LITES – Legal Issues in Transdisciplinary Environmental Studies

Volume 2

The 'Legal Issues in Transdisciplinary Environmental Studies' (LITES) Book Series is based on the assumption that the process of dialogue and cultural integration between law, life and earth sciences, and social and human sciences should be strengthened and updated, by relying on transdisciplinary research platforms such as agroecology, environmental studies, environmental science, and sustainability science. According to the new paradigm of social-ecological systems (SES), the concept of the environment is conceived as a complex system of relationships between ecological and social factors, including the cultural and economic ones. The primary purpose of law, in this conceptual framework, is to preside over the durability of the essential conditions for the survival of the social-ecological systems and the protection of life at all scales (of individuals, societies, ecosystems).

LITES Series aims to explore the relationships between legal and environmental sciences according to a transdisciplinary perspective. On the one hand, natural and social environmental sciences need to integrate the point of view of law: this entails to study the complexities of SES in the light of normative and institutional variables, with the lens of categories such as rights, duties, powers, responsibilities, and procedural safeguards. On the other hand, law is called upon to review its own internal geometries, confronting them with the holistic approach toward sustainability in the scientific debate. Accordingly, law should address the need of changing the approach that so far has led to both hypertrophy and disarticulation when regulating closely linked matters such as the environment, agriculture, forestry, landscape and cultural heritage, energy, and food.

LITES Series is addressed to a wide international and interdisciplinary readership, targeting academic researchers and scholars, experts and practitioner lawyers, public administrations, judges, and law-makers. Its volume editors and contributing authors have different backgrounds and come from all over the world in order to provide a forum for discussion and normative analysis about new legal frontiers of human-environment interactions across disciplinary barriers.

More information about this series at http://www.springer.com/series/15038

Alessandro Isoni • Michele Troisi • Maurizia Pierri
Editors

Food Diversity Between Rights, Duties and Autonomies

Legal Perspectives for a Scientific, Cultural and Social Debate on the Right to Food and Agroecology

Springer

Editors
Alessandro Isoni
Department of History, Society and Human Studies
University of Salento
Lecce, Italy

Michele Troisi
Department of Legal Studies
University of Salento
Lecce, Italy

Maurizia Pierri
Department of Legal Studies
University of Salento
Lecce, Italy

ISSN 2522-5049 ISSN 2522-5057 (electronic)
LITES – Legal Issues in Transdisciplinary Environmental Studies
ISBN 978-3-030-09167-5 ISBN 978-3-319-75196-2 (eBook)
https://doi.org/10.1007/978-3-319-75196-2

Printed on acid-free paper

This Springer imprint is published by the registered company Springer International Publishing AG part of Springer Nature.
The registered company address is: Gewerbestrasse 11, 6330 Cham, Switzerland

Introduction

This collective volume, entitled "*The Food Diversity Between Rights, Duties and Autonomies: Legal Perspectives for a Scientific, Cultural and Social Debate on the Right to Food and Agroecology*", contains the proceedings of a workshop held in Lecce (Italy) on 6–7 May 2015 with the same title.

The idea of analysing this topic arose from the opportunity given by the 2015 Milan Expo to reflect on the issues concerning, on the one hand, the difficulty in feeding an ever-increasing world population and, on the other hand, the need of building new productive systems able to preserve the planet from overexploitation.

For the first time, this volume brings together legal scholars, agroecologists, historians and sociologists coming from different countries to reflect on an issue that has never been studied until now with this wide range of cultural perspectives, in the hope of contributing to promote a transdisciplinary dialogue among different research experiences and scientific languages.

In fact, the concept of "food diversity" is a synthesis of multiple diversities: biodiversity of ecological sources of food supply; socio-territorial diversity; cultural diversity of food traditions. The role of law in the analysis of this theme is not prominent but serving: in fact, it is a necessary instrument for protecting the food diversity as an essential element of the right to food, understood as a human but also cultural right, connected to territory and conditioned by the progressive loss of biodiversity.

On a structural level, the volume is composed of three parts. The first part focuses on the theme of food diversity, agro-biodiversity and typicality, putting these issues in relation to territories, tourism and rural development. The second part highlights the multi-dimensionality of the right to food, which justifies the methodological choice of a trans-disciplinary research on this theme. The third and last part is devoted to analyse legal questions concerning regulations aiming to protect both the food diversity and the right to food, in the light of political, economic and social implications related to the problem of feeding world population respecting, at the same time, local communities' rights, especially in the developing and less developed countries.

In this line, according to a transdisciplinary perspective, the book collects 28 contributions, each of them devoted to a different feature concerning this complex and multifaceted issue, to create a polyocular scientific framework.

Contents

About the Editors and Contributors

About the Editors

Alessandro Isoni Associate Professor of Constitutional History, Department of History, Society and Human Studies, University of Salento (Italy).

Formerly Senior Researcher and Assistant Professor of Constitutional History and Public Law at the Department of History, Society and Human Studies, University of Salento. In 1999, he graduated in Political Sciences, final mark 110/110 *summa cum laude* at the University of Sassari. In 2001, he received an LLM degree in "National, Supranational and International Law of Market and Public Policies" by the Advanced School ISUFI (Institute of Advanced Interdisciplinary Studies) of the University of Salento. In 2005, he received the PhD degree with distinction in "Economics and Market Law" by the Advanced School ISUFI (Institute of Advanced Interdisciplinary Studies) of the University of Salento. Since 2005, he has lectured in courses for the University of Salento in Administrative Law and History of the Italian Administration and Public Law. On 24 December 2013, he obtained the National Scientific Qualification as Associate Professor in History of Political Institutions. He is a member of the editorial board of some reviews, like *Eunomia*, *Historia Magistra* and *Itinerari di Ricerca Storica*. He has been an invited speaker in national and international conferences. He is a cofounder of the Group LAIR "Law and Agroecology – Ius et Rus" established at the University of Salento. He has been an author of several articles and studies about the political and historical causes of the ECSC High Authority, together with other studies concerning the European integration institutional features and, in addition, other studies devoted to the parliamentary select committees in Italy under the Albertine Statute and the Republican Constitution. Among his recent publications are: Isoni A (2012) Farewell to the European Community. The Lisbon Treaty and the Conceptual Shifts of a Sui Generis Public Law Experience. Eunomia 1(1):85–98; Isoni A (2012) Federalismo e sussidiarietà nel processo di integrazione europea. In: Tedoldi L (ed) Alla ricerca della statualità. Un confronto storico-politico su Stato, federalismo e democrazia in Italia e in Europa.

QuiEdit, Verona, pp 181–200; Isoni A (2014) "Fonder la République". The French National Convention and the Revolutionary Government (1793–94). Parliaments, Estates and Representation 34 (1):40–54, Isoni A (2014), Jean Monnet, the UN Administrative System and the Creation of the ECSC High Authority. In: Mechi L, Migani G, Petrini F (eds.), Networks of Global Governance. International Organisations and European Integration in a Historical Perspective. Cambridge Scholars Publishing, Newcastle Upon Tyne, pp. 61–76; Isoni A (2015) The Common Agriculture Policy (CAP): Achievements and Future Prospects. In: Monteduro M. Buongiorno P. Di Benedetto S, Isoni A (eds.) Law and Agroecology. A Transdisciplinary Dialogue. Springer, Berlin Heidelberg, pp. 185–205; Isoni A (2017) La Royal Australian Navy e il blocco del Canale d'Otranto 1917–1918. In: Denitto A, Masella L (eds.) La Puglia e la Grande Guerra. Besa, Nardò.

Email: alessandro.isoni@unisalento.it

Web: https://www.unisalento.it/web/guest/scheda_personale/-/people/alessandro.isoni

Maurizia Pierri Full Researcher of Public Comparative Law, Department of Legal Studies, University of Salento (Italy).

She is currently Assistant Professor of Public Comparative Law at The Faculty of Social, Political and Educational Sciences, University of Salento. In 1989, she graduated in Law, final mark 110/110 *summa cum laude* at the Tor Vergata University of Rome. Since 1991 to 2004, she was Public Services Adviser (DPCM, 18/4/1995) for studies and researches in the Authority for Strikes in Public Services, ex lege 12/6/1990, n. 146. In 1998, she had the Admission to the Bar (Lecce) and in 2006 a contract for research in "Constitutional Transition in Latin-American countries, with a particular regard to Chilean situation" (University of Salento). In 2009, she received the international PhD degree in Comparative Political Social Systems (University of Salento). She is a member of the Editorial Board of *Eunomia* and of some scientific series. She has been an invited speaker in national and international conferences. In 2013, her paper "The limits of procedural democracy in intercultural conflicts about values" has been accepted and presented in the XXIII World Congress of Philosophy and is currently being published in the Proceedings of the Congress. She has been author of several articles and studies on administrative authorities, minorities rights and federalism. She is a member of the Group LAIR "Law and Agroecology – Ius et Rus" established at University of Salento. Among her recent publications are: Pierri M (2012) La funzione familiare della donna nella prospettiva delle gender theories. In: Various Authors, Studi in onore di Aldo Loiodice. Bari, Cacucci, pp 1021–1033; Pierri M (2012) Il Welfare State e la cultura dell'illegalità. In: Perrotta C, Sunna C (eds) L'arretratezza del Mezzogiorno. Le idee, l'economia, la storia. Bruno Mondadori, Milano-Torino, pp 210–229; Pierri M (2012) Regioni e welfare: l'impatto del federalismo fiscale sugli squilibri tra Nord e Sud d'Italia. In: Various Authors, Il regionalismo italiano dall'Unità alla Costituzione e alla sua riforma, vol. 2. Atti del Convegno ISSIRFA, "Il regionalismo italiano dall'Unità alla Costituzione e alla sua riforma", 20–21–22 ottobre 2011. Milano, Giuffrè, pp 687–704; Pierri M (2012) La tutela delle iperminoranze in India. In: Poggeschi G (ed) Le iperminoranze. Pensa, Lecce, pp 113–136; Pierri M (2012) La razionalizzazione del rapporto di fiducia nella forma di governo parlamentare. Nuova Rassegna di Legislazione, Dottrina e Giurisprudenza 16:1–9; Pierri (2013) Il welfare in Brasile. Diritto Pubblico Comparato ed Europeo 15(4):1457–1483; Pierri M (2013) Federalismo fiscale e sviluppo del territorio meridionale. In: Amendola A, Musella M (eds) Formazione e relazioni sociali: gli investimenti virtuosi per rimettere in moto il Mezzogiorno, vol. 7. Giappichelli, Torino, pp 135–144; Pierri M (2014) Boschi e foreste nel "nuevo constitucionalismo", ecologicamente orientato dei Paesi andini. In: Brocca M, Troisi M (eds) I boschi e le foreste come frontiere del dialogo tra scienze giuridiche e scienze della vita. Dalle radici storiche alle prospettive future. Editoriale Scientifica, Napoli, pp 155–162; Pierri M (2015) Agrobiodiversity, Intellectual Property Rights and Right to Food. In: Monteduro M, Buongiorno P, Di Benedetto S, Isoni A (eds) Law and Agroecology: A Transdisciplinary Dialogue. Springer, Heidelberg – Dordrecht – New York – London, pp 451–470.

Email: maurizia.pierri@unisalento.it

Web: https://www.scienzegiuridiche.unisalento.it/scheda_personale/-/people/maurizia.pierri

Michele Troisi Full Researcher of Constitutional Law, Department of Legal Studies, University of Salento (Italy).

He is currently Assistant Professor of Constitutional Justice at the Faculty of Law, University of Salento. In 2000, he graduated in Law, final mark 110/110 *summa cum laude*, at the "Aldo Moro" University of Bari. In 2005, he received the International PhD degree in Comparative Political Social Systems (University of Salento). Since 2008, he has lectured in courses for University of Salento in Public Law and Constitutional Law. He has been an invited speaker in national and international conferences. He is a cofounder of the Group LAIR "Law and Agroecology – Ius et Rus" established at the University of Salento. He has been an author of several articles and studies about the right to education, the school in the Constitutional law and regional electoral laws. Among his recent publications are: Troisi M (2012) Le assemblee elettive regionali, tra (perenne) crisi d'identità e (labili) segnali di ripresa. In: Mangiameli S (ed) Il Regionalismo italiano dall'Unità alla Costituzione e alla sua riforma. Giuffrè, Milano, pp 155–186; Troisi M (2013) La famiglia e misure di sostegno: la legislazione regionale. In: www.gruppodipisa.it; Troisi M (2014) Il diritto allo studio. In: Campanelli G, Carducci M, Tondi della Mura V (eds) Diritto costituzionale regionale della Puglia. Giappichelli, Torino, pp 1–10; Troisi M (2014) Boschi e foreste e ruolo della Corte costituzionale. In: Brocca M, Troisi M (eds) Boschi e foreste come frontiere del dialogo tra scienze giuridiche e scienze della vita: dalle radici storiche alle prospettive future. Editoriale Scientifica, Napoli, pp 163–173; Troisi M (2015) Environment, Landscape, Agriculture and Food in the Framework of State and Regional Legislative Powers as per Art. 117 of the Italian Constitution. In: Monteduro M, Buongiorno P, Di Benedetto S, Isoni A (eds) Law and Agroecology: A Transdisciplinary Dialogue. Springer, Heidelberg – Dordrecht – New York – London, pp 207–226.

Email: michele.troisi@unisalento.it

Web: http://www.giurisprudenza.unisalento.it/scheda_personale/-/people/michele.troisi

Contributors

Valentina Albanese Department of History and Cultures, University of Bologna, Bologna, Italy

Aida Giulia Arabia Institute for the Study of Regionalism, Federalism and Self-Government "Massimo Severo Giannini", National Research Council, Rome, Italy

Silvia Bolognini Department of Legal Sciences, University of Udine, Udine, Italy

Marco Brocca Department of Legal Studies, University of Salento, Lecce, Italy

Gianluca Brunori Department of Agronomy and Management of the Agroecosystems, University of Pisa, Pisa, Italy

Giuseppina Buia Department of Legal Studies, University of Salento, Lecce, Italy

Rosa Calderazzi Department of Economics, Management and Trade Law, University "Aldo Moro" of Bari, Bari, Italy

Allison Condra Davis Wright Tremaine LLP, Portland, OR, USA

Martina Cutazzo University of Salento, Lecce, Italy

Simona D'Amico Department of Agricultural, Food and Agro-Environmental Sciences, University of Pisa, Pisa, Italy

Antonello Denuzzo University of Salento, Lecce, Italy

Saverio Di Benedetto Department of Legal Studies, University of Salento, Lecce, Italy

Elizabeth Dooley University of Exeter, Exeter, UK

Roberto Franco Greco University of Salento, Lecce, Italy

Antonio Gusmai Department of Legal Studies, University of Salento, Lecce, Italy

Alessandro Isoni Department of History, Society and Human Studies, University of Salento, Lecce, Italy

Séverin Jean Institute of Judicial Studies for Urban Planning and Construction (IEJUC), Toulouse 1 Capitole University, Toulouse, France

Hope Johnson Queensland University of Technology School of Law, International Law and Global and Governance Research Program, Brisbane, QLD, Australia

Clelia Losavio Institute for the Study of Regionalism, Federalism and Self-Government "Massimo Severo Giannini", National Research Council, Rome, Italy

Serenella Luchena Department of Economic Sciences, University of Salento, Lecce, Italy

Nicola Lucifero Department of Agricultural, Food and Forestry Systems, University of Florence, Florence, Italy

Rowena Maguire Queensland University of Technology School of Law, International Law and Global and Governance Research Program, Brisbane, QLD, Australia

Massimo Monteduro Department of Legal Studies, University of Salento, Lecce, Italy

Rocco Palma Department of Economic Studies, Catholic University of Sacred Heart, Rome, Italy

Christine Parker School of Law, University of Melbourne, Carlton, VIC, Australia

Maurizia Pierri Department of Legal Studies, University of Salento, Lecce, Italy

Fabio Pollice Department of History, Society and Human Studies, University of Salento, Lecce, Italy

Adanella Rossi Department of Agronomy and Management of the Agroecosystems, University of Pisa, Pisa, Italy

Guido Ruivenkamp SADE, University of Wegeningen, Wegeningen, The Netherlands

Anjeza Sina Department of Legal Studies, University of Salento, Lecce, Italy

Giuliana Strambi Section on International and Comparative Agricultural Law, Institute for International Legal Studies, National Research Council, Florence, Italy

Virginia Tascagni University of Salento, Lecce, Italy

Michele Troisi Department of Legal Studies, University of Salento, Lecce, Italy

Francesco Fabrizio Tuccari Department of Legal Studies, University of Salento, Lecce, Italy

Giulia Urso Social Sciences Unit, Gran Sasso Science Institute, L'Aquila, Italy

Luca Zamparini Department of Legal Studies, University of Salento, Lecce, Italy

Part I
Food Diversity, Agrobiodiversity and Typicality: Territories, Tourism, Rural Development

A Comprehensive and Participatory Approach to the Valorisation of Biodiverse Products

Gianluca Brunori, Adanella Rossi, and Simona D'Amico

1 Introduction

The loss of biodiversity is currently underway and it has intensified over the past 20 years. The loss of biological diversity is the result of a combination of anthropological and ecological interrelated factors.[1] Studies mention habitat conversion, invasive alien species, resources overexploitation, climate change and pollution among the causes of species diversity reduction. Moreover, these factors are likely to be behind the contraction, or even homogenisation, of species variety, as they bring with them other environmental damage to ecosystems.[2]

Agriculture has historically had the largest impact on biodiversity, and it is expected to continue to be a factor of biodiversity loss in the future.[3] The way food production and consumption are organised in the mainstream industrial food system is indeed responsible of a high pressure on natural resources. This system has been relying on agricultural intensification, which has implied the conversion to predominantly monoculture farming of lands with an original high biodiversity value, and on the production of few major crops and staples, neglecting the diversity

[1] Ash and Fazel (2007) and Mayes et al. (2011).

[2] Pimentel et al. (1992), Ash and Fazel (2007) and Mayes et al. (2011).

[3] OECD (2008).

G. Brunori (✉) · A. Rossi
Department of Agronomy and Management of the Agroecosystems, University of Pisa, Pisa, Italy
e-mail: gianluca.brunori@unipi.it

S. D'Amico
Department of Agricultural, Food and Agro-Environmental Sciences, University of Pisa, Pisa, Italy

A. Isoni et al. (eds.), *Food Diversity Between Rights, Duties and Autonomies*,
LITES – Legal Issues in Transdisciplinary Environmental Studies 2,
https://doi.org/10.1007/978-3-319-75196-2_1

of traditional, place-specific, crops.[4] The intensive usage of resources by a growing human population will further exacerbate these negative dynamics.[5]

Besides reducing biological diversity this model has also undermined cultural diversity because of the shrinking space left to traditional methods of crops improvement, farming and food traditions that are rooted in specific cultural contexts.[6]

The transition to sustainable farming systems may contribute to the protection and enhancement of biodiversity, through the diversification of species and varieties in the productive cycles and the provision of agro-ecosystem services. Preserving agro-biodiversity, however, appears as a complex problem that requires systemic intervention. The particular and diverse nature of values involved in the conservation of agro-biodiversity and the complexity of the processes that underlie it demand to address the issue from different perspectives—technical, but also cultural, social and economic, as well as legal and political. There is the need for significant changes in the way of managing genetic resources and production activities, in the capacity of recognising and appreciating the values embodied in the final products, and in the role of institutional and political environment. Going deeper, this involves deepening the processes underlying the biodiversity value perception and the development of coherent practices, looking at the related enabling and hampering factors.

All this has significant implications for the definition of the approaches to the management of this issue, in all areas of intervention. This is particularly evident for the research, as shown by the recent EU funded project DIVERSIFOOD, aimed at studying the factors underlying the consolidation and market valorisation of biodiverse food.[7] This contribution draws on the experience gained in this project, considering it as a useful exemplification of the comprehensive and integrated approach that is needed.

2 Preserving and Enhancing Agro-Biodiversity

Acknowledging and managing the complexity of values connected to agro-biodiversity, translating them in coherent practices, are the key to its conservation and valorisation. These processes may take place at different stages of agro-biodiversity management and interact with other factors, dynamics and processes that intervene in it. After the illustration of the different values encompassed in

[4]Historically, humans have exploited thousands of plant species for food; today, however, most people on Earth depend on 20 types of plants and three staple crops (rice, wheat, and corn). See Laverty and Sterling (2003), Gruère et al. (2006) and Padulosi et al. (2013).

[5]Ash and Fazel (2007).

[6]Ceccarelli and Grando (2007), Morris and Bellon (2004) and Johns et al. (2013).

[7]DIVERSIFOOD is a four-years project funded by the European Union within the Horizon 2020 Programme under Grant Agreement no 633571. It involves 21 partners, belonging to 12 countries. The partners are public and private research institutes and various organisations engaged at regional level on issues of conservation and enhancement of agro-biodiversity. See: www.diversifood.eu.

(agro-)biodiversity and their role for the main actors involved, this section deals with: the role of the cultural and social dynamics; the meaning and implications of the different approaches to agro-biodiversity management, with reference to stakeholder involvement; the significance of strategies of market valorisation of biodiverse products, as specific space of alignment and coherence building amongst the several actors involved; the role of political frameworks within which interventions to maintain and enhance agro-biodiversity developed.

2.1 Building and Recognising Values of Biodiversity

Through its various components, biodiversity provides different goods and services; these, in their turns, offer different benefits—ecological, economic, socio-cultural—, not always easy to distinguish and to evaluate. Furthermore, biodiversity assumes the form of a public or of an open access good; in both cases, market has proved to be inefficient. In the first case, normal market process has for a long time failed to catch the values of some, or all, aspects connected to biodiversity, so not showing its scarcity. In the second case, the benefits of biodiversity are realised as private benefits, whereas the associated costs of using biodiversity resources are shared as social or public costs.[8]

All this complicates the process of preserving and enhancing biodiversity. However, the need to assign a value to biodiversity has grown, hand in hand with the necessity to face the issue of biodiversity increasing reduction through adequate measures and actions. Some studies identify two main kinds of biodiversity values: *intrinsic values*, which are inherent to biodiversity, to the right to exist of the life forms on which biodiversity builds up; *extrinsic values*, which grow out of the uses or applications of the life forms constituting biodiversity.[9] Another consolidated taxonomy, adopted in environmental economics, resolves the overall economic value of biodiversity into a set of values.[10] The first important distinction is between *use values* and *non-use values*. The first derive from the use that people make of goods and services that can be considered as the by-products of biodiversity. They are distinguished in *direct* and *indirect use values*. The former imply consumption of resources and are associated to market prices; they refer to goods, such as food, fibbers, fuel, or medicines. The latter do not imply consumption of resources and are characterised by a greater difficulty in evaluation; they are linked to services that mainly refer to the ecosystem services granted by the existence of biodiversity, such as soil and water conservation, pollination, nutrient cycling, aesthetic components,

[8] Government of Ireland (2008).

[9] Alho (2008) and Laverty and Sterling (2003).

[10] Turner et al. (1994) and OECD (2002).

recreation and other. A particular form of the use value is the *option value* that refers to the possibility of a future use of the resources, allowed by a current conservative management.[11] The *non-use values* refer to not utilitarian values. They encompass the *existence value* and the *bequest value*. The former is the intrinsic value recognised to the life forms constituting biodiversity, to the mere existence of a wide range of genetic resources, regardless the possibility of their use; the latter is the value attributed to the future societies chance to make use of the resources, according to intergenerational equity principles.

The different stakeholders involved in processes of biodiversity preservation may perceive these values differently and so may give different importance and attach different meanings to them.[12]

What described so far with respect to the value of biodiversity holds in the context of agro-biodiversity too. The value of agro-biodiversity expresses in different forms, which, although not all or not fully or easily translatable into economic value, all are crucial to the appreciation of agro-biodiversity by the different actors who play a role in the food system. They include actors involved in all the stages of the food chain, from breeders to consumers, as well as researchers and policy makers. In the case of farmers, *use value* of agro-biodiversity may refer to direct benefits, when the specific genetic resources allow creating a differentiated offer on the market, and to the functional benefits deriving from the greater resilience of the agro-ecosystems, which may have an effect in economic terms too. In the case of consumers, this value expresses in the opportunity to meet dietary needs (such as nutritional and health-related needs) or to enjoy sensorial or hedonistic benefits, according to the peculiarities of biodiverse products. Consumers also benefit of *indirect use value*, as in the case of utilisation of recreational or educational services provided by agro-tourism. More in general, *indirect use value* relates to the wider positive effects stemming from production systems and products building on biodiversity. These may include the benefits to society because of the greater resilience of the agro-ecosystems (which, in its turn, may translate in future use values by guaranteeing, for instance, food/environment conservation). At local level, they may include the benefits stemming from the inclusion of these products in broader programmes of valorisation of territorial capital (as in the case of many rural development projects). Regarding *non-use value*, the role that agriculture may play in conserving and managing biodiversity is recognised by growing segments of society. Within them, concerned consumers reward this role also through their consumption choices.

---

[11]It is the case of resources whose value has been untapped yet but can emerge from further research and experiences.

[12]Laverty and Sterling (2003).

2.2 The Role of Cultural and Social Dynamics

As said above, the acknowledgment of agro-biodiversity values is crucial to their production, management and enjoyment.

As more in general for biodiversity values, going from *direct use values*, through *indirect use values*, to potential or intrinsic values (as *bequest* and *existence values*) the tangibility of benefits and furthermore the possibility of their conversion into economic value decrease. Most of these values are expression of a less utilitarian and anthropocentric attitude, which relates to cultural and ethical value systems. These values may however be important in affecting the production and consumption choices and, more generally, decisions related to its management.

The importance attached to the (need for conservation of) agro-biodiversity is so closely dependent on the understanding of the interactions between biodiversity and the societal life conditions and, however, on the importance individually/socially assigned to the extrinsic and intrinsic values of biodiversity.

The relevance of the cultural determinants emerges also considering the role of knowledge, culture and skills in the handling of genetic resources (e.g. in farming and processing activities) and in the enjoyment of benefits stemming from biodiversity (e.g. product quality or aesthetic values of landscape in consuming activities). The process of co-evolution between the two systems makes this link even more meaningful. The evolutionary adaptation of ecological biodiversity and of the knowledge that is needed for its management go in fact together. The same is for the capacity to appreciate the values of the biodiverse food or the biodiverse natural environment. The decrease of genetic resources leads to the decrease of cultivation adaptability to the environment, but also implies a loss of knowledge of their use and management, in a sort of 'co-involution'. Furthermore, this affects the potential values of biodiversity (*option* and *bequest values*) twofold, as it loses most of its significance with the loss of human capabilities to use the genetic resources.

All this confirms how central is the issue of culture in the conservation and reproduction of (agro-)biodiversity.[13] It is evident, on the other hand, that these culturally mediated processes are dependent on the characteristics of the specific socio-economic contexts: the ways of looking at biodiversity, of grasping its values, of understanding the related problems and the need for solutions vary according to the cultural models and social organisations.[14]

In that regard, the contexts where the development of the agro-food system has led to the simplification of both agro-ecosystems and human diets need a relevant effort to recover knowledge and skills, both from the side of the producers and from that of consumers. Farmers need not only to adapt the farming activities but also to reorganise their relational network, to find new opportunities to exchange and share information. In most of the cases, taking responsibility of the management of agro-biodiversity also implies a real change of identity, not only to leave the model of

[13]Mathez-Stiefel et al. (2009).

[14]Nuijten et al. (2013) and Johns et al. (2013).

modern, specialised farmer, but also to move from a socially recognised and often economically rewarded role of conservation, to an active, community-based engagement in recovering and improving biodiversity. This process involves a considerable cultural shift. For consumers the needed changes in terms of sensitivity, preferences and practices is not less demanding, because of the more mediated relationship with natural resources. The growth of knowledge about the existence and properties of other species/varieties compared to the few usually marketed and the change of dietary habits are the first step to any initiative of conservation and valorisation of agro-biodiversity. Positioning this agro-biodiversity in the specific territorial contexts and identifying the specific production systems that manage it represent a further significant step. This connection is what has characterised the preference accorded to the typical and traditional food products and has frequently represented the central element of rural development projects. With an even greater emphasis on the relational dimension of the recognition of the product value, more recently it is what underlies the development of localised circuits of production-consumption and the related community-based engagement for biodiversity enhancement.[15]

In other contexts, as those of developing countries, where the pool of knowledge linked to agro-biodiversity has been eroded less because of the survival of the traditional farming systems and dietary habits, the challenge is to create the conditions to preserve and enhance it while trying to consolidate or improve food systems.[16] Here the role of cultural aspects, linked to social components and, in general, strongly context-dependent, is even more complex. They may include, for instance, gender, age, identity, cultural tradition, religious and ethnic or political conflict issues. These factors deeply shape the relationship with and the evolution of technologies (varieties, tools, techniques) that are significant to agro-biodiversity management.[17] The interdependencies of cultural and biological diversity and the need to enhance socio-cultural processes to foster agro-biodiversity is at the basis of the "biocultural approach" of some regional development programmes.[18] On the other hand, for the same interdependence, enhancing the use of specific genetic resources is seen as a means to keep food traditions, cultural values and community and social identities.[19]

All this points out the importance to work on the deep mechanisms underlying the cultural and social mediators of agro-biodiversity management. In that regard, learning processes that develop in the specific social environments and relational spaces are certainly to be explored.

[15]Brunori et al. (2011) and Simoncini (2015).

[16]Johns et al. (2013).

[17]Nuijten et al. (2013).

[18]Mathez-Stiefel et al. (2009).

[19]Johns and Eyzaguirre (2006).

2.3 The Approach to the Agro-Biodiversity Management

The development of learning processes and through them of new attitudes, knowledge and practices around agro-biodiversity management is closely linked to the degree of actors' involvement. This especially concerns farmers and consumers, for a long time not directly and actively engaged in managing agro-biodiversity.

In this regard, it is particularly meaningful the difference between the conventional approach to genetic improvement and the alternative approaches that have developed around the objective to tackle agro-biodiversity-related problems.

The conventional approach to crop improvement has been traditionally guided by the aim of meeting agro-industry needs in terms of crop stability, adaptability to a wide range of environments and high productivity (both relying on intensive use of chemicals), as well as wide acceptance of the related final products. Hence, this approach is effective when operating in uniform and stable environments and at the expense of biological and cultural diversity. Furthermore, it does not need much autonomy by farmers, but rather relies on control of know-out and a well organised marketing of reproductive material and related needed inputs.

The willing to overcome the limitations of this approach has led to seek a closer relationship with environmental, biological and cultural specificities of contexts. To that end, Participatory Crop Improvement (PCI) aims at linking the conventional, globalised crop improvement approach with localised and culture specific ones. It seeks combining productivity and cost-effectiveness with maintenance and enhancement of agro-biodiversity and cultural diversity.[20] The PCI-strategy is twofold: on one side, it inserts huge genetic diversity into the local farming systems; on the other, it relies on farmers' capacity to select, conserve, re-produce and exchange seeds that prove to be particularly suitable to the specific environment and to the local farming and food traditions.[21]

Central to these approaches is evidently the direct involvement of farmers in the various stages of agro-biodiversity management. PCI includes Participatory Variety Selection (PVS), Participatory Plant Breeding (PPB) and *in-situ* conservation of crops. PVS involves farmers and other stakeholders along the food chain together with researchers in the selection of varieties from formal and farmer-based collections and trials. The selection criteria give particular importance to the suitability of the varieties to local agro-ecosystems, as well as to needs, uses and preferences of the involved stakeholders. PPB sees the same approach applied to the different plant breeding stages of the process of plant breeding. Finally, *in-situ* conservation is a strategy for crops conservation that relies on farmers taking up, growing and exchanging seeds. It generates genetic varieties that, while being common in certain ecological and cultural environments, may be absent from conventional seed bank catalogues (the *ex-situ* conservation).[22]

---

[20]Hardon (1995).

[21]Almekinders and Elings (2001).

[22]Almekinders and Elings (2001).

PCI has been spreading in marginal and changing ecological environments of developing countries where small and traditional—sometime subsistence—farming is predominant. In these contexts, PCI worked to compensate the ineffectiveness of conventional crop improvement approaches.[23] Its more suitable approach relies on the use of locally adapted genetic resources and learning from traditional knowledge to deal with environmental variation and production risks.[24] Furthermore, in some experiences, special attention is given to gender issues. The involvement of women is prioritised for their relevant but also specific role in farming. Involving women is seen as a way to consider broader varieties of seeds, breeding, farming and conservation techniques that may be suitable to as much broader cultural and ecological environments.[25]

Hence, PCI results in valorising biological and cultural diversity, as well as in empowering rural communities, while responding to the needs of traditional farming systems in marginal environments. However, its positive impact highly depends on how it is implemented. In this regard, the degree of genetic diversity used and the level of involvement of farmers and other local stakeholders, besides scientists, are crucial.[26]

More recently, PCI has been adopted also in developed countries, where it again represents an alternative approach to the practices supported by the mainstream scientific and agro-industrial system. Also, in this case it refers to the guiding principles of reintroducing genetic variety in farming systems, by valorising the specific characteristics of local environments and the knowledge, skills and preferences of farmers and other stakeholders, with respect to the conventional powerful role of scientists and input-industry agents.

In these contexts, the participatory involvement of farmers and the other chain agents is even more significant, because of the force of the hegemonic culture that informs any aspects of technology of farming and processing stages and of the loss of autonomy, decision power, knowledge and skills that has interested their actors over the modernisation of agro-food system. We have already mentioned these aspects dealing with cultural factors. Likewise, the involvement of users and final consumers radically changes the condition in which they usually play their role, generally passive and disadvantage in terms of knowledge, skills and decision power. In this regard, there is an increasing attention to the opportunities stemming from processes of 'value co-production'.[27] In both cases, it thus emerges how the engagement of these stakeholders in the process of production, management and enjoyment of biodiversity is not just a matter of rights, but also implies the existence of enabling conditions for its actualisation.

[23]Tripp (1997).

[24]Clawson (1985) and Van Noordwijk et al. (1994).

[25]Farnworth and Jiggins (2003).

[26]Morris and Bellon (2004).

[27]Krucken (2005).

2.4 The Potential of Market Valorisation of Biodiverse Food

The conservation and enhancement of biodiversity finds important opportunities in the market valorisation of the final products.[28] The appreciation of these products on the market may generate income for the chain actors and indirectly for other actors involved, and at the same time spread further knowledge about the existence and properties of the products. In both cases, it may contribute to the maintenance of the production systems and the related environmental and social benefits. Furthermore, a market valorisation of these products may play an important role in broader strategies of enhancement and promotion of territorial resources, favouring the creation of synergies among different sectors.

This economic valorisation assumes specific characters in different socio-economic contexts.

In developing countries, structuring the marketing of these products, by the creation of proper value chains, is considered a way to overcome the limitations because of the subsistence level of farming or the scant knowledge out of the local contexts of production-utilisation or, in general, logistic difficulties. This may increase social welfare by generating income opportunities and new small economies.[29] It may also be seen as the needed condition to broaden the commercialisation of the products to catch other opportunities in niche or novelty markets linked to the increasing interest towards 'new varieties'.[30] It is recognised the role that these processes play in broader territorial development strategies aimed at increasing local livelihood.

In developed countries, marketing strategies focusing on products whose properties build on specific genetic resources and the related interactions with the local environments are not new. As said above, many European 'typical products' present this peculiarity, which has been also recognised through geographical indications since the early 1990s (as the Protected Designation of Origin (PDO) and the Protected Geographical Indication (PGI)). The market valorisation of these products has strongly contributed to their reintroduction in the local production-consumption systems and, more generally, in the food culture and in the eating practices. Again, also in these contexts, this has favoured processes of local development, thanks to the synergies created with other economic sectors.

The economic valorisation of products more specifically related to initiatives of biodiversity conservation—as underutilised or forgotten species—however represents a more recent opportunity. This is linked both to the new, increasing attention of public opinion to biodiversity among the other sustainability issues, and, in the case of food products, to the greater knowledge about nutritional aspects. The latter

[28]There are of course also the benefits stemming from the recognition of the ecosystem services provided by agro-biodiversity through market mechanisms. However, we refer here to the market valorisation of final biodiverse food products.

[29]Will (2008).

[30]Padulosi et al. (1999).

is supported by the spread of the findings of researches focusing on health benefits from underutilised products also out of the scientific environment (as, for example, in the case of properties of certain grain species/varieties to celiac disease). The opportunity to create new markets, based on these 'new products' and their multiple uses, is also at the base of the growing interest of public bodies and private economic actors, which results in multiple initiatives of promotion and valorisation.[31]

2.5 The Role of the Political Framework

Over the years, policy makers have recognised the importance of preserving and enhancing biodiversity. This has given rise to a variegated legislative framework that has contributed to shape the issue and has driven interventions, while at the same time defining the operating space of involved actors.

Political interventions have been promoted at international, regional, country and local levels. There have been three main categories of interventions: formal commitments to develop strategies to preserve biodiversity; measures to operationalise these strategies; legislative acts that allow the legal implementation of the strategies and operational measures. It is beyond the scope of this contribution to give a complete overview of the political interventions for biodiversity preservation and enhancement. However, some milestone cases, at the international and European level, are reported as examples of the political and legislative contributions to this matter and as source of insights on the most significant aspects at stake.

The first formal commitment to preserve and enhance biodiversity at the international level dates to the early 1990s: the Convention on Biological Diversity (CBD), signed by 150 Governments at the 1992 Earth Summit in Rio de Janeiro. Through it, Governments aimed at combining the human need for food, health, shelter and other human needs with the preservation and enhancement of biodiversity. The Convention covers key aspects of biodiversity conservation and management, including natural resource management, and the social, cultural and economic values of biodiversity.[32]

To fulfil the obligations deriving from the CBD, in 1998 the European Commission has adopted the EC Biodiversity Conservation Strategy, through which the European Union (EU) commits to reduce the rate of biodiversity loss. The theme of biodiversity conservation returns in the EU Sustainable Development Strategy, adopted in 2001. In the frame of this strategy, the EU Member States commit to halt biodiversity loss by the year 2010. To this end, four Biodiversity Action Plans have been adopted at EU level. They foster cooperation between States and between States and civil society organisations, as well as set responsibilities for European Institutions. Moreover, they promote a cross-sectoral approach with actions planned

[31]Padulosi et al. (1999).

[32]Alho (2008) and Government of Ireland (2008).

in the field of agriculture, fisheries, economic and development cooperation, and conservation of natural resources.[33]

A range of other multilateral environmental agreements were also established mainly during the 1970s to protect biodiversity.[34] These measures attempt to ensure a coordinated process for addressing biodiversity loss. Their implementation is generally at national level through policies that, after valuation, address the sources of impacts on biodiversity. Valuation, underlying prioritisation and setting of objectives, is to that end the crucial factor.[35]

As far as it concerns the measures to operationalise the general commitments towards biodiversity, a relevant contribution comes from the European agricultural policies, which have over time included the maintenance and enhancement of agro-biodiversity amongst their main objectives. The various policies offer significant support to interventions addressed both to maintain or improve agro-biodiversity looking at final product diversification (food and other products) and to strengthen functional agro-biodiversity, referring to its capacity to provide eco-system services. Furthermore, the new programme for rural development (Reg. 1305/2013) includes these objectives among the priorities of the forms of cooperation underlying the implementation of the European Innovation Partnership (COM (2012)79), stressing the importance of a multi-actor, integrated approach to the issue, aimed at actively involving all the stakeholders.

The support to biodiversity by the EU agricultural policy so focuses on two main areas of intervention. On the one hand, it addresses the conservation aspect, by funding interventions that are instrumental to the protection of local species/varieties; on the other hand, it pursues the way of economic valorisation of biodiversity, by supporting initiatives aimed at qualifying and defending on the market the specific quality products and at promoting services useful for the local communities.

Among the forms of market valorisation, the former tools are integrated with the existing EU regulatory system of Geographical Indications aimed at protecting typical and traditional products (PDO and PGI). As said above, these measures build on the close link between agro-biodiversity and the biological and cultural specificities of regional products. Afterwards, the need to intervene not only on the farming system but rather on the whole value chain has led to consider the key role of the short supply chains, where favourable conditions for the appreciation of the product quality can develop.[36]

As significant it appears the public support provided to the creation and valuation on the marketplace of environmental services, which is representing one of the most

---

[33]Government of Ireland (2008).

[34]They encompass, for example, the Convention on International Trade in Endangered Species, the Ramsar Convention on Wetlands, the Convention on Protection of the World Cultural and Natural Heritage, and the Convention on the Conservation of European Wildlife and Natural Habitats.

[35]OECD (2008).

[36]Naziri (2009) and Simoncini (2015).

promising way to attach economic value to agro-biodiversity.[37] Also in this case the adoption of an integrated approach is crucial. The growth of the environmental service supply in fact demands cooperation among different actors and related expertise (farmers and other enterprises, researchers, policy makers and administrators).

When moving to the regulatory framework, most of the acts concern the development, conservation and diffusion of seeds and plant varieties. The most significant aspect is here represented by the tendency to introduce recognitions and guarantees to the farmers' rights to produce, preserve and exchange seeds and the products, along to the breeders' property rights and the consumers' rights to nutritional safety. This move is meant to acknowledge farmers' contributions in developing and preserving seeds and plant varieties, which have become increasingly blatant with the diffusion of participatory crop improvement and market valorisation strategies.

In 1961, the International Convention for the Protection of New Varieties of Plants recognised the key role of plant breeders, creating the condition for the development of a system for plant variety protection working for registered varieties within the frame of conventional breeding approach. On the other hand, the more recent International Treaty on Plant Genetic Resource for Food and Agriculture (ITPGRFA) introduces some inputs to recognise farmers' rights to save, use, exchange seeds saved on farm and other propagating material. The Treaty encourages the contracting parties to promote policies, legal measures and strategies to acknowledge and encourage the contributions of farmers and traditional knowledge to maintain and reinforce the variety of genetic resources used for farming and food purposes.

The EU policies reflect the same dynamics. Some Directives, defined to regulate the production and diffusion of seeds, have been mostly focusing on guarantying varietal stability and safety. As an example, Dir. 2002/55/EC sets that only seeds varieties that prove to be distinct, stable and uniform can be registered in official catalogues and commercialised. Other Directives, such as Dir. 2009/145/EC, have been passed to open to the acceptance of traditional vegetable landraces that do not fit the above-mentioned criteria but are relevant to combat genetic erosion. Furthermore, the EU Seed and Plants Reproductive Material Marketing Law has been going through a process of reform under the pressure of civil society organisations. They claim that this legislation still fails to adequate the EU regulation to the indications of the ITPGRFA because it does not safeguard the informal systems of production, conservation and exchange of reproductive material, which contribute to the biodiversity protection and in which farmers and their traditional knowledge play a crucial role. On the other hand, it has becoming increasingly evident how complex may be a change of the current system. It in fact demands not only to consider the ethical issue of the recognition of farmers' contribution to the development of reproductive material, but also to face the questions of the legal responsibility for the submission

[37]OECD (2008).

and of the costs for the variety registration, the variety maintenance and basic seed production.[38]

Together with these more recent tendencies in dealing with the issue of biodiversity management, it is finally worth mentioning how there is a growing recognition that effective policies for biodiversity conservation need to focus on the mitigation of socio-economic pressures on biodiversity, either directly or through modification of their underlying driving forces.[39] This is thought to be the most effective and durable option to reduce the rate of biodiversity loss. Also, in this case, the fine-tuning of these measures demands an integrated research approach that integrate social sciences and economics with biodiversity research.

3 A Comprehensive Approach to Preserving and Enhancing Agro-Biodiversity: Diversifood

As showed in the previous sections, maintaining and enhancing agro-biodiversity is a complex process. It involves multiple actors, at the different stages of the food chain up to consumers; furthermore, it may also include other stakeholders, more in general interested in the promotion of territorial resources and food culture. The interactions amongst these actors around the meanings attached to agro-biodiversity, and the related learning processes and development of new practices that arise from them are central to agro-biodiversity management. They contribute to the definition, fine-tuning and running of the specific technical, organisational, cultural, social, economic and institutional components that, as seen above, are diversely involved in the various areas of agro-biodiversity management. The need to guarantee and optimise these interactions and to connect the different areas of action involved also highlights the importance of the adoption of proper approaches in promoting, analysing and supporting these processes.

The challenge of DIVERSIFOOD project is to consider all this. The following sections explain the goals and the approach of this project as an exemplification of how research can contribute to define and implement comprehensive approaches to preserving and enhancing agro-biodiversity.

The main objective of the project is to deepen the factors and processes of various nature that can support the reintroduction of biodiversity in cropping systems, improve its management and promote a social and economic valorisation of the related food products through proper communication and value chains.

Underlying this objective is the recognition of the urgent need to move towards farming systems able to recover and reproduce cultivated diversity, complementing the not fully effective efforts of *ex-situ* conservation (e.g. the gene banks) but also

[38]Almekinders and Elings (2001).

[39]Haberl et al. (2009).

overcoming the reductionist approach of the *in-situ* 'conservation'.[40] In addition, the project builds on the belief of the importance of the availability of a range of varieties suitable to organic, low-input and marginal agriculture, able to cope with particular/ more difficult environmental conditions.[41] As important is improving the economic performance of biodiversity-oriented food system and strengthening the social awareness and consensus around biodiverse food values.

The DIVERSIFOOD project faces the challenge to improve the understanding and capacity to manage biodiversity-oriented food systems by adopting an integrated approach both in the identification of the relevant processes involved and in the way to explore them.

3.1 The Approach of DIVERSIFOOD

The approach conceived for DIVESIFOOD follows two interrelated lines. On one hands, the project investigates all the stages involved in the processes of enhancement of agro-biodiversity and valorisation of biodiverse products. These stages include: breeding of genetic material, farming activities, processing of row products, marketing and consumption of the final products. On the other hand, DIVERSIFOOD investigates a combination of aspects that intervene in the process of valorisation of biodiverse products in each of the listed stages. These aspects are: technic-technological, organisational, economic, social, cultural, political-legislative. In both levels of analysis, the project explores how the various actors interact in the management of agro-biodiversity values and how they develop new consistent practices.

The project relies on insights emerging from empirical research to produce conceptual and methodological reflections for scientists working in the field of agro-biodiversity preservation, as well as recommendations for food chain practitioners, policy makers and civil society. To this end, its approach includes some space for theoretical reflections and dissemination activities.

A large part of the project activities is aimed at exploring the potential of underutilised genetic resources through multi-actor, on-farm evaluations. These activities involve experimentation on several agricultural and forestry cultivations aimed at testing different species and varieties in different agro-ecosystems, in the various European countries involved. The species and varieties examined belong to the local farming experience. However, their management is often to be improved, to respond both to the different environmental conditions and to the social changes in food requirements and uses.[42] They are so evaluated for their agronomic properties and for the results in terms of nutritional and organoleptic quality of the final

[40]Maxted et al. (2012) and Bonneuil et al. (2012).

[41]Wolfe et al. (2008).

[42]Fess et al. (2011).

products, directly involving farmers and consumers. Central in this stage is the attention paid to several aspects, such as: local traditions in cropping and food practices; the particular needs of local farming contexts or of small scales of production affecting the technical approaches to the crops; the farmers' knowledge and expertise in managing biodiversity to be mobilised (*in-situ management*); the sustainability and profitability of the production systems; the creation of new food quality through combining old genetic resources, new techniques, a renewed food culture.

The effort to revalorise underutilised species and varieties is accompanied by activities of participatory crop improvement. These activities are aimed at identifying the breeding process and the crops that are suitable to the conditions of different regions in Europe. The aim is here to create new diversity in several important crops in Europe, but also, and above all, to develop new approaches to the care of agro-biodiversity by fine-tuning suitable methods and tools, designed for on-farm decentralised participatory breeding. This allows complementing the work of *ex-situ* conservation to make available genetic resources more suitable to the current needs of production and consumption. The interaction amongst different sources of knowledge and expertise and amongst different fields of interests—by involving farmers, breeders, scientists, processors, users and final consumers—is here essential.

Other activities in DIVERSIFOOD focus on exploring the role social, institutional and political factors may play in enabling the participatory, multi-actors, on-farms approaches to genetic resources evaluation and crops improvement identified and tested during the project.

For many years a traditional, reductionist view considered the *in-situ* conservation of genetic resources for agriculture as a static function (the creation and running of a sort of open-air museum), not consistent with development goals.[43] Following more recent approaches, the project advocates the role of 'on-farm management', seen in its potential of dynamic utilisation of genetic resources, especially within a communitarian dimension.[44] In this perspective, the agro-biodiversity is seen as a common good and its conservation and reproduction (both of them not excluding evolutionary dynamics) is faced by farmers collectively and through close interaction with other agents. This function is recognised and supported by society, represented in the local communities.

Focusing on the management of seeds as crucial factor of biodiversity management, the project looks at seed systems as complex adaptive systems spatially and socially distributed, reflecting the immaterial diffusion of farmers' knowledge.[45] The community seed systems or seed localised networks become the space where to fine-tune the social and economic conditions needed to implement on-farm seed breeding and management. This includes identifying the key elements to strengthen

[43]De Boef et al. (2013).

[44]Louwaars et al. (2010), de Boef et al. (2010) and Thijssen et al. (2013).

[45]Leclerc and Coppens (2011) and Isaac et al. (2007).

community or network capacity, as well as to raise society awareness about the values of these resources.

Regarding enabling factors, another relevant area of (action-)research aims at conducting an in-depth analysis of the role of the international legal framework of agro-biodiversity management to increase the awareness of agents and policy world on its bottlenecks and possible improvements. As seen above, biodiversity management is affected by many national and international rules, laws and agreements dealing with access to genetic resources, the saving, reusing, exchanging and selling of seeds and other reproductive material, the improving of varieties and the conserving of landraces (from agro-biodiversity policies to seed laws, variety release procedures and intellectual property rights policies). In particular, the project focuses on the legal aspects that relate to on-farm management, aiming at highlighting problems and identifying ways to overcome them. To that end, it explores all the factors (of technical, organisational, economic, social and cultural nature) involved in the design of new solutions. Based on the findings of analyses and experimentations, at the end of the project the specific task aims at providing recommendations to policy makers, addressed to the creation of a new, enabling environment.

The enhancement of the agro-biodiversity through genetic resources evaluation, crops selection and other farming practices needs to integrate with a general acknowledgment and consensus by society and, more specifically, the appreciation of the final products on the market. Another area of activities in DIVERSIFOOD is so exploring the key elements underlying the building of a consistent valorisation strategy, aimed at strengthening and promoting the whole production-consumption system based on specific genetic resources, so guaranteeing their conservation and reproduction. According to the general approach, also the development of a valorisation strategy is considered as a multi-actor process, based on the role of the different actors involved from the breeding and farming stage to that of consumption of the final products. The interaction among these actors and the alignment of their attitudes and practices around the values embodied in the products are crucial aspects to explore. These aspects are important for the definition of the qualitative attributes of the final products, to which the genetic resources employed and their management in the fields, as well as the agronomic techniques to produce the raw material and the processing modes are determinant. Furthermore, they underlie the economic and cultural valorisation of these attributes through marketing and communication practices, in the local and extra-local systems. The last are essential to further spread the knowledge of the value of biodiverse products and so contribute to the consolidation of the production systems. Also, in this case, the project considers all the different dimensions—technical and organisational, economic, cultural and institutional—involved in the valorisation of the biodiverse products.

Going back to methodology, it is evident that the project attaches great importance to the contribution that can come from the different actors involved in the construction of systems of production and consumption of biodiverse products. Their active involvement is considered central for the evaluation of the performance of the various genetic resources on the field, of the economic aspects of seed management, of the properties of raw material for processing, of nutritional, healthy and organoleptic quality of the final products. As already emerged, this approach finds application through appropriate participatory methodologies in all the subtasks of the project.

At the basis of this participatory approach is the importance attached to the encounter amongst different forms of knowledge and expertise, considered potential sources of synergies, and to the development of a pool of new shared knowledge through the social learning that develop from interaction. More in general, the need to combine natural sciences, to catch the biological aspects of agro-biodiversity management, with social sciences, to also consider the cultural and social determinants that intervene, sees the project engaged in experimenting and conceptualising forms of inter-disciplinarity.[46] Together with finding effective modes of dialogue between academic and practitioners, between different chain agents or these and consumers, the integration amongst different disciplines represent the main methodological challenge of the project and one of the expected outcomes.

4 Some Final Remarks

There is a general social and political consensus on the importance of preserving and enhancing biodiversity. Nonetheless, several factors are causing a progressive loss of biological diversity. The process of agricultural intensification and specialisation that is the basis for the mainstream, industrial food system is identified as one of the threats to biodiversity. On the other hand, transitioning to sustainable farming is seen as a way to protect and enhance biodiversity, through the diversification of species and varieties in the production cycles, as well as through the provision of agro-ecosystem services. Conversely, preserving agro-biodiversity appears as a contribution to the broader goal of protecting and enhancing biodiversity.

There are however clear evidences of the complexity of the matter of agro-biodiversity management. This has firstly to do with the appreciation of the value of agro-biodiversity, which is crucial but inherently complex. There are in fact several values attached to agro-biodiversity. There are economic values, as well as other values, that may be decisive for agro-biodiversity management but are not always immediate and not easy to quantify in monetary terms, or subject to different interpretation by different stakeholders. The perception of agro-biodiversity values and their translation in coherent practices are, however, central to agro-biodiversity management. As illustrated by the study, these processes are related to other important aspects and processes, such as cultural and social dynamics, forms of stakeholders involvement, strategies of market valorisation of final products, role of the political frameworks within which strategies to maintain and enhance agro-biodiversity have to develop.

[46]Mathez-Stiefel et al. (2009), Nuijten (2011) and Johns et al. (2013).

This study has presented the DIVERSIFOOD project as exemplification of the comprehensive approach that is needed to effectively preserve and enhance agro-biodiversity. The structure of the project expresses the effort to explore the processes underlying the re-shaping of practices according to the objective of agro-biodiversity enhancement in all the areas involved in tackling with agro-biodiversity, from the farming activities to commercialisation of the biodiverse food products. The need to consider the different perspectives on the values and ways of preserving agro-biodiversity together with the social learning processes that lead to coherent practices underlies the adoption of participatory, multi-actors approaches to each of these stages. Breeder, farmers, processors, consumers and other actors of the food chains, scientists, civil society organisations and other relevant stakeholders are involved. Similarly, an inter-disciplinary approach, aimed at combining natural sciences and social sciences characterises the work of researchers. Looking at the mechanisms underlying and affecting the implementation of coherent practices, DIVERSIFOOD also investigates and raises awareness on the role that socio-cultural, legal and political frameworks may play in enabling socially recognised, participatory, community-based management of agro-biodiversity resources. This concerns all the areas of intervention analysed, but assumes special relevance in the specific case of on-farm management of seeds. This is the real and symbolic space in which the presence of different views, goals and approaches is more evident.

Finally, this contribution sheds light on the role that research may play in agro-biodiversity preservation. To be effective in this task, research projects shall look through a systemic approach at all the (environmental, technical, socio-cultural, economic, legal and political) aspects that intervene in the various areas of agro-biodiversity management. To that end, they shall involve both academics—from different disciplines—and practitioners and stakeholders—from different stages of the food chain and even beyond. Together, they should provide insights and recommendations on the many, connected entry points of agro-biodiversity enhancement.

References

Alho CJR (2008) The value of biodiversity. Braz J Biol 68:1115–1118

Almekinders CJM, Elings A (2001) Collaboration of farmers and breeders: participatory crop improvement in perspective. Euphytica 122:425–438

Ash N, Fazel A (2007) Biodiversity. Global Environmental Outlook, pp 160–192

Bonneuil C, Goffaux R, Bonnin I, Montalent P, Hamon C, Balfourier F, Goldringer I (2012) A new integrative indicator to assess crop genetic diversity. Ecol Indic 23:280–289

Brunori G, Rossi A, Malandrin V (2011) Co-producing transition: innovation processes in farms adhering to solidarity-based purchase groups (GAS) in Tuscany, Italy. Int J Sociol Agric Food 18(1):28–53

Ceccarelli S, Grando S (2007) Decentralized-participatory plant breeding: an example of demand driven research. Euphytica 155:349–360

Clawson DL (1985) Harvest security and intraspecific diversity in traditional tropical agriculture. Econ Bot 39:56–67

Convention on Biological Diversity (CBD) (1992) United Nations, retrieved from www.cbd.int
De Boef WS, Dempewolf H, Byakweli JM, Engels JMM (2010) Integrating genetic resource conservation and sustainable development into strategies to increase the robustness of seed systems. J Sustain Agric 34(5):504–531
De Boef WS, Subedi A, Peroni N, Thijssen M, O'Keeffe E (eds) (2013) Community biodiversity management: promoting resilience and the conservation of plant genetic resources. Issues in agricultural biodiversity. Routledge, London
European Commission (1998) EC Biodiversity Conservation Strategy (ECBS), retrieved from http://ec.europa.eu/environment/nature/biodiversity/strategy/index_en.htm
European Union (2001) Sustainable Development Strategy (SDS), retrieved from http://eur-lex.europa.eu/legal-content/EN/TXT/?uri=URISERV%3Al28117
Farnworth CR, Jiggins J (2003) Participatory plant breeding and gender analysis. PPB Monograph n. 4
Fess TL, Kotcon JB, Benedito VA (2011) Crop breeding for low input agriculture: a sustainable response to feed a growing world population. Sustainability 3:1742–1772
Government of Ireland (2008) The economic and social aspects of biodiversity. Benefits and costs of biodiversity in Ireland. The Stationary Office, Dublin
Gruère G, Smale M, Giuliani A (2006) Marketing underutilized species for the benefit of the poor: a conceptual framework. IFPRI, Washington, DC (EPT Discussion Paper 154)
Haberl H, Gaube V, Díaz-Delgado R, Krauze K, Neuner A, Peterseil J, Plutzar SC, Vadineanu ASJ (2009) Towards an integrated model of socioeconomic biodiversity drivers, pressures and impacts. A feasibility study based on three European long-term socio-ecological research platforms. Ecol Econ 68(6):1797–1812
Hardon J (1995) Participatory plant breeding. The outcome of a workshop on participatory plant breeding, 26–29 July 1995. Plant Genetic Resources No. 3. IPGRI, Rome
International Convention for the Protection of New Varieties of Plants (1961) Available at http://www.wipo.int/wipolex/en/other_treaties/details.jsp?treaty_id=27
Isaac ME, Erickson BH, Quashie-Sam SJ, Timmer VR (2007) Transfer of knowledge on agroforestry management practices: the structure of farmer advice networks. Ecol Soc 12:32
Johns T, Eyzaguirre PB (2006) Linking biodiversity, diet and health in policy and practice. In: Symposium on Wild-gathered plants: basic nutrition, health and survival. Proc Nutr Soc 65:182–189
Johns T, Powell B, Maundu P, Eyzaguirre PB (2013) Agricultural biodiversity as a link between traditional food systems and contemporary development, social integrity and ecological health. J Sci Food Agric 93(14):3433–3442
Krucken L (2005) Design and the valorisation of agricultural biodiversity products - a case study. In: Proceedings of the 6th international conference of the European academy of design. University of the Arts of Bremen, Bremen
Laverty MF, Sterling EJ (2003) Why is biodiversity important? Center for Biodiversity and Conservation of the American Museum of Natural History
Leclerc C, Coppens d'E G (2011) Social organization of crop genetic diversity. The G × E × S interaction model. Diversity 4(4):1–32
Louwaars N, Le Coënt P, Osborn T (2010) Seed systems and plant genetic resources for food and agriculture. Thematic background study for the Second Report on the State of the World's Plant Genetic Resources for Food and Agriculture. FAO, Rome. http://www.fao.org/fileadmin/templates/agphome/documents/PGR/SoW2/tbs_Seed_Systems_081209.pdf
Mathez-Stiefel SL, Gianella MC, Rist S (2009) Valorisation of Agrobiodiversity products and strengthening of local identities in the Peruvian Andes: experiences from the BioAndes Programme. Colloque "Localiser Les Produits"
Maxted N, Akparov ZI, Aronsson M, Asdal A, Avagyan A, Bartha B, Benedikova D et al (2012) Current and future threats and opportunities facing European crop wild relatives and landrace diversity. In: Maxted N, Iriondo MM, Pinheiro de Carvalho MMA (eds) Agrobiodiversity

conservation: securing the diversity of crop wild relatives and landraces. CABI Publishing, Wallingford, pp 333–353
Mayes S, Massawe FJ, Alderson PG, Roberts JA, Azam-Ali SN, Hermann M (2011) The potential for underutilized crops to improve security of food production. J Exp Bot 63(3):1075–1079
Morris ML, Bellon MR (2004) Participatory plant breeding research: opportunities and challenges for the international crop improvement system. Euphytica 136:21–35
Naziri D (2009) Direct sale as a means for promoting the sustainable use of plant genetic resources: the case of the Tuscany region. J Agric Environ Int Dev 103:65–80
Nuijten E (2011) Combining research styles of the natural and social sciences invagricultural research. NIAS-Wageningen J Life Sci 57:197–205
Nuijten E, Temudo M, Richards P, Okry F, Teeken B, Mokuwa A, Struik PC (2013) Towards a new approach for understanding interactions of technology with environment and society in smallscale rice farming. In: Wopereis MC, Johnson S, Ahmadi DE, Tollens N, Jalloh A (eds) Realizing Africa's rice promise, vol 29. CAB International, Wallingford, pp 355–366
OECD (2002) Handbook of biodiversity valuation. A guide for policy makers
OECD (2008) OECD environmental outlook to 2030
Padulosi S, Eyzaquirre P, Hodgkin T (1999) Challenges and strategies in promoting conservation and use of neglected and underutilized crop species. In: Janick J (ed) Perspectives on new crops and new uses. ASHS Press, Alexandria
Padulosi S, Thompson J, Rudebjer P (2013) Fighting poverty, hunger and malnutrition with neglected and underutilized species (NUS): needs, challenges and the way forward. Bioversity International, Rome
Pimentel D, Stachow U, Takacs DA, Brubaker HW, Dumas AR, Meaney JJ, O'Neil JAS, Onsi DE, Corzilius DB (1992) Conserving biological diversity in agricultural/forestry systems. Bioscience 42(5):354–362
Simoncini R (2015) Introducing territorial and historical contexts and critical thresholds in the analysis of conservation of agro-biodiversity by Alternative Food Networks in Tuscany, Italy. Land Use Policy 42:355–366
Thijssen M, De Boef WS, Subedi A, Peroni N, O'Keeffe E (2013) General introduction. In: Thijssen M, De Boef WS, Subedi A, Peroni N, O'Keeffe E (eds) Community biodiversity management: promoting resilience and the conservation of plant genetic resources. Routledge, London, pp 3–10
Tripp R (1997) The structure of the national seed systems. In: Tripp R (ed) New seeds and old laws. Regulatory reform and the diversification of national seed systems. IT Publications, London, pp 14–42
Turner RK, Pearce DW, Bateman I (1994) Economia ambientale – una introduzione elementare. Il Mulino
Van Noordwijk M, Spek LY, De Willingen P (1994) Proximal root diameters as predictors of total root system size for fractal branching models. I. Theory. Plant Soil 164:107–118
Will M (2008) Promoting value chains of neglected and underutilised species for pro-poor growth and biodiversity conservation. Guidelines and good practices. Global Facilitation Unit for Underutilised Species, Rome
Wolfe MS, Baresel JP, Desclaux D, Goldringer I, Hoad S, Kovacs G, Löschenberger F, Miedaner T, Østergård H, Lammertsvan Bueren E (2008) Developments in breeding cereals for organic agriculture. Euphytica 163(3):323–346

Remarks on Integrated Production (IP), Different Agricultural Systems and Coordinating Groups

Giuseppina Buia

1 The Definition of Integrated Production and Its Regulation: The Problem of Shared Failure

"*The Message of Ovronnaz*",[1] the first amongst the official scientific documents written on the theme of integrated production, has represented an important step in the progress of human society towards an "ecological culture".[2]

Since its first formulation in the *Message*, integrated production (IP) has shown itself to be perfectly adherent to a model of society not merely sensitive to environmental protection, but interested to interact with it.

The link between "ecological culture" and IP is even more evident if one considers their definitions.

"Ecological culture" describes "a life attitude that is based on the ecological paradigm, i.e., the systematic interpretation of reality, corroborated in scientific and philosophical terms, and pursues the objective of eco-development, namely the

PhD student in "Law of private, public and common goods. Tradition and innovation in the models of belonging and fruition" and Teaching Assistant ("Cultore della materia") of the Environmental Administrative Law Chair at the Faculty of Law—University of Salento (Italy).

[1]Published in 1977 in the Official Bulletin of OIBC/WPRS (International Organisation for Biological Control/West Palaeartic Regional Section), the *Message* represents the product of 30 years of experience of researchers in the field of integrated plant protection. It is a theoretical document but drawn from practical inspiration: it contains the regulatory bases and principles of the material formulated from the research but also the direct experience of its authors in the field. See Steiner et al. (1977). On the history prior to 1977 and, therefore, more widely on the concept of *integrated protection*, reference is made to the first part of the work by Principi (1992).

[2]Caporali et al. (2010), p. 3.

G. Buia (✉)
Department of Legal Studies, University of Salento, Lecce, Italy

A. Isoni et al. (eds.), *Food Diversity Between Rights, Duties and Autonomies*, LITES – Legal Issues in Transdisciplinary Environmental Studies 2,
https://doi.org/10.1007/978-3-319-75196-2_2

harmonic coexistence of man with the environment which evolves and where species multiply".[3]

Integrated production, however, is embodied by a different way of producing food, voluntary and ecosustainable, the essence of which lies in the simultaneous "*respect, integration and enhancement*" of positive factors of the ecosystem.

In practical terms, this translates into the use of techniques favouring production that does not contrast with ecological principles[4]: the agricultural sector is itself part of an ecosystem, and can only be best realised with the application of principles internal to the ecosystem itself.

The definition of the production model under analysis is not yet exhausted in attention and exaltation of the ecosystem, but consists of an additional element, so fundamental as to be preferred to others for the qualification of this production model: *integration*.

In addition to being respected and valued, the ecosystem should therefore be integrated with the best available scientific and technological techniques, able to compensate for the poor productivity and profitability often criticised of other production forms (organic production, for example). In this way, IP reproduces ancient farmer logic based on the mediation between economic and ecological needs.[5]

Consequently, it does not seem rash to argue that IP will contribute to the creation of a new definition of agroecosystem, in which natural factors potentially harmful to crops (e.g., some species of insects) are combatted, but through methodologies that interweave natural remedies with high-quality remedies offered by new technologies.[6]

Integrated production can therefore be said to express a new way of relating to the proper existence of the "ecological society": the agroecological approach. In fact, agroecology is nothing but the "the integrative study of the ecology of the entire food system, encompassing biophysical, economical and social dimensions".[7] The deviation into an integrated study of disciplines can be explained by virtue of the "need for cooperation between law and science in agroecology", but above all in consideration of the need to create a "second-order platform of agroecology to obtain a

---

[3]Caporali et al. (2010), pp. 3–4.

[4]For the definition of IP, *ivi* containing the quote on "*respect, integration and enhancement*" of positive factors of the agroecosystem, Cf. Steiner et al. (1977); Deguine et al. (2008), pp. 121 et seq.

[5]Boller et al. (1998), p. 5.

[6]Consider, for example, precision agriculture and in particular sensors built with the help of nanotechnological devices: the sensors detect the moisture level of the soil and, more generally, monitor the health status of the plant. In this way, it is possible to intervene *ex ante* on the plant in case of pathological symptoms, avoiding the risk of an *ex post* cure to be made through the use of pesticides. On the prospects of nanotechnology and on the inherent problems in their regulation, please refer to Monteduro and Buia (2014).

[7]Cf. Caporali (2015), p. 5, which summarises the version of Francis (2004).

polyocular view on the potentials for sustainable regulation and support of the development of agroecosystems".[8]

From the first formal elaboration in the *Message* of 1977, the concept of IP has been enriched in content, while maintaining unaltered those which can be considered its basic elements: exaltation of the quality of the entire ecosystem and the application of ecological principles.[9]

More specifically, these aspects are found today in the fixed objectives for IP: the idea of respect for and enhancement of positive factors of the ecosystem is materialised in the intent to reduce the use of synthetic pesticides as much as possible, replaced by the growing application of agronomic techniques. In this way, we will meet the increasingly looming need to meet the nutritional requirements of the entire planet and to promote protection of the environment. The remarkable success of IP in the field of pest management underlines the urgent need to address the problem of scarce food resources. It is estimated that the total absence of this type of management would cause a considerable decrease (up to 40%) in world production.[10] With IP, therefore, pest management is not abandoned but is rethought and practiced through methods that are environmentally friendly. The aforementioned techniques are based on a principle already known to science: the principle that chemistry and biology are not *alternative* but *complementary* to each other.[11]

Regarding integration, it corresponds to the aim of combining organic and biotechnology methods with good agricultural practices and the moderate use of chemicals.[12]

Finally, borrowing the definition offered by the International Organisation for Biological and Integrated Control (IOBC), IP can be defined as "a food production system that optimises the use of resources and natural regulatory mechanisms, ensuring long-term viable and sustainable agriculture". It carefully selects the organic methods, cultivation techniques and chemical processes to utilise, "looking for a balance between *the environment, profitability and social needs*".[13]

Conversely, integrated production makes use of the combination of principles, methods and instruments cultivated by various branches of science—biology, ecology, etc.—and sectors not strictly scientific—economics, ethics, sociology—relegating to

---

[8]Noe and Alrøe (2015), p. 42.

[9]On the evolution of the IP concept, Cf. Ferron (1999).

[10]Cravedi (2012), pp. 56–57.

[11]Cf Stern et al. (1959), p. 94: "Biological control and chemical control are not necessarily alternative methods; in many cases they may be complementary, and, with adequate understanding, can be made to augment one antoher. One reason for apparent incompatibility of biological and chemical control is our failure to recognize that the control of arthropod populations is a complex ecological problem. This leads to the error of imposing insecticides on the ecosystem, rather than fitting them into it".

[12]Boller et al. (1998), p. 14; Canali (2008a, b), p. 9.

[13]European Economic and Social Committee (2014/C 214/02), p. 2.

extrema ratio the use of harmful methods from all of the above mentioned fields.[14] In this way, IP allows for the satisfaction of environmental needs and, at the same time, needs expressed by the economy and society. Therefore, it could be argued that: if the agroecological approach is linked to organic production by virtue of the exclusion of the use of pesticides, and does not find margins of permeability in traditional production because it is based, amongst other things, on the massive use of pesticides, the connection between the agroecological approach and IP may be found in the balance and proportionality of the solutions it offers for environmental, economic and social needs.

At the national level, however, the accepted definition of IP in official texts of the Ministry of Agricultural, Food and Forestry Policy (in Italian, Ministero delle Politiche Agricole Alimentari e Forestali (MIPAAF)) (most recently from Law n. 4 of 3 February 2011,[15] establishing the "National quality system of integrated production") is the following: IP is the "agricultural production system that utilises all means of production and protection against adversity, aimed at minimising the use of synthetic chemicals and at rational fertilisation, in compliance with ecological, economic and toxicological principles".

A well-established definition of IP does not reflect an equal stability in the legal framework.

This fact may be seen at the European level where a uniform regulation for IP is missing, but it is also evident on the national and regional scale, where a series of guidelines[16] and various regional regulations crowd the legal landscape for IP.

Particularly, regional regulations were essential because of the need to guarantee that the techniques used are adaptable to the economic, biological and chemical conditions of the impacted territories.[17]

Regardless of the evaluation of opportunities for elaboration of a European legal framework for IP, the fragmentation of the current legal structure could lead to the danger of internal conflicts between different national guidelines and regional regulations (however essential) existing in all Member States.

The absence of EU legislation on the subject not only feeds the danger involved, but also affects a different level that is the *sharing* of a joint legal structure.

---

[14]MIPAAF (2008), p. 2.

[15]Law later implemented with Ministerial Decree 4890/2014.

[16]The National Guidelines for Integrated Production 2016, broken down into areas: integrated pest management (http://www.reterurale.it/flex/cm/pages/ServeAttachment.php/L/IT/D/u%252Fn%252Fi%252FD.7d471996196bd0287a04/P/BLOB%3AID%3D15591), agronomic techniques (http://www.reterurale.it/flex/cm/pages/ServeAttachment.php/L/IT/D/u%252Fn%252Fi%252FD.9da029e1485b0923d891/P/BLOB%3AID%3D15591); quality system for integrated production (http://www.reterurale.it/flex/cm/pages/ServeAttachment.php/L/IT/D/u%252Fn%252Fi%252FD.ca52796339a17b7dd981/P/BLOB%3AID%3D15591).

[17]To get an idea of the various regional agricultural practices present in Europe for IP, see the document created by AREFLH, the French association that brings together the leading fruit and vegetable producing regions of France, Spain, Italy and Portugal. Cf. AREFLH (2013).

The central role of sharing, and therefore the need to determine the problems related to its absence, has been implicitly confirmed by the European Economic and Social Committee in its recent opinion on IP and the relative notes.[18] In these documents, the Committee references sharing many times, referring to it in two different ways: on the one hand, it expresses their doubts at the moment that consumers are adequately informed on IP and that therefore actually know the basic concepts behind it; on the other hand, they speak about sharing when the Committee considers the opportunity to create a legal framework for IP that is common to all Member States.

Conversely, an *established* definition of IP seems to not correspond to a *shared* definition of it, and it is very likely that a scenario characterised by multiple legal disciplines such as today is still a symptom of the lack of sharing.

2 Relations Between the Different Production Chains in Light of the Principle of Coexistence

In the previous section, we have highlighted two of the focus points of the question concerning the definition and regulation of IP: the lack of consumer information about the basic concepts of IP and the European regulatory gap on these issues.

Both points are also relevant with reference to a further problem, which is the relationship between IP, conventional, organic and genetically modified (GMO) production.

Although IP and organic production express different forms (although in approach) of a single matrix that is sustainable agriculture, they are not jointly regulated at the European level. Indeed, while for organic production a European framework exists, the same cannot be said for IP.

The lack of a similar regulatory *background* could cause, as is obvious, some application difficulties and many problems in terms of the organisation of production.[19] Basically, it could be repeated for IP as happened in the past at the moment GMO production entered the European territory: the disorientation of producers and consumers in the choice between the different supply chains available and also in the rules to follow for each type of production (e.g., labelling and traceability).

In this sense, the lack of information would damage not only the consumer (as feared by the Economic and Social Committee), but also the producer, who *must* be put in the position to be able to choose which type of supply chain targets his

[18]European Economic and Social Committee (2014/C 214/02); European Economic and Social Committee (2014 NAT/596).

[19]In the United States, difficulties of this type were encountered especially about Integrated Production Management (IPM): the disappointing effects of the Integrated Crop Protection Program revealed the state of confusion on the definition of IPM and the consequent need to return to these issues. Cf. Ehler and Bottrell (2000).

production, given that "none of these forms of agriculture should be excluded in the European Union".[20]

The power of the choices' roots lies in the so-called *principle of coexistence*,[21] according to which each type of crop shall be carried out without hindering the other. Coexistence is explained by the need to preserve the "production peculiarities and specificities"[22] of each crop and for this reason it is important that the measures chosen to achieve coexistence ensure as much as possible the avoidance of mingling between the different crops.[23]

In this regard, the entry of IP in the legal framework could complicate the fulfilment of the strict separation between the productive models: as anticipated, not only the production but the whole organic agriculture is converging increasingly towards integrated production (and agriculture). Organic agriculture, in fact, despite having the advantage of not negatively impacting the environment, reveals a certain imbalance between demand (rising) and supply, consequently affecting the rise in sale prices. On the other hand, even traditional agriculture is often linked to one integrated with some elements, for example, the use (albeit regulated in IP) of pesticides.

Second, it is not clear if with IP the differentiation of supply chains is destined to fade over time, and with it the principle of coexistence.

In its opinion on IP, the position of the Economic and Social Committee is rather ambiguous. The recommendations and considerations of the Committee are mainly concerned with the idea that farming should be based on sustainability, namely on the rational use of available natural resources in accordance with preservation of the environment, profitability and attention to social context.

Thus, the coexistence of different production models is not excluded to achieve sustainable agriculture, which has become "a fundamental necessity of civil society". At the same time, however, the Committee sets up IP as a "goal to be reached" and claims that producers who choose integration are those that take us toward an "*even greater* compliance with sustainable production norms".[24]

---

[20]European Commission (C (2003) 2624), p. 39.

[21]On the application of the principle of coexistence with the precautionary principle in situations of scientific uncertainty, see Marini (2007). The Italian Constitutional Court, with judgment 116/2006, has described the principle of coexistence as "the point of synthesis between divergent interests of constitutional importance, made up on one hand by the freedom of economic initiative of the farmer and on the other by the requirement that such liberty is not exercised in conflict with the social utility, and in particular causing disproportionate damage to the environment and health." The Court does not explicitly refer to IP but it refers to GMO, organic and conventional crops; however, while the framing it provided also appears extensible to the above type of production. Interesting doctrinal comment on the judgment 116/2016 by Altili (2007).

[22]Law n. 5/2005.

[23]On the issue of coexistence and the danger of contamination between conventional, organic and genetically modified, Cf. in the end Rossi Dal Pozzo (2014).

[24]Quotations from the European Economic and Social Committee (2014/C 214/02).

To paraphrase, it could be said that all production models *present and future* are regarded as eligible provided they are able to guarantee sustainable agriculture, but currently the model that ensures the highest level of sustainability is precisely IP.

In this step, the Committee's position seems to favour the interpretation of the principle of sustainable development that tends as much as possible towards the protection of the environment and the social sphere. The latter retreat before economic requirements (profitability) only in exceptional cases (properly motivated) and provided that certain conditions are satisfied (the guarantee of quality and residual nature of the utilised scientific instruments).

Because of a similar trend in Europe, protection of the plurality of supply chains, whose existence is presently guaranteed by the principle of coexistence, may be somewhat compromised by the expansion of IP. The model is a winner by the balance between voluntary adhesion and strictness in the certification, but not just in economic terms, given that certification involves a heavy financial burden for producers.

The compromise may also be favoured by the compulsory integration into the relative field of crop protection[25] that was scheduled to begin in 2014, integration that instead remains voluntary for all other areas of incidence within the production model analysis (e.g., the one on fertilisation). The purpose of the decision is to ensure one minimum standard of sustainability in agriculture, whatever type of supply chain is chosen.

3 The Regulation of Integrated Production in Italy

Despite the absence of a European legal model of regulation, which could serve as a reference for all Member States, the Italian legislature felt the urgent need to develop an appropriate form of regulation for the sector in question. To this end, a particularly complex organisational structure was designed, although characterised by the establishment of coordinating bodies with mixed composition, functional to assist the MIPAAF in above-mentioned activity.

Before delving into the detailed examination of these bodies, it can be useful to carry out, beforehand, a brief analysis of the organisational structure.

According to the latest act of organisational regulation by MIPAAF (the Presidential Decree of the Council of Ministers 27 February 2013, n. 105), the IP sector is amongst those attributed to the "Department of competitive policies, food quality, horse racing and fisheries" (DIQPAI), to which, amongst others, the following tasks are assigned: (*i*) protection and enhancement of product *quality* (delegated by the Ministry); (*ii*) treatment of *communication and informational activity* in the matter of *quality* of agricultural and food products.

[25] On the importance of crop protection in agriculture, See Boller et al. (1995).

On the point of engaging two considerations: from regulatory provisions it may be inferred that the Italian legislature presents itself as perfectly in line with what can be considered the goal of IP, or incentivising an agricultural market whose food products are primarily qualitatively competitive before quantitatively present. Consistency is given by the constant references to quality in Decree n. 105/2013, particularly Article 3, which is entirely dedicated to DIQPAI, but also from the non-coincidental formula "food quality" inserted into the name chosen for the Department.

The second consideration rather relates to the connection between the content of point (*ii*) and the issue denounced by the European Economic and Social Committee in relation to awareness-raising activity to be carried out concerning consumers, especially for the dissemination of IP content. It is clear that, in Italy, the DIQPAI should avoid the shortcomings identified by the Committee, and it could be called upon to conduct a more insistent campaign in layman's terms regarding definitions, objectives and effects of the IP model.

The competences of the Directorate General for the promotion of food quality and horse racing, or one (of interest for the topic under discussion) of the three offices at the general management level that make up DIQPAI, appear to be more detailed.

Next to the qualification of general discipline and coordination in the field of quality of agricultural and food products, the aforementioned management office is assigned another competence of particular importance to the issue under discussion: the definition of the regime and administration of the "*National Quality System for Integrated Production*".

As the Directorate is entrusted with the competence on the System's discipline in general, the fundamental role of regulating the matter is carried out particularly by other types of bodies: coordinating bodies.

The bodies are essentially concerned with operating functions (such as the development and approval of criteria and general principles of IP) and control (such as monitoring the proper functioning of the System).

This is the case of the Committee on Integrated Pest Management (CDI), which in 2005 became the successor to the National Committee established in 1996.[26] The CDI was then chosen (*ratione materiae*) to assist the MIPAAF in regulating the National Quality System for Integrated Production: the Committee, in fact, existed prior to the first accord between the Ministry, the regions and the autonomous provinces that establishment it (dated 20 March 2008).[27]

The CDI was then absorbed into the Committee on Integrated Production (CPI), becoming one of the specialised technical groups that composed that body; the CPI, however, remained in operation only until 2014, the year in which a true and proper Technical Scientific Body on Integrated Production (OTS) was established.[28]

[26] Art. 2, Ministerial Decree 2722/2008.

[27] Cf. Ministerial Decree 2722/2008.

[28] Art. 3, Ministerial Decree 4890/2014, implemented by art. 2, para. 6, Law n. 4/2011 on "Measures on labeling and quality of food products" (published in the Gazetta Ufficiale 29 July 2014, n. 179).

4 Committee on Integrated Production and the Technical Scientific Body on Integrated Production Compared: Emergence of a Strong Technicalisation of the Coordinating Bodies

The changes that we have seen in terms of administrative organisation in the agricultural sector are even more evident if one looks at the composition and functions of the old and new coordinating bodies with which MIPAAF has collaborated.

Comparing, for example, the now repealed Committee on Integrated Production with the new Technical Scientific Body on Integrated Production results in the following observations.

In terms of their composition, the most important difference is the *obligatory* rather than optional presence of a competent representative of the national scientific institutions on IP.

Before 2014, the MIPAAF expressly required three representatives of the Council for Research and Experimentation in Agriculture (CRA)[29] "or other scientific institutions at the national level".[30] With the Decree establishing the OTS, however, the number of CRA representatives was set at two and simultaneously one representative is required from other important national institutions, which must be specifically competent in the field of IP.

With respect to the past, therefore, we see there is a requirement for persons with specialised training in the regulated sector.

The OTS, like the old CPI, is thus supported by three specialist groups: the then-autonomous Group on Integrated Pest Management (GDI), the Group on Agronomic Techniques (GTA) and the Technical Group on Quality (GTQ).

The groups in question were chosen as representatives of each region or autonomous province, MIPAAF and the CRA; also, in the OTS, as in the CPI, there are respective *experts* in integrated pest management, agronomic practices and the matter of quality systems.

In terms of the tasks performed by CPI and OTS, it is possible to distinguish "inherited" functions from the newly acquired functions.

Within the list of old functions certainly falls the approval criteria and the general principles of IP expressed with the national guidelines.

More precisely, the criteria and general principles of the guidelines are developed by these specialist groups within the OTS, broken down by area of expertise: the GDI addresses the guidelines related to integrated pest management, the GTA handles those linked to agronomic techniques and the GTQ defines the rules and

[29]The CRA is a national body under the supervision of MIPAAF, which enjoys a certain autonomy in research and experimentation activities. For thematic enquiries generally concerning the agricultural, industrial agriculture, fisheries and forestry sectors, http://sito.entecra.it/portale/index2.php?lingua=IT&access_flag=0. Accessed 6 Feb 2016.

[30]Art. 2, Ministerial Decree 2722/2008.

management procedures of the National Quality System for IP. Once the creative phase is completed, each group presents their guidelines to the OTS, which directly approves or initially integrates and then approves them as appropriate.

The approval and updating of the guidelines are crucial phases of the IP control process, as these lines serve as a benchmark for specifications (containing the "technical standard"),[31] that the regions are obliged to draft for the territory within their competence.

Coming now to the newly-acquired functions, the novelties to mention concern the activation of the coordinating body on the operational level: the body, from a mere theoretical creator of the IP regulatory model that was at the time CPI, also becomes a promoter of application techniques with a *modus operandi* that are indicated as a theoretical model to follow. Thus, the OTS is also given the task of fixing the *means of intervention* in the sectors impact by IP.[32]

The *modus operandi* and the means of intervention, now merged into the national guidelines, will then be subject to further verification based on performance indicators prepared by the OTS. The requirement for effectiveness could be satisfied presently when implementation of the guidelines has a positive impact on the social, economic and environmental sphere, with due respect for sustainable development.

Consequently, those specific guidelines could be used only if they are qualified by the OTS as best suited to achieve the same objective imposed by IP: balance between environment, profitability and social needs.[33]

Based on the concise analysis above, it is possible to encounter important changes in the regulation of IP and particularly find two aspects that are probably the consequence of each other.

First, the tasks attributed to the coordinating body called upon to support the MIPAAF in the investigative field are no longer within political-administrative policy activities, but extend to encompass practical functions such as the identification of intervention measures.

Secondly, there has been a strong and progressive "technicalisation" of these bodies, as, moreover, suggested by the name change to "technical scientific body".

The obligatory presence of persons who can be bearers of ideas pertaining to a specific scientific community (experts in IP), together with the incisiveness of the functions reserved by OTS, would seem to confirm this hypothesis.

---

[31] The inclusion of a "technical standard" that is drawn up at the regional level stems from the need to ensure it is "responsive to the particular soil and climate conditions of each region and autonomous province". Art. 2, Ministerial Decree 4890/2014.

[32] Art. 3, Ministerial Decree 4890/2014.

[33] Cf. Sect. 1 in this chapter.

5 Conclusions

The comments made thus far should have demonstrated how IP represents, by now, an indispensable basis for comparison for each of the productive models present in agriculture. This is in consequence of the fact that it incorporates an optimal system (at least for now) for the achievement of sustainable agriculture.[34]

This fact, albeit positive, may have side effects on the level of the plurality of supply chains, currently guaranteed by the existence of the principle of coexistence.

Making the concept in more explicit terms, it could be said that IP conceals in itself a paradox: on the one hand, it is an expression of a different way of feeding oneself,[35] which balances between environment, profitability and social needs, and thus contributes to the protection of the plurality of supply chains; on the other hand, the extreme level of the myth, of the "goal to be reached", could dent the same plurality and debase the principle of coexistence.

As often happens, the solution to the problem could be internal to the problem itself, and in this case, may lie, in fact, in the greatest gift of IP: balance.

In another area, the results of the investigation have revealed an aspect of IP that reverberates even in the bodies responsible for its regulation, or sharing.

Sharing is required with respect to the legal definition of IP, which would be more widespread and communicated to the producers,[36] as well as the consumers. But it is above all a constitutive element of IP and emerges when the analysed model is described as a system that utilises natural principles and instruments, integrated with devices that are the fruit of human ingenuity.

The attempt to communicate between natural and social sciences has also involved Italian bodies involved with the regulation of IP.

They now consist of highly specialised technicians that support policy on the regulatory development of IP, fostering dialogue between experts in disciplines distanced from each other for too long, such as law, economics, biology, chemistry. The propensity to dialogue together regarding the will to contribute to the creation of a "poli-ocular platform" capable of simultaneously responding to the pressures exerted on the environment, economy and social sphere could lead to the conclusion that the coordinating bodies achieve the key organisational agroecological perspective.

The change responds perfectively to the nature of the regulated object, also based on sharing and the agroecological approach, and tells of a continuous evolution of

[34]On the issue of sustainable agriculture in Europe in view of the targets set for 2020, See the Swiss Thematic Proposals for the 6th Framework programme for Research and Technological Development of the European Communities.

[35]On food diversity as a "value-system" and the new interplay between "nature and society", cf Monteduro in this volume. On other profiles inherent to food diversity, see the other contributions in this volume.

[36]The inadequacy and even the lack of technical assistance to farmers were already reported in 1992, a time when this inefficiency was traced to the non-transmission of the parties concerned of the results achieved with research on IP. Principi (1992), p. 94.

the administrative organisation, whose control centres are no longer only "multilevel", but also "multi-species".[37]

References

Albisinni F (2013) Prodotti alimentari o agroalimentari? Il TAR Lazio, giudice del mercato e law maker, smentisce il MIPAAF e l'AGCM. Riv dir alim (3):33–47

Altili P (2007) La coesistenza tra colture transgeniche e colture convenzionali nella sentenza della Corte Costituzionale n. 116/2007. Dir giur agr alim, e amb (2):96–100

Assemblée des Régions Européennes Fruitières, Légumières et Horticoles (AREFLH) (2013) Guide des pratiques européennes de production intégrée. http://www.areflh.org/images/stories/PDF/Dossiers/Protection_des_plantes/Production_integree/Production_int%C3%A9gr%C3%A9e_2013_FINAL_FR.pdf. Accessed 10 Feb 2016

Boller EF et al (1995) Pflanzenschutz als Teil einer nachhaltigen Produktion. Agrarforschung (2):504–507

Boller EF et al (1998) Integrated production in Europe: 20 years after the declaration of Ovrannaz. Bull IOBC/WPRS 21(1):1–33. https://www.iobc-wprs.org/ip_ipm/04011_IOBC_Ovrannaz.pdf. Accessed 12 Feb 2016

Canali G (2008a) Verso una strategia europea per i prodotti agroalimentari di qualità: il "Green Paper". Agriregionieuropa 4(15):7–9

Canali G (2008b) The role of "integrated production" schemes in the new fruit and vegetable CMO: a tool for competitiveness, sustainability or oligopsony by large retail chains. Paper presented at 109th seminary EAAE on "The CAP after the Fischler reform: national implementations, impact assessment and the agenda for future reforms", Viterbo (Italy), 20–21 novembre 2008

Caporali F (2015) History and development of agroecology and theory of agroecosystems. In: Monteduro M, Buongiorno P, Di Benedetto S, Isoni A (eds) Law and agroecology. A transdisciplinary dialogue, 1st edn. Springer, Heidelberg, pp 1–29

Caporali F, Campiglia E, Mancinelli R (2010) Premessa: il primato della cultura e la necessità del suo orientamento ecologico. In: Caporali F, Campiglia E, Mancinelli R (eds) Agroecologia. Teoria e pratica degli agroecosistemi, 1st edn. Città Studi Edizioni, Torino, pp 3–5

Cravedi P (2012) Misure per promuovere un'agricoltura a basso impatto ambientale. In: Cordini G, Alessandro V (eds) Agricoltura e ambiente. XIII Convegno annuale dell'Associazione "Club giuristi dell'ambiente" – Bobbio 10 settembre 2011, 1st edn. Aracne Editrice, Roma, pp 53–61

Deguine JP, Ferron P, Russell D (2008) Bases écologique d'une gestion de population. In: Deguine JP, Ferron P, Russell D (eds) Protection des cultures: de l'agrochimie à l'agroécologie. Edition Quae, Versailles, pp 111–123

Ehler LE, Bottrell DG (2000) L'illusion de la protection intégrée des cultures par Lester E. Ehler et Dale G. Bottrell. Courr. environ. INRA 40:85–88. http://www7.inra.fr/lecourrier/assets/C40Ehler.pdf. Accessed 12 Feb 2016

European Commission (C (2003) 2624) Recommendation of 23 July 2003 on guidelines for the development of national strategies and best practices to ensure the coexistence of genetically modified crops with conventional and organic farming. http://eur-lex.europa.eu/legal-content/EN/TXT/PDF/?uri=CELEX:32003H0556&from=EN. Accessed 9 Feb 2016

European Economic and Social Committee (2014 NAT/596) Integrated Production in the European Union (own-initiative opinion) – Nota d'informazione. https://www.google.it/search?q=produzione+integrata+comitato&ie=utf-8&oe=utf-8&gws_rd=cr&ei=D2q0VpSiFMn4Uubci8AJ#. Accessed 5 Feb 2016

[37] Albisinni (2013), p. 40.

European Economic and Social Committee (2014/C 214/02) Integrated Production in the European Union (own-initiative opinion). http://eur-lex.europa.eu/legal-content/EN/TXT/PDF/?uri=CELEX:52013IE2103&from=IT. Accessed 5 Feb 2016

Ferron P (1999) Protection intégrée des cultures: évolution du concept et de son application. Cah Agric (8):389–396. http://www.jle.com/download/--agr-269754-protection_integree_des_cultures_evolution_du_concept_et_de_son_application-Vr2ZeH8AAQEAAE7L3p8AAAAD.pdf. Accessed 12 Feb 2016

Francis CA (2004) Education in agroecology and integrated systems. J Crop Improv 11(1–2):21–43

Marini L (2007) OGM, precauzione e coesistenza: verso un approccio (bio)politicamente corretto? Riv. giur. ambiente (1):1–14

Ministero delle politiche agricole alimentari e forestali (MIPAAF) (2008) Istituzione del Comitato di produzione integrata ed i relativi gruppi tecnici specialistici di supporto, per la redazione delle linee guida nazionali di produzione integrata file:///C:/Users/utente/Downloads/DM_SR__Istituzione_Comitato_Produzione_Integrata%20(1).pdf. Accessed 1 Mar 2018

Monteduro M, Buia G (2014) Minima Curat Praetor? La sfida di una disciplina giuridica delle nanotecnologie. GiustAmm.it (7):1–60

Noe EB, Alrøe HF (2015) Regulation of agroecosystems: a social systems analysis of agroecology and law. In: Monteduro M, Buongiorno P, Di Benedetto S, Isoni A (eds) Law and agroecology. A transdisciplinary dialogue, 1st edn. Springer, Heidelberg, pp 31–45

Principi MM (1992) Protezione integrata e produzione integrata delle colture agrarie: realizzazioni e prospettive. Boll. Ist. Ent. "G. Grandi" Univ Bologna 47:79–100. http://www.bulletinofinsectology.org/pdfarticles/vol47-1993-079-100principi.pdf. Accessed 12 Feb 2016

Rossi Dal Pozzo F (2014) Profili recenti in tema di organismi geneticamente modificati nel settore agroalimentare fra procedure di Comitato e tutela giurisdizionale. Dir del Comm Internaz (2):339–396

Steiner H et al (1977) Vers la production agricole intégrée par la lutte intégrée. Bull OILB/SROP (4):1–153

Stern VM, Smith RF, Van Der Bosch R (1959) The integrated control concept. Hilgardia 29 (2):81–101. http://www.entsoc.org/PDF/2009/2009-IntegratedControlConcept.pdf. Accessed 12 Feb 2016

Genetically Modified Organisms as Politicizing Products?

Guido Ruivenkamp

1 Introduction

The past few decades have seen an impassioned scientific and social debate on the pros and cons of biotechnology with the supporters stressing the blessings of this new development and its opponents pointing out its risks. While these two groups were diametrically opposed to one another, they were agreed on one fundamental point. Both saw the development of biotechnology largely as "a given" and not as a field of battle in which various social and political aspects of this development has still to be ironed out by creative discussion between social and technical stakeholders concerned. Neither the supporters nor the opponents of biotechnology paid much attention to the attempts of multiple actors to change the social and technical dimensions, although an analysis of these attempts could have made an important contribution to an academic and social reflection on the development of biotechnology.

The debate focussed on the pros and cons while possibilities for a redesign of that technology almost got no attention. Both supporters and opponents of biotechnology—despite their different viewpoints on the value of biotechnology—agreed in making the theoretical assumption that biotechnology can be seen "as an instrument" with positive or negative consequences for society. Conversely, both detach biotechnology form the social context (and from the social struggles) in which it was developed. When they talk about the social dimensions of biotechnology they generally refer to the (good or bad) *consequences of this technology for society*, without enquiring why the technology was developed in this particular way. I therefore call both these groups '*splitters*', since they discuss biotechnology as a

G. Ruivenkamp (✉)
SADE, University of Wegeningen, Wegeningen, The Netherlands
e-mail: guido.ruivenkamp@wur.nl

A. Isoni et al. (eds.), *Food Diversity Between Rights, Duties and Autonomies*,
LITES – Legal Issues in Transdisciplinary Environmental Studies 2,
https://doi.org/10.1007/978-3-319-75196-2_3

factor A, which has a certain influence on factor B (society), where A and B are assumed to be essentially independent of one another.[1]

Another group, the '*weavers*', take the viewpoint that biotechnology not only causes social change but is itself a form of social change. They stress that the development of biotechnology is not a random event that can be viewed in isolation but is closely interwoven with other social trends. They therefore speak of the *co-creation* of biotechnology and these other social trends. According to this view, biotechnology should not be regarded simply as a group of techniques but rather as a complex of social and technical dimensions.

Yet another group is that of the *re-designers*. They agree with the weavers about the interaction between and co-creation of biotechnology and other social trends but differ from them in wishing to influence this interaction. The re-designers are the optimists among the weavers, trying to find room for manoeuvre within which other socio-technical ensembles, other forms of biotechnology, can be developed. The re-designers are however also realists. While they know that biotechnology (and genomics) develop within global power structures, they stress that this does not inevitably mean that biotechnology must be developed primarily to serve the interests of the multinationals.

The re-designers regard it as a challenge to investigate to what extent the development of biotechnology can be aligned with other social interests. This makes them networkers, always looking for like-minded spirits within various institutions with whom they can work together to bring about social and technological change. They place the scientific development of biotechnology and genomics directly within the context of social change processes and try to uncouple biotechnology from the global biopower system and to transform it into a catalyst for specific local developments (conversely, to create a system of what I call tailor-made biotechnology).

In line with this, this contribution attempts to indicate ways in which agro/food biotechnology can be reconstructed for which purpose a three-step argument is presented. In the first paragraph we contextualize the development of biotechnology from within the social conflict regarding the social relations of post-Fordist production processes, indicating that biotechnology is related to three historical processes of separation in food production, to a strengthening of a spatial transformation of food production and to establishing new techniques of governing humans and particularly the social (scientific workers (immaterial labor).

The argument continues with a description of the new power relations within the changing food system. This indicates that the political system is increasingly shifting from the domain of policy into the domain of S&T through the development of *politicizing products* on which a political system of biopower is being established. The final step in the arguing line comprises a response to this through *biopolitics* and *tailor-made biotechnologies*. This paragraph discusses the perspective and concrete practices of resistance, of subjectivation and/or subtraction from biopower through

[1]Ruivenkamp (1997).

reinvention, particularly of the politicizing products of biotechnology, by challenging the embodied political content these contain and transforming them into catalysts for *multiple, location-specific developments.*

The article concludes by stressing the social relevancy of criticizing the ideological dominance of the instrumental-anthropocentric paradigm of technology that has characterized the pro-con biotech debate for so many years and to explore possibilities for strengthening concrete practices of agrobiodiversity and food diversity by transforming the politicizing products of agro-industrial biotechnology into tailor-made biotechnologies becoming catalysts for location-specific developments.

2 Contextualization of Biotechnology: A Spatial Reorganization of Food Production

The development and application of biotechnology also does not take place in a historical vacuum. Here, we consider the social conflict (struggle) regarding (new) social relations in post-Fordist production processes as the context out of which biotechnological developments have taken shape.[2] Expressed as the "*creative destruction of the peasantry*",[3] the modernization of agriculture in the nineteenth and twentieth centuries was realized through increases in scale and mechanization for the supply to industrial workers of cheap food.[4] This transformation has occurred and is ongoing throughout much of the world and with similar social consequences: agricultural labor became largely redundant and farms were converted to enterprises. As a function of 'development', the depopulation of the rurality with the growth of the urban and the marginalization and disappearance of the peasant, landless laborer, and smallholder was at the same time the birth of the farmer, a modern entrepreneur integrated first into the market and then into the social organization of agro-industrial production chains (followed by his demise, or complete incorporation). Within and as a part of this—actually teleological and always uncompleted—history, the emergence of biotechnology has been strongly entwined with (scripted by) two long-term processes summarized by the terms 'appropriation' and 'substitution'. *Appropriation* refers to the gradual take-over of controllable biological activities from farming practice by external institutions, in particular industry. Examples include the breeding of new cultivars and the maintenance and propagation of basic seeds (once a common good maintained by peasants in communities, now increasingly controlled by public research institutions and private life-science companies) and the regulation of soil fertility (previously performed by the peasant herself with, e.g., crop rotation, but now more usually achieved using artificial fertilizers supplied by agrochemical conglomerates). *Substitution*, meanwhile, refers to the historical development by

---

[2]Ruivenkamp (1989, 2003).

[3]Jongerden (2008) citing Hobsbawm (1995), p. 289.

[4]Vroom (2009).

which the agrarian origin of food sources is gradually being replaced by the products with an industrial-biochemical base. Foods are replaced by foodstuffs, a development that undermines the direct relation between agriculture and food production. Prime examples of this are the exchange of butter (made from milk) for margarine (manufactured from vegetable oils) and of beet and cane sugar by maize fructose syrup and synthetic sweeteners, such as aspartame.[5]

This dual process of appropriating and substituting peasants' activities and 'natural' foods has been accompanied by the (co-)development of specific biotechnological artifacts, such as seeds for sowing, hybrid seeds, and fertilizers, as well as new products to extract food components for recombinations from a broad spectrum of agricultural and synthetic resources. Through the development of various new bio-techniques, the life-science companies increase their capacity to intervene more radically and efficiently in the genetic structure of plants and food-processing and to determine where, when, and how a crop should be sown, harvested and processed. This agro/food biotechnology development has run alongside and been intercausally connected with the spatial transformation of food production. In the move from geographically concentrated units of peasant production towards distant and dispersed foci in interconnecting lines and zones of production-distribution-consumption, the socio-technical code of biotechnology has been particularly oriented to three processes of separation. Comprising a *de-territorialization*, these are contributing to the emergence of a qualitatively changed organization of food production.

3 The De-Territorialization of Food Production

The externalization and scientification of agricultural research co-determinate with appropriation and substitution has contributed to a dual process of 'liberating' agricultural produce from its territorial constraints and creating a dependence on the inscribed scientific information regulating cultivation and processing. This occurred in the context and through a strengthening of three disconnection processes.

3.1 Separation of Agriculture from Its Agro-Ecological Environment

That plants thrive in specific types of climate illustrates that they have internalized the environmentally-specified requirements for good growth in their genetic programs via processes of evolutionary selection. Traditional plant improvement and

[5]See Ruivenkamp (1986) for an analysis of the characteristic historical phases of this agricultural-to-synthetic substitution process.

cross-breeding techniques have already been used for many centuries to broaden the boundaries of this close relationship between growth for crops and the natural environment, a process that leaped forward with the Green Revolution introduction of new information into the genetic programs of various cereal grains. While the new plants gave higher yields, however, they needed external protection (manufactured weed-killers, pesticides, and fertilizers) to flourish in what was now a 'foreign' environment for them, thus adding a third dimension to the original plant-environment relationship and initiating the "*biochemicalization of agriculture*".[6] Advances in agro-industrial biotechnology then enabled the development of crops with many additional properties—especially herbicide-, pest-, disease-, drought- and salt-tolerant varieties—which (further) 'freed' plants from their relationship with their (somewhat) natural environment and made them (yet more) dependent on externalized agricultural research (the information incorporated in the seeds).

New biotechnologies are likely to further affect the already shifting climatological boundaries of crop cultivation, effectively causing the regulation of agricultural production to move to the domain of product development. The information inscribed in new seed products will further discipline cultivation practices, establishing new systems of power relationships based on control of agricultural production *at a distance* by the research, particularly the immaterial (genetic informational) content resulting from the work of researchers active in the complex international network of cooperative and commoditized knowledge development. This uncoupling of seed production from its surroundings and recoupling to scientific design has already become a core element of a new struggle for food sovereignty.

3.2 Separation of Agriculture and Food Production

New information is being introduced not only into seeds but also into products that play a crucial role in the processing of agricultural products to foodstuffs. This further contributes to the long-standing historical process of loosening and even completely breaking from the basic principle of food production according to which farm products are processed to make (regionally specific) foods. This cleavage is supported by biotechnology via the development of new enzymatic techniques for deriving food components from an ever-wider range of cultivated crops and of techniques for the factory production of these components using micro-organisms. The replacement of the direct linkage of specific farm products to specific foodstuffs involves a new form of productive cooperation in which an increasing *flexibility* of biochemical, as well as agricultural raw materials in the food processing industry, plays a pivotal role. This involves an expanding *multilevel interchangeability* of product groups and producer combinations—such as, for sweetening, in which corn

[6] Ruivenkamp (1989).

(high-fructose corn syrup) is used alongside beet sugar and cane sugar while also various microbial sources such as aspartame and other intensive sweeteners function as competitive raw materials for the beverage industries.[7]

The informatization of enzymes and microbiological fermentation processes is increasingly influencing the social organization of agriculture and food production. Advances in enzyme production can be decisive in prompting a move to the use of crops grown in a given region in the production of a wider range of foodstuffs; the use of corn as a raw material for sweeteners is a good example of this. The international flow in the trade of agricultural products can also undergo radical changes (e.g. the slump in Philippian cane sugar exports because of the American use of corn fructose), with a variety of economic consequences that directly impact all those individuals and institutions involved. Thus, enzyme technology and microbiological fermentation technologies, like that of seeds, carry an *intrinsically political ability to bring about social reorganizations*, with a potency inherent in the specific combination of immaterial and material components they embody—and research institutions that produce these combinations can be regarded as key actors on the political stage.

Because of these developments, food products can be increasingly regarded as assemblages from several basic nutrients (carbohydrates, proteins, and fats with amino-acid and vitamin supplements, etc.) and additives (flavors, colorants, etc.) extracted from a variety of interchangeable products and processes (agricultural, biotechnological, etc.). The entire mix is typically made up to resemble a traditional food as closely as possible and finally presented to the consumer in an attractive packaging (which may imply another range of biotechnical inputs, e.g. involving recycling). This process of uncoupling of the food product from the agricultural product and recoupling food components assembled at an international level has certainly not been completed everywhere yet and is sometimes rejected by consumers. There is thus a pressing need to investigate the extent to which the informatization of food processing methods could also be reprogrammed to follow other restructuring scenarios and stimulate other regional development strategies (see Section IV).[8]

3.3 Uncoupling of Agricultural Products from Their Intrinsic Nutritional Quality

Another consequence of the transition to a food production system based on the international assembly of food components is that agricultural products are increasingly uncoupled from their intrinsic nutritional quality. When agricultural products are regarded simply as 'components,' the food industry can obtain these components

[7]Ruivenkamp (1986).

[8]*Infra.*

from the industrial sector instead of the traditional source (farms)—the biochemical production of amino-acids as building blocks for artificial sweeteners is a good example of this. At the same time, agricultural products can also serve as inputs for the manufacture of non-food products—examples are the use of cane sugar and rapeseed to produce ethanol and lead substitutes for petrol. This increasing uncoupling of agricultural products from their intrinsic nutritional quality means that they only continue in part to function as food products, so part of the agricultural sector moves outside the social organization of the food chain.[9] The 'liberation' of the food industry from specific agricultural products and of agricultural products from their nutritional quality is at the same time their dependence on the complex network of developments around the delivery of the informationalized products, including the labor implications of this.

This uncoupling can also lead to an increasing demand for the creation of new *symbolic* values. As the (relatively) natural content is gradually hollowed out of the new internationally assembled food products, strenuous efforts are made to refill them with a new synthetic content that has a significance that can be attached to the products. The new assembled food components are increasingly presented and marketed as brands claimed to contribute to the health, beauty, and sporting abilities of consumers and to ethical norms (fair-trade, sustainability, etc.). This process is accompanied by checks carried out to determine the justification of the claims made, as well as monitoring claimed exclusivities—are they really made in GMO-free production chains, do they live up to the regional *Appellation Contrôlée* claimed for them? The ability of the immaterial laborer to create a specific emotional and symbolic significance for products will only reinforce his hegemony, while the informatization of products plays an increasingly important role in the creation of new social relationships.

4 Biopower Through the Politicization of Biotechnological Products

The qualitative changed organization of food production with its territorial disconnection combined with increasing dependency on the complex research network that inscribes specific scientific information in the biotech products equates to spatially reorganized food production systems incorporating an emergent biopower. An important characteristic of the new biopower is a redistribution of political influence, increasingly shifted to the *social organization of production*. Agriculture ministries (public policy) are being replaced by research institutions (technological development) as regulators of agricultural production at the expense of farmers and their lobbies and cooperatives. The development of new knowledge-intensive inputs and

[9]These processes have been going on since time immemorial, but the changing face and especially biotechnologization of agriculture has given them a qualitatively new significance.

food processing methods enable the research institutions to control at a distance the agricultural production systems and make agricultural products, producers and even regional production systems interchangeable production units for the food companies that assembled the food components at an international level. Moreover, there is an ongoing, global seismic *shift in the power balance* between different actors at the social institutional level that is, I would argue, steadily *reducing the democratic regulation of food production*. The depoliticizing of policy and patenting of politicizing products are two perfectly compatible arms of a movement that functions to create a new (productive) bio-power system. Further to eroding the capacity of social organizations grounded in agriculture to influence change strategies, the new bio-power system also strengthens its position by *eroding regional political identity*. The initial strategy of maintaining political-economic power primarily by marketing the same 'miracle seeds' for development on a global scale—e.g., the global distribution of hybrid corn varieties seems to be increasingly changing into a bio-power approach based upon a strategy of *segmenting regionally differentiated agricultural production systems*: along with de-territorialization, there is also a *re-territorialization*, driven by capital in combination with biotechnology.[10] To understand this trend properly, it is important to note that the same (patented) basic techniques and products can be used in the development of different varieties of agricultural crops (e.g., for herbicide-, pest-, and disease-resistance but also for drought-resistance, nitrogen-binding, or raised nutritional value). Consequently, regionally differentiated agricultural systems become dependent on the same basic techniques or products patented by a limited number of life-science companies. This implies a non-autonomous development of regionally differentiated agricultural systems; the regions thus lose their particular *political identity*, which was based on a historical development of their diversified food production systems. Just like colonial political regimes caused some parts of the world to develop into agricultural commodity production zones (e.g., West Africa for cocoa), so also should we expect to see things like life-science companies developing specific crops for designated regions through their supplies of specific politicizing products.

The replacement of regional political identities with regional production systems (of food components) creates possibilities for increasing regional food supplies, since the desired food components can now be derived from a wider range of local raw materials. However, the increasing flexibilization of food production also leads to *increasing competition* between the different regional methods of producing the corresponding basic nutrients (or the social forms in which these production methods are organized), and the combination of these two developments—regional food component production alongside an intensified competition among these regional extraction and production units—tends to transform (and reduce the political identity of) the specific regional forms of food-component production (either small-scale or large-scale, either on a cooperative or private basis) into (just economic, competitive) production forms that are interchangeable within the companies that assemble

[10] Kloppenburg (1988).

the food components at a global level. Hence, the regional productive diversity no longer gives a clear indication of whether a particular regional form of development is still politically autonomous. On the contrary, the different, apparently independent, regional production systems now compete for representation among the various international food-component assembly companies, leading to yet further dependency of the regional production forms on the scientific research that provides the basis for the development of the politicizing products.

Apart from de-politicizing of policy, political erosion of social organizations and the loss of a regional political identity, the new bio-power system is also maintained through the *complex social organization of the immaterial scientific work* that provides the basis for the preparation of the politicizing products. The collection, appropriation, and incorporation of genetic information in the politicizing products takes place in very complex, flexible knowledge development environments. Knowledge is increasingly developed within all kinds of relatively short-lived cooperative structures built up around permeable research bubbles involving a variety of public-private and fundamental-applied information flows, among others. The overall trend is for individual researchers to retain less and less insight into or control over the contribution they make to the way in which their labor is ultimately incorporated, even as this scientific work makes a steadily greater contribution to the development of products that themselves harness a greater transformative power. Scientists thus become *disconnected* from the social significance of their work—just at a time when their work is impinging increasingly on society.

Moreover, scientists are expected to show complete personal dedication to their restricted field of research, as illustrated by the increasing blurring of the boundaries between working time and free time, work and hobby, production and training in scientific research. There is no longer a clean, well-defined threshold separating labor-time from non-labor time.[11] For the immaterial laborer, her work is her life, and his life is his work. This close personal involvement in a steadily more limited field of research means that those immaterial laborers (enzyme researchers, plant breeders, etc.) who lay the foundation for the incorporation of politicizing products into the food chain are *intimately bound up with the new bio-power system,* making that power system hegemonic. For the immaterial laborer all aspects of his/her individual human life become a potential to be used to produce immaterial products. The ability to learn, to adapt to different networks, to integrate different scientific disciplines, to be accustomed to mobility, to be affective and communicative; all these elements, which cannot be separated from the person, increasingly contribute to the weight of the individual contribution—outside the workplace—to the production of the politicizing products. The infusion of the hegemonic biopower system in the personal life of the researcher is based upon the voluntary 'self-chastisement' of the immaterial worker, however, which does imply an affordance on the part of the researchers to reflect critically on their work and initiate various forms of resistance—one of which involves the exploration of a new biotechnology pathway, that

[11]Virno (2004).

of 'tailor-made biotechnologies' in the company of the various social organizations that share an interest in this field.

5 Biopolitics and the Development of Tailor-Made Biotechnologies (TMBT)

Multiple forms and trajectories of various struggles and alternative practices of resistance to the biopower in global food production are appearing. Here, there follow four overviews of domains of practice of such resistance in the shape of efforts to recodify the politicizing biotechnological products.

5.1 Reuniting the Elements Separated in the Biopower System

Because products such as seeds, enzymes and microbiologically produced food components contain the immaterial politicizing content enabling and impelling an increasing control at a distance over agricultural production systems and interchangeability of agricultural (and synthetic) raw materials (and their producers), they cannot be considered as just economic products. The transformation in the social organization of global food production blurs the difference between 'politics' and 'economics,' realizing a transfer of (policy) politics into the domain of S&T. As crucial links in the social organization of global food production, the products become key politicizing agents in deciding which plants (varieties) are cultivated and where, when and how they are sown, tended, harvested and processed as all inscribed in the products being a composition of immaterial (scientific information) and material components (seeds, plants). As described the development and application of these politicizing biotech products are interwoven with three historical processes of separation in global food chains through which the agricultural production becomes *de-territorialized* and new power relations emerge. Thus, the bio-power system develops through the proliferation of politicizing products, with agricultural production increasingly 'liberated' from the agro-ecological environment, foodstuffs from the supply of specific agricultural products, and agricultural products from their intrinsic nutritional qualities.

The collection and incorporation of genetic information in seeds, enzymes and fermentation processes need not be irretrievably cast, however, in the role of handmaiden to the separation processes. It is also possible to use this technology based on the life processes of plants, micro-organisms, cells and parts thereof precisely to *reunite what has been separated in global food chains* and thereby to *recreate and strengthen sustainable local food networks of agro- and food diversity*. Various examples of concrete practices can be referred to in which such attempts are being made. For example, just as farmers establish new seed-sharing commons in resistance to the private and public appropriation of the seeds scientists can also concentrate on collecting genetic information and incorporating it in seeds to reinforce the *regional quality* of agricultural crops.[12] This will involve them liaising with local groups of farmers to ascertain the additional information that can improve the regional quality and diversity of farm or food products and speed up the transition to more sustainable, democratic production methods. It also implies that farmers' groups can become key actors in progressive decision-making about the ways that food crops are cultivated and establish their agency in the contemporary design of biotech agro/food products (see Agency, below). Attempts can also be made to introduce genetic information into cultivars to refine (and renew) the (traditional) *rotation of crops*. For example, dual-purpose and early-maturing sorghum varieties have been developed in India that do not only contribute to restoring soil fertility but also deliver multiple social benefits, such as improved nutrition, higher income for farmers, and lower consumption of herbicides/pesticides).[13]

Another way of loosening the grip of the bio-power system is by *restoring the relationship between food production and agricultural production* through the establishment of networks linking critical actors in the field of enzyme-technology research. These may be directed to develop, for example, enzymes inscribed with genetic information supportive of the diversity of regional agricultural products and farming methods that are represented in the end-products, such as region-specific cheeses.[14] Instead of supporting increased interchangeability of agricultural raw materials and destroying the linear integration of agricultural with food production, research in this critical research network concentrates on *recreating links between agricultural and food production in a given region* or locality or even on a given farm.[15]

---

[12]Öztürk et al. (2014) and Patnaik (2016).

[13]Rana (2000).

[14]Ruivenkamp (2008).

[15]Since 1999 an international network (Tailor-Made biotechnologies for Endogenous developments) has been working on facilitating an information exchange among various practices of tailoring biotechnologies to location-specific networks in various countries. One of the initiatives has been a common research project of twelve PhD researchers from Ecuador, Ghana and India to investigate opportunities for strengthening the social transformative power of location-specific food products as lupine (Ecuador), cowpea (Ghana, Benin) and mungbean (India) by attuned tailor-made biotechnological developments. Some results of these initiatives are represented in four documentaries included in the book "Biotechnology-in-development: Experiences form the South" Ruivenkamp (2008).

5.2 Rivitalizing Place-Oriented Developments

The spatially reorganized food production—agro-industry's de-and re-territorialization importantly and increasingly informed by biotechnology—illustrates that the biopower of food companies is anew in conflict with the emergent composition of social movements formed now rather on basis of place- instead of class-consciousness.[16] It also illustrates that the researchers developing the biocatalysts become key political actors, although they may not always identify themselves as such. In view of this crucial political role, the formation of (alternative), international networks to facilitate the efforts of those critical scientists also becomes vital—including networks of those endeavoring to incorporate scientific information into biocatalysts that aim to challenge these power relations by *revitalizing place-oriented developments that go beyond the interchangeability of regional methods of producing food components.*

For example, there are innumerable initiatives that refer to the richness of local culinary practices and cultural habits embedded in specific food products and which try to design those biocatalysts that stimulate these local food practices and products; and there are also location-specific catalysts to establish home-based fermentation units for household development as a food sovereignty initiative. In Kenya, for instance, a research network working with households has developed new techniques to make different end-products from cassava (cakes, snacks, soap, etc.), while other households are making snacks to be eaten by their children during school time.[17] In West Africa, meanwhile, research networks have improved fermentation techniques for the many different fermented foods such as *kenkey*, *waachi, banku*, and *tuo zaafi* often consumed as street foods so providing a popular form of nourishment. Research networks and civil society organizations are also looking for opportunities to convert waste products in resources. In Ghana, cocoa husk waste products are made economically relevant for smallholder farmers by converting the cocoa husk waste into resources to produce fertilizers and end products like the traditional soap, *alata samena*.

These different initiatives to revitalize location-specific resources have nothing to do with conserving traditional practices or a romantic return to the past; on the contrary, this *alternative re-territorialism* represents a future-oriented way of re-organizing open local spaces from within *endogenous developments* that redesign agro/food products. Apart from the various initiatives that challenge the separation processes on which the hegemonic position of the emerging biopower system is establishing, that is, there are also these various initiatives inscribing biotechnological artifacts in a way that is grounded in place rather than abstracting from place. Both groups of initiatives are focusing on changing the scientific information that is

[16] Magnaghi (2000), p. 78.

[17] Ruivenkamp (2008).

incorporated in the politicizing products manufactured within complicated networks of research collaboration. This complex organization of knowledge production is itself another contested arena.

5.3 *Creation of Maneuver Space in the Complex Organization of Knowledge Production*

With the international production of politicizing agro/food products characterized by an increasing concentration of conglomerates in tandem with countless fluid and hybrid networks of public/private, fundamental/applied research institutions, it is near impossible for individual researchers to accurately direct their contributions toward the political content of the designed end-products. Although these end-products are characterized by their contributions to the creation of new social relations, the non-transparency—because of this complexity of the production system—causes researchers to become alienated from the social significance of their work. While there are immense organizational opportunities for individual researchers to develop their work at the technical level, the same organizational form constrains them regarding critical reflection on the social content of their work. Even outside of corporate politics, this highly specialized type of work even within public research environments implies that the technical potentials of the immaterial production are increasingly separated from a critical-reflexive attitude to the politicizing content precisely while the politicizing content is a major characteristic of the products being delivered by the immaterial laborers. This contrary development may, however, stimulate researchers to liberate themselves from their interwovenness with the biopower of the spatially reorganized food production.

The examples cited to show efforts to create maneuver space and to reunite elements separated in the biopower of food production and revitalize endogenous developments of agr- and food diversity. All involve individual researchers indeed playing a constructive role to change the political content of biotechnological artifacts. By making concrete choices about the specific information to be incorporated in the immaterial products, individual researchers do challenge the hegemony of the biopower system and actively contribute to the reshaping of alternative, new informationalized seeds and biocatalysts within alternative networks, as well as making use of common pool frameworks, illustrating that the immaterial laborer can also be related to democratic social change and become a protagonists for the development of what Beck et al. (1994) called 'sub-politics'.

5.4 Restoring Agency

Alongside the various efforts to contribute to a revitalization of place-bounded food developments and to create a maneuver space in the complex organization of knowledge production there are also efforts to restore agency in the development of technology as such—meaning, not creating relations of submission and dependency but contributing to processes in which people become the subjects of their own history. In the context of an advanced technological society, the agency issue may be keenly contested. It may be suggested that in an advanced technological society, the expansion of technological means as a whole becomes an actant in historical development reducing people's activities to a constant functional adaptation to instrumental technological rationality. In this context, man may even become an integral part of the technological apparatus and people's lives may increasingly be characterized by enabling and furthering the functioning of this apparatus changing life as an end-in-itself into life as a means for the maintenance and development of the new abstract technological developments (a new abstractum).[18] There may be a trend that leads to an increasing *technologization of life* that is based on using life as the center of potentiality for the future development of that instrumental technological rationality.

In this context also, a new system of domination appears in which the potentiality of life, the potentiality of creativity and imagination become forces for the further development of that abstract technological development. Indeed, these phenomena are already present in the work of the individual biotech researcher, where all aspects of her/his individual human life have become a potential to be used to produce the politicized products. However, just as the emergence of the modern farms and global food chains have not resulted in a complete banishment of the peasantry, so also the social dominance of the instrumental rationality and the technologization of life of the individual researcher have not led to a complete disappearance of ends and values other than the instrumental efficiency.[19] On the contrary, there is an emergence of location-specific agro-techno developments in a redefined social space wherein researchers cooperate with other actors to *redesign* the biotechnological artifacts, and researchers working in the networks of biopower do end up *questioning their personal involvement* in the design of politicizing products. This critical reflection on their relationship with the bio-power system may be further reinforced by consideration of how they are, in fact, exploited and their work misused.

Such concerns may lead to action in many different ways. For example, microbiologists and other researchers may decide to learn more about the social

[18]Galimberti (2003), p. 685.

[19]Van der Ploeg (2008).

implications of biotechnology, thus training themselves to be 'weavers' rather than 'splitters'.[20] One risk involved in this approach is that they may feel that their concerns entitle them to wear the mantle of social scientists, a move that may, however, turn them into 'splitters' of a new kind, those who exclusively attack the social risks of biotechnology, perceived as 'immutable.' But researchers *can* look for ways in which they can deploy their technical knowledge for the development of products of a different kind; they *can* team up with other actors to use their work in a directed way that serves other strategies of change. Conversely, they can turn into 're-designers,' looking for areas of work and life that do not reinforce the bio-power system but rather lead to a 'subpoliticization from below'[21] and a multiplicity of work-styles and lifestyles beyond the bio-power system, more or less.[22]

Some of these new approaches to immaterial work have been indicated here with possibilities for a *consciously alternative* politicized inscription of biotechnological artifacts. These initiatives may further be strengthened by the establishment of new social relations with other actors. Indeed, within an alternative re-territorialist perspective, platforms of researchers and post-modern peasants are formed in which efforts are made *to reclaim* the renewed biological activities previously appropriated and changed by external agencies. Thus, there are multiple initiatives attempting to bring part of the crop improvement work back to the field and under the control of the contemporary smallholder. One example concerns the cloning of teak and neem trees in India. Initial trials of this technique involved rooting and growing tissue-cultured clones of these trees in laboratories in the town, but these phases of production have now been moved to field laboratories in the countryside. This simple adjustment has not only reduced costs (since the shoots from the tissue-cultured clones transported to the country by truck take up much less room than the young trees that used to have to be transported) but also encourages the farmers (many of them women) to take an active role again in strategies for the determination of their environment.[23] Similarly, groups of local people from the village of Gaddipalli in the southern Indian state of Andhra Pradesh have developed a potting soil from a mixture of cow dung, cellulose waste and worm culture that is used for rooting the tissue-cultured clones in the field laboratories and as manure to promote

[20]Ruivenkamp (1997, 2008).
[21]Ruivenkamp (2005).
[22]Hardt and Negri (2004).
[23]Ruivenkamp (2008).

the growth of the young cloned trees in the field.[24] Other initiatives focus on a re-appropriation of specific phases of plant propagation, aiming, for example, to circumvent the biological protection of hybrid (corn) varieties that forces farmers to buy new seed each year instead of laying aside part of the harvest for sowing in the traditional way. This is done by applying specific technical solutions to remove this politicizing content while maintaining other advantageous traits of the hybrid varieties. Thus, despite the limited space of maneuver in the spatially reorganized agro/food production systems of biopower, a great many successful initiatives are constantly springing up to restore regional, local and individual agency and thereby launching tailor-made biotechnologies for endogenous developments of agro- and food diversity.

6 Some Reflexive Remarks

Within the context of a spatially reorganizing food production and an emergent biopower system, new domains of social conflict appear that offer not only constraints but also new opportunities for redesigning biotechnological developments attuned to agro/food diversity by disconnecting the research from the social context of food multinationals and the social organization of global food chains and by re-territorializazing the biotech research to location-specific initiatives of agro and food diversity leading diversified forms of tailor-made biotechnologies.

Various practices of resistance described above referred to the initiatives of researchers in cooperation with other actors aimed at redesigning biotechnological products by unravelling the political content they contain and inscribing alternatives into products that can thus become catalysts for a democratic and agro/food diversified productions. This movement implies multiple social struggles in various domains in which both deconstructive and reconstructive activities must be carried. Firstly it implies a social struggle against the ideological dominance of the instrumental-anthropocentric technology paradigm, going beyond and substituting the instrumental pro-con biotech debate with a critical-reconstructivist approach that focuses on the rewriting of the socio-technical code in biotechnology.[25] Secondly it implies strategic actions to unravel the social relations from which the specific socio-technical code in the agro-industrial biotechnology emerge and determine their specific political character inscribed in the various biotechnological artifacts as

[24]Ruivenkamp (2008).

[25]In the sociology and philosophy of technology, there is an intensive debate on the concept of "socio-technical code" in technology that has been stimulated among others by the article of Langdon Winner (1985) "Do artefacts have politics" and followed by debates and analyses of Noble, Feenberg, Ruivenkamp, etc.. In general terms the code concept indicates that technologies are developed within a cultural horizon (Feenberg 1999), specified by the socio-cultural assumptions grounding and directing their development and from which 'the politics in technologies' is practiced.

seeds, enzymes and microbiologically produced food components. Thirdly it implies various concrete tactical actions to rewrite the socio-technical code of biotechnology by changing the (im)material form of the biotechnological artifacts, transforming the concrete political content the biotech products contain and transforming them into catalysts for agr- and food diversity for location-specific developments.

While these practices of resisting agro-industrial biotechnology and recodifying it into tailor-made biotechnologies are strongly interwoven with actions to redirect the social organization of food production towards agro- and food diversity, there are also initiatives that challenge the expertocracy and high-tech character of biotechnology. Do-it-yourself biology (DIYbio) or garage biology is a set of practices through which lay people practice biotechnology and thus also challenge the exclusive control exercised on biotech R&D by Big Bio. DIYbio translates into practice the assumption of material semiotics that objects (non-humans) have the capacity to act or participate in systems and networks and can thus be regarded as actants[26] or even as politicizing products.[27]

Acknowledging that practical means are the mediators through which domination is produced, the BIYbio practitioners aspire to deliver actionable knowledge regarding the use of these means and go beyond the dichotomy between the designers and users of these tools by realizing praxis-oriented interventions built on sharing, participation, and creativity.[28] BIOYBio may be viewed as a participatory practice through which amateur biologists explore the possibilities for fostering non-hierarchical peer production models that have proved successful in the IT industry[29] (into the biological domain of lab, organisms, and cells.[30] Indeed, there are concrete actions of garage biologists reversing the plant-growing process of GM cut flowers by tissue-culture techniques, stimulating simple fermentation experiments with local beverages, and exploring possibilities for obtaining alternative bioenergy sources.[31] The general aim of these activities is to develop a praxis-oriented engagement with science and technology that enables citizens to exercise their citizenships rights in a technological culture by actively and autonomously shaping techno-science developments instead of being mastered by the technological means and thus directed into passive adaptation.

A social re-appropriation of (bio)technology development does not only take place by organizing an equal access to and control over the technological tools but also requires a commoning of the knowledge production in which the (socio-) technological tools are produced. In that sense, it becomes crucial to understand the ways in which science workers can create maneuver space in international research networks to common the knowledge production of various agro/food

---

[26] Latour (2005).

[27] Ruivenkamp (1989, 2005).

[28] Latour (2005).

[29] Benkler (2006).

[30] Delfanti (2013), p. 112.

[31] Golinelli and Ruivenkamp (2015).

biotechnological artefacts. In view of the practice of commodifying knowledge and patenting bio-products, it will become increasingly important to understand how the enclosure of knowledge commons by state and market can be prevented and through which social relations a new way of producing knowledge can be realized. Various research projects are under way to develop new insights in these trajectories of communizing agro/food production. A social re-appropriation of biotechnology development may be closely bound up with this new way of producing knowledge and designing technology, commonly. Therefore, this contribution concludes by emphasizing the relevance of creating specific networks of science workers and other stakeholders to challenge the political content of emerging politicizing products. For example, many research activities are programed to hybridize two of the world's most important food crops, rice and potato, which may again lead to a reinforcement of relations of dependency, as in the history of hybrid corn (above). Therefore, it seems imperative that we intervene and investigate how the research activities are organized and whether and in which ways the socio-technical code of the hybridization techniques can be changed and related to an *agro-ecological production*. It may be very difficult to intervene in the design process of these and other politicizing product (some personal efforts have not proved successful, yet). However, it may be worthwhile to continue trying. Instead of passively accepting or rejecting technological artefacts, it may be worthwhile to restore agency in the development of biotechnology and develop both deconstructive and reconstructive activities; deconstructive in the sense of unravelling the social relations from which the specific social-technical code in the biotechnology emerge and determine the specific political character of the various politicizing products; reconstructive in the sense that concrete efforts can be made to rewrite social-technical code of biotechnology by changing the embodied political content biotechnological products contain and transform the immaterial content towards an empowering force for agro-food diversity.

References

Beck U, Giddens A, Lash S (1994) Reflexive modernization. Politics, tradition and aesthetics in the modern social order. Polity Press, Cambridge

Benkler Y (2006) The wealth of networks how social production transforms markets and freedom. Yale University Press, London

Delfanti A (2013) Biohackers. The politics of open science. Pluto Press, London

Feenberg A (1999) Questioning technology. Routledge, Oxford

Galimberti U (2003) Psiche e techne. L'uomo nell'eta della tecnica. Giangiacomo Feltrinelli Editore, Milano

Golinelli S, Ruivenkamp G (2015) Do-it-yourself biology: action research within the life sciences? Action Research, 05/06/2015

Hardt M, Negri A (2004) Multitude: war and democracy in the age of empire. The Penguin Press, New York

Hobsbawm E (1995) The age of extremes: the short twentieth century, 1914–1991. Penguin, New York

Jongerden J (2008) First the peasant? Some reflections on modernity, technology and reconstruction. In: Ruivenkamp G, Hisano S, Jongerden J (eds) Reconstructing biotechnologies: critical social analyses. Wageningen Academic Publishers, Wageningen, pp 119–142
Kloppenburg JR (1988) First the seed: the political economy of plant biotechnology, 1492–2000. Cambridge University Press, Cambridge
Latour B (2005) Reassembling the social. An introduction to actor-network-theory. Oxford University Press, New York
Magnaghi A (2000) Il Progetto Local. Bollati Boringhieri, Torino
Öztürk M, Jongerden J, Hilton A (2014) Commodification and the social commons: Smallholder autonomy and urban relations in Turkey. Agrar South J Polit Econ 3:337–367
Patnaik A (2016) Seeds as biosocial commons. An analysis of various practices in India. PhD thesis, Wageningen University
Rana BS (2000) Annual Report National Research Centre for Sorghum. Indian Council of Agricultural Research, Hyderabad
Ruivenkamp G (1986) The impact of biotechnology on international development: competition between sugars and sweeteners. Vierteljahresberichte des Forschungsinstituts der Friedrich Ebert Stiftung, 103 (Special issue on 'New technologies and Third World Development')
Ruivenkamp G (1989) De Invoering van Biotechnologie in de Agro-Industriele Productieketen. De overgang naar een nieuwe arbeidsorganisatie. Jan van Arkel Uitgeverij, Utrecht
Ruivenkamp G (1997) Biotechnology as a socio-technical ensemble – closings remarks and reflections. In: Wirtz J, Lammerts van Bueren ET (eds) The future of DNA. Kluwer Academic Publishers, Dordrecht
Ruivenkamp G (2003) Biotechnology: the production of new identities. In: Koot W, Leisink P, Verweel P (eds) Organizational relationships in the networking age. Edward Elgar, Northampton, pp 265–290
Ruivenkamp G (2005) Tailor-made biotechnologies: between biopower and subpolitics. Tail Biotechnol 1:11–33
Ruivenkamp G (2008) Biotechnology in development. Experiences from the South. Wageningen Academic Publishers, Wageningen
Van der Ploeg JD (2008) The new peasantries, struggles for autonomy and sustainability in an era of empire and globalization. Earthscan, London
Virno P (2004) A grammar of the multitude. An analysis of contemporary forms of life. Semiotext, Cambridge
Vroom W (2009) Reflexive biotechnology development. Studying plant breeding technologies and genomics for agriculture in the developing world. Wageningen Academic Press, Wageningen
Winner L (1985) Do artifacts have politics? In: MacKenzie D, Wajcman J (eds) The social shaping of technology. Open University Press, Milton Keynes, pp 26–38

Agroecology and Geographical Indications at the WTO and in the EU Between Magic and Rationality: 'Reinventing' Marketing Designations to Preserve Rural Economy, Cultural Heritage and the Environment

Rocco Palma

1 Introduction. Good Fate of 'Agroecology' and the 'Rediscovery of the Law of Geographical Indications'. Scope of This Inquiry: GIs Between Magic and Rationality

'Agro-ecology' and the law of geographical indications (GIs) have become particularly popular in the last 30 years between scholars and policy makers. 'Agro-ecology' might be defined as the science of sustainable management of the environment, inclusive of natural, social and human resources used in agricultural production. In this regard, scholars indeed speak of the 'birth' of a new paradigm, whose main feature is the detachment from the matrix provided by 'ecology' of an autonomous scientific method that investigates the 'environment', both in its 'natural' and 'socio-economic' aspects, by focusing on the impact of agriculture on the environment and promoting sustainable traditional agriculture as a guiding principle of the movements for preservation of local peasant economies, especially in Developing Countries (DCs) and Least Developed Countries (LDCs).[1] The fame of newly adopted GI laws stems, on the opposite, from a mere 'rediscovery' of pre-existing legal concepts such as *'indications of provenance'* ('IP') and *'appellations of origin'*

[1] Altieri (1971/2002), pp. 1–24.

R. Palma (✉)
Department of Economic Studies, Catholic University of Sacred Heart, Rome, Italy

A. Isoni et al. (eds.), *Food Diversity Between Rights, Duties and Autonomies*, LITES – Legal Issues in Transdisciplinary Environmental Studies 2,
https://doi.org/10.1007/978-3-319-75196-2_4

('AO' or 'AOC'),[2] *i.e.* the revision—in a significantly changed context—of a branch of law protecting geographical names of products with a natural and a cultural link with the places of production, which knew a long-lasting tradition both in national (France and Mediterranean Europe Countries, including Italy) and International law.[3] New GI law has added to the picture new types of designations such as, in Europe, '*Protected Designations of Origin*' ('PDOs') and '*Protected Geographical Indication*' ('PGIs')[4] and, in the multilateral context, '*Geographical Indications*' ('GIs') in Section 3 of the TRIPS Agreement.[5] The qualifying element of this new wave of law is the call for a wider set of justifications for GIs. At the TRIPS Council and in the WTO forum, the EU, Switzerland, India and some African and Asian countries, traditionally in favor of stronger GIs protection ('*Demandeurs*'), support the view that non-wine GIs, now protected under the 'misleading standard', should be equipped with an absolute right to use, *i.e.* 'objective' because independent from the opinion of the public (the same way of GIs for wines and spirits in TRIPS

[2]This has been a classic distinction in International law and in domestic law until the 1990s of the last century: simple '*indications of provenance*' (IPs) inform the consumer about the mere geographical origin of products ('*Made in Italy*', '*product of France*'), while '*appellations of origin*' (AOs) are distinctive geographical names guaranteeing not only product origin but also territorial quality because of natural and human factors available in the area ('*Parma*', '*Barolo*', '*Roquefort*'): cf. Ladas (1975), p. 1574 et seqq. The expression '*appellations d'origin controlèe*' (AOCs) refers to registered geographical names for traditional products for which systems of monitoring and compliance of quality standards are provided and, indeed, this is largely the way the notion is used during this inquiry, even when the expression '*geographical indications*' is used.

[3]A first set of international rules on GIs protection, albeit incomplete, was adopted within the Unions of Geneva for the protection of industrial property in the second half of the twentieth century, such as Articles 1, paragraph 2, 9 and 10 of the 1883 Convention of the Paris Union for the Protection of Industrial Property ('PC'), the 1891 Madrid Agreement on the suppression of false indications of origin ('MA') and the 1958 Lisbon Agreement on the Protection of Appellations of Origin and their International Register ('LA'): cf. the texts at http://www.wipo.int/geo_indications/en/#laws.

[4]Cf. Regulation (EEC) 2081/92 of the Council of 14 July 1992 on the protection of geographical indications and designations of origin for agricultural products and foodstuffs, OJ [1992] L 156/1, Regulation (EC) 510/2006 of the Council of 20 March 2006 on the protection of geographical indications and denominations of origin for agricultural products and foodstuffs, in OJ [2006] L93/12, Regulation (EU) 1151/2012 of the European Parliament and the Council of 21 November 2012 on quality schemes for agricultural products and foodstuffs, OJ [2012] L 343/1, the last currently in force. The first Regulation was mirrored by provisions negotiated on request of the EC during the Uruguay Round and later incorporated into the TRIPS Agreement.

[5]From Section 3 of the TRIPS Agreement has then derived a conspicuous number of WTO Members national implementing GI laws: cf. the TRIPS text at https://www.wto.org/english/docs_e/legal_e/legal_e.htm. For an overall picture of national implementing laws and diversity of legal conceptions, cf. TRIPS Council IP/C/W/85 of 17 November 1997, p. 12 et seqq, and IP/C/W/85/Add. 1 of 2 July 1999, p. 22 et seqq.

Section 3, Art. 23).[6] For *Demandeurs*, in addition to informing the public about the origin and territorial quality, GIs would be, in fact, a tool for 'economic development', both for advanced countries and DCs or LDCs. GIs would enable local communities to preserve economic revenues, cultures and lifestyles, local identities and traditional methods of production, as well as agrobiodiversity and the environment,[7] preventing the homologation of products induced by food industry, intensive agriculture and monocultures.[8] USA, Canada, Australia and some South American Countries (so-called '*Opponents*') question the effectiveness of GIs as 'income generators' and their ability to preserve the authenticity of traditions and 'non-trade interests'. GIs naturalist and spiritualist narrative would be a sort of magic, a mythology lacking any rational basis and effects in the face of epochal changes brought by market forces: advertising, economic convenience, product innovation have, in fact, irresistibly changed both the 'culture of production' (offer) and 'consumption habits' (demand), as well as the same processes by which individual and collective identities are shaped and self-perceived.[9] The *Opponents* emphasize the monopolistic effect of GIs and the costs for governments and individuals of *sui generis* GIs law and conclude that the 'misleading' standard of protection is sufficient.[10] Twenty years after the 'rediscovery' of GI law, it is reasonable to subject the opposing views, often stiffened in the iteration of arguments not always empirically substantiated, to a first, albeit provisional review on the background of available experience. Borrowing from anthropology the distinction between 'magic' and 'rationality', the crucial issue is whether GIs law is to be placed in the realm of 'magic', as a mere rhetorical device devoid of any economic or practical effect, or in that of 'rationality', being a concrete instrument with significant impact on the realities of the market and social life.[11] This inquiry then aims at ascertaining—through the lenses of IP law—on what terms, at least as a general trend, in the last decades GIs have been effective in promoting 'economic development', 'culture' and the 'environment', and if so, to what extent this would have happened. Bearing

[6]GIs of non-wine products are now protected in Art. 22 of Section 3 by reference to the ordinary prohibition of misleading statements and Art. 10bis of the Paris Convention against unfair competition.

[7]WTO, doc. TN/C/W/25 cit, p. 3.

[8]Fink and Smarzynska (2002), p. 403 et seqq. GIs would allow local communities to safeguard a 'traditional' model of agriculture, alternative to mass production, thus avoiding the risk of a '*McWorld*': cf. Broude (2005a), p. 21.

[9]Broude (2005b), p. 623, and especially Hughes (2006), p. 299 et seqq., where a critical review is made of the same founding premises of GIs and AO theory, seen as a sort of rhetoric device, not correspondent to economic and geo-physic realities. For a positive account—from an historian perspective—of industrially-processed foods against the rhetoric of traditional products, cf. Laudan (2013).

[10]On the monopolistic effects of GIs Mosoti and Gobena (2007), p. 202 et seqq.

[11]De Martino (2015), p. 7 et seqq.

in mind the intrinsic double nature, theoretical and practical, of agroecology and IP law,[12] this work is structured in a first short preliminary reflection aimed at detecting symmetries and interactions between 'agroecology' and the 'theory of GIs and *terroir*' and a more thorough review of empirical findings so as to verify to what extent policy declaration have been confirmed by market practice, also with a view to identify loopholes in the law and propose ameliorations. This appears particularly relevant in a context where unprecedented transformations are occurring, including grave threats to agricultural communities and the environment (such as the '*Xylella*' bacterium in the South of Italy), that seem to require a deeper theoretic reflection on the relationship between the 'environment' (both natural and social) and 'economic development' to properly guide public and private action.[13]

2 Common Assumptions Between Agro-Ecology and GIs Theory: Cultural and Environmental Value of Traditional Agriculture

'Agro-ecology' on the one hand, the theory and practice of AOs and GIs on the other, present several contacts in terms both of contents and addressees and indeed each of the two seems at first sight instrumental to the objectives of the other. In very simplified terms, 'agro-ecology' prescribes a holistic perspective on the environment and subjects it to investigation in its 'natural', 'social', 'cultural' and 'economic' dimensions, including 'agricultural production' where the role of men is prominent ('agro-system').[14] In this perspective, agroecology privileges traditional agriculture, *i.e.* productions based on small or medium-size organizations and traditional intergenerational knowledge rooted in customs ('TK'), as opposed to the paradigm of mass food industry and intensive agriculture (monocultures), with a view to keep a balanced relationship between agriculture and the environment and preserve the self-determination of local communities and their right to food from market pressures and established economic interests.[15] While being not a science but a mere

[12]Ladas (1975), p. 1574. For a view of the theory of property rights on intangibles as an 'arbitrary cognitive act' aimed at unifying under the same umbrella diverse juridical institutions and practices, previously covered under a different set of scientific studies or branches of law (labor law, public law of economy), cf. Spada (2012), pp. 3–5 et seqq.

[13]This would place agro-ecological studies into a long and robust perspective of thought, deeply rooted in a tradition of legal studies of Southern Italy dating back to that particular version of the '*Enlightenment*' that took place in Naples and to one of its most prominent representatives, Antonio Genovesi, with his idea of an advanced relationship between theory and practice and, particularly, of the role of theoretic reflection as an activity that should not limit itself to speculation, but be pragmatic and inspired to the public good, so as to find new paradigms to guide public and collective action: cf. de Nitto (2002), p. 164.

[14]Altieri (1971/2002), pp. 5–7.

[15]Altieri (1971/2002), p. 16.

doctrine underpinning a very specific branch of law, the theory of AOs and GIs-which despite some theoretic uncertainty, stands out for a significant generality of principles and independent features from the law of trademarks—shows some complementarity with 'agroecology'.[16] GI law grants exclusive rights on geographic names to circles of producers settled in a delimited area only when the quality of the goods has an exclusive link with the territory, *i.e.* with the natural ('natural resources') and human factors ('methods of production') ('*terroir*' theory).[17] Both doctrines thus identify a clear distinction between 'industrial or mass production', which is the subject of a negative judgment, and 'traditional production' based on small scale organizations and common knowledge, which is the subject of a positive judgment for the very considerable implications it involves in terms of supporting rural economy, as well as promoting respect for the environment and the natural and social landscape. This common perspective, which is both symbolic and practical, suggests a clear attitude of GIs to become proper tools to concretely apply the outcome of agro-ecological research.

[16]GIs are usually given a heterogeneous protection in different countries depending on the interest in protecting agro-food products. An Anglo-German model is traditionally opposed to the French-Romanistic one: the former is based on the perception of the public as a 'subjective' standard for repression of misleading under unfair competition doctrine, the latter on *sui generis* laws which provide (a) preventive registration by public authorities on the premise of a preexisting exclusive 'link-with-the-*terroir*' (*i.e.* between the quality of products and natural and human factors therein available) and (b) a strengthened 'objective' and 'quasi-absolute' protection which results in an exclusive right to use, conferred to the circle of producers settled in the designated area, independent from the perception of the public, coupled with the prohibition to become generic. Despite a prolonged uncertainty about the legal nature of rights, especially on AOs [still attested in 1992 by the Resolution OIV on the legal nature of AOs: cf. Audier (1993), p. 28 et seqq], GIs and AOs are today placed, with full merit, within IP law, namely as collective rights of property on intangibles (in that particular meaning that is given to property rights in IP law, as 'rights on reputations').

[17]'*Terroir*' theory has been differently molded in course of time: it might be roughly stated that, in a first phase, it was privileged a spiritual and symbolic conception rooted in the need for reconstructing French national identity after the collapse of the '*Ancient regime*' (nineteenth century); at a second stage, it was given priority to a 'naturalist' approach, based on natural factors of production (land, climate, raw materials), thus limiting protection only to '*produits naturels*' (French law on AOs of 1919 and Decree on AOCs of wines and cheeses of the 1930s); later the conception shifted—as early as the late 1950s—to a more nuanced and holistic view which resulted in a so called "humanistic approach", open to the contribution of both natural and human factors (local knowledge, traditional methods of production), so as to allow—because of the pressure of the industrial '*milieu*' (*Limoges*)—a wider protection, extended to '*les produits industriels et artesanals*' (Lisbon Agreement, from the 1950s). For a mature 'naturalist' approach to *terroir*, cf. Roubier (1954), p. 762 et seqq. For an account of '*terroir*' as a tripartite notion ('spiritual', 'natural' and both 'natural and cultural') and as it related to the different stages of formative politics in the post-Revolutionary France, cf. Gangjee (2012a), p. 86 et seqq.

3 Debate at the WTO and in the EU on the strenghtening of protection: GI Law Protected Interests Between 'Trade' and 'Non-Trade Concerns'

3.1 WTO Debate on the 'Extension' of GI Protection and the Establishment of a 'Multilateral Register': GIs as Tools for Economic Development and 'Cultural' or 'Environmental Guardians'

The enhancement of GI protection in TRIPS Section 3 is one of the most controversial issues that has been debated during the '*Doha Round*'. The debate is currently focused on the so-called 'extension' issue, *i.e.* whether the stronger 'objective' protection for wines and spirits[18] should be extended to non-wine products, and on the establishment of a 'multilateral register', binding on all Members.[19] The EU and some DCs and LDCs have called for strengthened GIs protection arguing that this would improve global market access—and therefore remuneration of goods and amelioration of the balance of payments—for traditional quality products coming from the South of the world.[20] GIs would serve as a circuit of collective commercial exploitation, alternative to large retail channels and mass distribution.[21]

---

[18]Non-wine GIs are given in TRIPS Art. 22 'subjective' protection, *i.e.* the traditional common law standard depending on the extent to which the use of the geographic sign for products not originating from the designated area would result in a mislead of the public: which is to be excluded when the geographical meaning of the sign is obscure, is hardly connectable with an originating production, or is perceived as the common name for a type of product ('*Kolnisch Wasser*'), or is used in connection with 'delocalisers' (additions such as 'style', 'type', 'imitation', 'like' or the indication of the true origin, so as to exclude mislead). The evolution of GIs law from simple IPs in the Paris Convention or Madrid Agreement to AOs in the Lisbon Agreement and wine GIs in TRIPS might be accounted as the attempt to make protection 'objective', *i.e.* detached from the perception of the public, also by prohibiting delocalizing additions (cf. Madrid Agreement Art. 4 and TRIPS Art. 23).

[19]WTO Doc. n. TN/C/W/25, of 18 May 2005 on 'extension' and TN/IP/W/12, of 14 September 2005, on the 'multilateral register', both (not undisputedly) recognized as '*outstanding implementation issues*' by the Doha Declaration: cfr. *WTO Ministerial Declaration*', WT/MIN(01)/DEC/1, of 14 November 2001, par. 18.

[20]*"Several new Members and many developing countries had a long tradition of products grown or manufactured with special characteristics due to their origin which, in spite of the limited amount to which they were traded internationally, had become known and reputed. For These countries, ensuring market access for their 'star' products, might be the only way to keep their agricultural sectors alive. As a result of increased international competition in generic basic agricultural commodities, prices might plummet and severely diminish income in a number of new Members. Therefore, it was vital for countries to ensure those alternative sources of income, which could be done through market entry for ensuring their quality products"* (TRIPS Council, Minutes of meeting, 5–7 March 2002, IP/C/M/35, par. 159).

[21]*"In a time of globalizing and highly-competitive trade, the added-value of a GI creates new and better opportunities for quality products, especially for smaller and developing Member States and a valuable alternative to the benefit of economies of scale"* (WTO doc. cit., TN/C/W/14, 9 July 2003, par. I.10).

Many DCs and LDCs are weak in high added value sectors (IT, services), but rich in raw materials and traditional products (food and crafts, textiles) whose excellence and quality originate from their wide local biodiversity and natural resources, while reflecting cultural identities and TK[22] ('*Roibos*', '*Darjeeling*', '*Basmati*').[23] Methods for transformation of agro-food products are sometimes used in 'culturally and spiritually significant contexts', often separated by the market, and use rare or special raw materials (such as seeds or crops, plant varieties, breeds), selected from generation to generation through processes of trial and error, with a view to identify the best available variety or race, suitable for cultivation in a specific *terroir.*[24] Production knowledge, such as techniques for cultivation and processing, is often associated with specific biomaterial ('*biodiversity-related knowledge*') and is usually 'informal' and 'customary' and, thus, hedged in positive law and easily appropriable for industrial application by third parties unrelated to the originating community ('*bio-agro-piracy*').[25] GIs would be able to shield these 'TK-intensive products' from free-riding and gain them a certain position on the global markets, so safeguarding—in one expression—the economic, social and natural environment. Certain traditional productions do meet prevalent 'consumption habits', but are uneasy and costly and may move out from the market for social or economic reasons.[26] GIs would signal to the buying public the added value of these

---

[22]The argument had already been raised during the 1970s and the 1980s at the time of talks for a WIPO treaty on GIs and a new PC Art. 10*quater*, which were widely affected by the debate on the 'New World Economic Order' and 'the right to development'. As to the debate on TRIPS reforms in favor of DCs and LDCs, cf. Land (2014), p. 142 et seqq, and Deere (2009), particularly 303 et seqq. WIPO has published a 'Model Law for Developing Countries on Appellations of Origin and Indications of Source' in the 1970s: cf. WIPO document TAO/I/INF. 1, of 30 October 1974.

[23]'*Roibos*' is a South Africa root used as a beverage and for healing purposes, while '*Darjeeling*' tea and '*Basmati*' rice are major Indian traditional quality products, whose production and export is rapidly growing in terms of quantity and value. '*Blue Mountain*' is a fine traditional coffee grown in Jamaica: cf. Downes and Laird (1999).

[24]Cf. Correa (2001), pp. 3–5 et seqq.

[25]The landmark essay is Shiva (1996), pp. 141–170.

[26]A traditional product may disappear, for instance, because of: (a) changes in in trends and habits of consumption by virtue of competition by cheaper alternatives or substitutes, eventually more 'desirable', if only for the novelty or the attractive force of fashion and aggressive advertising campaigns; (b) a rise in costs of raw materials, available techniques, labor force or other inputs eroding margins and causing lower revenues; (c) events such as a reduction of income or investment in the areas of origin and production, the degradation of living conditions, also because of economic crises or catastrophic events (war, drought or famine, epidemic events, migrations), causing erosion and loss of natural resources for significant climate changes, lack of human or intellectual resources, financial support or economic viability or the abandonment of traditional techniques (or their failure to adapt to the needs of present time) for the lack of an incentivizing context; (d) a choice of industrial policy (such as financial incentives for mass agriculture or industry), draining away resources from traditional agriculture. An example of public choice is in Gramsci (1966), p. 6: *"(T) he protective duty on wheat has prompted many in the countryside at sowing in the middle of barren lands in the certainty of an easy gain artificially provided by the State, for the usual reason of increasing domestic production. The State monopoly created by the war, which from 29 francs brought the grain to more than 40 francs, is to create the illusion that even spreading on the sand there is always enough to gain"*(a.t.).

products, thereby preserving not only their diversity but also the methods by which they are produced,[27] the 'contexts of consumption,' as well as certain 'lifestyles', that surrounds the products themselves, put at risk by market forces and homologation induced by mass production.[28] From this angle, GI protection is to intersect with the 2003 UNESCO Convention on the protection of '*intangible cultural heritage*' (CICH) and with the 2005 UNESCO Convention on the protection and promotion of '*diversity of cultural expressions*' (CDC)[29] and by the angle of the right to a healthy environment, is accounted as a tool for safe guarding agrobiodiversity according to the Rio Convention on Biological Diversity of 1992 (CBD).[30] In this last regard, by promoting TK agriculture (mostly family-based), GIs would enhance the use of alternatives to chemicals and pesticides and preserve local plant species[31] and the

---

[27]The aging method of '*infossatura*' for the '*Fossa di Sogliano*' cheese of Emilia Romagna, Italy, or the 'filatura in bronzo' ('extruded through bronze') used for certain types of pasta made in '*Gragnano*', Naples.

[28]International trade law contributes to homologate consumption styles by requiring harmonized standards for the production and marketing of products and an equal treatment of goods and services to allow their free circulation across borders. On mass consumption and the role of media cf. Schiller (1976). Habermas (2001) highlights the erosion of local traditions and habits brought by global trade. Contexts of consumption include the set of social practices, places and contexts where products are sold and consumed, which are a significant portion of processes of self-perception of individuals and social groups' own identities, reinforcing communitarian ties and sense of belonging. Consumption of traditional quality products often takes place in 'cultural' and 'social' places (small markets, cafes, specialized shops), or in particular occasions (feasts, festivals): cf. '*der Zur Blauen Frasche*', accounted as the first 'cafè', opened in Vienna by *Georg Koslchitzky* after the retreat of the Turks in 1683, which inspired the '*Lloyd's Coffee House*' in London in 1687, a 'must' for public relations and commerce: Niola (2015). On the role of food in building up cultural and communitarian identities, cf. Montanari and Sabban (2006), p. XI et seqq.

[29]Respectively, Articles 2 and 4: both texts are available at http://portal.unesco.org/en/ev.php-URL_ID=12025&URL_DO=DO_TOPIC&URL_SECTION=-471.html. The dual nature of GIs paves the way for considering the interrelations between GIs law and international cultural law, as well as the fundamental right to culture under the UN Covenant on Economic, Social and Cultural Rights: an interesting approach in this regard is in Gangjee (2008), pp. 383–395. On IP law and fundamental rights cf. Intellectual Property and Human Rights, WIPO, Geneva, 1998. For a general framework of the interrelations and conflicts between general International law and special International law (or between its special branches), cf. International Law Commission, Fragmentation of International Law: Difficulties Arising from the Diversification and Expansion of International Law: Report of the Study Group of the International Law Commission, 58th Session, May 1-June 9, July 3-Aug. 11, U.N. Doc. A/CN.4/L.682 (Apr. 13, 2006), as corrected U.N. Doc. A/CN.4/L.682/ Corr.1 (Aug. 11, 2006) (finalized by Martti Koskenniemi).

[30]CBD text is available at https://www.cbd.int/convention/text/. FAO publications support the view that, apart from being "*of great economic benefit to rural communities or local niche producing or widely popular products*", GIs may serve as a tool for implementation of the CBD, whose Art. 8 (j) requires States Parties to promote a "*wider application*" of TK and communitarian practices for the conservation of BD encouraging an "*equitable benefit sharing*" and whose Art. 11 requires to adopt "*economically and socially sound measures that act as incentives for the conservation and sustainable use of biological diversity*": cf. Mosoti and Gobena (2007), p. 202 et seqq.

[31]Grafting techniques, seed selection, rotation of species and use of biologic or organic substances with regenerative, anti-bacterial and anti-viral purposes, with a view to maintain biologic cycles and

landscape, otherwise bound to suffer the unaesthetic prevalence of monocultures, as well as public health, affected in the last decades by a substantial reduction of food diversity, *i.e.* the same number of nutrients that are available for human alimentation.[32]

3.2 GIs Attraction Into the CAP as Instruments for Rural Development and to Protect the Gastronomic Heritage and the Environment

Such a set of underlying justifications coincides to a large extent with the arguments provided by the EU to adopt GIs regulations from 1992 until 2012, as a cornerstone of EU quality policies under the CAP.[33] EU law had first recalled the economic dimension of GIs protected interests and expected outcomes in Reg. 2081/92 and later in Reg. 501/2006,[34] and only in Reg. 1151/2012 their social, cultural and environmental dimension by placing GIs economic functions in a broader framework that moves well beyond the narrow limits of the protection of business goodwill and customer information and includes a significant spectrum of public and general interests in between private and public law.[35] The preamble of the latter Regulation highlights the connection between market promotion and the preservation of the cultural heritage and agrobiodiversity,[36] as well as the economic value of

the organic generative capacity of fields, as well as a balanced use of resources, reduction of sulphites in wine and additives in general.

[32]Khoury (2014), p. 4001 et seqq.

[33]PDOs and PGIs Regulations were designed in the late 1980s and early 1990s in times of overproduction and falling agricultural commodity prices as a counterbalance at the time the EU was dismissing the unrealistic policy of so called 'vertical harmonization' of products, leaving room to the principle of mutual recognition ('*Cassis de Dijon*'): cf. Communication of the Commission on the *'Realization of the Internal Market: community regulation of agricultural products and foodstuff'*, 6 November 1985, par. 28, COM(1985) 603def.

[34]Cf. Reg. 2081/92: GIs are capable to ensure "*the improvement of farmers incomes*" and of "*promoting the diversification of agricultural production in order to achieve a better balance between supply and demand on the market*", as well as "*the promotion of products having certain characteristics could be of considerable benefit to the rural economy, particularly in disadvantaged and outlying areas*".

[35]GIs are supported by a complex administrative and self-professional monitoring system and a wide scope of protection, compared to simple IPs and trademarks: cf. as to the mixed nature of GIs law cf. Olszak (2001), p. 13: "*...ce système de protection s'est organisé suivant des principles assez différents de ceux des autres branches de la propriété industrielle et on se trouve parfois face à un "conflit de logiques", (...) c'est-à-dire un conflit entre unelogique de droit public, d'essence collective, et celle de droit privé, éminentement individuelle*".

[36]Reg. 1151/2012: "*(T)he quality and variety of agricultural production (...) of the Union represent a strength and an important competitive advantage for producers of the Union and are an integral part of its living cultural and gastronomic heritage*".

European farmers TK, strongly rooted in tradition but open to innovation and development,[37] and the GIs potential in protecting the environment, landscape and animal welfare.[38] At the outcome of a long-standing normative process, the EU thus acknowledged what, after all, was implicit in the same '*humanist*' conception of *terroir*[39] since the very first delimitation of the IP law protected 'intangible' in the 1992 Regulation,[40] namely that GI products are not only assets providing a comparative advantage to EU producers in international trade, but also significant pieces of the social and economic life of EU local communities, whose regulation transcends the boundaries of commerce and aims at protecting EU cultural heritage and the natural environment.[41]

[37]Cf. Reg. 1151/2012: "*This is due to the skills and determination of Union farmers and producers, who were able to preserve traditions while taking into account the development of new materials and production methods*".

[38]"*An agricultural product or foodstuff bearing such a geographical description should meet certain conditions laid down in a specification, such as specific requirements aimed at protecting the natural resources or landscape of the production area, or to improve animal welfare*".

[39]As to the 'naturalist' conception, often opposed to a 'humanist' conception of GIs, cf. note n. 17 above.

[40]As known, IP law does not simply accommodate pre-legal intellectual entities into legislation as they are in the real word or in human thought, but it qualifies as deserving protection a specific portion of a given intellectual entity (*e.g.* an 'original' solution to a technical problem which has not been 'published or make known to the scientific community and the public', is susceptible of 'serial industrial application' and has been 'deposited and registered at IP specialized offices'), carved out in accordance with pre-established legal functions and protected interests ('original ideas for patents' and not any idea, 'distinctive signs' for trademarks and not any sign). Regarding GIs in Reg. 1151/2012, protection is limited to those geographical names designating products originating from a specific area whose characteristics, qualities or reputation are exclusively due (or attributable) to the environment, including natural and human factors (Art. 5): this making clear that not all signs are protected, but only those signs suggesting an essential link with 'origin' and 'quality'.

[41]Some hints for a cultural conception of GIs as 'national cultural heritage' are in La Pergola, Conclusions, Case C-289/96, C-293/96, C-299/96, *Denmark et al. v. Commission* [1998] *ECR* I-1541, para 8; Ruiz-Jarabo Colomer, Conclusions, Case C-317/95, *Canadane Cheese Trading* [1997], ECR p. I-4681, para 9 et seqq.

4 Criticism. Arguments Against Strengthened Protection for GIs: Alleged Inadequacy in Preserving 'Non-Trade Concerns'

4.1 GI Ineffectiveness as 'Income-Generators' in Face of Decreasing EU Wine World Market Shares. Transformative Effects of 'Market Forces' and 'Consumption Habits'

These positions have been subject to harsh criticism within the TRIPS Council at the WTO. According to important scholars, GIs ability to bring significant increases in international trade would be completely unproven, while burdens on government sand producers to align their system to GI standards are undoubtedly high.[42] According to Broude, in the last 20 years Europe has lost considerable world wine market shares to the benefit of California, Chile, South Africa and Australia.[43] The rhetoric of GIs as '*cultural guardians*' against homologation would be baseless. GIs have not prevented in Piemonte and Toscana, Italy, the so-called '*Chianti wars*', *i.e.* innovative winemakers from adopting new styles and aging methods, closer to international tastes ('*barrels*' and '*trellising*'), and leaving the DOCG '*Chianti Classico*' to achieve greater simplification by using grape names as exclusive identifiers of their wines ('*Supertuscans*').[44] Similarly, in the area of '*Parmigiano Reggiano*', traditional local cattle breeds have been replaced first with the Swiss breed '*Bruna-Alpina*' and later with North American '*Holsteins*' and Dutch '*Fresian*'.[45] GIs, by establishing strong monopolies oriented to exports, would also increase prices restricting access to food for local communities, while unified quality standards, drawn up in accordance with power relations among producers based on commercial interests, would end up undermining authenticity and dispersing the diversity of nuances of traditional products.[46]

---

[42]GIs law implies to put in place new institutions: offices for the registration and quality monitoring, alignment of quality standards, administrative protection, education and training: cf.WTO doc. n. IP/C/W/360, p. 14 et seqq.

[43]In 2000, Australian exports to the UK were for the first time higher than the French, so unveiling a change in taste of the British consumer who turned his attention to new wine producing countries. This has caused even the '*Ministere de l'Agriculture*' to question the need for an *AOC* reform: cf. Broude (2005b), p. 670 et seqq. and Broude (2005a), p. 21.

[44]Under the modest but less rigid 'table wine' label, before the adoption of IGTs: cf. Broude (2005b), p. 664 et seqq. This led the Consortium '*Chianti Classico*' to dismiss traditional varieties according to the 'Barone Ricasoli' recipe (such as '*Malvasia*' and '*Trebbiano*') and increase new grapes until 10% and later 15%.

[45]Hughes (2006), p. 360. Similar changes have occurred as it concerns rules on grazing fields and pasture.

[46]Cf. FAO publication Mosoti and Gobena (2007), p. 202 et seqq.

4.2 Criticism to the 'Terroir' Theory on the Background of the 'Reality Principle': GI Products as 'Invented Traditions' and a Loose Basis to Overrule the 'Misleading Standard'

Anglo-American scholars subjected to criticism the same foundations of *terroir* in its rigidly '*naturalist*' conception, questioning the admissibility of an exclusive link between product and territory on the background of the 'reality principle', where conceived in terms of strict impossibility to reproduce the product elsewhere ('*exclusivité d'origin*').[47] In Hughes opinion, the rhetoric of '*terroir*' would not correspond to the geological nature of most of PDOs production areas.[48] Distinctive traits of famous wines would be partially reproducible in any place with similar soil or climatic conditions: the impossibility to reproduce a product in another place—historically relied to justify claims to exclusive use of AOs and GIs—would then constitute a mere dogma that distorts reality for the sake of hegemonic interests.[49] The dichotomy '*industrial v. traditional or familiar*' would be the outcome of Eurocentric perceptions not adhering to the realities of production.[50] In other terms, PDOs and PGIs would be mostly what Hobsbawm called an '*invented tradition*', *i.e.* the labelling or reinterpretation of a chain of activities as typical or connected with the past or custom even if it reflects only partially authentic historical conditions, so as to invoke the authority of the past for specific political purposes (in this case, a purely protectionist and expansive trade policy). '*Terroir*' would then be a very loose justification for the 'extension' of GIs protection. According to the '*Opponents*', making GIs protection 'objective' and 'quasi-absolute', *i.e.* independent from the opinion of the public, would result in the loosening of a fundamental concept of trademark law (the '*confusion*

---

[47]Cf. Roubier (1954), p. 762 et seqq.

[48]Zoning operations would be largely fictitious and would rarely correspond to homogeneous geologic conditions and to the extension of customary uses, showing inconsistency with the real organization of production (procurement of raw materials is often made on a larger scale than the delimited area), because of the pressures of lobbies, interested in the commercial value of names: cf. Broude (2005b), p. 664; Hughes (2006), p. 360. For a recent account of disputes concerning the delimitation of the geographical area, cf. Gangjee (2006), p. 291.

[49]Such as retaining the evocative power of geographical signs on the international markets and preserving EU monopolistic rents in agriculture on the face of political and electoral pressures: cf. Hughes (2006), p. 360.

[50]Many PDO/PGI champions produce significant quantities of products and apply technologies far from being traditional: cf. Hughes (2006), p. 339; Broude (2005b), p. 651. In addition, GIs law would drive local communities to emphasize the most marketable aspects of traditions: Broude (2005a, b), respectively pp. 21 and 677. The success of '*Champagne*', made according to the codification of *Dom Perignon*, is reported as an example of a particularly palatable wine defeating traditional wines.

or misleading standard'),[51] with significant trade-restrictive effects.[52] Exceptions to free trade should be carefully circumscribed to prevent that the defense of tradition and biodiversity are in fact used to shield veiled forms of protectionism.[53]

5 Reconsidering GIs Protected Interests and Functions Through the Lenses of IP Law: Scope of Rights and Expected Effects

5.1 Dual Nature, Commercial and Cultural, of GI Products. Cultural Protection as *'as an end in Itself'*

The vagueness of certain statements in support of GIs, on the one hand, the nature of certain criticism often not lacking arguments, but seemingly biased by not always well founded or significantly proven premises, on the other, suggest to review the opinions so far exposed based on available literature and practice. Taking the angle of IP doctrine, it is rather useful to 'order' the entire set of policy justifications of GIs and unpack protected interests as well as underlying functions,[54] with a view to define both expected benefits and interactions with International cultural and environmental law.[55] A multitude of GI products, especially food and artisanal productions ('*Gorgonzola*' cheese, '*Chianti'* wine, '*Murano'* glass, '*Delft*' and '*Limoges*' pottery), have a dual nature being at one time 'commodities produced in significant quantities for trade' and 'cultural expressions': thus they are potential

[51]WTO Doc. IP/C/W/360, 26 July 2002, par. 16 e 20: "*To require the trademark system to begin using a different standard - one not based on consumer confusion - would require, at least in the trademark regimes of some members, a fundamental philosophical change*".

[52]Dismissals of names (previously used as generic)to the advantage of GIs holders with consequent confusion of the public, advertising of new names and fine-tuning of sign less known by the public, limitation of freely available trade descriptions, unfair competitive advantage conferred to GIs holders, not based on any initial effort or goodwill: cf. WTO IP/C/W/360, par. 14 et seqq, and 23; WTO, doc. TN/IP/W/12/Add.1, par. 14, 75 and 100): cf. also WTO IP/C/W/360, cit., par. 23–26; in literature, cf. Landes and Posner (2003), p. 169 et seqq. and p. 189 et seqq.

[53]Cf. Broude (2005b), p. 678; Hughes (2006), p. 366; Voon (2009), p. 311.

[54]The Italian contribution to the legal analysis of protected interests and functions of IP rights is well established in Europe: for a functional theory of commercial law, cf. Ascarelli (1949), p. 101; in IP and trademark law, cf. Vanzetti (1961), p. 16 et seqq. and Spada (2012), pp. 3–5 and 15 et seqq; for an overview of the different theoretic approaches to IP rights, cf. Are (1959), p. 244 et seqq.

[55]Having particular regard to the UNESCO CICH and CD and the CBD. For a comment to the UNESCO Conventions cf. Von Schorlemer and Stoll (2012) and Blake (2006). As to CBD and its critical interrelation with biotechnologies, cf. Pavoni (2006), p. 29, and Ullrich (2006), p. 201 et seqq.

objects of legal protection both under supranational GIs law and the UNESCO CCD and CICH.[56] This conclusion also holds true for TK used in the production processes, as well as the contexts in which products are produced and consumed, particularly suitable to qualify as '*intangible heritage*' as of Art. 2 of the CICH, since they comply with 'objective' and 'subjective' requirements and 'space and time' conditions provided therein.[57] However, in no way this symmetry in substance corresponds to identity of protected interests and type of protection conferred. While CCD and CICH do consider the commercial value of cultural activities ('complementarity' principle),[58] as a general rule cultural law moves from the assumption that cultural activities *"may be an end in themselves"*.[59] Beyond a vision limited to the patrimonial value of property interests, protection of cultural assets aims at identifying, preserving or recovering any single entity (or group of entities) that have a cultural relevance, especially when at risk of extinction, and the overall social environment they live in on the assumption of their social, educational and aesthetic value, also in terms of perception of one's own 'identity'. This becomes particularly relevant in those cases where the invisible hand of the market does not 'self-sustain' cultural activities and allocate human and financial resources on more rentable activities, so putting at risk the *humus* for cultural expressions to survive.[60] From

[56]Both in their tangible dimension, *i.e.* entities in their physical features, and intangible dimension, *i.e.* the 'knowledge' and the 'information' used in the processes, intended as 'ideas' or 'intellectual entities' regardless of their material support.

[57]GIs incorporate '*practices*', '*expressions*', '*knowledge*', '*skills*' that local '*communities*', '*groups*' and '*individuals*' recognize as part of their culture, transmitted '*from generation to generation*' and constantly '*re-created*' in interaction with the '*place*', the '*environment*', '*nature*' and '*history*', to provide a sense of identity and continuity based on respect for diversity and creativity. Nevertheless, the same products and techniques also reflect the identifying features of the notions of '*cultural expressions*' and '*cultural activities, goods and services*', as required by Art. 4 of the CCD, *i.e.* expressions, goods, activities that incorporate or are the result of '*expressions of the creativity of individuals and social groups with cultural content*' as it relates to '*symbolic meaning, artistic dimension and cultural values that originate or express cultural identities*'. Most natural resources used for producing GIs products ('*plant and animal varieties*', '*vines*', '*micro-organisms*') and the knowledge used to treat them, may qualify under '*biodiversity*' and '*biodiversity-related knowledge*' definitions as of Articles 2 and 8 (j) of the CBD.

[58]Art. 2.5 of the CCD (as to measures aimed at promoting '*cultural industries*' cf. Art. 6.2).

[59]Art. 4.4 of the CCD.

[60]Starting from these assumptions, CCD and CICH establish the legal framework for the adoption—and legitimacy *vis-à-vis* other branches of International law—of domestic measures and international cooperation aimed at safeguarding the basic conditions (including the economic ones) for the continuous reproduction, reinterpretation, promotion and further growth of cultural expressions and intangible heritage. This kind of protection is both 'conservative' and 'evolutionary', because is committed to continuously identifying entities and assets worth of protection (through inventories, databases or periodic plans linked to specific sectors) and to provide the means, incentives and awareness (including financial support, appropriate communication campaigns) to support persistence and continuity of traditions and cultural expression (*e.g.* depending on the type of entity, investigation, research and dissemination of knowledge, creation of spaces for the free development, revitalization or continuous renewal, contamination and '*re-creation*' of cultural and artistic contents, laying down the infrastructures for '*creative and cultural industries*'):

this angle, while it stands out more sharply the difference in approach between the two branches of law, it is possible to remark overlaps of cultural law with IP law (promotion of aesthetical values, identities and good living), tensions with WTO law (ambition to 'special treatment' of cultural goods and activities v. the unifying approach of the International law of trade)[61] and a certain degree of instrumentality of GI law to cultural policy objectives.[62]

5.2 *Difference in Nature of GI Law: The Distinctive Function as Protection of Producers' (Reputation) and Consumers' Interests (Information). The 'Premium Price Theory'*

According to the method of IP law, GI provisions (especially those on AOs), in defining the portion of pre-legal reality that is qualified by the relevant norms as an IP intangible, establish a right to exclusive use limited to a geographical name (not covering quality regulations) to the benefit of producers, so as to distinguish products whose quality or characteristics derive from origin and the natural and human resources therein available.[63] As shown by economic analysis of law, the legally protected function of GIs—by a sort of partial 'trademark analogy'—is to identify products by making reference to the dual link between product and territory

cf. in this regard CCD Art. 4, para 7 and 6 (principles, regulatory measures and incentives), from 7 to 19 (interventions, transparency and information sharing, awareness campaigns, international cooperation, international fund, monitoring and reporting); and CICH Articles from 11 to 18 (regulatory, incentives, awareness and planning), from 19 to 24 (international cooperation), from 25 to 28 (ICH International Fund) and 29–30 (monitoring and reporting).

[61]Cf. general exception in GATT Art. XX, lett. b and, as to the audio-visual sector, the so called 'cultural exception', establishing quotas and percentage for domestic products, GATT Art. IV. There is no exception in GATT Art. XX which justifies restrictions of trade to preserve national culture, a part from the protection of 'public morals' (Art. XX, lett. a).

[62]In particular: (a) points of overlap with IP law where the promotion of aesthetic value, cultural identities and good living which are enshrined in artistic creations is one of the 'underlying assumptions' which justifies at one time '*droit d'auteur*', GI law and the legal protection of cultural expressions and intangible heritage; (b) potential points of tension with the law of international trade, where cultural law, while recognizing the commercial value of certain cultural assets, sets out the conditions for their 'special treatment', which might collide with the unifying and standardizing and non-discriminatory approach which lies at the heart of free trade principles, raising the far more difficult issue of the limits within which a special regime for circulation of cultural goods and services (or preferential aids), which might result trade-restrictive, is likely to alter ordinary conditions of competition or imply veiled restrictions to international trade; (c) relationships of instrumentality between GI law and cultural law where, assuming the complementarity between economic and cultural protection, GI law might serve as a tool to achieve cultural policy aims.

[63]The double nexus (geography and quality) has been highlighted in French and Spanish doctrine: cf. Roubier (1954), p. 762 et seqq., p. 762; Fernandez Novoa (1970), p. 146; Botana Agra (2001), p. 22.

(origin and quality),[64] rather than—as in the case of trademarks—to the uniform entrepreneurial source (the producing firm), so as to differentiate territorial products from the genus of serially processed product to which they belong. Therefore, the GI performs a dual economic purpose that is: (a) to reduce consumers search costs by providing crucial information for purchase decisions; and (b) to enhance the image of products polarizing on the signs a 'well-established reputation', dependent on the quality of the product, so as to turn the designated goods in '*niche*' products, of particular excellence, for which the public is generally willing to pay a '*premium price*' in exchange for quality.[65] Thus, GIs primary protected interests are not 'cultural' in the strictest sense, but rather: (1) the commercial and patrimonial interest of producers to the recognition of 'geographical market reputations'[66]; (2) the interest of consumers to receive truthful information to make informed purchase decisions.[67]

5.3 How GIs Law Protects 'General Interests': GIs as a Settler of Incentives to protect 'Living Traditions', Fully Subjected to the Laws of the Market

This statement does not mean that GI provisions, as trademark and unfair competition law, cannot indirectly pursue general or public interests.[68] More simply, it is

[64]Better, geographical origin and the guarantee of territorial quality, inclusive of the '*conditions naturels des terroirs*' (natural factors) and the '*savoir faires*' of producers (human factors).

[65]This is particularly true for 'experience' and 'credence' products, whose quality cannot be weighted before consumption and after consumption: cf. Akerlof (1970), pp. 488–500, and, regarding 'reputation', 'asymmetrical information' on the markets and 'premium price theory', Shapiro (1982) and Shapiro (1983), pp. 659–679. For a review of economic literature on GIs, cf. OECD (2000), p. 7 et seqq.

[66]In particular, reputation determined by territorial quality, natural (purity of water, soil, type of plants, climatic conditions) and human (craft machinery, hands harvesting, traditional recipes): as to the connection between quality and reputation today cf. CJEU, Case C-388/95 Belgium v. Spain (Rioja II) [2000], ECR I-3123, paras 54–56 and 59; CJEU, Case C-108/01 Parma et alteri v. Asda Stores [2003], ECR I-5121, paras 64–66 and 80. Quality reputation is protected by conferring an exclusive right to use on the club of producers who comply with quality regulations (and so gained reputation to the products) so as to prevent any interference (or free-riding) from third competitors in the channel of communication established through the sign with the public. Since the very beginning, as early as it was adopted the first French *AOC* law in 1919, protecting reputation from fraud was the function of quality regulation, as a response of quality producers to the '*phylloxera*' epidemic, which destroyed high quality grapes and caused lesser quality wines exported from Algeria and Tunisia be sold under famous geographical labels.

[67]Protecting consumers from misleading and shielding producers from unfair competition, as primary justifications of GI law, are firmly stated from the 70s of the past century: cf. Fernandez Novoa (1970), p. 10 et seqq.; Ladas (1975), p. 1574, and in US law McCarthy (2006), ch. 3; cf. also, CJEU, Case C-12/74, *Commission v. Germany* (Sekt), ECR 1975/181, para 7.

[68]It is well-established in general theory of law, as well as in the theory of property rights and IP that, by conferring rights of actions to individuals, the legal order may be willing to achieve not only

useful to clarify that in GIs law the protection of interests of greater range takes place on a level far different from that on which individual commercial interests and the general interests internal to the market are protected. More precisely, this is a causally 'indirect' protection performed through the selective function of the market. The exclusive right to use the geographical name for origin and quality description[69]—except what is highlighted below—cannot ensure 'by itself' and 'in isolation from the market' persistence of productions or any presumed 'traditional purity' of products.[70] It rather works by signaling the added value of products to establish *de facto* incentives to invest in quality, excellence and tradition and receive a reward. Although it is correct to point at the risk of dilution of the purity of '*Chianti*' or the replacement of traditional breeds in the '*Parmigiano Reggiano*' area, it is not the sole GI law that can be entrusted with the task of preserving the traditional features of certain products. Such protection, being exposed to the fluctuations of public willingness to buy, is real and actual only on condition that producers maintain the traditional added value of GI products, consumers appreciate the latter and exercise their preferences for those GI-designated products and the relationship between costs of production and revenues remains efficient, in relation to the expectations of a significant number of producers who are willing to remain engaged in production. As regards the 'link-with-the-*terroir*' as a loose basis for GIs reinforced protection, while the exclusive or essential link with the territory should not be given for grant, there should not be much discussion about the principle that any natural product is mainly the outcome—both in texture, color, organoleptic consistence, taste and collective 'cultural recognition'—of the place and the knowledge that gave it birth and originality.[71] From this angle, it is hardly deniable that GIs, if internationally endowed with an exclusive right to use reserved to local producers, would better prevent any interference (imitation or free-riding) by third competitors in the channel of communication between producers and the public established by the sign, while

---

individual interests, but also a wide range of superior general or collective interests (transparency of the market, free competition, stimulus for innovation, public disclosure of new technologies for the sake of future free exploitation).

[69]As to the double nexus between products and the place of production (geography and quality), cf. Roubier (1954), p. 762 et seqq., p. 762; Fernandez Novoa (1970), p. 146, Botana Agra (2001), p. 22.

[70]At a deeper glance, this feature of GIs is coherent with the premise of the 'living' character of traditions, also recognized by the UNESCO Conventions: traditions should not be closed in unlikely juridical 'museums', but rather be deemed as a changing and evolving phenomenon, inclusive of modifications induced by consumers tastes.

[71]Even those authors that deeply criticise the '*terroir*' theory do not really doubt about the existence of a deterministic link between products and territory, limiting themselves to argue that some territorial traits might be reproduced elsewhere. In any case, even if some features of a product might be reproducible anywhere, with analogous results, any presumed 'natural' product is always also 'cultural', *i.e.* it is the result of a process of historical and cultural recognition in a given originating community as something edible and culturally worthwhile that gives it originality and uniqueness, also when partially or totally reproducible elsewhere: as to the notion of 'cultural recognition' cf. Montanari and Sabban (2006), p. XII et seqq.

ensuring the conditions for a clearer perception of the cultural and environmental added value of GI products, with minimal monopolistic effects.[72] At the same time, it seems rather confirmed that GI law, while posing very considerable market incentives for traditional productions, is just one of the factors that enhances GIs value and contribution to authenticity of traditions, being the other common cultural perceptions, the general conditions of the markets and the whole set of regulatory conditions set out to support culturally relevant activities and to develop the entire earning and conservative potential of GIs.

5.4 Elements Suggesting a Direct, Albeit 'Incomplete', Protection of Culture, Tradition, Food Diversity and the Environment: A Function in Progress?

Certain structural and functional elements in GI law leave a glimpse of a 'direct' legal protection of the guarantee for cultural and environmental value incorporated into GI products, which might go beyond simple '*de facto*' incentives to quality production. First, GIs protection consists both of a private right to use and an administrative protection, performed by public authorities and professional organizations,[73] and its contents and scope depart from traditional protection of trademarks

[72]As to GIs protection against free-riding and dilution, cf. Lehmann (1986), pp. 746–767. The anti-competitive effects of IP restrictions on distinctive signs (as in the case of GIs) are not comparable in quality and nature with restrictions on use of original knowledge (as in the case of patents): GI exclusive rights do not cover quality specifications, which remain freely available, but simply prohibit the use of geographical name for non-complying products or products not originating from the designated area. Thus, there is a limited monopolistic effect on language if generic and geographic alternatives are available ('sparkling wine', '*Champagne*' and '*Spumante*', '*Cava*' and '*Prosecco*'). While restrictions on signs based on origin may well favor national preferences, contrary to the principle of free circulations of goods[this is mirrored in CJEC, Case 249/81 Commission v. Ireland ('*Buy irish*')(1982), ECR 04005, Case 207/83, Commission v. UK and Northern Ireland (1985), ECR 01201: all these decisions prohibiting use of national geographical labels in the single market and reflecting the traditional conceptual hostility of the principles of free trade to commercial use of geographical names, dating back to the British Commonwealth], volumes of GI products are usually not massive since GIs regulations provide for limitation of yields. In addition, monopolistic effects of GIs protection are not so different in scope if compared to the effects of the rights conferred by the TRIPS on (reputed) trademarks.

[73]Inspectorates against fraud, judicial authorities, consortia and professional associations: in France, GI judicial protection has prevalently been the realm of public prosecutors (followed by constitution of private parties in criminal proceedings), while the '*Institute National des Appellations d'Origine*' (INAO), '*Syndicats de défense*' and '*Comités Interprofessionnels*', as well as some offices at the '*Ministère de l'Agriculture*' and '*de l'Economie et des Finances*' have been given the powers of inspection, monitoring and quality control(similarly in Italy). Public *ex officio* protection by public authorities of Member States (MS) has been provided—with some misunderstanding and uncertainty among States (cf.CJEU, Case C-132/05 *Commission v. Germany* (Parmesan) [2008], *ECR* I-957, para 75-by Reg. 2081/92 and later placed within the system of controls on health and composition of food products provided in Reg. (EC) 882/2004 by Reg. 510/2006 (Articles

by conferring an 'objective'('quasi-absolute') right to use which stretches in several significant directions.[74] This 'special status', public and private, is a first clear recognition of the 'extra-commercial' value of GI products.[75] Second, the symbolic message conveyed by GIs, more and more, does not simply indicate geographical origin and quality, but tends to attach to the main message a communicative content that, in a nearly pre-established form, points at the cultural, social and environmental value of the designated products, to which consumers tend to give relevance at the time they make their purchasing decisions.[76] Third, the same existence in GI law of quality regulations (and a system of monitoring their compliance), by whose respect depends the authorization to make use of the GI-label, may directly guarantee 'conservation of the traditional and historical content of products', to the extent that culturally relevant or environment-friendly methods are actually codified into quality regulations. In contrast to this view, it is to be pointed out that, while remaining constantly in the background of the public discourse as underlying justifications of GI law, the 'historical' or 'environmental' character and even the 'reputation' of a typical product (apart from 'reputed' PGIs) are not formal preconditions for PDOs registration,[77] except where they are imposed by a MS legislation.[78] It could then be inferred that EU GI law contains embryonal elements that

10–11), to then find full and final recognition (although serious problems of implementation) in Articles 36–40 of Reg. 1151/2012.

[74]Prohibition of declaring registered GIs generic, independence from the perception of the public, wide protection against usurpations, imitation and evocation, even beyond similarity of products. Reinforced protection of GIs against usurpation, imitation and evocation, prohibiting the simple use of any simple trait resembling the protected name, reflects a logic like the '*droit d'auteur*', that protects artistic works from any alteration or encroachment of the moral rights of the author and prevent any copying or imitation of the artistic expression: cf. Chen (1996), p. 5.

[75]EU law confers to GI holders special rights of/to protection, with a mixed public-private nature, that cannot be explained but not by virtue of a pronounced will to protect the 'extra-commercial' value of GIs. This trend is rooted in a French reaction in the early twentieth century to some court decisions in the US declaring French famous GIs generic, which was favored by their having been declared generic in the country of origin ('*La République française v. Saratoga Vichy Spring Company*', 191 U.S. 427, pp. 435–437, 24 S. Ct. 145, 1903). Consequently, the need for a GIs reinforced domestic protection as a prerequisite for the international one and their qualification in terms of national public property or heritage: cf. Roubier (1954), p. 754.

[76]As to the widening of the communicative content of GIs today, beyond tradition and quality (suggesting consumers to buy for *"motivations de nature hygiéniste, patriotique ouethique"*), cf. Olszak (2001), p. 1.

[77]Absence of traditional, cultural or environmental value are therefore hardly 'justiciable', that is to say they are unlikely to be the basis of an actionable claim in court, claiming the annulment or hetero-integration of PDOs or PGIs registration.

[78]As it is the case of France, Italy and Spain: cf. below notes 109 and 110. The absence of a traditional character may however be found, indirectly, as non-compliance with the requisite of a 'link-with-the-*terroir*': to establish a link, especially for PDOs, implies time to consolidate production, a certain level of recognition by the public, coordination and organization by the circle of producers. The Lisbon Agreement, Art. 2.2, seems to regard 'reputation' as an implied feature of International AOs: *"Le pays d'origine est celui dont le nom, ou dans le quell est située la régionou la localité dont le nom, constituel'appellation d'origine qui a donné au produits a notoriété"*.

underlie a function of legal protection of traditional cultures and the environment still *'waiting to realize'*. Such a function is in fact only 'eventual', its most rigorous fulfillment depending on the actual use that is made of product regulations and on the willingness of national authorities and producers to intensify the quality guarantee by adding cultural and environmental *'plus-elements'*. The extent and content of this last proposition will be the subject of a more thorough examination in paragraph 7, by means of the examination of a sample of EU PDOs and PGIs quality regulations.

6 Arguments in Support of GIs Law: EU Wine Agreements and the Establishment of Independent Commercial Identities

6.1 EU Agreements on trade in wine and the establishment of international reputations for local geographical designations in Australia and South Africa

The considerations exposed so far, while highlighting the 'commercial' nature of GI protection, suggest at one time to critically re-examine those opinions according to which GIs ability to generate income and protect culture would be unproven. In this perspective,[79] the decrease of the world exports of European wines in favor of the Australian ones from the mid-1990s might be read—rather than as a failure of EU GIs—as the consequence of the signing of the EU-Australia Wine Agreement that marked a turning point for the Australian wine industry, paving the way for the establishment of local wine denominations and the dismissal of EU imitations (*'Coonawarra'*, *'Barossa'*, *'Shiraz'*).[80] Analogous wine and free trade agreements have certainly contributed to the growth of quality wine production in Hungary,

[79] It is not an easy task to establish unequivocally the weight of each of the factors that determine the performance of complex markets such as the international ones, all the factors leading to increase or decrease of world market shares for a given industry being interdependent. Such an operation is even more difficult where the question is how to measure the economic effects of a legislative measure of IP law, which is notably difficult to measure in terms of economic analysis of law: cf. Landes and Posner (2003), p. 10. Bearing this in mind, any reading should be made being aware that available data do not provide objective truths but, at the most, some guidance on the impact of any single factor and that those data should be interpreted with caution according to the given historical context.

[80] 'Agreement between Australia and the European Community on Trade in Wine and Protocol, done at Brussels and Canberra (26-31 January 1994)'. By stopping the use of French names (such as Burgundy and Chablis) and relying on their own, Australians have built the world's most dynamic wine industry: Calboli (2006), p. 201. Commercial trends between 1992 and 2000, right after the signing of the Wine Agreement, show that the reputation of Australian local GIs, as well as the extra-price paid from the international public in exchange of well-known geographic wine brands, have significantly grown in the reference period: cf. Schamel and Anderson (2003), pp. 357–369.

South Africa and in some South American countries like Chile and Argentina[81] since, by prohibiting usurpations of EU wine names, they contributed to the creation of local GIs reputations ('*Paarl*' and '*Western Cape*' in South Africa, '*Maipo*' and '*Elqui Valley*' in Chile).[82] The latest case studies and numerous inquiries measure GIs capacity to generate income in very positive terms. Correa detects increases in market prices of some PDO/PGIs products at 30 to 100% of the price of generics ('*Poulet de Bresse*', '*Riviera Ligure*' olive oil).[83] Outside Europe, a survey on '*Boseong*' green tea in South Korea found, because of registration of a name, an increase in production between 1997 and 2005 from 500 to 1200 tons and a growth rate of 90% between 2002 and 2006, in a context in which the prices of similar generic tea grown in the same country remained unchanged.[84] The 'partial' character of some of these studies, often carried out at the request of the EC, might raise some doubts about the reliability of results.[85] However, the general consensus that GIs gained among producers and the continued use of those designations by a large part

[81]EU-US agreement of 2005, EU-Chile Agreement of 2002, EU-South Africa agreement of 2002. The 2005 EU-US Agreement prohibited the use of EU names in US (Champagne, Chianti, Chablis, Burgundy, Madeira), previously deemed as 'semi-generics', and pushed US vintners to use local GI names (consequently, the debate about the origin of Zinfandel, claimed by local vintners as a traditional grape of California).

[82]Cf. Agreement in the form of an Exchange of Letters between the European Community and the Republic of South Africa on trade in wine, OJ [2002] L28, 30.1.2002, pp. 134–135; Agreement on Trade in Wines annexed to the Agreement establishing an Association between the European Community and its Member States, of the one part, and the Republic of Chile, of the other part, OJ [2002] L352, pp. 3–1450. A similar agreement, especially devoted to the protection of European GIs, the phasing out of European GIs used as semi-generic in the US and the mutual recognition of oenological practices, was concluded by the EU and the US in 2005, but never endorsed by the Congress (cf. 'Agreement in the form of an Exchange of Letters between the EC and USA on matters related to Trade in Wine', OJ [2005] L 301, p. 16). The EU-US Agreement was meant to prohibit the use of an important set of EU wine names in the United States (Champagne, Chianti, Chablis, Burgundy, and Madeira). In any case, in the US, a flourishing viticulturist movement has consolidated—alongside the continued use of European IGs as 'semi-generics'—the protection and reputation on the domestic and international markets of Napa Valley wines, thus moving with small step towards the use of local geographic designations (see, in this regard, as a symptomatic element of a new trend, the debate on the origin of 'Zinfandel' vine that Californian vintners claim to be a traditional California vine, though it is accounted to have Austrian origins and shares the same DNA with the 'Primitivo' vine from Salento, Puglia, Italy). In the US, the 'Alcohol and Tobacco Tax and Trade Bureau' is entitled to issue authorizations to use in commerce the so-called 'American Viticultural Areas' (AVAs), geographical labels whose structure is in essence similar to European designations of origin, but whose extension and prescriptive contents are well far from being so specific and detailed such as those in place in Europe (the mandatory content of grapes grown and harvested in the area is set for AVA wines at a threshold around 80/85%).

[83]Correa (2001), p. 17.

[84]Paus and Reviron (2011), p. 16, with special regard to the case-study by Suh, Macpherson, The impact of geographical indication on the revitalization of a regional economy: a case study of '*Boseong*' green tea. This case study also reports the full panoply of advantages brought by GIs regulations, including increases in prices and sales, in land value and investments or the creation of new jobs, as well as reduction of polluting practices and greater sustainability of productions.

[85]Studies at http://ec.europa.eu/agriculture/quality/reports/index_en.htm , accessed 7 July 2015.

of producers are the most decisive arguments in support of GIs effectiveness: increasing costs of quality compliance and monitoring can in fact be tolerated only if the costs are largely outweighed by revenues.[86]

6.2 'Rioja', 'Parma' and 'Grana Padano' at the ECJ: Benefits for Side-Industries, Investments Attraction and 'Job-Relocation' in GIs Production Areas

GIs are also accounted of benefitting not only the market of the registered product, but also related segments.[87] EU law provides under certain conditions for a normative reserve of collateral activities for the area of production, such as packaging and products conditioning.[88] In '*Rioja II*' and later in '*Parma*' and '*Grana Padano*', the ECJ stated that certain operations, if necessary to obtain or safeguard the quality of products, shall remain under the control of producers in the delimited area, thereby retaining in its territorial sphere entire economic activities and job-places.[89] At the same time, the economic viability of GI products are also a multiplier of foreign investments and value of lands.[90] While these arguments are not supposed to prove the infallibility of GI *sui generis* law, they may well serve to balance unilateral perceptions and to identify best practices and further ameliorations. They provide reliable indications about GI law ability to contribute or co-determine increased profitability of quality products, even in highly competitive scenarios and in conditions of economic downturn, without claiming that this is the only process that triggers growth in rural areas.

[86]'Commission Staff Working Paper' (2010) Impact Assessment on Geographical Indications, 2010, pp. 4 and 28, at http://ec.europa.eu/agriculture/quality/policy/quality-package-2010/ia-gi_en.pdf.

[87]Agro-tourism and catering, connected activities such as agro-food technology, communication and marketing, consulting, logistics and transport, packaging and the getup of products.

[88]Typically the bottling, grating, slicing of products.

[89]CJEU, Case C-388/95 *Belgium v. Spain* (Rioja II) [2000], ECR I-3123, paras 54–56 and 59; CJEU, Case C-108/01 *Consorzio del Prosciutto di Parma et alteri v. Asda Stores* [2003], ECR I-5121, paras 64–66 and 80; CJEU, Case C-469/00 *Ravil v. Bellon and Biraghi* [2003], ECR I-5053, paras 90 and 49.

[90]GIs have historically favored attraction of foreign investments: not by chance '*Bordeaux*', '*Champagne*', '*Jerez*' and 'Porto' main houses of productions were established by British investors in the nineteenth century.

7 Ambivalent Content of PDOs/PGIs Quality Regulations as Regards Tradition, Food Diversity and the Environment

7.1 EU Register as an Inventory of Cultural Diversity and TK. Elements for the Preservation of Traditions and the Environment in PDOs and PGIs Quality Regulations

It now remains to be seen whether GIs quality regulations registered from the early 1990s confirm the concrete capacity of GIs to contribute to keep into the market a wide range of traditional products or to determine positive environmental effects, by ascertaining whether 'plus-elements' are present into quality regulations, designed to preserve general interests. It first comes to mind that the same public records of EU GIs registrations, apart from performing the primary function of making known to the public IP entitlements to exclusive use, do collect all quality regulations and thus codify the natural and cultural diversity of traditional products in Europe (and their producing methods consolidated over time according to traditional, fair and consistent uses).[91] A sample examination of quality regulations included in the EU databases confirms—with the *caveats* that will be made below—this assumption. The PDO '*Fiore Sardo*' historical does prescribe the use of fresh whole milk from sheep of Sardinia breeds, raised on natural wild pasturages, and of traditional manufacturing methods requiring that the milk is heated in containers of cork filled with hot stones and fermented with spontaneous thistles of the genus '*cynara*'.[92] The PDO '*Formaggio di fossa di Sogliano*' provides for the historical method of curing cheese in pits dig in the ground ('*infossatura*') at various moments of the year ('*spring and summer*'), lists the characteristics of the soil ('*sandstone rock with veins of clay and marl*') and the measures of the pits, as well as instructions to make them.[93] Surveys testify, on the other hand, the belief of producers that GI registration has safeguarded products such as red and white rice old varieties of '*Riz de Camargue*' from extinction[94] and report that GI registrations have allowed to keep competitiveness, put at risk from widespread counterfeiting and imitations[95] and

[91]DOOR, E-BACCHUS and E-SPIRIT DRINKS, on line databases, available at www.europa.eu/qualitypolicy.

[92]DOOR, dossier IT/PDO/0017/0007, p. 2.

[93]DOOR, dossier IT/PDO/0005/0580.

[94]A similar view is expressed by manufacturers of the PDOs '*Fromage de Herve*' and '*Voilaille de Bresse*' (France) and '*Szeged Fuserpaprika*' (Hungary): cf. London Economics et alteri (2008), p. 226.

[95]This is the opinion of PDO '*Tuscan olive oil*' (Italy) and of '*Jersey Royal Potatoes*' (England) producers: cf. London Economics et alteri (2008), p. 226.

spurred diversification.[96] As regards the environment,[97] there is evidence of significant traditional methods for sustainable land management, such as the '*alumbra*' ('*serpia*' or '*deserpia*') in the area of '*Jerez*',[98] that consists of creating in the land small tanks 20 or 30 cm deep around the stump, to store rainwater and prevent erosion, or the use of local acorn varieties for pigs in the area PDO '*Dehesa de Extremadura*'.[99] The French cheese '*Comtè*' stands out for an entire set of environment-friendly practices (inventory of grazing areas subject to audits and periodic checks; quasi-absolute ban of chemical fertilizers; duty to use local feedings on the spot, coupled with upper limitations to feed storing so as to prevent mass silos storage) having a direct impact on maintenance of local crops and thus foster both agrobiodiversity and the overall profile of the landscape.[100]

[96]E.g. '*Gala*', '*Morgenduft*' and '*Fuji*' apple varieties in '*Melinda*' in Italy, or '*White Stilton with Apricot*' cheese in the UK and a special type of '*Esrom*' cheese in Denmark: cf. London Economics etalteri (2008), p. 226 et seqq.

[97]The 'environment-oriented' use of GIs builds on French *AOC* law that has provided the normative model for the EU. *AOC* quality regulations in France limit the use of pesticides and chemicals, established what lands are to be used for grazing and what foods for breeding, define exactly the arable land and usable plant varieties and fix quantities for crops yields, preventing the modification of traditional cultivations and intensive farming, which would lead to low quality and the degradation of the landscape: cf. Lorvellec (1996), p. 78 et seqq. Environment is also relevant by a prospective 'external' to GIs law, as legal treatment conferred to areas hosting GIs, at the moment of taking administrative decisions about urban and countryside planning, or regulating dangerous economic activities: cf. Lorvellec (1996), p. 79. Dangerous economic activities might alter the landscape or expose the geographical area to accidents and potential risk of erosion and loss of natural resources (groundwater, land or air pollution, reduction of biodiversity, acid rain). Special treatment as to building or dangerous economic activities licenses would then prevent that GI productions no longer comply with quality rules and immemorial productive customs—turning into 'misleading' the message of quality conveyed to the markets—and ensure that local environmental conditions are objectively better than the average.

[98]Also known as '*Xeres*' or '*Sherry*': for an accurate explanation of this method, cf. Garcia de Lujan (1988), pp. 55–56.

[99]Cf. DOOR, dossier ES/PDO/0017/0076, para 5, lett. (c) (e f), registered with Reg. (EC) n. 1107/96 of the Commission of 12 June 1996 on the registration of geographical indication and appellations of origin within the framework of the procedure provided in Art. 17 of Reg. n. (EEC) n. 2081/92 of the Council, OJ L 148, of 21 June 1996, pp. 1–10.

[100]Quality regulations provide for exams and checks to verify the presence of chemicals, metals and germs on the land, and contemplates the use of local hay for feeding animals on the spot: cf. DOOR, dossier FR/PDO/0117/0116, as amended under the implementing Regulation (EC) n. 256/2015 of 13 February 2015 approving non-minor amendments to the specification for a name entered in the register of PDOs and PGIs [Comté (PDO)', OJ L43 18.02.2015, containing the '*Cahier des charges de l'appellation d'origine Comtè*': cf. Annex 1].

7.2 Under-Consideration of and Generic or Apodictic Reference to Cultural and Environmental Methods in Quality Regulations

In other GI quality regulations, the reference to local traditions appears to a large extent a general statement, lacking any description of cultural methods. A higher degree of generic statements is frequent for products from those countries (usually in Northern Europe) that do not know advanced 'agri-food' traditions, for natural, historical or cultural reasons. However, this is a recurring feature also for some Southern Europe-located GI products. Certain quality regulations, for instance in the olive oil sector, define traditional methods 'circularly' as those *"traditional and characteristics of the area and, however, able to provide the olives and olive oil with the specific quality characteristics"*.[101] Loose wording might be explained by the fact that for certain types of products natural factors are determinant (if compared to other sectors where transformation prevails) and thus a stronger emphasis is placed on the bio-geological link with the environment.[102] It is true, however, that similar problems also arise for processed products for which, on the contrary, prevailing contribution comes from the 'human factors',[103] usually lighting up never-ending litigation on 'traditional recipes' and their correspondence with historic usage in a certain delimited area.[104] Environmental standards in PDOs e PGIs quality regulations show at first sight similar patterns. In fact, a thorough examination of production rules does not always confirm the inclusion of plus-elements at different stages of production, nor shows much change after the entry into force of the first 'environmental' Regulation n. 1151/2012. Quality rules often refer to the 'use of traditional techniques' without specifying features, fixing the frequency, nor identifying types of plants or pastures for breeds or establishing thresholds or particular conditions for use of local resources.[105] Many regulations simply reduce the use of

---

[101]Likewise, for planting layouts, types of cultivation and pruning systems which *"must be those generally used or at least not in any way change the characteristics of the olives and oil"*: cf. 4.1 of PGI quality regulation of '*Olio Toscano*'. Art. 4.1 of PDO quality regulations of '*Olio di Terra d'Otranto*' provides similarly, while art 5.3 specifies: *"For the extraction of extra-virgin olive oil as of art. 1 only physical and mechanical processes are admitted apt to ensure that olive oil is obtained without any alteration or of the qualitative features of the fruit"*.

[102]Texture and composition of soil, exposure to climatic agents, plants varieties, rather than processing techniques.

[103]Bakery and confectionery, beer, sausages, sweets: for these products processes are normally more relevant than natural factors and they are usually accommodated within the looser definition of PGI.

[104]Complaints concern especially the identification of the exact extension of territory—normally expanding over time—where the historical production has taken place. Political pressures often determine the overexpansion of the producing area and the loosening of quality specifications: Gangjee (2006), p. 301 and in ECJ case law, cf. CJEU, Case C-269/99, Kuhne ('*Spreewalder Gurken*') [2001], and TFI order of 15 September 1998, T-109/97, 'Molkerei v. Commission'.

[105]For an in-depth analysis, cf. 'European Forum on Nature, Conservation and Pastoralism' (2015), pp. 5–6.

pesticides and chemicals with high impact, but do not provide for alternative agents, nor specify terms of use of grazing lands, density of cultivations or allowed fertilizers.[106] This is reflected in a limited capacity to preserve the landscape and the environment[107] and might confirm—within the limits of this inquiry and the cross-section of quality regulations under examination—that the enhancement of the cultural and environmental function, although openly declared by EU law, is left to the actua and concrete will of producers to voluntarily add 'plus-elements' designed to protect general interests.

8 Tendential Lack of 'Justiciability' of Enhanced TK Requirements in EU Law: The Case for 'CSR' or 'soft law'?

8.1 'Historical' or 'Traditional Character' and 'Environment-Friendly Elements' in PDOs and PGIs Quality Regulations as 'Non-Strictly Mandatory Requirements' in EU law

Unlike what often emerges in public discourse, at the state of the art of EU GI law, the 'historical content' and 'traditional nature' of a product or even a significant 'reputation' (except for 'reputed' PGIs) are not formal requirements for registration. The EU legislator in the Council is likely to have compromised not to disadvantage those MS alien to GI traditional agriculture.[108] The consequence is that, as a general rule, lacking requirements of historicity or tradition are not strictly 'justiciable', *i.e.*

[106]Para 5, lett. (f) of the '*Camembert de Normandie*' unique document, if compared with that of '*Comtè*' cheese, appears to vaguely indicate grass pasture without any further detail. The '*Cahier des charges*' lists some criteria for pasture (density) and foodstuff for animals (types) without any reserve for local plants or prohibitions of non-natural substances or elements (cf.DOOR, '*Camembert de Normandie*', dossier FR/PDO/0017/0112 and '*Décret no 2008-984 du 18 septembre 2008 relatif à l'appellation d'origine contrôlée Camembert de Normandie*', JORF, 21 September 2008, 11, 60: annex 3).

[107]The area of '*Camembert de Normandie*' is alleged to have been intensively cultivated so that it has lost much of its traditional landscape and plants diversity: cf. European Forum on Nature, Conservation and Pastoralism' (2015), p. 6.

[108]This choice, while diluting the strictly 'naturalist' (or 'traditionalist') conception of GIs law and paving the way to the registration of 'new' products, is coherent with legal and philosophical premises. The notion of tradition is in itself a 'living' concept, subject to constant change and transformation (cf. Art. 2 of the CICH). Thus, a stricter standard would have limited protection to historical methods without allowing evolution and making it more difficult for those MS where an agricultural tradition to obtain PDO/PGI protection is lacking ('at what point in time and with which patterns would the beginning of a tradition be fixed?'). Among protected GIs there are both reputed designations of long-lasting tradition with great international value ('*Parma*', '*Parmigiano Reggiano*') and designations whose reputation is limited in time and space, especially to the State of origin (*e.g.*, '*Kraškizašink*', a particular type of Slovak salami).

they cannot easily provide the basis for claims of annulment or hetero-integration of quality regulations before courts,[109] unless the lack of 'historicity' requirement is expressly provided by national law[110] or is tantamount to 'absence or weakness of the link-with-the-*terroir*'.[111] Thus the correspondence of quality rules to authentic traditions and the *genius loci* is left to the discretion of MS and even more to the bargaining among producers and the transactions within their associations, rather than to public authorities.[112] This finding is even truer as regards environment-friendly methods. References in Preamble of Reg. 1151/2012 to 'sustainability' point at an 'implied-value' and are worded in 'non-compulsory' terms, as 'optional' criteria that producers may adopt as business ethics.[113] They have an 'incentive' function, aimed at enhancing producers' ability to make ethic choices at the time of drafting production rules, while considering the positive impact of these choices on the public ('premium price in exchange of environmental value'). Loose wording and optionality of environmental quality regulations should not be deemed negative elements since they allow to keep regulatory flexibility[114] so as to preserve the

[109]This is not to say that 'production requirements' and their connection with 'established traditions' are not justiciable 'in absolute'. Examples of pleas invoking non-correspondence with the 'traditional receipt' and—more often—the revision of the borders of the geographical area (normally, asking for larger areas than those agreed by producers' associations) are not unknown [cf., among others, '*Northern Foods Plc v. DEFRA, Melton Mowbray Pork Pie Association*' (2005) EWHC 2971]. In addition, quality regulations can be opposed (where still at the stage of application) or amended (if approved and registered), respectively, under Art. 10–51 and 53 of Reg. 1151/2012. Simply, in general terms, substantial and formal conditions to suit or apply for administrative modifications are restrictive and they often imply grounds for action very difficult to prove.

[110]MS may provide for additional requirements in the national phase of registration according to a 'variable geometry' pattern that enhances the individual choices of each MS. Stricter requirements for registration, such as a certain quality or duration in time of previous use, might be provided so as to select designation with a higher degree of quality: cf. in Italy, Art. 5 of the Decree of the Ministry of Agricultural policies n. 5442, of 21 May 2007, requires a *"report attesting that production has lasted at least 25 years, even in a non-continuous way, and constant and traditional use of the appellation in commerce and common language"*; in France, Art. L. 115-5 of the *'Code de la consommatiòn'* provides that '*AOC*' products shall have a *"notoriété dumentétablie"*.

[111]Somehow, a certain degree of local reputation or historicity is intrinsic to the structure of GI law and to a GI to be registered. In fact, it is hardly conceivable the registration of a product with a 'link-with-the-*terroir*' without a collective production with a certain recognition on the market, consolidated over time: registration is not an instantaneous act as it implies coordination among producers, identification and setting of quality regulations as 'traditional' and a certain preexisting reputation making profitable to engage in a collective costly production.

[112]Cf. Gangjee (2012b), p. 102.

[113]Cf. 'Commission Staff Working Paper, Impact Assessment on Geographical Indications', 2010, at http://ec.europa.eu/agriculture/quality/policy/quality-package-2010/ia-gi_en.pdf, accessed 7 July 2015, where it is clarified that GI producers rejected the EC request for compulsory environmental standards, obtaining by large majority that their addition be voluntary.

[114]Flexibility of language, however, appears to some extent fictitious since the discretion left to producers is sometimes filled up by national and regional (or professional) normative sub-levels, allowing the continuous adaptation of production standards within the normative limits posed by the registered regulation (self-regulation by consortia and producers associations).

necessary room for producers' economic initiative.[115] There is however no doubt that vague formulation might result in significant uncertainty at the moment to monitor production and as to clarity of the statements made to the market, so that it might not be effectively ensured that the GI is connected with authentic traditions and that applied methods are sustainable.

8.2 *Enhancing 'Cultural and Environmental Added Value' in GIs: Fostering Normative Integration Through National Law or CSR and Conditionality Policies*

The elements so far exposed suggest to consider how to further enhance the objectives set out in the main Regulation to widen the range of traditional and environmental 'plus-elements' in quality specifications. A suitable path is that resulting from the 'state of the art', *i.e.* the existing condition based in the main Regulation for which MS are free to impose stricter domestic legal standards on their producers and better detail cultural, social and environmental criteria as prerequisites for the access to the PDO or PGI. 'Inviting' in a systematic form MS to self-determine by domestic legislation additional requirements, as it already happens for certain countries, would have the advantage for each single adhering state of origin to mark the difference with other MS GI products with an average value-added[116] so as to create the conditions for a 'competition between legal systems', indeed able to raise—where the public would come to give value to national plus-elements—the qualitative standard of national products.[117] However, it is more

[115]According to '*Jamòn de Teruel*' and Danish cheese '*Esrom*' producers, GI standardization may hamper creativity and diversity inducing specialization and jeopardizing the different 'nuances' of the final product. This might also depend on difference in political weight of producers within GIs associations, since big firms normally vest different interests than SMEs, mainly, the interest in having more standardized conditions of production, which might be suitable to achieve economies of scale, with possible dispersion of TK and diversity.

[116]So, as it is already the case for France, Italy and Spain. In a context of enhanced competition and gradual saturation even of GIs markets, further differentiation through an enhanced guarantee on tradition and social value may further raise the market profile for certain GI products claiming enhanced authenticity and environmental sustainability.

[117]Organic sub-segments for traditional products might be established within the horizon of a single protected GI, conveying to the public authenticity and social value by means of specific indications coupled with sub-regulations ('*traditional*', '*original recipe*'), to obtain 'additional *premium-prices*'. This solution would build on existing practices, such as the '*Aceto balsamico di Modena*' GI, that however has brought to the registration of two different GIs, '*Aceto balsamico di Modena*' under a PGI and the PDO version marketed under the label '*traditional*' ('*Aceto balsamico tradizionale di Modena*') and might be applied also to a particular type of '*Parmigiano Reggiano*', prepared according to ancient recipe of Benedictine monks and today at risk of extinction. Hyper-differentiation based on origin and quality is criticized by those who think that this would create confusion in the average consumer [cf. Broude (2005b), p. 670 et seqq.]. However, the positive experience of certain raw commodities markets experimenting for the first time differentiation (craft

likely that plus-elements would be properly implemented both through coordinated action at the national and local level or by resorting to non-binding standards such as soft law and Corporate Social Responsibility (CSR). Capacity-building in drafting of quality regulations might be achieved by selecting best practices to be brought to the attention of GI applicants[118] under specific conditionality policies.[119] The pattern of GIs as clubs for production through common quality rules would be particularly significant in exceptional cases of increasing pollution, natural epidemics, urbanization and depletion of agrobiodiversity, as CSR may serve as a voluntary tool to implement (with the instruments of market reward) special environmental policy objectives, such as in the '*Xylella*' case in the South of Italy the integration of best practices.[120] EU quality policy from 2012 and the objectives set out in the Quality Package and Agenda 2020 seem to suggest that the EU institutions have timidly begun this last path.

beers, coffee, chocolate) suggests a favorable trend, at least for growing markets where generic sale still prevails.

[118]Greater selection during registration might be of impact, but it could be prevented by political relations among MS and the Commission. ECJ case law has clearly distributed competence in the two-tiers procedure of registration to MS (national phase) and the Commission (EU phase), under the judicial control of national courts and the ECJ, being the competence on substance mainly attributed to national authorities: cf. CJEU, Case C-269/99 *Kuhne* [2001], *ECR* I-9517, paras 57–58; CFI order, Case C-447/98 P *Molkerei* [2001], para 72; CFI order, Case T-114/99 *Pampryl* [1999], ECR 1999, p. II-3331; CJEU, Case C-343/07, *Bavaria et alteri v. Bayerischer Brauerbund* [2009], para 71 e 82.

[119]Incentives and conditionality usually reinforce CSR policy objectives by providing hooks in other public policies (trade promotion, support to SMEs, cooperation among producers, collective promotion and advertising).

[120]'*Xylella fastidiosa*' is a bacterium that, originally identified in California, from 2009 has affected olive trees in the South of Puglia, Salento, Italy, causing an environmental disaster and seriously harming local olive oil economy, by reducing production volumes and spreading a risky (wrong) perception that the epidemic might have affected the quality of olive oil. Among the practices that are being identified as a response there are substitution of plants with species more resistant to the bacterium, plowings and cleaning of the ground, hooding of infected trees or other mechanical barriers to contamination, and similar. Hetero-integration of quality regulations, that may extend also to collective marks and simple GIs, may also serve (a) to counter-balance the negative effects of the epidemic in terms of public perception so as to make known to the buying public policy actions and prevent the erosion of the market image of local products and (b) to push fragmented prducers to collectively adopt coordinated quality processes fragmented producers, so as to better exploit local resources and face international competition.

9 Conclusions. GI Law as a 'Rational Piece of the Institutions of the Market', Designed to Foster 'Quality', 'Culture' and 'Sustainable Rural Development'

This inquiry leads to conclude that the assumption that GIs would be a factor distorting tradition—more than a tool to preserve income, authenticity and agrobiodiversity—is at least misplaced. Traditional products, protected under EU and TRIPs GIs, are 'living' traditions whose conservation cannot be stiffened within the patterns of an 'authenticity prescribed by the law', separated from transactions among producers and social transformations, including changes in tastes and consumption trends.[121] Their effectiveness, more than other, is confirmed from the very circumstance that producers keep on using GIs and associating to establish new signs. Even where GIs were deemed to be a sort of (re)*'invention of tradition'*, this '(re)invention' should not be deemed as inherently negative because of a non-compliance with a code of formalistic 'authenticity', *a priori* established in relation to an alleged 'golden age'.[122] The GI model is fully subjected to the law of demand and offer and it represents, within the law of distinctive signs, a 'rational' piece of the institutions of the market pursuing a dual purpose: to protect producers 'local reputations' and consumers 'information'. Thus, if GIs are 'magic', they are in a way which is not so distant from the way market forces are. In addition, GIs also envisage a 'collective transactional organization for quality production'. As stated by the DSB at the WTO, EU GI law is a legitimate and even desirable body of rules for quality enforcement on condition that it complies with the basic principle of non-discrimination that underpins the law of International trade.[123] This inquiry

[121]Probably, the 'traditional purity' of the '*Barone Ricasoli*' recipe for Chianti wine should not be overstated: finally, adding increasing percentages of '*Malvasia*' and '*Trebbiano*' to the strong black '*Sangiovese*' wine of that time had no other meaning than creating a 'Supertuscan' *ante litteram*, intended to make it more palatable to intercept changing tastes.

[122]On the contrary, forms of recovery and reinterpretation of tradition are frequent and socially significant and appear as a sort of 'social engineering', hardly distinguishable from 'evolution' or 'contamination' that are typical attributes of living traditions and, somehow, a qualifying sign of a certain modernity: for a positive evaluation of 'inventions of traditions' as innovative processes, cf. Hobsbawm (2002), p. 16. Leaving aside radical distortions of traditions (often politically-driven), the success of genuine reinterpretations of the past is more simply explainable as a mass reaction to the advancement of globalization and homologation, which brings people to fold to the 'small' and the 'known' and, in market terms, to switch to products rooted in a communitarian, traditional production, as something familiar and sustainable, often original and distinctive, perceived as diverse from serial products, identical at every latitude. A successful example of re-interpretation of productive traditions in the viticultural sector is the experience of certain Italian wine-producers in the Romagna region leading the "Ritorno al Sangiovese", i.e. the attempt to revitalize and conjugate the cultivation of traditional Sangiovese grapes with the production of high quality excellency wines (cf. f.i. http://www.noeliaricci.com/identita/).

[123]Cf. WTO, 'European Communities – Protection of trademarks and geographical indications or agricultural products and foodstuffs', Reports of the Panel, 15 march 2005, doc. WT/DS174/R (complaint by the United States), as to inspection and prescriptive objective requirements for quality

tried to demonstrate that EU GI law is shifting with some significant consequences from mere 'quality enforcement' to 'cultural policy' and 'environmental protection'. Normative domestic adjustments and CSR initiatives may enhance the impact of culturally relevant and sustainable productions and show effective also in facing exceptional circumstances, such as the '*Xylella*' case, and try to turn them into chances for economic growth.[124] Thus, criticism against GIs effectiveness seems rather to confirm what is actually logical for IP lawyers and policy-makers, *i.e.* that it is not convincing to rely solely on GIs to foster income and preserve traditional agriculture, authenticity and the environment, being GIs only a part of an entire set of public policies (as the cultural ones under the UNESCO Conventions) designed to enhance social and economic welfare in a given market context.

References

Akerlof AG (1970) The market for lemons: quality, uncertainty and the market mechanism. Q J Econ 84:488–500

Altieri M (1971/2002) Agroecology: the science of natural resource management for poor farmers in marginal environments Agric Ecosyst Environ 93:1–24

Are M (1959) Beni immateriali. In: Enciclopedia del diritto. V. Giuffrè, Milano, pp 244–269

Ascarelli T (1949) Funzioni economiche ed istituti giuridici nella tecnica dell'interpretazione. In: Ascarelli T (ed) Saggi giuridici. Giuffrè, Milano

Audier J (1993) De la nature juridique de l'appellation of origin. Bulletin de l'OIV, pp 21–37

Blake J (2006) Commentary on the 2003 UNESCO Convention on the safeguarding of the intangible cultural heritage. Institute of Art and Law, Leicester

Botana Agra MJ (2001) Las Denominaciones de Origen. Marcial Pons, Madrid-Barcelona, 2001

Broude T (2005a) Culture, trade and additional protection for geographical indications. Bridges 9:21. www.ictsd.org. Accessed 7 July 2015

Broude T (2005b) Taking trade and culture seriously: geographical indications and cultural protection in WTO law. U Pa J Int Econ Law 26:623–692

Calboli I (2006) Symposium: the first ten years of the trips agreement: expanding the protection of geographical indications of origin under TRIPs: "old" debate or "new" opportunity? Marq Intell Prop Law Rev 10:181–203

Chen J (1996) A sober second look at appellations of origin: how the United States will crash France's wine and cheese party. Minn J Glob Trade 5:29–64

Correa CM (2001) Traditional Knowledge and Intellectual Property, Issues and options surrounding the protection of traditional knowledge. The Quaker United Nations Office (QUNO). https://www.humanrights.ch/upload/pdf/061127_correa_trad_knowledge.pdf. Accessed 7 July 2015

---

under GI Regulations: *"The Panel agrees with the European Communities that WTO Members are entitled to aim for objective assessment of product conformity, provided that they implement this objective in a WTO consistent manner"*.

[124]Finally, the birth of the monumental *AOC* law in France (particularly the 'provenance' and 'quality' discipline as embodied in the law of 1918) was no more than the reaction of French '*vignerons*'—somehow not differently from today—to 'a plant epidemic that came from the Americas', *i.e.* the '*phylloxera crisis*', affecting European countryside in between the nineteenth and the twentieth century.

de Nitto A (2002) Giuseppe Pisanelli e la legge. In: Diritto dei giudici e diritto dei legislatori. Argo, Lecce, pp 146–203
De Martino E (2015) Sud e magia, Milano
Deere C (2009) The implementation game. The TRIPS and the global politics of intellectual property reform in developing countries. Oxford University Press, Oxford
Downes DR, Laird SA (1999) Innovative mechanisms for Sharing Benefits of Biodiversity and Related Knowledge, UNCTAD biotrade initiative. http://www.ciel.org/Publications/InnovativeMechanisms.pdf. Accessed 7 July 2015
European Forum on Nature, Conservation and Pastoralism' (2015) EFNCP response to the EC, Green paper on agricultural product quality: product standards, farming requirements and quality schemes. http://www.efncp.org/download/EFNCPresponseGreenPaper.pdf. Accessed 7 July 2015
Fernandez Novoa C (1970) La Protecciòn Internacional de las Denominaciones Geograficas de los Productos. Editorial Tecnos, Madrid
Fink C, Smarzynska BK (2002) Trademarks, geographical indications, and developing countries. In: Hoekman B, Mattoo A, English P (eds) Development trade and WTO. A handbook. The World Bank, Washington DC, pp 403–412
Gangjee D (2006) Melton Mowbray and the GI Pie in the sky: exploring cartographies of protection. Intell Prop Q 3:291–309
Gangjee D (2008) Geographical indications and human rights. In: Torremans P (ed) Intellectual property and human rights. Kluwer Law International, The Hague, pp 383–395
Gangjee D (2012a) Geographical indications and cultural heritage. WIPO J 4:92–102
Gangjee D (2012b) Relocating the law of geographical indications. Cambridge University Press, New York
Garcia de Lujan A (1988) Singularidades de la Viticoltura del Jerez como base de su denominaciòn de origen. In: Proceedings of the Historic Appellation of Origin Symposium, Consejo regulador de la Denominaciòn de Origen 'Jerez-Xeres-Sherry', Jerez, pp 55–66
Gramsci A (1966) La questione meridionale. In: Dalla Chiesa N (ed) (2014) Melampo, Roma
Habermas J (2001) The postnational constellation: political essays, 1st edn. MIT Press, Cambridge
Hobsbawm EJ (2002) Introduzione: Come si inventa una tradizione. In: Hobsbawm EJ, Ranger T (eds) L'invenzione della tradizione. Einaudi, Torino
Hughes J (2006) Champagne, feta, and bourbon: the spirited debate about geographical indications. Hastings Law J 58:299–386
Khoury CK (2014) Increasing homogeneity in global food supplies and the implications for food security. Proc Acad Sci USA 11:4001–4006. http://www.pnas.org/content/111/11/4001.full.pdf. Accessed 7 July 2015
Ladas SP (1975) Patents, trademarks, and related rights. National and international protection, vol III. Harvard University Press, Cambridge
Land M (2014) Adjudicating TRIPS for development. In: Ghidini G, Peritz RG, Ricolfi M (eds) TRIPS and developing countries. Towards a new IP world order? Edward Elgar, Cheltenham-Northampton, pp 142–162
Landes WM, Posner RA (2003) The economic structure of intellectual property. Harvard University Press, Cambridge
Laudan R (2013) Cuisine and empire. Cooking in world history. University of California Press, Berkeley
Lehmann M (1986) Unfair use of and damage to the reputation of well-known marks, names and indications of source in Germany. Some aspects of law and economics. Int Rev of Intell Prop Comp Law 6:746–767
London Economics, ADAS, Ecologic (2008) Evaluation of the CAP policy on protected designations of origin (PDO) and protected geographical indications (PGI), Final report. http://ec.europa.eu/agriculture/eval/reports/pdopgi/report_en.pdf. Accessed 7 July 2015
Lorvellec L (1996) You've got to fight for your right to party: a response to Professor Jim Chen. Minn J Glob Trade 5:65–81
McCarthy JT (2006) McCarthy on trademarks and unfair competition. Thomson West, Eagan

Montanari M, Sabban F (eds) (2006) Storia e geografia dell'alimentazione. Risorse, scambi, consumi. I. Utet, Torino
Mosoti V, Gobena A (2007) International trade rules and the agriculture sector. Selected implementation issues. FAO Legislative Study 98, Rome
Niola M (6 August 2015) series 'Mangiare I simboli: il caffè', La Repubblica
OECD (2000) Appellations of Origin and Geographical Indications in OECD Member Countries: Economic and Legal Implications, COM/AGR/APM/TDWP(2000)15/FINAL
Olszak N (2001) Des appellations d'origineet indications de provenance. Tec&Doc, Londres-Paris-New York
Paus M, Reviron S (2011) Evaluating the effects of protecting geographical indications: scientific context and case studies. In: Bettelli G, Marescotti A, Paus M, Reviron S, Deppeler A, Stamm H, Thévenod-Mottet E (eds) The effects of protecting geographical indications, ways and means of their evaluation. Swiss Federal Institute of Intellectual Property, Bern. http://www.ige.ch. Accessed 7 July 2015
Pavoni R (2006) Biodiversity and biotechnology: consolidation and strains in the emerging international legal regimes. In: Francioni F, Scovazzi T (eds) Biotechnology and international law. Hart, Oxford, pp 29–57
Roubier P (1954) Le droit de la propriété industrielle, Partie speciale II. Sirey, Paris
Schamel G, Anderson K (2003) Wine quality and varietal, regional and winery reputations: Hedonic prices for Australia and New Zealand. Econ Rec 79:357–369
Schiller H (1976) Communication and cultural domination. M.E. Sharp, New York
Shapiro C (1982) Consumer information, products quality, and seller reputation. Bell J Econ 13:20–35
Shapiro C (1983) Premiums for high quality products as a return to reputations. Q J Econ 98:659–679
Shiva V (1996) Protecting our biological and intellectual heritage. In: Drahos P (ed) (1999) Intellectual Property, International library of essays in law and legal theory. Second series. Aldershot, Ashgate Dartmouth, pp 141–170
Spada P (2012) Il diritto industriale. Parte generale. In: Auteri P, Floridia G, Mangini V, Olivieri G, Ricolfi M, Spada P (eds) Diritto Industriale, Proprietà Intellettuale e Concorrenza. Giappichelli, Torino, pp 3–53
Ullrich H (2006) Traditional knowledge, biodiversity, benefit sharing and the patent system: romantic v. economics? In: Francioni F, Scovazzi T (eds) Biotechnology and international law. Hart, Oxford, pp 201–229
Vanzetti A (1961) Funzione e natura giuridica del marchio. Riv di dir comm, I:17–88
Von Schorlemer S, Stoll P (2012) The UNESCO Convention on the protection and promotion of the diversity of cultural expressions: explanatory notes. Springer, Berlin
Voon T (2009) Geographical indications, culture and the WTO. In: Ubertazzi B, Muniz Espada E (eds) Le indicazioni di qualità degli alimenti. Diritto internazionale ed europeo. Giuffrè, Milano, pp 300–311

Rocco Palma is a former contract lecturer of International Trade Law at the Catholic University of Sacred Heart, Piacenza, Italy, and "ad hoc" lecturer of International Intellectual Property Law at the same university in Rome. He is currently serving at the Intellectual Property Unit of the Italian Ministry of Foreign Affairs and International Cooperation. The opinions expressed in this article are those of the author and do not reflect the official policy or position of the Italian Government.

Food Diversity and Typicality in EU and in Italian Law: Protected Designations of Origin (PDOs); Protected Geographical Indications (PGIs); Traditional Speciality Guaranteed (TSGs)

Silvia Bolognini

1 Introduction

On 14 December 2012, Regulation 1151/2012 (hereinafter referred to as the "Quality Schemes Regulation"), was published in the EU Official Journal.[1] The adoption process formally began on 10 December 2010 when the European Commission presented its proposal and ended on 21 November 2012.

Immediately following the adoption of Regulations 509/2006[2] and 510/2006,[3] a review process concerning the provisions of the PDOs, PGIs and TSGs resulted in the adoption of the Quality Schemes Regulation.

As is known, Regulations 509/2006 and 510/2006 replaced Regulations 2081/1992[4] and 2082/1992,[5] introducing some significant changes, mainly aimed at extending - in compliance with a 2005 WTO Panel - the protection afforded by

[1]Reg. (EU) 1151/2012 of the European Parliament and of the Council of 21 November 2012 on quality schemes for agricultural products and foodstuffs, OJ [2012] L 343/1.

[2]Reg. (EC) 509/2006 of the Council of 20 March 2006 on agricultural products and foodstuffs as traditional specialties guaranteed, OJ [2006] L 93/1.

[3]Reg. (EC) 510/2006 of the Council of 20 March on the protection of geographical indications and designations of origin for agricultural products and foodstuffs, OJ [2006] L 93/12.

[4]Reg. (ECC) 2081/1992 of the Council of 14 July 1992 on the protection of geographical indications and designations of origin for agricultural products and foodstuffs, OJ [1992] L 208/1.

[5]Reg. (EEC) 2082/1992 of the Council of 14 July 1992 on certificates of specific character for agricultural products and foodstuffs, OJ [1992] L 208/9.

S. Bolognini (✉)
Department of Legal Sciences, University of Udine, Udine, Italy
e-mail: silvia.bolognini@uniud.it

A. Isoni et al. (eds.), *Food Diversity Between Rights, Duties and Autonomies*,
LITES – Legal Issues in Transdisciplinary Environmental Studies 2,
https://doi.org/10.1007/978-3-319-75196-2_5

those Regulations to the geographical indications of third countries (where those were protected in their country of origin).[6]

However, one look at the title and the structure of the Quality Schemes Regulation is enough to understand that its scope is not limited to the rules concerning PDOs, PGIs and TSGs. First, the Regulation refers to "quality schemes for agricultural products and foodstuffs" (and no longer, as in the past, to "the protection of geographical indications and designations of origin for agricultural products and foodstuffs" or to "agricultural products and foodstuffs as traditional specialties guaranteed"). Secondly, the Regulation is composed of six Titles, of which only Titles II, III and V provide the new rules for PDOs, PGIs and TSGs.

At the same time, we must be careful not to be misled by the generality of the title of the Quality Schemes Regulation. Even if its title refers to the "quality schemes for agricultural products and foodstuffs", the Regulation does not consider the provisions relating to all the quality systems present in the European agri-food market. On the contrary, the Regulation concerns only the protected geographical indications, the traditional specialties guaranteed and "a second tier of quality systems"[7] newly introduced as that of "optional quality terms", to which the Quality Schemes Regulation dedicates Title IV.[8]

However, there is no doubt that the adoption of the Quality Schemes Regulation places itself in the context of a broader revision regarding the entire quality policy for European agri-food products to enhance this policy's potential contribution to achieving the objectives of the CAP (Common Agricultural Policy).

As is known, the day after the adoption of Regulations 509/2006 and 510/2006, the European Commission expressed its intention to review not only the rules of PDOs, PGIs and TSGs to address the critical issues which emerged during its application, but also the entire system of quality schemes for European agricultural products and foodstuffs. This review was programmed because over the years the European quality schemes evolved in a fragmented way, often preventing the producers to be rewarded for their efforts in producing a diverse range of quality products.

As also pointed out by the European Commission, since the nineties the quality policy for European agri-food products has based itself on PDO and PGI, organic production and TSG (originally called "certificates of specific character") schemes.[9] In addition, since it was established, the CAP has imposed the observance of

[6]Costato (2006), p. 351 et seqq; Coppola (2013), p. 31; Pisanello (2006), p. 556 et seqq; Gencarelli (2006), p. 609 et seqq; Petrelli (2012), p. 285 et seqq.

[7]See recital 44 of the Quality Schemes Regulation.

[8]See Articles 27–34 of the Quality Schemes Regulation.

[9]See the Proposal for a Regulation of the European Parliament and of the Council on agricultural product quality schemes, of 10 December 2010, COM(2010) 733 fin., point 1.2., p. 3. See also the Proposal of a Regulation of the European Parliament and of the Council amending Council Regulation (EC) 1234/2007 regarding marketing standards, of 10 December 2010, COM(2010) 738 fin., point 1.2, p. 4.

marketing standards to promote fair competition and to ensure the proper functioning of the market.

Over the past decade, especially in the private sector, certification schemes for food quality have been widely proliferated. These schemes have been used both to inform buyers and consumers about the added value of certain products, and to reassure them regarding the compliance with technical rules, in some cases imposed by specific regulations and in others voluntarily imposed by the producers.[10]

The result is a complex picture - not always transparent and easily understood –, which has hindered the achievement of the set goals. The excess of certification schemes decreases the possibility for producers to operate in fair competition and, therefore, to be reasonably rewarded for their efforts. At the same time, the opportunity for buyers and consumers to easily and correctly identify (and choose) the products that best meet their needs is reduced.

Hence, the need - stressed also by European institutions - to improve the rules for food quality schemes, by also paying close attention to national and private certification schemes. Enhancing the agri-food quality policy, which is one of the major strengths of European agriculture, permits the emergence of a more competitive economy, in line with the "Europe 2020" policy priorities, as set out in the Commission Communication of 3 March 2010 entitled "Europe 2020: A strategy for smart, sustainable and inclusive growth".[11] The European agri-food sector, thanks to the high quality of its products, enjoys an excellent reputation in the international community. For this reason, it is necessary to improve the way in which this quality is communicated. Nevertheless, since the agri-food quality policy is strictly connected to the CAP (the first, as we know, is an integral part of the second), the European institutions have recognised that it can play a decisive role in contributing to and in complementing not only market and income support policies, but also the rural development policy. Enabling producers to obtain a financial advantage for the added value of their products, quality schemes can contribute to the realisation of an economy with a high employment rate, able to encourage both social and territorial cohesion. From this perspective, the quality schemes are the ideal tool to reward the efforts made by producers in producing a diverse range of quality products, especially in areas - such as disadvantaged areas, mountain areas and most remote regions - in which the farming sector is of greater economic importance, but it has to reckon with higher production costs.[12]

[10]Albisinni (2000), p. 109 et seqq; Altili (2012), p. 95 et seqq.

[11]COM(2010) 2020 fin.

[12]See recital 4 of the Quality Schemes Regulation.

2 From Food Diversity to Food Diversities

To specify the focus of my topic, I must clearly define the meaning of the expressions "diversity" and "typicality" when used in combination with "food". First, the expression "food diversity" can be used with two different meanings, to indicate a quantitative or a qualitative profile.

The expression "diversity" can indeed be used as a synonym for multiplicity, to emphasise the richness of the European agri-food production, which guarantees a wide choice for the consumers. Here the quantitative aspect of diversity is being emphasised.

Otherwise, the expression "diversity" can be used as a synonym for variety, to underline the wide range of quality products offered by the European agricultural and food production. Here the qualitative aspect of diversity is being emphasised. In this case, the expression "diversity" helps to highlight that some foods, thanks to their characteristics, are different from other products of the same category.

Whereas, in the past, the use of the expression "food diversity" to emphasise the qualitative aspect has been predominant, at present frequently the EU legal acts refer to a diversity concerning the entire European agricultural and food production, and not specific products. This also applies to the Quality Schemes Regulation: consider, for example, the first and the second recitals of the Regulation, in which there is a reference to "the quality and diversity of the Union's agricultural, fisheries and aquaculture production" and to "the diversity of the agricultural production in the Union".

Therefore, it seems appropriate to try to categorise food diversity or, better yet, food *diversities*. It could be useful to use the expression "external food diversity" to refer to the difference between European agricultural products and foodstuffs and those which are non-European. On the other hand, the expression "internal food diversity" may refer to the presence of European agricultural products and foodstuffs, which distinguish themselves thanks to their uniqueness.

At the same time, through the years, the use of the expression food diversity to indicate the qualitative profile of certain foods has also evolved. Because of the changes that occurred in knowledge, in the agri-food market dynamics and in the scale of values on which the European system was and is based, new food diversities have emerged. While the "traditional" diversities refer to the identity and/or the composition of certain foods - regarding, for example, physical, chemical, organoleptic properties –, the new diversities often capture the ability of some foods to comply - thanks to their unique properties - with the development of environmental, moral, ethical, even religious, consumers' sensitivities.[13]

For this reason, it is useful to distinguish between "intrinsic" and "extrinsic" food diversity, where "intrinsic" food diversity concerns elements, which make up a food's identity or composition, while "extrinsic" diversity regards the capacity of

[13]Jannarelli (2011), p. 199 et seqq; Adornato (2010), p. 19 et seqq.

some foods to best meet the environmental, moral, ethical, even religious consumers' needs.

However, also the extrinsic diversities can affect consumer purchasing decisions and thus the competitiveness of foods. For this reason, extrinsic diversities must be considered - first by the European legislator - when the goal is to enhance the ability of European foods to find a place in the market.

At the same time, we must keep in mind that these new extrinsic diversities are able to contaminate intrinsic food diversity. It is very often no longer enough that the food is ontologically different from similar products belonging to the same category. Consumer's changing sensitivities must also be considered. In other words food competitiveness is not only a question of agreement to tastes, but also of conformity to feelings.

In this regard, it is interesting to note that the twenty-third recital of the Quality Schemes Regulation stresses that also a PDO and PGI product "should meet certain conditions set out in a specification, such as specific requirements aimed at protecting the natural resources or landscape of the production area or improving the welfare of farm animals".

3 Food Typicality. Traditional Food as an Enriched Typical Food

Similar considerations can be made, however, about the "typicality". "Typicality", indeed, is nothing more than a form of food diversity, whose main characteristic lies in the fact that it indicates not only that a given agri-food product possesses specific features, but also that those properties strictly depend on its production. A "typical" food is a food which, thanks to its characteristics, differs to the other foods of the same category, or because it comes from a specific geographical area, or because it has been obtained using a particular method of production. This implies that only the products made in "that" particular geographical area or obtained with "that" particular method of production are allowed to be classified as "typical".

It also follows that the "traditional" food may be considered a "typical" food, in a certain sense an "enriched" typical food. The expression "traditional food" indicates the ability of certain foods to express one of the most important functions delegated to agriculture, in the context of its multifunctional character, which is its cultural function. By keeping alive and transmitting to future generations certain production methods, the primary sector operators transmit, in fact, the culture that characterises certain areas.

Diversity and typicality are concepts, which have always existed in the food industry, because agricultural products and foodstuffs in themselves are unable to transmit their specific properties without the help of a communication tool, such as, for example, a label.[14]

Since everyone must eat to survive, producers are not required to convince consumers to eat, but rather, to entice them to buy their products. From this perspective, food quality is of strategic importance.[15]

It is also for this reason that the European legislator - and before him the legislators of some Member States, among which also ours can be counted - has given special attention to "typicality". Meaning, how a particular geographical environment, with its inherent natural and human factors, and traditional production methods influence the qualities possessed by a food.

PDOs, PGIs and TSGs can surely be considered a form of typicality and thus, also of food diversity: indeed, they represent a clear example of food diversity "certified at European level".

In this regard, it is interesting to recall that in the nineties some national legislations in force, especially, in France and in Italy, inspired the introduction of the first European harmonised discipline on PDOs and PGIs. The goal of these national legislations was the promotion of "food typicality", both as a geographical typicality, and as a typicality linked to a traditional production method. Our country was among the first to adopt some national provisions aimed at protecting certain typical foodstuffs, such as cheeses, hams and various cured meats and wines.[16]

It is clearly not a coincidence that at present Italy and France are the two Member States with the highest number of registered PDOs and PGIs. More specifically, Italy is the Member State with the widest number of registered PDOs and PGIs. This fact proves that our country has always given great attention to the quality of its agri-food production and to its connection with its geographical areas of origin.

[14]Rook Basile (1996), p. 182; Losavio (2007), p. 19; Lucifero (2011), p. 326; Tommasini (2011), p. 493.

[15]Bolognini (2012), p. 47 et seqq.

[16]We can mention here, for example: Law 125/1954, referred to the protection of the typical designations of cheese (whose implementation was regulated by the Decree of the President of Republic 667/1955); Law 506/1970, related to the provisions on the protection of designation of origin of Parma ham, the delimitation of its production area and its characteristics; Law 507/1970, related to the protection of designations of origin and typicality of the San Daniele ham; Law 224/1989, laying down the protection of the designation of origin of the Varzi salami, the delimitation of its production area and its characteristics. Some of these regulations, subsequently amended, have been repealed, while others are still in force.

4 The Context of the Revision of Quality Schemes for European Agricultural Products and Foodstuffs

Someone has defined - wrongly or rightly, it is hard to say - the Quality Schemes Regulation as the cornerstone of the revision of the quality schemes for European agricultural products and foodstuffs.[17] However, since the Quality Schemes Regulation is only one of the steps of the quality policy's revision, we can properly understand its content only by also considering the context of its adoption.

The revision of the quality policy for European agri-food products started in 2006 with a stakeholder consultation, which was followed by a conference significantly entitled "Food quality certification - adding value to farm produce", held in Brussels in early February of 2007. During this conference the quality schemes used in the European agri-food market were closely examined, in their many different forms (public, private, European, national and even local).

Based on feedback provided by this conference, on 15 October 2008 the Commission released a Green Paper - known as "The Green Paper consultation on agricultural product quality policy" - asking open questions about the different issues related to quality.[18] At the close of the consultation, 560 written responses were received from a broad range of stakeholders.

The comments made by stakeholders, as well as the results of the High Level Quality Policy Conference, organised in Prague on 12 and 13 March 2009, were taken into account by the European Commission for the drafting of "The Communication on agricultural product quality policy", adopted in May 2009.[19]

This Communication considered the strong support expressed by stakeholders for the existing main quality schemes for agricultural products and foodstuffs, including that of protected geographical indications. Furthermore, the Communication placed emphasis on the need for the European producers to better communicate to consumers the qualities of their products, to retain fair share of the added value, while at the same time giving to consumers the opportunity to make informed choices.

After "The Communication on agricultural product quality policy", the European Commission adopted the so-called "Quality Package 2010", a comprehensive policy on certification schemes, value-adding terms for agricultural product qualities and product standards. The "Quality Package 2010" covered the different facets of quality, from compliance with minimum standards to the production of highly specific products. The Package comprised: the proposal of a new "Agricultural Product Quality Schemes Regulation"[20]; the proposal of a new general base line Marketing Standard for all agricultural products and a specific power to adopt

[17]Gragnani (2013a), p. 376.
[18]COM(2008) 641 fin.
[19]COM(2009) 234 fin.
[20]COM(2010) 733 fin.

place-of-farming and other sectoral rules for marketing products[21]; new Guidelines of best practices on voluntary certification schemes[22] and on the labelling of products using PDO-PGI ingredients.[23]

The European Commission drafted the Proposal for the Quality Schemes Regulation to bring coherence and clarity to the EU quality schemes for agricultural products and foodstuffs and to reinforce the flagship schemes for protected designations of origin and geographical indications (PDOs and PGIs). Other goals of the new Regulation were to overhaul the traditional specialities guaranteed scheme (TSGs) and to introduce a new framework for the development of Optional Quality Terms, such as feeding methods and production methods.

What is of most interest here is to recall that "The Communication on agricultural product quality policy" highlighted the need to review and simplify the existing schemes of the geographically protected indications (both PDOs and PGIs). In view of this, the Commission initially postulated the possibility of merging the PDO and PGI schemes, notwithstanding the need to differ the level of protection guaranteed by the EU. At the same time, the European Commission considered a merger of the existing systems - respectively - for wine, spirits, aromatised wines and agricultural products and foodstuffs, which should have led to a common regulation, able, however, to preserve the specificities of each system. Finally, the Commission put forward the idea of eliminating the TSGs because of lack of success.

As reported by the "Explanatory Memorandum" accompanying the Proposal for the Quality Schemes Regulation, after the adoption of "The Communication on agricultural product quality policy", stakeholders were again consulted. In this case, they opposed the solutions proposed by the European Commission. First, stakeholders did not welcome the idea of merging the two schemes of PDOs and PGIs and the existing systems for wine, spirits, aromatised wines and agricultural products and foodstuffs. Secondly, stakeholders expressed almost unanimous support for continuation of the TSG scheme. On the other hand, stakeholders asked for the simplification of the bureaucratic procedures for accessing PDOs, PGIs and TSGs schemes and highlighted the need to consider the difficulties of small-scale producers, for whom those quality schemes were too burdensome.[24]

Based on the outcome of the stakeholders' consultation, the European institutions adopted the Quality Scheme Regulation, maintaining the previously existing quality schemes. At the same time, the Quality Schemes Regulation introduced a new quality scheme concerning the optional quality terms.

In the context of the revision of the quality schemes for European agricultural products and foodstuffs, the Quality Schemes Regulation "aims to help producers of agricultural products and foodstuffs to communicate the product characteristics and

[21]COM(2010) 738 fin.

[22]OJ [2010] C 341/3.

[23]OJ [2010] C 341/5.

[24]See the Proposal for a Regulation of the European Parliament and of the Council on agricultural product quality schemes, of 10 December 2010, COM(2010) 733 fin., point 2.2., p. 5.

farming attributes of those products and foodstuffs to buyers and consumers".[25] More specifically, the Quality Schemes Regulation aims to ensure fair competition for farmers and producers of agricultural products and foodstuffs having value-adding characteristics, the availability to consumers of reliable information pertaining to such products, the respect for intellectual property rights and the integrity of internal market.

At the same time, "the measures set out in this Regulation are intended to support agricultural and processing activities and the farming systems associated with high quality products, thereby contributing to the achievement of rural development policy objectives".[26]

5 The Main Novelties Introduced by the Quality Schemes Regulation in Relation to the Rules on PDO and PGI Schemes

The Quality Schemes Regulation brings into one legislative instrument the rules for PDO, PGI and TSG schemes.[27] Regarding the PDO and PGI schemes, the Quality Schemes Regulation combines - where possible - rules of application process and controls, ending current divergences in procedures, with benefits in terms of coherence of rules across schemes. Furthermore, the Quality Schemes Regulation shortens and streamlines procedures and introduces some clarifications, particularly in relation to intellectual property rights. In this regard, one of the main novelties introduced by the Quality Schemes Regulation is a mandatory requirement for Member States to put in place administrative and judicial procedures to prevent or stop ex officio the unlawful use of a protected geographical indication.[28]

---

[25]See Art. 1, para 1, of the Quality Schemes Regulation.

[26]See Art. 1, para 1, of the Quality Schemes Regulation.

[27]See Costato (2012), p. 648 et seqq; Capelli (2014), p. 52 et seqq and Coppola (2013), at p. 31 et seqq.

[28]See Art. 13 of the Quality Schemes Regulation and especially its para 3. This provision fills a gap present in the previous regulation, which, in the famous controversy "Parmesan", allowed Germany to refuse to prosecute on its territory the placement on the market of cheese under the designation 'Parmesan', which did not comply with the specification for the protected designation of origin ('PDO') 'Parmigiano Reggiano'. The behaviour of German national authorities was justified by the lack of a specific European provision, which required Germany to prosecute ex officio the marketing under the designation 'Parmesan' on its territory of cheese which did not comply with the specification for the PDO 'Parmigiano Reggiano'. In particular, according to the Court of Justice, the Commission failed to demonstrate that Germany was under an obligation to take such action in the absence of sufficient and appropriate external impetus to do so. See CJUE, Case C-132/05, *Commission of the European Communities v Federal Republic of Germany* [2008] *ECR* I-957. The new rule provided by Art. 13, para 3, aims at pursuing one of the most important objectives of the Quality Schemes Regulation, which is to increase the effectiveness of the PDOs and PGIs protection. In this regard, the Quality Schemes Regulation also better defines the roles assigned to

According to the Quality Schemes Regulation, a designation of origin is a name which identifies a product originating in a specific place, region or, in exceptional cases, a country, whose quality or characteristics are essentially or exclusively due to a particular geographical environment with its inherent natural and human factors. In this case, the production steps of the PDO product shall all take place in the defined geographical area.[29] A geographical indication is a name which identifies a product originating in a specific place, region or country, whose given quality, reputation or characteristics is essentially attributable to its geographical origin. In this case, instead, at least one of the production steps of the PGI product shall take place in the defined geographical area.[30]

The protection of PDOs and PGIs depends on the registration. The right to use a designation of origin or a geographical indication is subordinated indeed to a procedure of registration. The Quality Schemes Regulation has confirmed that only a group - defined as "any association, irrespective of its legal form, mainly composed of producers or processors working with the same product" - is entitled to apply for registration.[31] The procedure of registration is regulated by the Quality Schemes Regulation.[32] It is interesting to recall here that the procedure of registration starts with an application addressed to a Member State, which shall contain the product specification and a single document setting out the main points of the specification.[33] The Member State, to whom the request of registration is addressed, shall verify that the application meets the conditions of the Regulation. After a reasonable period, within which any natural or legal person may lodge an objection to the application, the Member States forwards the request to the Commission through a decision that can be appealed on a national level. From this moment, the

the so-called groups in supervising the correct use of the quality schemes. The importance of this objective emerges also from the new legal basis of the Quality Scheme Regulation, such as Art. 118, para 1, TFEU. The legal bases of the Regulation are, in fact, Art. 43, para 2, and Art. 118, para 1, TFEU. As known, Art. 43, para 2, TFEU is the agricultural legal basis. Art. 118, para1, TFEU is, on the contrary, a new provision introduced by the Lisbon Treaty, under which "In the context of the establishment and functioning of the internal market, the European Parliament and the Council, acting in accordance with the ordinary legislative procedure, shall establish measures for the creation of European intellectual property rights to provide uniform protection of intellectual property rights throughout the Union and for the setting up of centralised Union-wide authorisation, coordination and supervision arrangements". It is evident that the presence of Art. 118, para 1, TFEU as one of the legal bases of the Quality Schemes Regulation aims to permit the EU to exercise all the actions required to protect intellectual property rights in a uniform way within the European Union. See Albisinni (2015), p. 251 et seqq; Capelli (2014), p. 53.

[29]See Art. 5, para 1, of the Quality Schemes Regulation.

[30]See Art. 5, para 2, of the Quality Schemes Regulation.

[31]See Art. 3, no 2, of the Quality Schemes Regulation.

[32]See Art. 7 et seqq of the Quality Schemes Regulation.

[33]See Masini (2012), p. 354, who underlines that "Product specification plays a crucial role in the procedure of registration, as it represents a *self regulation code* or *negotiation regulation*, that every single producer must adopt to use the geographical indication".

Member State can grant transitional national protection to the geographical indication.

Then, the Commission will scrutinise the application sent by National authorities to check that it is justified and meets the conditions laid down in the Quality Schemes Regulation.

When the Commission considers that the conditions laid down in the Regulation are met, it shall publish the single document and the reference to the products' specification in the Official Journal of the European Union. Within 6 months from the date of publication in the Official Journal of the European Union, any Member State or third country may object to the proposed registration. The Commission shall invite the interested parties to engage in appropriate consultations. If no agreement is reached, the Commission shall take a final decision and publish it in the Official Journal of the European Union.

After registration, all the producers who respect the conditions set out in the product specification have the right to use the geographical indication. At the same time, after the registration, PDOs and PGIs are protected against any unfair use by third parties, who attempt to exploit its reputation but do not respect the geographical indication product specification. PDOs and PGIs are also protected against any misuse, imitation or evocation, even if the true origin of the product is indicated or if the protected name is translated or accompanied by an expression such as "style", "type", "method", "as produced in", "imitation" or similar.[34]

PDOs and PGIs are also protected against any other false or misleading indication as to the provenance, origin, nature or essential qualities of the product on the inner or outer packaging, advertising material or documents relating to the product concerned, and the packing of the product in a container liable to convey a false impression as to its origin, as well as any other practice liable to mislead the consumer as to the true origin of the product.[35]

6 The Revision of the TSG Scheme

In relation to the TSG scheme, the Quality Schemes Regulation introduces different changes. To fully understand the scope of these changes, we must be aware that the revision was a response to the criticisms that emerged when the former regulatory frameworks were in force. When the TSGs were introduced in Europe (originally with the name "certificates of specific character"), the goal of European institutions was the introduction of a specific communication tool, able to promote products obtained by a traditional method of production. However, in the minds of the European legislator, the traditional method of production in question could be

[34]See Art. 13, para 1, lit (a) and lit (b), of the Quality Schemes Regulation.

[35]See Art. 13, para 2, lit (c), of the Quality Schemes Regulation.

"relocated", that is freely used even by operators located in places different from that of its origin.[36]

Nevertheless, as is widely known, the system was a complete failure. From 1992 to the present, more than one thousand PDOs and PGIs have been registered, while only a little over thirty names have been registered as TGSs, most of which have failed to meet its potential.[37]

The previous rules can be criticised from different points of view. First, criteria and conditions for the registration were not clear. In particular, the definitions of "specific character" and "traditional" were not understandable and this gap made it difficult to verify their existence. Moreover, this lack of clarity allowed names of non-genuine traditional products to be registered under the scheme.[38]

However, the most controversial aspect of the former scheme was the possibility of registering a TGS without name reservation. Indeed, the producers could choose between two different types of registrations, which corresponded to two different types of protection in the EU: under the old TSG scheme, producers could apply for a TSG registration without or with name reservation.[39]

If they applied for a TSG registration without name reservation, the name registered as TSG could continue to be used freely even in relation to products which did not comply with the registered TSG specification. Consequently, anyone could place on the market products identified with the same name, even if obtained otherwise. The only limit was the ban of accompanying the product with the TSG expression or with its European logo.

On the contrary, if a TSG was registered with name reservation, the name could not be used for similar agricultural products and foodstuffs. The name, even where unaccompanied by the indication "traditional speciality guaranteed", the abbreviation "TSG" or the associated European symbol, may no longer be used on the labelling of similar agricultural products or foodstuffs not satisfying the registered specification. In other words, the registration with name reservation permitted the use of the name registered as TSG exclusively for marketing a product obtained by following the corresponding product specification.

Unfortunately, the registration with name reservation was difficult to obtain. To avoid it, it was sufficient to present an objection, aimed at demonstrating that the name had already been used previously in a lawful, renowned and economically significant manner for similar agricultural products and foodstuffs.

Furthermore, the application for the registration of a TSG was too complex and expensive. Considering that those who might be interested in registering a TSG were medium- and small-scale producers, it seems that the entire TSG scheme was designed in a counterproductive way.

---

[36]Galli (2004), p. 71.

[37]Capelli (2014), p. 55; Strambi (2014), p. 367; Gragnani (2013b), p. 156.

[38]See Strambi (2014), p. 368 et seq. See also recital 36 of the Quality Schemes Regulation.

[39]Canfora (2011), p. 75 et seqq.

Concerning the TSG scheme, the most relevant innovation introduced by the Quality Schemes Regulation lies in the fact that under the new scheme the registration of a TSG without name reservation is no longer possible. In fact, the Quality Schemes Regulation maintains only the possibility of registering a TSG with name reservation. It follows that from now on the registration of a TSG automatically entails the ban of using the registered name for products obtained following a production method different from that described in the TSG product specification.

At the same time, the Quality Schemes Regulation, to make the TSG scheme more understandable, operational and attractive to potential applicants, introduces a new definition of "specific character" and tries to clarify what "traditional" means. In accordance with the Quality Schemes Regulation, the expression "specific character" - in relation to a product - means "the characteristic production attributes which distinguish a product clearly from other similar products of the same category".[40] In this regard, it seems that the concept of "specific character", on which the current TSG scheme rests, is more respectful of the fact that only products with an added value must be allowed to access food quality schemes.

In relation to the traditional character, the Quality Schemes Regulation specifies that "traditional" means "proven usage on the domestic market for a period that allows transmission between generations"; this period should be at least 30 years (and not more 25 years).[41]

A further novelty introduced by the Quality Schemes Regulation concerns the removal of the rule previously provided by Regulation 509/2006, under which the TSG registration should not be permitted when the specific character of the agricultural product or foodstuff was due to its provenance or geographic origin.[42] In the past, this rule was justified by the desire to draw a clear dividing line between PDO and PGI schemes, on the one hand, and the TSG scheme, on the other. PDO and PGI schemes should enhance the typicality of agricultural products and foodstuffs connected to their geographic origin. The TSG scheme, instead, should highlight the added value of an agricultural product or foodstuff produced using traditional raw materials or characterised by a traditional composition or method of production and/or processing reflecting a traditional type of production and/or processing. For this reason, the name of an agricultural product and foodstuff was allowed to be registered as a TSG only if the specific characteristics of the product did not depend on its provenance or its geographic origin. However, this negative requirement was extremely difficult to meet: as a rule, in fact, the traditional agri-food products are intrinsically linked to their provenance or geographic origin.

The Quality Schemes Regulation, not re-proposing that provision, seems to have given up the idea that it is necessary to keep the TSG scheme separate from those of the PDO's and PGI's. Under the Quality Schemes Regulation, it is possible to register as TSG the name of an agricultural product or foodstuff whose specific

[40]See Art. 3, no 5, of the Quality Schemes Regulation.

[41]See Art. 3, no 5, of the Quality Schemes Regulation.

[42]See Art. 4, par. 2, second provision, of Reg. (EC) 509/2006.

characteristics depend on its provenance and geographic origin. This makes possible an overlap of PDO, PGI and TSG schemes.

In this regard, the literature asks if the European legislator wants to create a sort of third category of protected geographical indications, less qualified than PDOs and PGIs and intended, for that reason, to afford less protection.[43]

In my opinion, this question is certainly important for the systematic interpretation of the Quality Schemes Regulation, but it is not the main question to pose regarding the new TSG scheme. In my opinion, the most important question is whether changes introduced by the Quality Schemes Regulation are an appropriate remedy (at least potentially) for the lack of success recorded by the old TSG scheme. Can the new TSG scheme contribute to realising the Quality Schemes Regulation objectives?

To answer to this question, it is necessary to consider that the main reason for the failure of the old TSG scheme was that it was unattractive to producers for a variety of reasons. In fact, it was not clear what advantages - in terms of allocation of their products on the marketplace - the producers could have by registering a TSG.

We have seen that, in the past, when a TSG was registered without name reservation, the advantage was strongly reduced by the possibility of finding similar products on the market with the same name, even if not obtained following the product specification associated with the registered TSG. This possibility led to confusion and, consequently, consumers were unable to recognise the added value of TSG products.[44]

On the other hand, in the case of the registration of a TSG with name reservation, the product specification, which is the traditional method of production, is published. It follows that, after the registration, all competing producers can use the traditional method of production, whenever and wherever they want, which means that they can relocate and standardise the production.[45]

Unfortunately, in drafting the Quality Schemes Regulation, European institutions did not pay enough attention to this problematic profile and the current situation does not differ so much from the previous one. Since the Regulation draws attention to the fact that producers can continue to produce a diverse range of quality products only if their efforts are fairly rewarded, we must conclude that the ability of TSG scheme to facilitate the achievement of this objective is significantly affected by the way in which it operates.[46]

Even if now the registration of a TSG automatically implies the name reservation, TSG products can be obtained everywhere, resulting in the quality scheme benefiting only those producers who can both locate their activities in places where production costs are lower, and standardise the production.[47]

---

[43]Gragnani (2013a), p. 383.

[44]Gragnani (2013b), p. 155.

[45]Gragnani (2013b), p. 156; Masini (2006), p. 490; Canfora (2011), p. 79.

[46]See recital 3 of the Quality Schemes Regulation.

[47]Masini (2015), p. 279; Gragnani (2013b), p. 156.

For this reason, I completely agree with the literature when it emphasises that the weakness of the TSG scheme lies in the fact that it ignores the link between agricultural products and foodstuffs and its provenance and geographic origin.[48] From this perspective, since it is completely indifferent where the TSG product is produced, the contribution that the new TSG scheme can offer in enhancing the development of rural areas is minimal.

7 Concluding Remarks: Is the Quality Schemes Regulation a Panacea for All Its Critics?

Since one of the main objectives of the Quality Schemes Regulation is to help European producers to better communicate the quality of their agricultural products and foodstuffs, it is also necessary to ponder if the new Regulation has made PDO, PGI and TSG schemes more understandable and clearer to consumers.

The Quality Schemes Regulation provides some specific rules directly intended to make more operational the communication tools referring to PDO, PGI and TSG schemes.

First, Title V, Chapter I, of the Regulation concerns the official controls, which must be carried out to verify if products which apply for PDO, PGI and TSG schemes, before being placed on the market, comply with legal requirements related to these quality schemes. In this context, official controls shall also ensure that indications, abbreviations and symbols referring to such quality schemes are used only in connection with products produced in conformity with the rules.[49]

Secondly, the Regulation allows the groups that can legally apply for registration to develop information and promotion activities aiming at communicating the value-adding attributes of the products to consumers.[50]

Controls and promotion are undoubtedly the most important tools to ensure an effective protection to PDO, PGI and TSG schemes. Controls decisively contribute to preserve the distinguishing character and the attractiveness of the quality schemes. By ensuring the access to quality schemes only to products with an added value, controls impede the loss of credibility of the quality schemes.

Otherwise, through information and promotion activities it is possible to make consumers aware of the specific characteristics possessed by agricultural products and foodstuffs registered under these schemes.

The rules provided by the Regulation are based on the conviction that the quality schemes were not successful because of the wide proliferation of certification schemes for food quality and the ineffectiveness of the communication tools. However, there are also further criticisms that affect PDO, PGI and TSG schemes.

[48]Strambi (2014), p. 372 et seq; Gragnani (2013b), p. 158; Canfora (2011), p. 80.

[49]See Articles 35–40 of the Quality Schemes Regulation.

[50]See Art. 45, para 1, lit. (d), of the Quality Schemes Regulation.

As previously stated, at present, more than one thousand PDOs and PGIs and a little over thirty TGSs have been registered. But, some market surveys have shown that very often European consumers are struggling to distinguish PDOs from PGIs, although they tend to place the first in a hierarchically higher position than the latter. Even if the Quality Schemes Regulation tries to further stress that PDO products are more strongly linked to the geographical area of its origin than PGI products, the dividing line between the two schemes is still far from being clear or well defined. The adoption of a uniform regulation for both PDO and PGI schemes obscures the differences between these two quality schemes. At the same time, the fact that the European symbols used to identify them are strikingly similar (the two logos, in fact, differ only in colour) reduces their capacity to efficiently communicate to consumers the differences between the two quality schemes.

Moreover, also under the Quality Schemes Regulation, in some cases, certain names are treated as PDOs although the raw materials for the products concerned come from a geographical area larger than, or different from, the defined geographical area. This rule has been introduced in the past to allow producers to meet the high demand of their PDO products, when it was almost impossible to meet the production needs with raw materials coming only from the defined geographical area. However, this rule has certainly not helped consumers to correctly understand the PDO and PGI schemes.[51]

Furthermore, only some registered PDOs and PGIs are widely recognised and, consequently, in high demand in the marketplace. On the contrary, most of PDO and PGI products are almost unknown, sometimes even by consumers of their country of origin.[52]

As is known, the European regulation on PDO, PGI and TSG schemes was adopted to give to producers the chance to use some distinguishing marks playing a similar role to individual brands, which are mostly employed by agri-food companies. However, the achievement of this goal seems to be possible only if the products bearing these distinguishing marks reach wide dissemination, able to allow the producers to amortise the high costs both of production and of promotion. Unfortunately, except for very few cases, PDOs, PGIs and TSGs have not been able to reach this kind of visibility. In terms of market penetration, overall, the objectives of the protected geographical indications and of the traditional specialities guaranteed have not been met. Producers, instead of taking advantage in terms of an increased competitiveness of their products, in almost all cases have not been rewarded for their efforts, especially regarding the high economic costs. In relation to the most famous PDOs and PGIs, it must also be considered that often official controls are not effective. As shown by recent market surveys, the counterfeiting of these PDOs and PGIs is constantly expanding, which causes a substantial economic loss to producers who have decided to invest in the quality of their products.

---

[51]Di Lauro (2005), p. 219.

[52]Strambi (2010), p. 113.

Furthermore, the entry into force of the Marrakech Treaty and thus, the simultaneous establishment of the WTO has worsened the situation. First, increasingly European agricultural products and foodstuffs have been called to compete with the low cost of similar non-European products. Secondly, under the obligations assumed by the European Union as a member of the WTO, the access to PDO, PGI and TSG schemes has also been opened to non-European products.

Finally, many names, regularly registered as PDOs and PGIs in the European Union, had been earlier registered as individual brands in some Member States of the WTO. In accordance with the Agreement on Trade Related Aspects of Intellectual Property Rights (TRIPS Agreement), the protection afforded to earlier registered brands has prevented the use of the European distinguishing marks in a large part of the non-European markets.

The criticisms emerged at the international level in relation to PDO and PGI schemes should have led the European legislator to also consider the possibility of giving up on these quality schemes or, at least, of rethinking and strengthening the function played by the business to consumer communication in the enhancement of the European agricultural products and foodstuffs quality.

At the same time, the EU should have paid more attention to its food quality schemes in the negotiations of the FTA US-EU. As is known, the European Union and the United States are almost at the end of the negotiations on a new Transatlantic Trade and Investment Partnership (TTIP), a free trade agreement that aims to "harmonise" standards and regulations in both regions to expand corporate profits. In this regard, the European Commission has affirmed that "the Transatlantic Trade and Investment Partnership (TTIP) will create new trade and investment opportunities for companies, big and small, and new jobs. At the same time, the TTIP will cut prices and widen choice, while keeping the EU's high standards for consumer protection, social rights and environmental rules. It will also boost Europe's influence in the world – by shaping global trade, projecting our values and attracting more investment".[53]

But the TTIP is particularly insidious for the European quality schemes for agricultural products and foodstuffs. To prove this, it is sufficient to recall that on 18 May 2015 the European Commission, questioned about this topic by the European Parliament, answered that it is aware of the economic impact resulting from the lack of protection experienced in the U.S. market by an important number of EU quality products bearing the protected designation of origin (PDO) or protected denomination of origin (PGI) protection in the EU territory, and of the fact that the U.S. may consider some of these EU names as common names/generic in the U.S. territory, particularly regarding some cheese names.[54] For this reason, the Commission has also declared that it is trying to insert into TTIP rules guaranteeing an appropriate level of protection and the satisfactory enforcement of that protection

---

[53]http://ec.europa.eu/priorities/balanced-eu-us-free-trade-agreement_en.

[54]http://www.europarl.europa.eu/sides/getAllAnswers.do?reference=E-2015-004165&language=EN.

in the U.S. for a selected list of EU geographical indications. But, as specified by the European Commission, the status of this selected list of EU geographical indications in the U.S. territory will have to be analysed by the negotiators based on concrete evidence.

Presently, it is not possible to know what the final content of TTIP will be, but, in my opinion, there are serious doubts that the TTIP will be able to ensure to all the registered PDOs, PGIs and TSGs the protection they need at the international level. On the contrary, the TTIP will probably reveal itself to be a Trojan horse and the European rules and, especially, the Quality Schemes Regulation will prove to be ill-equipped and unprepared to deal with its surprises.

References

Adornato F (2010) Le "declinazioni della qualità": una nota introduttiva. In: Adornato F, Albisinni F, Germanò A (eds) Agricoltura e alimentazione. Principi e regole della qualità. Disciplina internazionale, comunitaria, nazionale. Atti del Convegno internazionale I.D.A.I.C. di Macerata, 9–10 ottobre 2009. Giuffrè, Milano, pp 19–27

Albisinni F (2000) Azienda multifunzionale, mercato, territorio. Nuove regole in agricoltura. Giuffrè, Milano

Albisinni F (2015) Strumentario di diritto alimentare europeo, 2nd edn. Utet, Torino

Altili P (2012) Certificazione: nuova disciplina in materia di accreditamento e di organismi di controllo. Agri Isti Merc 1:95–105

Bolognini S (2012) La disciplina della comunicazione business to consumer nel mercato agro-alimentare europeo. Giappichelli, Torino

Canfora I (2011) Le "specialità tradizionali garantite". In: Costato L, Germanò A, Rook Basile E (eds) Trattato di diritto agrario, vol III. Diritto agroalimentare. Utet, Torino, pp 73–80

Capelli F (2014) Il Regolamento (UE) n. 1151/2012 sui regimi di qualità dei prodotti agro-alimentari: luci ed ombre. Riv dir alim 1:52–60

Coppola N (2013) Regolamento UE n. 1151/2012 su DOP, IGP E STG: una occasione mancata? Alimenta 2:31–35

Costato L (2006) DOP, IGP, STG nei regolamenti del 2006 adottati anche in relazione ai negoziati WTO. Riv dir agr 1:351–359

Costato L (2012) Il regolamento n. 1151/2012 del Parlamento europeo e del Consiglio sui regimi di qualità dei prodotti agricoli e alimentari. Riv dir agr 1:648–667

Di Lauro A (2005) Comunicazione pubblicitaria e informazione nel settore agro-alimentare. Giuffrè, Milano

Galli C (2004) Globalizzazione dell'economia e tutela delle denominazioni di origine dei prodotti agro–alimentari. Riv dir ind 2:60–80

Gencarelli F (2006) La politica comunitaria di qualità alimentare: origine ed evoluzione. Dir com sc intern 3:609–622

Gragnani M (2013a) The EU regulation 1151/2012 on quality schemes for agricultural products and foodstuffs. Eur Food Feed Law 6:376–385

Gragnani M (2013b) Le Specialità Tradizionali Garantite (STG) secondo il Regolamento UE n. 1151/2012. Alimenta 7–8:155–159

Jannarelli A (2011) La qualità dei prodotti agricoli: considerazioni introduttive ad un approccio sistemico. In: Jannarelli A (ed) Profili giuridici del sistema agroalimentare tra ascesa e crisi della globalizzazione. Cacucci editore, Bari, pp 199–216

Losavio C (2007) Il consumatore di alimenti nell'Unione europea e il suo diritto a essere informato. Giuffrè, Milano

Lucifero N (2011) La comunicazione simbolica nel mercato alimentare: marchi e segni del territorio. In: Costato L, Germanò A, Rook Basile E (eds) Trattato di diritto agrario, III. Diritto agroalimentare. Utet, Torino, pp 321–412

Masini S (2006) Sulla funzione delle specialità tradizionali garantite: una nomenclatura tra tradizione e delocalizzazione. Dir giur agr alim amb 1:490–495

Masini S (2012) PDO, PGI and TSG. In: Costato L, Albisinni F (eds) European food law. Cedam, Padova, pp 351–367

Masini S (2015) Corso di diritto alimentare, 3rd edn. Giuffré, Milano

Petrelli L (2012) I regimi di qualità nel diritto alimentare dell'Unione Europea. Editoriale Scientifica, Napoli

Pisanello D (2006) La riforma del sistema comunitario di tutela delle denominazioni d'origine e delle indicazioni geografiche dei prodotti agricoli e alimentari. Contr impr/Europa 1:556–574

Rook Basile E (1996) Prodotti agricoli, mercato di massa e comunicazione simbolica. In: Casadei E, Germanò A, Rook Basile E (eds), Gli attuali confini del diritto agrario. Atti del Convegno "Enrico Bassanelli", Firenze, 28–30 aprile 1994. Giuffrè, Milano, pp 181–186

Strambi G (2010) Gli strumenti di promozione dei prodotti agro-alimentari tradizionali ad alta vocazione territoriale alla luce della Comunicazione sulla politica di qualità dei prodotti agricoli. Riv dir agr 1:109–127

Strambi G (2014) La tutela delle "specialità tradizionali garantite" alla luce del reg. UE n. 1151/ 2012. In: Studi in onore di Luigi Costato, vol II. Diritto alimentare. Diritto dell'Unione europea, Jovene, Napoli, pp 365–373

Tommasini A (2011) L'etichetta intesa come regola tecnica e strumento di veicolazione di informazioni dal produttore al consumatore. In: Costato L, Germanò A, Rook Basile E (eds) Trattato di diritto agrario, vol III. Diritto agroalimentare. Utet, Torino, pp 493–497

Regional Development of Rural Areas and Promotion of Local Foods: Comparing the EU and US Approach

Elizabeth Dooley and Allison Condra

1 Introduction

The United States of America (US) and European Union (EU) are both developed economies that have witnessed similar trends in relation to the role of rural areas and agriculture in their societies. Rural areas have decreased in population because of increasing urbanisation and declining participation and employment in the agricultural sector (USDA 2014a). A consequence of this shift is that urban populations have become increasingly disconnected with where their food comes from and what it takes to get it onto their plate (USFRA 2011). Consumer demand has risen for different options within the food system and a reconnection with foods' origins, placing a higher importance on components such as how products were produced (e.g., environmental management of the land), whether chemicals were used or the product is organic, whether social values were respected (e.g., workers' rights, animal welfare), etc. Alongside consumer demand, institutional support and governance has developed to further the viability, credibility, and uniformity of local food systems, thereby contributing to rural development of the communities where these farmers and small- and medium-sized enterprises (SMEs) are operating.

This chapter will discuss approaches toward rural development and local food promotion in the EU and US. It is based upon research conducted in 2015; thus, the reader should be aware that changes may have occurred, which are not reflected here. Following the UK referendum to exit the EU (i.e., Brexit) and US presidential and

E. Dooley (✉)
University of Exeter, Exeter, UK
e-mail: ed419@exeter.ac.uk

A. Condra
Davis Wright Tremaine LLP, Portland, OR, USA
e-mail: allisoncondra@dwt.com

A. Isoni et al. (eds.), *Food Diversity Between Rights, Duties and Autonomies*,
LITES – Legal Issues in Transdisciplinary Environmental Studies 2,
https://doi.org/10.1007/978-3-319-75196-2_6

congressional elections in 2016, uncertainty is widespread as to the continuation of laws, policies, and funding championing local food and rural development under the Trump administration, and Britain's role as a leader in local food and rural development within the EU is effectively no longer relevant. Regardless, the development of strategic public policy schemes to fit the context, needs, and limitations of each system has been noteworthy and should be instructive for continued support to local producers and creation of opportunities within rural areas.

In considering promotion of local foods, the authors adopt a broad approach to the term and understand it to include the following concepts:

- 'local farming' means the production of agricultural products and foodstuffs with the aim of selling them in an area reasonably close to the farm of production;
- 'direct sales' means sales by a farmer directly to a consumer, without intermediaries on the selling side;
- 'short food supply chains' means sales from a farmer to a consumer with a reduced number of intermediaries;
- 'local food systems' means that production, processing, trading and consumption of food occur in a relatively small geographical area" (European Commission 2013a, p. 5).[1]

These terms correspond with on-farm (e.g., Community-Supported Agriculture (CSA), "farm shops, farm-based hospitality, roadside sales, pick-your-own schemes") and off-farm activities (e.g., "farmers' markets, shops owned by farmers, food festivals and fairs, farm-based delivery schemes, or through one single trade intermediary (cooperative shops, specialist shops, supermarkets, etc.)" (Kneafsey et al. 2013, p. 13). In addition, direct sales to public institutions (e.g., hospitals, schools, nursing homes, etc.) in the form of public procurement and even non-short distance sales through the internet are possible alternatives for local food sales. Thus, we will consider how regulations, policies, and incentive structures contribute to these varying forms of local food promotion.

The chapter begins with an overview of the EU and US systems of government to provide context for comparison between their various levels of laws and policies. Subsequently, the chapter provides five approaches toward promotion of local food and rural development that are utilised within both systems, in some cases similarly or with key differences, and some additional examples of local mechanisms that have been successfully implemented. Finally, conclusions are provided regarding the approaches taken by the EU and US and their contribution to the objective of increasing local food systems and rural development.

[1] *See also* Kneafsey et al. (2013) for a discussion of the proliferation of terms in academia, policy, technical and civil debates about the concept of "re-localising" food systems.

2 Overview of the EU and US Governance Structures

To effectively compare the EU and US systems for rural development and from where local food support flows and why, it is important to understand the governance structures that exist on both sides of the Atlantic. Under the US Constitution, Congress' authority to act is limited to the powers enumerated in the Constitution (U.S. Constitution Art. 1 § 8). For rural development and local food production, Congress' primary authorities are the ability to levy taxes (clause 1) and to regulate interstate commerce (clause 3), which includes goods that may flow across state borders such as food products. Any powers Congress was not granted are reserved for the states.[2] Thus, states can regulate when Congress does not have the power to regulate and in cases where Congress has the Constitutional authority to act but has not yet done so. States, in turn, can delegate to their local governments authority to act on issues of local concern (Briffault 1990; Weissert and Ice 2014). Policies around rural development and the production of local food are therefore possible and can be identified at all levels of government in the United States.

The EU is similarly restricted to certain competences, which it has been granted under the Treaty on the European Union (TEU, Article 5(2)) and the Treaty on the Functioning of the European Union (TFEU, Article 1). These competences can be exclusive,[3] shared,[4] parallel and supporting[5] in relation to the Member States' (MS) competences. The EU must also respect the principles of subsidiarity (except for exclusive competences) and proportionality in regulating, which means that regulation should be done at the appropriate level of government and only according to what is necessary to achieve the Treaties' objectives, respectively (TEU, Article 5 (3)-(4)). With regards to rural development and local food production, "economic, social and territorial cohesion" and agriculture are two principal areas that are shared competences between the EU and the MS (TFEU, Article 4(2)). Additionally, any competences that have not been conferred to the EU in the Treaties remain with the MS (TEU, Article 5(2)). Thus, the MS can regulate "to the extent that the Union has not exercised its competence", as well as where the EU has not been given competence to regulate (TFEU, Article 2(2)). The MS are sovereign nations, so their governance structures and how much authority is delegated to local governments may differ, but there is a trend toward more decentralisation and delegation of

[2]"The powers not delegated to the United States by the Constitution, nor prohibited by it to the States, are reserved to the States respectively, or to the people." U.S. Constitution Amendment 10.

[3]Only the EU can legislate within that area. TFEU, Articles 2(1) and 3.

[4]Member States may exercise their competence on the issue if the EU has not legislated. TFEU, Articles 2(2) and 4.

[5]The EU has the right to legislate on the issue, as well as the Member States. TFEU, Articles 2(5), 4 (3)-(4), and 6.

responsibilities and authority to subnational governments by MS.[6] Policies promoting rural development and local food production therefore exist on all levels of government in the EU.

3 Comparing the EU and US Approaches

Thus, for the sake of comparison, the EU and US have an overarching level of government legislating on specific issues, which the lower levels of government may not contradict with their own legislation. Then, there are multiple territorial entities (currently, 50 states in the US and 28 MS in the EU) whose governments must abide by the overarching government's regulations but also have a lot of power to set their own regulations. A major difference of course in including states and MS in the same category is that states are subnational entities and MS are sovereign national entities. However, in ceding power to the EU in the Treaties, the MS can be compared with US states for our purposes as the subsequent level of authority under an umbrella form of government. Finally, there are many regional or local governments that may hold authority over issues that are best regulated at the lowest level.

Based on the comparative assessment, the EU and US levels of government have a similar focus on local food systems in their legislative and policy frameworks and have a broad array of activities being carried out within the local food landscape. Their approaches differ slightly given the overarching framework and context, but in general, these two systems utilise many of the same tools to promote local food and rural development.

Centralised omnibus farm policy and programming is a top-down strategy utilised in both systems to provide financial and technical assistance to on-the-ground stakeholders. As will be discussed in more detail below, however, the US scheme is more diffuse in programs and funding sources than the EU approach that stems primarily from a central policy structure for programming tied to funds to be allocated throughout the regional and local levels.

- Health and food safety regulations recognising potential barriers (e.g., size, scale) and providing flexibility for direct sales (cottage foods) and local food processing facilities are present within both systems.
- Labeling schemes specifically promoting local foods are also present in the US and EU, but in the US, these are cooperatively instituted and not protected in the same way as the EU geographic protections established through a strictly enforceable regime.
- Public procurement policies are used in both the US and the EU to increase local sourcing of products by public entities. The US system is often structured as a requirement though and the EU protection of free movement of goods means that

[6]This has been influenced in part by the Cohesion Policy and requirements for allocation and use of structural funds, which have triggered regionalisation reforms in some MS. Dhéret et al. (2012).

often procurement policies must focus prioritised purchasing on other characteristics besides geographic proximity.
- Increasing focus on farmer-to-consumer relationships is evident through direct marketing opportunities such as farmers markets, which have grown in number in both jurisdictions because of consumer demand and policy promotion.
- Local initiatives that contribute to building an effective, comprehensive food system vary according to the local context, but certain mechanisms are used in one or both the US and EU.

3.1 *Omnibus Farm Policy: The Farm Bill and the CAP Primary Source of Agricultural Policy and Funding*

The US federal government's primary way of supporting rural development and local food production is through financial assistance in the form of grants and low-interest loans (money raised from federal taxes), and technical assistance. As the makeup of rural America has shifted over the years, with more people moving to urban centers, this support for rural development and local food production has changed and developed as well (Fluharty 2011).

The US federal government takes up issues of rural development and local food production every 5–7 years with the Farm Bill (USDA 2015a). This omnibus piece of federal legislation addresses a wide range of issues, and includes sections specifically on rural development, specialty crops (fruits, vegetables, and nuts), organic agriculture, local and regional agriculture, and beginning farmers and ranchers. It is in these titles of the Farm Bill that Congress decides what type of and how much support to give to rural America and local food production. The most recent Farm Bill (2014) added more support for local and regional foods, for example, by providing more financial support to programs like the Farmers' Market and Local Food Promotion Program and the Healthy Food Financing Initiative (USDA 2015b).

The EU also provides financial and technical assistance for local food and rural development through an omnibus farm policy, the Common Agricultural Policy (CAP) (TFEU, Art. 38). Despite the declining numbers of people employed in the agricultural sector and increasing urbanisation, rural areas are still a very important component of the EU landscape (comprising 52% of the EU territory and 23% of the population in 2013) and rural development policies aim to increase economic opportunities and quality of life in rural areas (European Commission 2013b; United Nations 2014).

Adopted every 7 years, the CAP is divided into two pillars: the first pillar provides direct payments to farmers and the second pillar deals with rural development within the EU. Rural Development Programmes (RDPs) must be adopted by the MS at national or regional level, through which they provide grants, technical assistance, knowledge sharing, and research and innovation to eligible beneficiaries (Regulation 1305/2013, Art. 6). This regulation sets the framework for the Member States to adopt RDPs that reflect their local conditions through the measures they choose to include and allocate funding. Thus, at MS level there are opportunities to specifically tailor solutions to the national, regional or local context through the RDP, especially in light of the principle of subsidiarity, and provide grants to actors and projects that promote local food systems, e.g., farmers markets, processing facilities, labeling and certification schemes.

The RDPs to be adopted during the 2014–2020 programming period must be based on at least four of the six common EU priorities: (1) knowledge transfer; (2) competitiveness; (3) food chain; (4) ecosystems; (5) resource efficiency; and (6) social inclusion. They must also reflect the strategic objectives set by the MS and agreed to in the Partnership Agreements with the EU (Ionescu 2015). Based on assessment of the issues and needs of the territory covered by the RDP, the region selects the priorities it needs to promote and chooses measures to include in the RDP from those outlined in the EAFRD regulation to achieve those priorities. Thus, the EU sets the broad objectives for rural development and local food promotion and the RDPs determine what type and how much support to provide for local food systems.

One of the priorities, "promoting food chain organisation", contains a focus area for "improving competitiveness of primary producers by better integrating them into the agri-food chain through quality schemes, adding value to agricultural products, promotion in local markets and short supply circuits, producer groups and organisations and inter-branch organisations" (Regulation 1305/2013, Art. 5(3)(a)). The EAFRD regulation refers to "short supply chains" as the mechanism for local food promotion. Short supply chain is defined as "a supply chain involving a limited number of economic operators, committed to cooperation, local economic development, and close geographical and social relations between producers, processors and consumers" (Regulation 1305/2013, Art. 2(1)(m)).

Thematic sub-programmes may also be developed within RDPs to specifically address certain needs and achieve the six priorities. Short supply chains is one of the possible themes (Art. 7). This effectively involves a targeted selection of measures that would contribute toward shortening supply chains (e.g., setting up producer groups, quality schemes for agricultural products and foodstuffs, basic services and village renewal in rural areas, knowledge transfer and information actions, etc.) and allocation of more funding toward those measures because they are part of a thematic sub-programme (Regulation 1305/2013, Art. 7, Annexes II, IV). Under Article 15, the RDPs may also include advisory support services that "help farmers, young farmers, forest holders, other land managers and SMEs in rural areas" improve the economic and environmental performance of their holding, enterprise and/or investment, with development of short supply chains as one of the possible topics for advice. Short food supply chains are typically composed of small farms or enterprises with small-scale producers who often practice organic farming as well, which is another measure supported under the RDPs and available for bundling into a thematic sub-programme on short supply chains (Regulation 1305/2013, Art. 29, Annex IV). Moreover, the RDPs may contain a measure supporting investments in physical assets to improve farm and rural enterprise performance, which specifically related to local food production may include "setting up of small scale processing and marketing facilities" (Regulation 1305/2013, Recital 15, Art. 17).

4 Alternative Programs and Support

The Farm Bill provides strong central agricultural and rural development legislation in the US, but consumer demand, changing societal priorities, and political will has also led to the development of additional executive branch programs. The United States Department of Agriculture (USDA) is the primary executive branch department that focuses on rural development and local food production. However, other executive branch departments and agencies have started playing a larger role.

During the Obama Administration and under the leadership of US Secretary of Agriculture Tom Vilsack, new programs were created to prioritise rural development, local food production, and local food production. First, Know Your Farmer, Know Your Food (KYF2) began in 2009 as an initiative of the USDA "to begin a national conversation to help develop local and regional food systems and spur economic opportunity" (USDA 2009). U.S. Secretary of Agriculture Tom Vilsack asserted that "reconnecting consumers and institutions with local producers will stimulate economies in rural communities, improve access to healthy, nutritious food for our families, and decrease the amount of resources to transport our food" (Ibid). KYF2 brings together various agencies within the USDA to pursue policies and programs that support and develop local and regional food systems (USDA 2015c). USDA programs within the KYF2 initiative provide grant money for farm-to-school projects, specialty crop (fruits, vegetables, and nuts) research and development, and rural business enterprises and opportunities, among others (USDA 2015d). USDA also offers loans to farmers and food producers, as well as financial assistance to individuals and families to help them purchase food (e.g., the Supplemental Nutrition Assistance Program (SNAP), formerly known as food stamps), and has created special programs to help those individuals and families purchase more fruits and vegetables from farmers markets.

In 2011, President Obama signed an Executive Order creating the White House Rural Council (The White House 2011). The Rural Council was established "to better coordinate federal programs and maximise the impact of federal investment to promote economic prosperity and quality of life in our rural communities" (Ibid). Chaired by the Secretary of Agriculture, the Rural Council is made up of heads of nearly all of the cabinet level executive branch agencies (except for the Department of State and the Department of Education), as well as heads of other offices and departments within the executive branch (such as the Small Business Administration, the Office of Management and Budget, and the National Economic Council). The policy initiatives of the White House Rural Council include: increasing capital access for rural small businesses; supporting rural health care; accelerating broadband infrastructure deployment; investing in rural schools; advancing water quality conservation across the United States; developing rural health information technology workforce; and expanding health information technology in rural communities (White House Rural Council 2015a). Many of the policy initiatives provide financial assistance and/or technical assistance; some of the initiatives seek to reduce regulatory barriers and make policy change to improve life in rural America.

Finally, in the summer of 2014, the White House announced a new grant program for local and regional food production in rural and other disadvantaged parts of the United States. The Local Food, Local Places grant program is an inter-agency initiative of the USDA, the United States Environmental Protection Agency (EPA), the United States Department of Transportation (DOT), the Appalachian Regional Commission (ARC), and the Delta Regional Authority (DRA) (McKalip 2014; The White House 2014). This is the first grant program to focus specifically on using local and regional food production to enhance rural development. In December 2014, twenty-six communities across the United States were selected to receive grant funding (USDA 2015e). Examples of the projects include the creation of a local food hub in an abandoned grocery store in downtown Fallon, Nevada; the creation of the Eastern Kentucky Food and Farm Hub, a local food aggregation and distribution center that aims to serve communities within a 50-mile radius of its downtown Hazard, Kentucky location; and, the establishment of "a job training program based on vegetable farming and community gardens that will supply food for a farmers market and cafe" in Clarksdale, Mississippi; among many others (White House Rural Council 2015b).

Therefore, many different grants are available to local actors in the US to improve and promote local food systems (e.g., to incorporate local processing or develop marketing for local products). This integrated approach to rural issues and local food production involving different US federal agencies attempts to leverage each agency's expertise and find better ways to support regional development overall through the vein of local food production.

Rural development and local food policy at the EU level is primarily supported through the strong central CAP rather than various additional top-down programs. Because of the supranational nature of the EU, additional programs targeting local issues would stem from the national MS or regional and local levels in accordance with the subsidiarity principle. Unlike the US executive branch, the EU-level Directorate-General for Agriculture and Rural Development of the European Commission primarily works on implementation of the CAP. Nonetheless, the European Network for Rural Development (ENRD) is an EU-level program aimed at raising awareness and supporting implementation and evaluation of rural development activities within the EU (Regulation 1305/2013, Art. 52). It highlights projects conducted through the LEADER+ measure of the RDPs—LEADER fosters bottom-up action for rural development through local action groups (LAGs). A Working Group on short supply chains was established by interested National Rural Networks in 2011 as well to build exchange of knowledge and best practice examples between the MS about producer and consumer awareness raising and capacity building, promotion of local food, increasing markets and market access, and value-added opportunities (ENRD Contact Point 2011).

Additionally, as of 23 March 2015, the European Commission and the European Investment Bank signed a Memorandum of Understanding to increase financial instruments (e.g., low-interest loans) for agriculture supported by the European Agricultural Fund for Rural Development (EAFRD) (European Commission 2015). Such alternate financing options may allow small and beginning farmers to access credit more readily and thereby make capital investments into farms that build up the local food system.

Another mechanism that was introduced in the 2014–2020 CAP programming period is the European Innovation Partnership (EIP) for Agricultural Productivity and Sustainability. A Focus Group was established under the EIP on Short Food Supply Chain Management. It focuses on identifying innovative solutions to help scale up the number of farmers involved in short food supply chains and products sold to improve farmer income (EIP-AGRI 2014). Operational groups made up of farmers, researchers, advisors, and businesses within the agricultural and food sector may also be formed under the EIP to develop, test, adapt or implement innovation projects (Regulation 1305/2013, Art. 56-57). This collaborative type of approach toward implementation of the CAP measures is an overarching fundamental characteristic of the EU's omnibus farm policy (Regulation 1305/2013, Art. 35). Particularly relevant to local food systems, cooperation between local actors and partnership approaches can help to strengthen opportunities, such as farmer cooperatives sharing costs and pooling resources, local businesses coordinating promotion and increasing their customer base, and sectors building cross-linkages (e.g., tourism, food supply, agriculture) and enhancing local economies overall. Thus, the RDP measure specifically provides that developing short supply chains and local markets and promoting them are eligible cooperation actions if carried out between at least two different actors under Article 35(2)(d) and (e).[7] A comparable emphasis on collaboration in the US may be seen in the Regional Conservation Partnership Program introduced in the 2014 version of the Farm Bill, but the EU emphasises and institutionalises cooperation actions more than in the US local food policies, such as through transnational projects with knowledge exchange and capacity building.

4.1 *Flexible Health and Food Safety Regulations*

Both the EU and US have regulations for hygiene and food safety for processed foods (e.g., industrial kitchens, inspections), which may be too burdensome and exclude SMEs from participating in local food processing and value-added sales. Thus, some MS and US states have tried to address this issue by passing regulations with more flexibility in requirements while maintaining the focus on food safety (i.e., cottage food laws).

US states' primary authority is to regulate for the health, safety, and welfare of their residents; this gives states a lot of flexibility in legislating on a variety of topics. Most commonly, the federal government and the states will work together on issues

[7] *See also* Commission Delegated Regulation 807/2014, Art. 11 (defining that short supply chain means that there is no more than one intermediary between the producer and the consumer and that local markets means the product originated, was processed and was sold to the final consumer within a certain kilometer radius set by the RDP).

of rural development and local food production. As discussed above, the federal government provides a significant amount of financial and technical assistance for rural development and local food production. The states often play a role in helping the federal government implement those projects. For example, the Specialty Crop Block Grant program is a federal level program that provides block grants to state departments of agriculture for specialty crop (e.g., fruits, vegetables, and nuts) research and promotion (USDA 2014b). Although it is federal money, the states are responsible for applying with proposed projects to the federal government, and then distributing the money from the federal government to those projects within their state. However, states have also taken their own steps to support rural development and local food production.

State governments have supported local food with policies encouraging local food processing. Local food processing increases the value of local products and benefits local economies by creating job opportunities and contributing to the community tax base (Buck et al. 2007). One example of innovative state government support for local food processing is the Vermont state legislature's support for the creation of mobile poultry slaughtering units and a mobile flash-freeze unit for berries (Vt. Stat. Ann. tit. 6, §§ 17-18). Similarly, states also widely support local food processing in the form of cottage food laws, which allow for the in-home processing of non-potentially hazardous foods (e.g., baked goods, granolas, bread, dried fruit, pasta, herb blends, etc.). As of 2013, 42 states had some form of cottage food law (Harvard Food Law & Policy Clinic 2013, pp. 6, 10). These products are then available for sale to consumers without having to comply with expensive and often prohibitive permitting and licensing procedures, equipment requirements (e.g., commercial kitchens), inspections, etc. of what have been described as "one-size-fits-all" food processing regulations.

When analysing the approaches toward rural development and local food production at the Member State level in the EU, wide variances are observed because of the differences in climate, soil type, production systems and methods, traditional varieties, size of landholdings, etc. (Kneafsey et al. 2013). Overall, local food production is quite economically important for farms in the EU, with estimates of direct sales, for example, showing that over 20% of farms in the EU-15 engage in some form of direct sales (European Commission 2013a).[8] The percentage of farms differs dramatically between the MS, however, with some as low as Ireland, where only 0.5% of farms participate in direct sales, compared to around 35% in Italy (Ibid). In addition to prioritising measures important to the national context in their 2014–2020 RDPs and implementing the CAP measures from 2007 to 2013, Member States have taken additional action to grow local food systems.

The Austrian Standing Committee of the Ministry of Health, a cross-stakeholder initiative of ministries, professional organisations and consumers, identified the flexibility available under the EU hygiene and food safety regulation (Regulation

[8]Citing the European research project IMPACT: The socio-economic impact of rural development policies: realities and potentials (CT-4288), 4th Framework FAIR programme, 2002.

178/2002) in light of the barriers Austrian SMEs face and concluded that flexibility could be implemented in Austria without compromising food safety (Markuszewska et al. 2012). Thus, national guidelines were developed involving less documentation and less complicated monitoring requirements for food processing, simplified laboratory testing based on a risk-based approach, and a training project to train advisors for web trainings of practitioners.

4.2 Labeling Schemes Promoting Local Foods

Food labels that encourage consumers to purchase products for a premium price because they have special characteristic exist in many different forms (e.g., all-natural, heart healthy, sustainably produced, etc.). Local food labeling schemes aim to achieve the same objective. The EU has a well-established system of labeling for foods from certain geographic locations or produced in a traditional way, but also national and local/regional labeling and certification schemes have been developed to promote consumer awareness and targeted purchasing to support local food producers. While examples of such local food labeling schemes exist in the US,[9] they are not as systematically instituted or strictly exclusive and enforceable as the EU quality food labeling scheme.

Under the EU quality labeling scheme, special labels may be placed on products that are certified as eligible for "protected designations of origin", "protected geographical indications", "traditional specialities guaranteed" and optional quality terms such as "mountain product" (Regulation 1151/2012). This structure of recognisable labels for products that meet the strict criteria provides a way for local food producers and producers of traditional specialities to enhance promotion of their products (given that consumers appreciate and seek the qualities associated with products from certain localities or which are produced in a traditional way) and demand a premium price. With the introduction of the new regulatory framework for quality schemes in the EU in 2012, the Commission was also required to explore the possibility of a local farming and direct sales labeling scheme (European Commission 2013a). In line with recommendations from a conference on the topic and expert opinions, it was determined that a labeling scheme would be useful to prevent abuse by imitation labels that are not certified and create unfair competition, and to reduce consumer confusion as long as it is clear, optional for producers and without additional costs for certification and accreditation.

National labeling schemes throughout the EU Member States also exist to certify local products, which enable producers to secure value-added prices and increase consumer confidence. Labeling schemes are likely not as important for locally sold products where there is a direct relationship between the seller and buyer and the

[9]See e.g., "Indiana Grown", Purdue Extension, https://www.purdue.edu/dffs/localfood/2016/07/07/indiana-grown/.

buyer already knows from whom they are buying and generally how the product was raised/prepared. However, regional labeling schemes can be valuable for expanding the reach of products beyond the local area and communicating certain characteristics of the product (e.g., landscape and environmental characteristics, culture, traditions of the area, etc.) (Markuszewska et al. 2012). They can also prevent exploitation of consumer demand for local products by farmers or companies who do not use traditional practices special to the area, etc. (European Commission 2013a). For example, Austria has a logo "Gutes vom Bauernhof" trademarked by the Austrian Chamber of Agriculture, which indicates that the product has come directly from a small-scale family farm (Gutes vom Bauernhof 2016). It was adopted as a voluntary national standard in 2001 and it aims to promote authentic, traditional, traceable products.

Local branding of products in the EU has been a way in which producers gained value-added prices because of the association between production in a certain area by certain growers and the products' level of quality. Another example can be found in the region of Southern Tyrol, Italy, where a group of eight farmers came together to form a cooperative and market the traditional breeds from that area, as well as to improve animal welfare, increase sales and profits, and shorten the food chain by developing the local brand "Laugenrind" for organic beef (ENRD 2015). Additionally, a labeling and certification scheme for the EU outermost regions' products has benefited local producers, e.g., banana producers on the Canary Islands, by increasing their marketing of exports to EU consumers (Strano et al. 2011).

4.3 Public Procurement Targeting Increased Purchase of Local Products

Public procurement is one of the ways that local food can be systematically integrated into supply chains for local public entities, offering a large market, stable income and potentially value-added prices for special forms of production or varieties. Both the US and EU lower levels of government have local procurement examples that contribute toward increasing local food promotion. At the overarching level of government, the EU regulatory framework for procurement has a threshold limit for it to apply to a procurement contract, which is quite high in comparison to the often small-scale contracts that local food producers may enter into. Thus, the MS and local level procurement policies are applicable to promoting local food, but given the restrictions on limiting free movement of goods, preferences that promote local foods are designed in terms of freshness or production style requirements. This differs from the US system wherein some procurement policies require public entities to source a specific percentage of food locally; however, examples of explicit local preferences can also still be found in the EU despite the non-discrimination principle.

Local procurement and/or farm-to-institution laws throughout US states provide some form of preference for purchasing locally made goods by state institutions (e.g., agencies, schools, hospitals) (see, e.g., R.I. Gen. Laws Ann. § 37-2-8). The term local procurement and farm-to-institution are closely related; farm-to-institution is broader and encompasses a wider range of activities (such as farm-to-school education programs), while local procurement focuses specifically on the purchasing of local food. In the US, 37 states have local procurement laws to increase purchasing of local products, although not all these laws explicitly mention food; however, 14 states have laws that focus specifically on purchasing locally grown and produced food (Condra and Broad Leib 2015). For example, Massachusetts state agencies must purchase food grown or produced in Massachusetts if it is not more than 10% more expensive than out-of-state food (price preference law) (Mass. Gen. Laws Ann. ch. 7, § 23B).

Much of the focus placed on local food has been through farm-to-school initiatives, which aim to increase sourcing of fresh farm products from local producers by surrounding schools. This alternative aims to replace large amounts of processed, packaged, and/or frozen food from being used in school cafeteria meals (potentially transported over long distances), and it is typically considered a healthier option because of the focus on sourcing fresh fruits and vegetables. Additionally, these initiatives provide a steady buyer to whom local producers can market their products and may increase stability of farmer income, as well as eliminate supply chain intermediary costs. Examples of states promoting farm-to-school programs can be found in both Maine and New Hampshire, which passed legislation creating farm-to-school task forces to grow and strengthen farm-to-school programs (see Me. HP784 (2009); N.H. S.B. 141 (2014)). Additionally, Vermont awards small grants to schools working to start farm-to-school programs and school gardens (Vt. Stat. Ann. tit. 6, § 4721).

Local areas may also choose to institute local procurement requirements. The Memphis City Schools give a 5% price preference to local businesses who submit bids to supply products in Memphis and Shelby County, Tennessee (Shelby Co. Bd. of Ed. Procurement Serv. 2014). The city of Cleveland also provides a 2–4% bid preference for food items sourced locally, as well as bid discounts for local companies that source at least 20% of their food contracts from regional growers (City of Cleveland, Ohio 2015).

As described above, no exclusionary preference for local food products in the EU may be included in the selection criteria because of the prohibition against restrictions on free flow of goods and thus no discrimination between economic operators in other MS (Art. 18).[10] However, the EU level framework was reformed in 2014 to simplify procedures and provide more flexibility for SMEs to participate in and potentially win public procurement contracts (European Parliament 2015). Consid-

[10]The 2014 revised EU public procurement policy did not adopt the suggestion of the Committee of the Regions to specifically provide for a "locally produced" preference in food procurement to develop local food systems. Committee of the Regions (2011), paras. 53–58.

eration for social and environmental criteria is also included within the process (Directive 2014/24/EU, Art. 67). Implementing national policies in accordance with the EU standards, MS can include tender specifications for freshness (e.g., no more than 3 days between harvest and consumption) or production circumstances that effectively allow local areas to present the most economically advantageous tender, e.g., price-quality ratio (Committee of the Regions 2011, para. 55; European Commission 2016).

France, for example, has a nationally recognised definition for short food supply chains under the term "circuit courts", where there is no more than one intermediary between the producer and the consumer (Chiffoleau 2010). Importantly, this does not encompass distance, so if a local element is added, those are termed "circuit court de proximité". National-level policies have demonstrated a commitment to supporting circuit courts, e.g., the Barnier Plan with four key objectives: (1) transfer knowledge about circuit courts, (2) support producers in creating circuit courts, (3) train producers, and (4) organise and promote circuit courts (Kneafsey et al. 2013). Circuit courts have also been adopted by the French Rural Network as a priority action, and both the National Food Programme and Agricultural Modernisation Law have supported them by emphasising collective catering (e.g., in schools) and authorising cafeterias to purchase food directly from farmers or producer groups (Ibid).

Public procurement such as the French example for farm-to-school sourcing aims to benefit local areas by targeting purchase preferences of public institutions. The approaches vary between the MS, with some requiring open, fair and competitive procedures like those for the larger EU contracts and some only issue administrative guidance rather than legislation (Markuszewska et al. 2012). Scotland, for example, has a Sustainable Procurement Action Plan under the National Food and Drink Policy (The Scottish Government 2010). England's School Fruit and Vegetable Scheme has contributed to greatly expanding the number of food supplier contracts to schools throughout the country (Morgan and Morgan 2009). Hungary also has a legal framework, which has flexibility built into the procurement process for local sourcing as a derogation from the lowest price criterion and exemptions for many farm products (Kneafsey et al. 2013).

Local procurement policies in the EU can also promote uptake of more sustainable practices by local producers to meet heightened criteria for green and socially responsible public procurement, implemented widely in relation to food at the MS level (Kahlenborn et al. 2010).[11] Multiple regional laws exist in Italy, which regulate the public procurement into school cafeterias, requiring organic, typical and traditional products (Spigarolo et al. 2010). For example, the Consortium of Valle d'Aosta Local Authorities (CELVA) has established tender specifications that

[11] National, regional and local authorities can further define criteria for procurement to help achieve their public policy objectives. Additionally, "food and catering are among priority sectors for green public procurement", which encourages greening provisions to be included within sourcing contracts. European Commission (2013a).

require 70% local produce, in which case "local" is broadly defined to encompass traditional agricultural practices and breeds (Markuszewska et al. 2012). These types of policies aim to benefit children's health and food education, as well as local farmers.

4.4 Farmers Markets and Other Forms of Direct Marketing

More local markets provide opportunities for producers to sell their products directly to consumers who seek out food with certain characteristics, e.g., produced locally, organic, processed on-farm. Local market options in the US have increased in addition to direct on-farm sales (e.g., the Know Your Farmer, Know Your Food program, discussed above in section III.A.2). The 2012 US agricultural census revealed that over 12,000 farms marketed products through some type of a community-supported agriculture (CSA) scheme (USDA 2014c). This provides farmers with the financial security of definite buyers of the farm's products, often advance capital to use for planning and purchasing the growing season's materials, better prices and reduced risk from losses or "non-conventional" products in terms of shape and size, and less marketing burden. The number of farmers markets hosted throughout the US have greatly expanded within the past decades (growing from only 1755 recorded markets in 1994 to 8268 in 2014) (USDA 2014d). This is due in part to federal grants available under the Farmers Market Promotion Program, but also increasing consumer demand for farmers markets has driven expansion of these opportunities.

Farmers markets are also very well established in various MS throughout the EU, but additional policy support for increasing such direct marketing opportunities could be included in MS and regions' RDPs, such as Poland's for 2007–2013. Poland allowed support for improving existing markets' infrastructure, as well as buildings and equipment for new markets under the measure 321—basic services for the economy and rural inhabitants (ARMA 2016; Markuszewska et al. 2012). Local food systems have also fostered increases in local food purchasing through promotion and awareness raising, such as the NEBUS Network in Belgium, which developed a regional distribution network and promotional campaign with an online store (ENRD 2014).

4.5 Local Mechanisms to Promote Local Food Systems

Throughout the EU, bottom-up rural development and local food system promotion is significantly driven forward by the LEADER program under pillar two of the CAP, which is specifically intended to foster local responses to rural development needs. Local actors develop projects that are contextually appropriate and community-led and apply for support to local action groups (LAGs) (ENRD

2016). The aim of the project must fit within the scope of the Local Development Strategy, which LAGs create to target the specific needs of their regions. Cooperation initiatives, as mentioned above, are strongly emphasised and may be utilised within projects targeting local food systems. For example, a transnational cooperation project between Polish and Scottish LAGs focused on sharing experiences and promoting local foods by regional producers (TORE 2016). In addition, a project in France was initiated by the LAG Mayenne to raise awareness about what types of products are available locally, how farmers can build short supply chains, and how to increase use of local food by public entities (Pays de Haute Mayenne 2016).

Local food policy councils are a mechanism that have become very popular in cities throughout the US (e.g., Pittsburgh, PA, Knoxville, TN, Minneapolis, MN). Formally established within a community or region, food policy councils are groups of representatives and stakeholders from various sectors and interests in the local food system. They may take a broad approach toward issues within the food system, or they may focus on specific problems, such as food waste, community gardens, farm-to-school, etc. For instance, food policy councils have been established in Winston and Montgomery counties in Mississippi to strategically assess opportunities for collaborations between local producers, businesses, community institutions, and consumers and build thriving local food systems (e.g., Block et al. 2015).

Land use and planning through zoning ordinances can provide a way to promote local food production by facilitating small-scale land uses within communities that increase the availability, visibility, and convenience for producers and consumers to connect. For example, Cuyahoga County in Ohio allows residential lots to keep up to six chickens in a coop and to have one beehive (Cleveland – Cuyahoga Co. Food Policy Coalition 2010). Lots that are one-acre or larger can keep a small number of larger animals, including pigs, sheep, goats, and geese. On the vast former Tempelhof airport grounds in Berlin, Germany, the city's zoning has designated it as an open public space, which allows for a community garden to have been established and thrive (The Local 2011).

Finally, state-level loan programs have been instituted that may contribute to local food promotion. Pennsylvania, for example, adopted the Next Generation Farmer Loan Program that targets beginning or first-time farmers (Pa. Dept. of Community & Econ. Dev. 2013). Under the program, lending institutions that provide mortgages to beginning or first-time farmers for capital investments qualify for federal and state income tax-exempt status on the interest earned. The intention is that the lender then passes on a lower interest rate to its borrower, thereby providing financing opportunities to these specific populations that may not have been available otherwise. This program may particularly benefit applicants without access to land and/or equipment who would like to become small-scale producers and operate within the local food system or short supply chain.

5 Conclusions About EU and US Local Food Promotion

The EU and US have both placed significant emphasis on developing legislative and policy environments wherein local food systems can thrive. Many approaches and mechanisms that may be used to support local food and rural development are utilised within both systems, although in some cases there is stronger emphasis on institutionalised rules rather than voluntary schemes and project funding. Both the US and EU legislative and policy landscapes for local food and rural development are dominated by their respective omnibus farm policies—the Farm Bill and the CAP. Financial and technical assistance are primarily the form of support, with the US providing even additional support through separate grant programs, e.g., KYF2, the White House Rural Council, and the Local Food, Local Places program. The EU CAP requires development of Rural Development Programmes by the MS or smaller regions, in which they prioritise their measures toward issues within their local context and may allocate funding to actors and projects that promote local food systems. Therefore, many different grant programs are available to local actors on both sides of the pond to incorporate local processing or develop marketing for local products, for instance, to improve and promote local food systems. Within the overarching farm policy structure, however, the EU emphasises and institutionalises cooperation actions more than the US does, such as through transnational cooperation projects with knowledge exchange and capacity building.

Both the EU and US have regulations for hygiene and food safety for processed foods (e.g., industrial kitchens, inspections), which may be too burdensome and exclude SMEs from participating in local food sales. Thus, some MS and US states have tried to address this issue by passing regulations with more flexibility in requirements while maintaining the focus on food safety (i.e., cottage food laws). In addition to the EU's well-established system of labeling for foods from certain geographic locations or produced in a traditional way, national and local/regional labeling and certification schemes have been developed to promote consumer awareness and targeted purchasing to support local food producers. While examples of such cooperative labeling schemes exist in the US, they are not as systematically instituted or strictly exclusive and enforceable as the EU quality food labeling scheme.

Public procurement is one of the ways that local food can be systematically integrated into supply chains for local public entities, offering a large market, stable income and potentially value-added prices for special forms of production or varieties. Both the US and EU lower levels of government have local procurement examples that contribute toward increasing local food promotion. However, the EU prohibition against discrimination between EU producers means that the qualitative focus of the purchasing preference may be on other elements, e.g., freshness, rather than simple geographic proximity. This differs from the US system wherein some procurement policies require public entities to source a specific percentage of food locally.

Finally, local areas in either one or both the EU and US support local food promotion with mechanisms such as bottom-up local action groups that select strategic local projects under the LEADER program, local food policy councils, and land use planning and zoning aiming to facilitate access and participation in the local food system.

Throughout this chapter, multiple effective approaches toward supporting local food systems and the stakeholders within them (farmers, suppliers, processors, distributers, consumers, etc.) have been highlighted for their development and implementation within the US and EU. The idea that local consumers can access fresh and healthy food from producers they know and gain an understanding of the process for it to get to their plate is a positive step toward fixing an elongated and highly processed food chain. Nevertheless, there is more work to be done in terms of not only that the food is produced locally, but that it is qualitatively the type of local food desired or needed. Examples above of EU local procurement being specific to cultural and traditional foods allude to the question of what type of local food should be promoted. This question was a key component of the discussion surrounding "what is the right to food?" and whether that right encompasses a right to food diversity and cultural/traditional foods at the University of Salento conference on the Right to Food in May 2015. The authors argue that such elements are inherent within the right to food and reflect the fact that the right to food will look different to different individuals, communities, and cultures. For instance, the Indigenous Food and Agriculture Initiative at the University of Arkansas is aimed at addressing Native American tribes' extreme levels of food insecurity and receipt of federal food provisions devoid of traditional and culturally appropriate foods. Resulting health impacts seen within local populations raise this question as to what their right to food encompasses and whether it is being met (Univ. of Arkansas 2017). Therefore, policies and laws aimed at developing robust local food systems should not only ensure that there is local food that is accessible, but that it is also good quality food that is appropriate for that local area.

References

ARMA (Agency for Restructuring and Modernisation of Agriculture) (2016) Rural Development Programme for 2007–2013 – Poland, http://www.arimr.gov.pl/o-arimr/information-about-the-agency/forms-of-aid-managed-by-arma/rural-development-programme-for-2007-2013-poland.html (last visited 13 Mar 2016)

Block S, Guo E, Balkus O (2015) Good laws, good food for our schools: strategies to improve school food in Montgomery County and Winston County, Mississippi. Report by the Harvard Law School Mississippi Delta Project and Harvard Food Law and Policy Clinic, http://www.chlpi.org/wp-content/uploads/2013/12/Winston-Montgomery-Counties-ADC-School-Foods-Report-January-2015.pdf (last visited 12 Mar 2016)

Briffault R (1990) Our localism: Part I—the structure of local government law. Columbia Law Rev 90:1–115

Buck K, Kaminski LE, Stockmann DP, Vail AJ (2007) Investigating opportunities to strengthen the local food system in Southeastern Michigan. Report for Master of Science (Natural Resources and Environment) project, University of Michigan

Chiffoleau Y (2010) Formes et enjeux des circuits courts de commercialisation en agriculture. INRA, Licence Professionnelle Gestion des espaces naturels agricoles, CEP Florac, Montpellier SupAgro CEP Florac, Montpellier SupAgro, 5 Mar 2010, http://www1.montpellier.inra.fr/PSDR/doc/coxinel/florac2010-formation-05032010.pdf (last visited 12 Mar 2016)

City of Cleveland, Ohio (2015) Local foods and sustainable business, http://www.city.cleveland.oh.us/CityofCleveland/Home/Government/CityAgencies/OfficeOfSustainability/LocalFoodsAndSustainableBusiness (last visited 12 Mar 2016)

Cleveland – Cuyahoga County Food Policy Coalition (2010) Land use & planning, http://cccfoodpolicy.org/working-group/land-use-planning (last visited 12 Mar 2016)

Commission Delegated Regulation (EU) No 807/2014 of 11 March 2014 supplementing Regulation (EU) No 1305/2013 of the European Parliament and of the Council on support for rural development by the European Agricultural Fund for Rural Development (EAFRD) and introducing transitional provisions, OJ L 227/1

Committee of the Regions (2011) Opinion of the Committee of the Regions on 'Local food systems' (outlook opinion). Official Journal of the European Union C 104/1

Condra A, Broad Leib E (2015) Increasing local food procurement by Massachusetts state agencies. Harvard Food Law and Policy Clinic, http://www.chlpi.org/wp-content/uploads/2013/12/Increasing-Local-Procurement-by-Massachusetts-State-Agencies_FINAL_FULL-REPORT.pdf (last visited 12 Mar 2016)

Dhéret C, Martinovici A, Zuleeg F (2012) Division of powers between the European Union, the Member States and local and regional authorities – additional fiscal/budgetary elements. Report by the European Policy Centre, http://cor.europa.eu/en/documentation/studies/Documents/division-of-powers_additional-fiscal-budgetary-elements.pdf (last visited 11 Mar 2016)

Directive 2014/24/EU of the European Parliament and of the Council of 26 February 2014 on public procurement and repealing Directive 2004/18/EC

EIP-AGRI (European Innovation Partnership for Agricultural Productivity and Sustainability) (2014) Discussion Paper from the 1st Meeting of the Focus Group on Short Food Supply Chain Management, Prague, 10–11 December 2014, http://ec.europa.eu/eip/agriculture/sites/agri-eip/files/fg12_short_food_chains_discussion_paper_2014_en.pdf (last visited 11 Mar 2016)

ENRD (2014) Authentic local rural product distribution network in Central West Flanders, http://enrd.ec.europa.eu/enrd-static/policy-in-action/rdp_view/fr/view_project_3500_fr.html (last visited 13 Mar 2016)

ENRD (2015) "Laugenrind" – organic beef and local ownership, https://enrd.ec.europa.eu/en/policy-in-action/rdp_view/laugenrind-organic-beef-and-local-ownership (last visited 13 Mar 2016)

ENRD (2016) LEADER Gateway, https://enrd.ec.europa.eu/en/leader (last visited 12 Mar 2016)

ENRD Contact Point (2011) Report of the Outcomes of the 11th NRN Meeting, Bad Schandau, Germany, 27 April 2011, Annex I, http://www.landsbygdsnatverket.se/download/18.e01569712f24e2ca09800012821/1370172568804/11th+NRN+meeting+report+09_05_11_Final+%282%29.pdf (last visited 11 Mar 2016)

European Commission (2013a) Report from the Commission to the European Parliament and the Council on the case for a local farming and direct sales labelling scheme, COM(2013) 866 final

European Commission (2013b) Rural Development in the European Union Statistical and Economic Information, Directorate-General Agriculture, http://ec.europa.eu/agriculture/statistics/rural-development/2013/full-text_en.pdf (last visited 11 Mar 2016)

European Commission (2015) Factsheet on 2014–2020 Rural Development Programme for Slovakia, http://ec.europa.eu/agriculture/rural-development-2014-2020/country-files/sk/factsheet_en.pdf (last visited 12 Mar 2016)

European Commission (2016) EU Public Procurement Legislation: Delivering Results Summary of Evaluation Report, http://ec.europa.eu/internal_market/publicprocurement/docs/modernising_rules/executive-summary_en.pdf (last visited 12 Mar 2016)
European Commission and European Investment Bank (2015) Questions & Answers in relation to the Memorandum of Understanding in respect of cooperation in agriculture and rural development within the EU between the European Commission and the European Investment Bank, http://ec.europa.eu/agriculture/events/2015/ec-eib-coop/qa-mou-ec-eib_en.pdf (last visited 11 Mar 2016)
European Parliament (2015) Public Procurement Contracts, Factsheet, http://www.europarl.europa.eu/ftu/pdf/en/FTU_3.2.2.pdf (last visited 12 Mar 2016)
Fluharty CW (2011) Why rural policy now matters to agriculture: rural development, regional innovation, and the next Farm Bill. Drake J Agric Law 16:31–67
Gutes vom Bauernhof (2016) Willkommen bei den bäuerlichen Direktvermarktern!, http://www.gutesvombauernhof.at/oesterreich.html (last visited 12 Mar 2016)
Harvard Food Law & Policy Clinic (2013) Cottage food laws in the United States, http://www.chlpi.org/wp-content/uploads/2013/12/FINAL_Cottage-Food-Laws-Report_2013.pdf (last visited 12 Mar 2016)
Ionescu SS (2015) Partnership agreements within cohesion policy 2014–2020. European Parliamentary Research Service, http://epthinktank.eu/2015/02/11/partnership-agreements-within-cohesion-policy-2014-2020/ (last visited 11 Mar 2016)
Kahlenborn W, Moser C, Frijdal J, Essig M (2010) Strategic use of public procurement in Europe. Final report to the European Commission MARKT/2010/02/C, http://ec.europa.eu/internal_market/publicprocurement/docs/modernising_rules/strategic-use-public-procurement-europe_en.pdf (last visited 13 Mar 2016)
Kneafsey M, Venn L, Schmutz U, Bálint B, Trenchard L, Eyden-Wood T, Bos E, Sutton G, Blackett M (2013) Short food supply chains and local food systems in the EU: a state of play of their socio-economic characteristics. European Commission Joint Research Centre (JRC)
Maine House Paper 784 (2009)
Markuszewska A, Prior A, Strano A, Bálint B, Midoux B, Bros C, Koutsaftaki C, Jochum C, Buffet C, McGlynn D, del Bravo F, Valtari H, Czaja J, Saalasto P, Töyli P, Kokovkin R, Redman M, Mazili SR, Silm S, Watson S, Leporati S, Marran T, Hudson T (2012) In: Peters R (ed) Local food and short supply chains. ENRD, EU Rural Review No. 12
Massachusetts General Laws Annotated ch. 7, § 23B
McKalip D (2014) Local food, local places: a federal partnership to help rural America use local food and build local economies. White House Rural Council, https://www.whitehouse.gov/blog/2014/06/09/local-food-local-places-federal-partnership-help-rural-america-use-local-food-and-bu (last visited 12 Mar 2016)
Morgan S, Morgan K (2009) Supply chain models and public sector local food procurement. The Centre for Business Relationships, Accountability, Sustainability and Society, http://www.academia.edu/3121870/Supply_Chain_Models_and_Public_Sector_Local_Food_Procurement (last visited 12 Mar 2016)
New Hampshire Senate Bill 141 (2014)
Pays de Haute Mayenne (2016) Qualité et proximité en restauration collective, http://www.hautemayenne.org/mission_leader_actions.html (last visited 13 Mar 2016)
Pennsylvania Department of Community & Economic Development (2013) The Next Generation Farmer Loan Program, http://www.newpa.com/find-and-apply-for-funding/funding-and-program-finder/next-generation-farmer-loan-program (last visited 12 Mar 2016)
Regulation (EC) No 178/2002 of the European Parliament and of the Council of 28 January 2002 laying down the general principles and requirements of food law, establishing the European Food Safety Authority and laying down procedures in matters of food safety, OJ L 31/1
Regulation (EU) No 1151/2012 of the European Parliament and of the Council of 21 November 2012 on quality schemes for agricultural products and foodstuffs, OJ L 343/1

Regulation (EU) No 1305/2013 of the European Parliament and of the Council of 17 December 2013 on support for rural development by the European Agricultural Fund for Rural Development (EAFRD) and repealing Council Regulation (EC) No 1698/2005

Rhode Island General Laws Annotated § 37-2-8

Shelby County Board of Education Procurement Services (2014) Notice to Bidders, http://www.scsk12.org/uf/procurement/uploads/bids/CNC%20Fresh%20Fruits%20and%20Vegetable%20Quarterly%20Bid%20Language%202014-2015%20-%20Geographic%20Preference%20%28Central%20Kitchen%29.pdf (last visited 12 Mar 2016)

Spigarolo R, Sarti V, Giorgi G (2010) Organic and conventional public food procurement for youth in Italy. Bioforsk Report, vol. 5, no. 109/2010

Strano A, Hudson T, Redman M, Meskell J, Stopes C, Schalenbourg C, Angiolini S, Ockso E, Baliuleviciene R, Bloch E, O'Hara E, Parissaki M, Hayes A, Gardner S, Toland J (2011) Agricultural product quality: a success factor for EU rural areas. EU Rural Review No. 8

The Local (2011) Urban gardening takes off at Berlin's fabled airport Tempelhof, http://www.thelocal.de/20110829/37247 (last visited 12 Mar 2016)

The Scottish Government (2010) Sustainable food procurement in the public sector, http://www.gov.scot/Publications/2010/11/19111212/5 (last visited 12 Mar 2016)

The White House (2011) Executive Order - Establishment of the White House Rural Council, https://www.whitehouse.gov/administration/eop/rural-council/executive-order (last visited 11 Mar 2016)

The White House (2014) Local foods, local places: announcement of federal assistance for sustainable communities, https://www.whitehouse.gov/sites/default/files/docs/announcement_-_local_foods_local_places_2014_final.pdf (last visited 12 Mar 2016)

TORE (The Taste of Rural Europe) (2016) TORE Project, http://www.tasteofruraleurope.eu/gb/00083,Project (last visited 13 Mar 2016)

Treaty on European Union (TEU), Consolidated Version published in the Official Journal of the European Union, C 326/13, 26.10.2012, Article 5(2)

Treaty on the Functioning of the European Union (TFEU), Consolidated Version published in the Official Journal of the European Union, C 326/47, 26.10.2012, Article 1

United Nations (2014) World urbanization prospects: the 2014 revision, highlights (ST/ESA/SER.A/352), http://esa.un.org/unpd/wup/Highlights/WUP2014-Highlights.pdf (last visited 11 Mar 2016)

University of Arkansas (2017) School of law, indigenous food and agriculture initiative, http://indigenousfoodandag.com/ (last visited 16 May 2017)

USDA (2014a) Rural America at a Glance: 2014 Edition. Economic Research Service, Economic Brief No. 26, http://www.ers.usda.gov/media/1697681/eb26.pdf (last visited 12 Mar 2016)

USDA (2014b) Specialty Crop Block Grant: fiscal year 2014 request for applications. Agricultural Marketing Service, https://www.ams.usda.gov/services/grants/scbgp (last visited 12 Mar 2016)

USDA (2014c) Community supported agriculture. National Agricultural Library, http://www.nal.usda.gov/afsic/pubs/csa/csa.shtml (last visited 12 Mar 2016)

USDA (2014d) Local food research & development. Agricultural Marketing Service, https://www.ams.usda.gov/services/local-regional (last visited 12 Mar 2016)

USDA (2015a) Agricultural act of 2014: highlights and implications. Economic Research Service, http://www.ers.usda.gov/agricultural-act-of-2014-highlights-and-implications.aspx (last visited 11 Mar 2016)

USDA (2015b) Local and regional foods: Title IV (Nutrition), Title V (Credit), Title VI (Rural Development), Title X (Horticulture). Economic Research Service, http://www.ers.usda.gov/agricultural-act-of-2014-highlights-and-implications/local-and-regional-foods.aspx (last visited 11 Mar 2016)

USDA (2015c) Know Your Farmer, Know Your Food: Our Mission, http://www.usda.gov/wps/portal/usda/usdahome?navid=KYF_MISSION (last visited 11 Mar 2016)

USDA (2015d) Grants, loans & support, http://www.usda.gov/wps/portal/usda/usdahome?navid=KYF_GRANTS (last visited 11 Mar 2016)

USDA (2015e) News Release, Obama administration selects communities to develop local food projects, encourage economic expansion: local foods, local places initiative encourages creative economic development in 26 communities in 19 states, http://www.usda.gov/wps/portal/usda/usdamediafb?contentid=2014/12/0262.xml&printable=true&contentidonly=true (last visited 12 Mar 2016)

(USDA) United States Department of Agriculture (2009) Press Release, USDA launches "Know Your Farmer, Know Your Food" initiative to connect consumers with local producers to create new economic opportunities for communities. Release No. 0440.09, http://www.usda.gov/wps/portal/usda/usdahome?contentidonly=true&contentid=2009/09/0440.xml (last visited 11 Mar 2016)

(USFRA) United States Farmers & Ranchers Alliance (2011) Nationwide surveys reveal disconnect between Americans and their food. The Food Dialogues, http://www.fooddialogues.com/press-release/antibiotics/nationwide-surveys-reveal-disconnect-between-americans-and-their-food (accessed on 7 Mar 2016)

Vermont Statutes Annotated title 6, §§ 17, 18, 4721

Weissert CS, Ice JL (2014) Relations between state and local governments. In: Haider-Markel DP (ed) The Oxford handbook of state and local government. Oxford University Press

White House Rural Council (2015a) Policy initiatives, https://www.whitehouse.gov/administration/eop/rural-council/policy-initiatives (last visited 11 Mar 2016)

White House Rural Council (2015b) Local food, local places community grant awards, https://www.whitehouse.gov/sites/default/files/docs/lflp_community_list.pdf (last visited 12 Mar 2016)

"Short Food Supply Chain" and Promotion of Local Food in EU and Italian Law

Giuliana Strambi

1 Introduction

In recent years, the European Union (EU) and Member States (MSs) have shown an increasing interest in short food supply chains as a tool to promote both local products and their food systems.[1] Short food supply chains and local production (or local farming), however, are not necessarily associated, at least not for all forms of the short supply chain.

This section analyzes this relationship, starting with a brief overview of the economic notions of "short supply chain" and "local products". In fact, although EU and Italian legislation have taken this terminology from economic literature they have never agreed on a legal definition of the two terms, and it is imperative to clarify and delimitate the boundaries of these notions to ensure the legislative field of application is clear and uniform, especially concerning incentive measures.[2]

[1]See, for example, Commission Staff Working Document on various aspects of short food supply chains accompanying the document: Report from the Commission to the European Parliament and the Council on the case for a local farming and direct sales labelling scheme, SWD (2013) 501 final, Brussels, 6 Dec 2013, especially para 3, where the Document talks about the "Revival of short food supply chains and local food systems".

[2]At EU level, see, for example, the Opinion of the Committee of the Regions on 'Local Food Systems' (outlook opinion), OJ [2011] C 104/01, where the Committee of Regions considers that the European Commission should "adopt definitions of 'Local Food Products' and 'Local Food Systems', and introduce a new logo and identify a common symbol and scheme identity for local products, to be added to the Agriculture Product Quality Policy regulation" (p. 1). At Italian level, see ISMEA, Capitolati di gara e best practices per le forniture di alimenti a filiera corta nelle mense

G. Strambi (✉)
Section on International and Comparative Agricultural Law, Institute for International Legal Studies, National Research Council, Florence, Italy
e-mail: giuliana.strambi@cnr.it

A. Isoni et al. (eds.), *Food Diversity Between Rights, Duties and Autonomies*, LITES – Legal Issues in Transdisciplinary Environmental Studies 2,
https://doi.org/10.1007/978-3-319-75196-2_7

Moreover, the evolution of the reasons behind EU and Italian promotion of different forms of short supply chains over time in the food system is examined, including a description of the current reasons for local food product promotion. The 2014–2020 EU Rural Development Policy and Italian Regional Law will be analyzed as significant examples of legislation addressing short food supply chains as an instrument instead of the object of promotion.

2 From "Short" to "Local": The Evolution of Policy Support for Food Supply Chains

The adjective "local" is always used to designate all that is bound to a place in its physical-spatial sense.[3] As regards food, it generally means that there is geographical proximity between farmer and consumer, so that food is produced close to where it is consumed, but there is no standardization of the distance that defines it, as it mainly depends on whether the food is fresh or processed, as well as the different geo-morphological contexts, agro-food market structures, and cultural heritage, i.e. factors influencing consumer perception of the proximity concept.[4] For example, if "local" in the US generally includes an area of hundreds of kilometres, in the EU and especially from an Italian point of view, this kind of distance seems excessive.

Increasing consumer sensitivity to food provenance is because of many reasons involving social, economic, environmental and health concerns[5] and is one of the main causes of the growing focus of EU and national Governments on local food markets and, in general, on alternative local food production and distribution systems.[6]

pubbliche, December 2011. On the specific problem of local food definition, see Sirsi (2014), p. 499.

[3]Sirsi (2014), p. 501.

[4]See Ansaloni (2012), p. 1; Galli and Brunori (2013), p. 4. The problem of "the absence of an established definition of what constitutes a 'local food'" is highlighted by US Congressional Research Service Report on "The Role of Local Food Systems in U.S. Farm Policy" of 12 March 2013, where "The lack of a universally agreed-upon definition" is also considered as a potential cause of "fraud in the marketplace with the sale of foods that are marketed as 'local' when they cannot be determined to be local" (p. 4). See the Report also for further references.

[5]See Sirsi (2014), especially pp. 502–503. On the specific philosophic view of "local" as the other side of the coin to the on-going process of globalization, see Certomà (2010), p. 113 et seqq.

[6]Recent studies have demonstrated that a short distance between producers and consumers and the absence or scarcity of intermediaries are beneficial also for farmers and local communities. In this regard, see SWD (2013) 501 final/2, pp. 5–6: "With regard to farmers, their role is enhanced as consumers understand better their function not only with regard to food production but also with regard to providing public goods like maintaining ecosystems, landscapes, and culture and tradition. Short food supply chains can result in a higher share of the value of the final sales price. This can result in a higher farm income and creates the opportunity to expand and modernize agricultural and non-agricultural activity on the farm as well as to improve viability of the farms". About local communities, they "may furthermore benefit from increased economic activity and a revival – or at the very least maintenance - of local services.".

It is against this background that the economic concept of short supply chain can be seen as a form of re-organization of the food supply chain that "aim[s] at re-connecting producers and consumers and re-localising agricultural and food production", although "short" doesn't always means "local".[7] Regarding this role specifically, the short supply chain has recently entered into the scope of institutional activities directly supporting the food sector.

In economic literature, in fact, the concept of short supply chain (also called "short circuit") is generally associated with the identification of alternate product distribution channels different from those of the current industrial and standardized distribution system, dominated by retailers.[8] Its hallmark is the reduction of one or more steps in the supply chain up to the elimination of all intermediaries, as in the case of direct access of farmers to final markets, i.e. to final consumers.[9] Therefore, a short supply chain can take various forms. Those with a single intermediary include, but are not limited to, restaurants, public food procurement for institutional catering services, sales through joint purchasing and distribution groups or through specialized shops. Regarding, instead, forms of direct distribution channels, i.e. without intermediaries, there are various forms of sale, for example: on-farm, in farmers' markets, in farm shops, online (with either home delivery or by post), vending machines and harvesting of products on the ground by the consumer on the farm (the so-called "pick-your-own").

Only when the chain is also "short" in a physical-spatial sense, namely when the distance of product transportation from production site to consumer is reduced, may consumers actually buy "local" products in the meaning of food produced in the same territory where it is consumed; only in this case, does "short" mean "local".[10] Conversely, a physical distance between producer and consumer, whatever it might

---

[7]EC Joint Research Center (2013), p. 13. However, "local" is not to be confused with quality production expressed by the protected designations of origin and geographical indications in EU territory. In this regard, see Kebir and Torre (2013), p. 1: "the local' appears as the vehicle of some form of intrinsic quality that can be objectified, described and most of all marketed on the food market (freshness, trust, information, etc.). This "local" quality provides (or at least is expected to) competitiveness to food products and most of all to generic food products, such as carrots, cabbage and leeks whose production do not require the use of specific assets such as terroir, traditional and local know-how, etc. As such it challenges the means of regional competitiveness which have pointed the importance of specificity in the paradigm of competition through differentiation and competitive advantages. Local food is in this sense a very different scheme than PGI/PDO, AOC (French) or even slow food movements that tend to valorise specificity and protect terroir and know-how (Torre, 2006). Of course both can overlap, products sold under PGI are often sold through SSFC (on the farm, at the cellar, farmers market) – in connection for example with tourism - but their main purpose is exportation out of the local area".

[8]On Alternative Food Networks, see Renting, Marsden and Banks (2003) p. 393 et seqq. The issue of the inclusion of short supply chains among Alternative Food Networks is complex, because a conventional, large scale farmer can sell some of his production in a farmer's market and, on the other hand, alternative producers can sell part of their production in international markets (Kebir and Torre 2013, p. 2).

[9]The literature is vast. See, for example, Giuca (2012), p. 11 et seqq. and Galli and Brunori (2013).

[10]Sirsi (2014), p. 501; Galli and Brunori (2013).

be, automatically makes the product not local, even if there are no intermediaries (for instance, in the case of online sales) and the chain is therefore short.

3 The Shortening of Food Supply Chains for Social and Collective Objectives. The Case of Farmers' Markets in Italy

Short chains as a tool for policies aiming to promote the consumption of local products is thus the result of reasons overlapping those that have traditionally driven supply chain economic studies and initial national legislation supporting food sold directly by farmers.

In this respect, direct sale, as a shorter form of the agro-food chain, has long had an independent legal significance in the Italian system with *ad hoc* regulation.[11] For example, the 1882 Commercial Code excluded direct selling of agricultural products from acts of trade because they were considered of minor importance, limited to farmers' needs to allocate their field products.[12]

More recently, special legislation has been still inspired by allocation purposes and by related opportunities to integrate farmers' income. It can thus be considered an expression of the specific regime of privileges that Italian law reserves for agricultural entrepreneurs.[13] In fact, the direct sale of food produced by farmers prevalently on their own farm is subject to a simplified administrative procedure.[14]

These "traditional" objectives have progressively taken on a further modern connotation: within the current industrial agro-food system, thanks to the shortening of the supply chain, a farmer can appropriate the profit margin that otherwise would

[11] Alabrese (2008), p. 3.

[12] Alabrese (2008), p. 3. On Art. 5 of Commercial Code, see Albisinni (2011), p. 263, according to whom (note 2, p. 263) "such exclusion reflects the choice of the location of the agricultural activities outside the area of enterprise, rather than the recognition of its specialties" (author's translation). In addition, also referring to special legislation on selling products directly by farmers, see Canfora (2008), p. 25 et seqq.; Bolognini (2011) p. 595 et seqq.; Germanò (2015), p. 161 et seqq.

[13] The literature on the status of favor that the Italian law reserves for agricultural entrepreneurs ("imprenditore agricolo"), now found outside the Italian Civil Code in special legislation (e.g. Art. 4 d. lgs. 228/2001), is vast. See, e.g., Jannarelli and Vecchioni (2009), p. 333 et seqq.; Germanò (2010), chapter 1.

[14] See Germanò (2002), p. 282 et seqq.; Masini (2007), p 289 et seqq.; Alabrese (2008), p. 3 et seqq.; Canfora (2008) p. 159 et seqq.; Albisinni (2001), p. 754 et seqq.; Albisinni (2011) p. 264 et seqq.; Albisinni (2013), p. 9. et seqq. Italian Law qualifies the sale of agricultural products directly by farmers as an activity "connected" to agricultural activities *stricto sensu* (Article 2135 Code Civil) and allows an agricultural entrepreneur to directly sell not only products in a natural state produced prevalently on his own farm, "but also products obtained as a result of treatment or processing on their own farm as well as purchased products (in the amount not exceeding the amount of farmer's own products)", see Kapala (2012), p. 53.

be spread over other stages of the food supply chain, and thusly receive a "fair income".[15]

Different and "complementary" objectives recently overlapped those cited above, which are no longer related to "individually important economic aspects", as were the previous ones, but to emerging issues of "collective" importance that can be classified into two categories.[16] The first one is related to environmental issues, such as the reduction of transportation emissions or food waste[17]; the second one is related to socio-economic issues, such as the development of local production, the recovery of the citizen-farmer relationship as a factor of social cohesion, increased surveillance of food hygiene because of direct contact between purchaser and seller, and the renewed interest in fresh and seasonal products for health reasons.[18]

Therefore, in recent Italian promotional measures, the "short" and "local" dimensions of food chains combine farmers' interests with those of consumers, who are "indirect addressees of social and economic advantage".[19] The case of farmers' market regulation is emblematic. This legislation, provided for in Agricultural Ministry decree of 20th November 2007,[20] establishes guidelines for the creation of markets reserved for farmers[21] selling directly to consumers in a clearly delimited place.[22] It aims to meet consumer requests to purchase agricultural products directly linked to the territory of production"[23] (using local ingredients) by providing measures and conditions supporting farmers' access to these markets. To sum up, this legislation, based on the existence of simplified rules for selling, which were not modified, reveals the new goals of "territorial policy" that substitute the traditional ones of merely supporting farmers through specific administrative rules.[24]

---

[15]Canfora (2008), p. 34.

[16]Canfora (2008), p. 34.

[17]The promotion of short chains is mentioned among the measures of the Italian National Programme of waste prevention 2013, III measure.

[18]Brunori and Bartolini (2013). On the short food chain in the organic sector, see Cristiani (2008), p. 15 et seqq.; Tommasini (2014), p. 33 et seqq.

[19]Canfora (2008), p. 34.

[20]More precisely, the first Italian legislation was established by Article 1, c. 1065, of Finance Act 2007, No 296/2006, which was implemented in 2007 by Agricultural Ministry decree. Each Italian Region has also implemented the national law, and specific administrative procedures have been adopted at a local level.

[21]According to Italian regulation, farmers are defined as agricultural entrepreneurs by Code Civil Article 2135. Farmers' cooperatives may access farmers' markets as well.

[22]Masini (2007), p. 289 et seqq; Alabrese (2008), pp. 6–7; Canfora (2008) p. 163 et seqq.; Canfora (2012) pp. 34–35; Albisinni (2013) pp. 30–33.

[23]Art. 1, c. 4, D. m. 20th November 2007.

[24]Albisinni (2011), p. 290 (author's translation).

4 The Definition of Supply Food Chain Within EU Rural Development Policy

The presence of environmental and social-economic motivations also emerged in the approach adopted at Union level for supporting short food supply chains, as the 2014–2020 Rural Development Policy reveals.

The promotion of short supply chains, in this context, is emancipated from the role of mere "instrument to support forms of marginal and non-competitive agriculture to become a key tool for achieving general goals of rural development, thus preserving the socio-economic vitality of rural areas", and of urban food policy.[25] It is against this background that it is interesting to exam the definition of "short supply chain" laid down by EU Regulation 1305/2013 on support for rural development; it "means a supply chain involving a limited number of economic operators, committed to co-operation, local economic development, and close geographical and social relations between producers, processors and consumers" (Art. 2, letter m).[26]

Here, the physical-spatial connotation is introduced into the short supply chain definition itself by referring to the territory and the local context. Moreover, in the specific measures dedicated to supporting "co-operation", the development of short supply chains is mentioned jointly with the development of "local markets",[27] according to an innovative approach that—as was remarked on at the beginning of this chapter—is the result of increased sensitivity of EU legislators towards the promotion of local forms of production and consumption, whose positive effects on the area involved, from an environmental and socio-economic point of view, have been established.[28]

However, the definition of the supply chain provided by Regulation 1305/2013 is still too general regarding subjects and quantities. Thus, in the subsequent Delegated Regulation 807/2014 supplementing regulation 1305/2013,[29] the Commission has introduced specifications to delimit and clarify the scope of incentive measures. Specifically, it aims to clearly indicate the short supply chains and local markets to which support may be given. Therefore, the number of intermediaries is established as a criterion defining the short supply chain, and the kilometric distance from the farm (considering specific geographic features of the area concerned) is established

---

[25]Brunori and Bartolini (2013), p. 4 (author's translation).

[26]Regulation (EU) No 1305/2013 of the European Parliament and of the Council of 17 Dec. 2013 on support for rural development by the European Agricultural Fund for Rural Development (EAFRD) and repealing Council Regulation (EC) No 1698/2005, OJ [2013] L 347/487.

[27]Regulation 1305/2013, Art. 35.2, (d) and (e).

[28]Cf. COM(2011) 236 final, Green Paper on promotion measures and information provisions for agricultural products: a reinforced value-added European strategy for promoting the tastes of Europe, 14 Sep 2011, Bruxelles.

[29]Commission Delegated Regulation (EU) No 807/2014 of 11 March 2014 supplementing Regulation (EU) No 1305/2013 of the European Parliament and of the Council on support for rural development by the European Agricultural Fund for Rural Development (EAFRD) and introducing transitional provisions, OJ [2014] L 227/1.

as a criterion defining the local market, unless a "convincing alternative criterion is presented" by a national rural development plan.[30] Regarding short supply chains, thus, cooperation measures shall cover only those involving no more than one intermediary between farmer and consumer.[31]

The two notions underpinning this approach have also been received in successive EU "Guidelines for State aid in the agricultural and forestry sectors and in rural areas 2014-2020" of July 2014.[32] Referring to short supply chains, although the definition continues to be broad,[33] the Commission text reproduces the same criterion of Regulation on rural development, stating that aid for the establishment and development of short supply chains must cover only those involving no more than one intermediary between farmer and consumer.[34] While regarding local markets, the Guidelines specify that the definition includes "markets in radius of 75 km from the farm of origin of the product, within which the activities of processing and sales to the final consumer have to take place".[35] Conversely, the Commission considers State aid promoting local markets placed in radius of 75 km from farm of product origin compatible with internal EU market rules, unless national rural development programs set out their own kilometric radius within which the activities of processing and sale to the final consumer must take place, or "a convincing alternative definition".[36]

In Italy, rural development programs are drawn up by the Regions and there are inevitably progressive adjustments to Commission Guidelines regarding promotion policies for local development. For example, the Campania Region has implemented the detailed definition of local market stated in EU Guidelines on State aid in its Regional Law No 6/2015 on rules for the support of Purchasing Groups and the distribution of short chain and quality food products,[37] where the sale of "zero-kilometre" products is defined considering the distance from the farm of origin, which must be within a maximum of 75 km.[38]

These supporting measures, even if intended to promote activities connected to short food supply chains and local markets development, should not be construed as tools to support only regional food, otherwise they would breach the principle of free

[30]Whereas 11 of Delegated Regulation 807/2014.

[31]Art. 1 (1) Delegated Regulation 807/2014.

[32]204/C 204/01, OJ [2014] C204/1.

[33]Part I, para 2.4 n. 56, of the Guidelines.

[34]Part II, para 1.1 "Rural development measures", n. 319, of the Guidelines.

[35]Part I, para 2.4, n. 60, (a), of the Guidelines.

[36]Part I, para 2.4. n. 60, (b) and (c) of the Guidelines.

[37]Campania Regional Law 6 March 2015, No 6, "Law on the support of joint purchasing groups and the distribution of food products from short chains and amendments to Regional Law 20/2014" (Campania Regional Bullettin of 9 March 2015, No 16) (author's translation).

[38]On farmers' markets and "food miles" (or "Zero-Kilometer food") and the issue of ambiguity of definitions, see Losavio (2011), p. 93 e seqq.

movement of goods and expose the State to complaints by EU Commission or to a declaration of illegality by EU Court of Justice.[39]

5 Towards an EU Labelling Scheme?

At EU level, the attention to the short food supply chain as a tool for promoting local agriculture and local food also emerged in the food information sector. Indeed, European Institutions have examined the issue of adopting a labelling scheme, considering that a growing number of consumers tend to give value to food bought directly in the place of production or produced in proximity of the place of purchase, without any considerations of how it was produced.

The labelling issue came up during the drafting of the new Regulation on quality schemes for agricultural products and foodstuffs, which establishes a scheme for protected designations of origin (PDO) and protected geographical indications (PDO) and a scheme for optional quality terms.[40] Therefore, in the final version of the Regulation in November 2012, the Commission was instructed to submit "a report to the European Parliament and to the Council on the case for a new local farming and direct sales labelling scheme to assist producers in marketing their produce locally".[41] The report "shall focus on the ability of the farmer to add value to his produce through the new label", and "should take into account other criteria, such as the possibilities of reducing carbon emissions and waste through short production and distribution chains".[42] Conversely, the report should consider environmental criteria as well.

In December 2013, the report was drawn up.[43] It refers to a definition of "short food supply chain" that appears most generally established at Union level, including

[39]See Losavio (2011), p. 99, especially about the case of Veneto Regional Law on the promotion of regional food. Many recent regional laws reveal critical elements, because they declare as their goal the promotion of products obtained through short chains or "zero-kms" but then they provide measures that support mainly regional products. Recently, on this issue, see Tommasini (2014), p. 55 et seqq.; Masini (2014), p. 54. Regarding the specific issue of State Law or regional Law supporting national or regional food through distinctive marks that promote regional short food chains, see Losavio, in this book.

[40]Regulation (EU) 1151/2012 of the European Parliament and of the Council of 21 November 2012 on quality schemes for agricultural products and foodstuffs, OJ [2012] L 343/1.

[41]Art. 55 of reg. 1151/2012 (italics is added).

[42]Art. 55 reg. 1151/2012, which concludes that "That report shall, if necessary, be accompanied by appropriate legislative proposals on the creation of a local farming and direct sales labelling scheme".

[43]COM (2013) 866 final, Report from the Commission to the European Parliament and the Council on the case for a local farming and direct sales labelling scheme, Bruxelles, 13.12.2013.

"sales from a farmer to a consumer with a reduced number of intermediaries".[44] However, in this case the vagueness of the definition is inevitable, as the report considers several studies conducted on the socio-economic and environmental implications of local farming and direct sales and on the existing labelling scheme throughout the EU, showing that there is a large variety of current short food supply chain models supported by Member States and that there are "disparities in the development and support of this type of sales"[45] because of national and regional differences in farm structures, distribution channels and culture.[46] For this reason, socio-economic and environmental impacts[47] have different value and weight regarding the forms of short supply chains subject to national measures.[48] Conversely, this also means that the choice of different parameters qualifying short food supply chains, regardless of the number of intermediaries, significantly affects the scope of promotional measures.

The EU Commission's report concludes that local farming and direct sales are certainly present within the European Union and "will continue to be part of European agriculture", and "there is a demand for a genuine farm product sold in short food supply chains, as well as the need to identify it", but a possible new EU label would actually be useful only if it was "simple and unburdensome for producers" while being controllable and ensuring sufficient credibility to consumers. It should also aim to reduce the risk of consumer confusion although existing EU legislation, if correctly enforced, allows action to be taken against misleading practices.[49] Moreover, it states that "a labelling scheme restricted to direct sales would have limited impact", because in this case consumers would be able to access all the information considered relevant for purchasing thanks to the direct relationship with the farmer. On the contrary, labelling information is important where there

[44]Para 3, p. 4, of the Report. An analogous consideration of the vagueness of the definition could be expressed referring to "local food systems" (which is defined as that one where "production, processing, trading and consumption of food occur in a relatively small geographical area") and to "local farming" (which is defined referring to "the production of agricultural products and foodstuffs with the aim of selling them in an area reasonably close to the farm of production"). On the contrary, the definition of "direct sales" is very precise, because there is no doubt throughout MSs legislations that it "means sales by a farmer directly to a consumer, without intermediaries on the selling side" (para 3, p. 4 of the Report).

[45]Para 4, p. 8, of the Report.

[46]Para 6, p. 10, of the Report.

[47]The environmental impact of short food supply chains should be examined, for example, in terms of reduction of food waste or of carbon dioxide emissions. See COM (2013) 866 final, para 3.2.

[48]It is useful to consider that the study described in the 2013 Report reveals "that to minimize the negative impact on the environment, short food supply chains should at the same time be local, be seasonal, use ecologically sound production methods and consider a low carbon footprint. Combining local and seasonal characteristics reduces storage needs, while ecologically sound production methods may also contribute to reduced use of pesticides, soil and water pollution and soil degradation, and enhance biodiversity and sustainable water usage" (para 3.2, p. 7).

[49]Para 6, p. 11, of the Report.

are intermediaries between the producer and the consumer because it replaces the direct communication in cases where this is not possible.[50]

6 Concluding Remarks

The current multilevel legal framework, which is developing around short food supply chain support for promoting local agro-food products, still requires clarification and legal certainty in terms of definition and, consequently, of application.

However, in conclusion, it can be noted that, as regards national promotion legislation, the framework tends to become uniform since it bases its legitimacy on rural development policy; while, as regards EU food law, the framework has the characteristics of "flexibility", and "exclusion" or "derogation" provisions,[51] either because of the specific features of direct contact and a relationship of trust between the producer and the consumer, or to the small quantity sold, or to the traditional method of local food production, as in the case of regulation of food information to consumers or EU food hygiene legislation.[52]

References

Alabrese M (2008) La vendita diretta dei prodotti agricoli. Riv di dir alim 3:3–7

Albisinni F (2001) Commento all'art. 4 d. lgs. n. 228/2001. Le nuove leggi civili commentate 3–4:754–763

Albisinni F (2011) La vendita diretta dei prodotti agricoli. In: Costato L, Germanò A, Rook Basile E (eds) Trattato di diritto agrario, vol I. Utet Giuridica, Torino, pp 263–294

Albisinni F (2013) Nel decennale dei decreti di orientamento: la vendita diretta. In: Germanò A (ed) Sul diritto agrario comunitario e nazionale. Rubettino editore, Catanzaro, pp 9–36

Ansaloni F (2012) Prodotto locale e sistemi alternativi di vendita. Agriregionieuropa 30:1–7

---

[50] Para 5.1, p. 9., of the Report. The conclusion on the ineffectiveness of labelling scheme restricted to direct sales is in line with the EU approach in general regulation on food information to consumer, where a direct relationship between producer and consumer, and the sale of small quantitative of product (which is typical of several forms of food supply chains) justify derogation or exemption from the general rules. For example, reg. UE n. 1169/2011 provides for an exemption from mandatory provision of nutrition information in foods "directly supplied by the manufacturer of small quantities of products to the final consumer or to local retail establishments directly supplying the final consumer" (Annex V, no 19). In addition, it states that "Where foods are offered for sale to the final consumer or to mass caterers without prepackaging, or where foods are packed on the sales premises at the consumer's request or prepacked for direct sale", only the provision related to ingredients or substances causing allergies or intolerance (listed in Annex II) are mandatory, unless Member States adopt national measures requiring other provisions of some or all of the particulars which are mandatory for pre-packed foods in general (Art. 9.1). On exemption rules on food information to consumers for unpacked products, see Jannarelli (2012), p. 44; Ragusa (2012), p. 457 et seqq.

[51] SWD (2013) 501 final/2, especially Annex II. Cf. Sirsi (2014), p. 517 et seqq.

[52] On relevant EU hygiene legislation applying to short food supply chains, see Strambi (2008), p. 1 et seqq.

Bolognini S (2011) Vendita diretta dei prodotti agricoli. Digesto italiano quarto. Sez. Civ. Agg. IV. Utet, Torino, pp 503–619
Brunori G, Bartolini F (2013) La filiera corta: le opportunità offerte dalla nuova Pac. Agriregionieuropa 35:1–10
Canfora I (2008) La commercializzazione dei prodotti agricoli nel diritto italiano e comunitario. Cacucci editore, Bari
Canfora I (2012), Dalla terra al territorio: Il ruolo dell'agricoltore nella filiera Corta. In: Giarè F, Sabrina G (eds) Agricoltura e filiera corta. Profili giuridici e dinamiche socio-economiche. INEA, Roma, pp 31–44
Cristiani E (2008) La filiera corta in agricoltura biologica. Riv di dir alim 3:15–20
Certomà C (2010) La retorica del locale. In: Cristiani E, Sirsi E, Strambi G (eds) Regole delle produzioni locali e mercato globale. III Giornate sulla Sicurezza alimentare. Edizioni ETS, Pisa, pp 113–121
EU Joint Research Center (2013) Short food supply chains and local food systems in the EU. A state of play of their socio-economic characteristics. Publications Office of the European Union, Luxembourg
Galli F, Brunori G (eds) (2013) Short Food Supply Chains as drivers of sustainable development. Evidence Document. Document developed in the framework of the FP7 project FOODLINKS (GA No. 265287). Laboratorio di studi rurali Sismondi. http://www.foodlinkscommunity.net/fileadmin/documents_organicresearch/foodlinks/CoPs/evidence-document-sfsc-cop.pdf. Accessed 29 Oct 2015
Germanò A (2002) Commento all'art. 4 d. lgs. n. 228/2001. Riv di dir agr (I):275–287
Germanò A (2010) Manuale di diritto agrario, 7th edn. Giappichelli, Torino
Germanò A (2015) I contratti di cessione dei prodotti agricoli. In: Germanò A, Rook Basile E (eds) I contratti agrari. Utet Giuridica, Milano, pp 161–217
Giuca S (2012) Conoscere la filiera corta. In: Giarè F, Sabrina G (eds) Agricoltura e filiera corta. Profili giuridici e dinamiche socio-economiche. INEA, Roma, pp 11–29
Jannarelli A (2012) La fornitura di informazioni sugli alimenti ai consumatori nel nuovo reg. n. 1169/2011 tra l'onnicomprensività dell'approccio e l'articolazione delle tecniche performative. Riv di dir agr (I):38–46
Jannarelli A, Vecchioni A (2009) L'impresa agricola. Section I, vol II.II. Trattato di diritto commerciale. Giappichelli, Torino
Kapala A (2012) Direct sale of agricultural products in Poland - general regulations. Riv di dir alim 4:51–54
Kebir L, Torre A (2013) Geographical proximity and new short supply food chains. In: Lazzeretti L (ed) Creative industries and innovation in Europe, concepts, measures, and comparative case studies. Routledge, New York, pp 194–202
Losavio C (2011) I prodotti agricoli "a chilometri zero" nelle leggi regionali. Agri Ist Merc 3:92–107
Masini S (2007) I mercatali (mercati degli imprenditori agricoli a vendita diretta). Dir e giur agr alim e dell'amb 5:289–294
Masini S (2014) Regole della concorrenza e differenziali di regolazione regionale e di garanzia di diritti individuali e di coesione sociale. Dir e giur agr alim e dell'amb 1/2:51–55
Ragusa VE (2012) I prodotti alimentari "preincartati" e gli "alimenti preimballati per la vendita diretta", alla luce del regolamento n. 1169/2011, "relativo alla fornitura di informazioni sugli alimenti ai consumatori". Riv di dir agr (I):457–487
Renting H, Marsden TK, Banks J (2003) Understanding alternative food networks: exploring the role of short food supply chains in rural development. Reg Stud Reg Sci 1:393–411
Sirsi E (2014) Regole e implicazioni giuridiche della produzione e del consumo del cibo locale. In: Nappi P, Sgarbanti G et al (eds) Studi in onore di Luigi Costato, vol III. Jovene editore, Napoli, pp 499–518
Strambi G (2008) I *farmers' markets* e la normativa sull'igiene degli alimenti. Riv di dir alim 3:1–12
Tommasini A (2014) Produzioni biologiche e filiera corta in funzione di un'alimentazione sostenibile. Riv di dir agr (I):33–69

Rural Development and Food Diversity in France

Séverin Jean

1 Introduction

Feeding and being fed, in English like in French, is a verbal construction from the noun "food". No need in English, unlike French, to use the adjective "dietary"—defined as "relating to your diet",[1] since the pre-modifier "food" associated with another noun "diversity", i.e. that which is "different; of various kinds",[2] will lead us to our topic of interest. Conversely, food diversity is nothing more than the diverse range of foodstuffs we consume. Nonetheless, food diversity is not the only theme in this contribution; it interacts with another expression: "rural development". If we look at the language behind these concepts one more time, this is about "the action of developing, deploying or unfolding"[3] "that which belongs to the fields, the countryside".[4] Rural development can therefore be considered an action, an instrument whose scope of investigation is very wide, as demonstrated by the definition given by Jean Morize: "Rural development involves improving the entire environment of the farmer, considered the primary beneficiary. It deals with roads, villages, health, education, and all the economic and social services likely to improve not only production, but also social wellbeing".[5] In fact, rural development as an action, as an instrument, does not have as sole beneficiary he who is the source of food

Published with the support of the Toulouse Faculty of Law.

[1]Littré, Dictionnaire de la langue française, tome 1, v° *alimentaire* (sens 1).

[2]Littré, *ibid.*, tome 2, v° *diversité* (sens 1).

[3]Littré, *ibid.*, v° *développement* (sens 1).

[4]Littré, *ibid.*, tome 5, v° *rural* (sens 1).

[5]Morize (1992).

S. Jean (✉)
Institute of Judicial Studies for Urban Planning and Construction (IEJUC), Toulouse 1 Capitole University, Toulouse, France
e-mail: severin.jean@ut-capitole.fr

A. Isoni et al. (eds.), *Food Diversity Between Rights, Duties and Autonomies*,
LITES – Legal Issues in Transdisciplinary Environmental Studies 2,
https://doi.org/10.1007/978-3-319-75196-2_8

diversity, meaning, the producer. Rather, rural development is also about land, the roots of a population in a rural environment often thought of as underprivileged in comparison to the "city". That said, it would be of no use, not to mention off-topic, to solely focus on rural development, since the expression "food diversity" does not impose a caesura, but establishes a link with rural development in France.

After the end of the Second World War, France was more interested in productivity; namely, producing in sufficient quantities to feed the population. This concern, in the context of European integration, found new expression in the creation of the Common Agricultural Policy (CAP), since the Treaty of Rome of 25 March 1957 includes specifically in its former Article 39[6] that "The objectives of the common agricultural policy shall be: to increase agricultural productivity (...), thus to ensure a fair standard of living for the agricultural community". It was not until 1999, after productivity issues had been resolved, that the expression "rural development" made a noted entry into community law—as the second pillar of the CAP through Council Regulation (EC) No 1257/1999 of 17 May 1999.[7] This Regulation inspired French law with a series of similar-sounding multi-annual "programmes" (anaphora) both at the national and regional levels, since the regional level is the most suitable for promoting local agricultural, agro-food, and forestry development, and to drive implementation of the second pillar of the CAP locally. This is because of the proximity to the land of the French administrative Regions and the leading role they play in economic development and spatial planning. In this context, rural development in France has included and still includes financial subsidies. Some are direct such as support to organise agricultural activity with the aim of securing agriculture through the maintenance of farms. Others are direct and indirect, and involve subsidised revenue for farms, by supporting the prices of agricultural products.[8] Although financial support, particularly via funds such as the European Agricultural Fund for Rural Development (EAFRD), is essential to reach the objectives of policies for rural development, it remains true when we look more closely, that food diversity is inseparable from policies regarding rural development, both at the European and French national levels.

It has become clear that rural development—as an instrument of "vertical" policy that sometimes makes the head spin—now attempts to ensure food diversity by focusing on the notion of "quality", along with the notion of "quantity". For instance, at the European level, one can read that "the purpose of Community farm investment aid is to modernise agricultural holdings, to improve their economic performance through better use of the production factors including the introduction

[6]Now Article 33.

[7]Council Regulation (EC) No 1257/1999 of 17 May 1999 on support for rural development from the European Agricultural Guidance and Guarantee Fund (EAGGF) and amending and repealing certain Regulations. This regulation was subsequently amended many times.

[8]In this sense, Krajeski (2009), No 67 *et seq.*

of new technologies and innovation, targeting quality, organic products and on/off-farm diversification".[9] Food diversity as an objective of rural development policy is even more explicit in French legislation since the law for the future of agriculture, food and forestry of 13 October 2014,[10] which inserted a preliminary "Book" in the Rural and Maritime Fishing Code, stating in Article L. 1.-I, 1 that this rural development policy is aimed at "ensuring that the population has access to safe, healthy, and diversified food, of good quality and in sufficient quantity, produced in economically and socially acceptable conditions, promoting employment, protection of the environment and the countryside, and contributing to mitigating the effects of climate change and adapting to them". As such, if we ask whether food diversity is ensured by rural development, clearly a positive answer must follow since rural development as an instrument tends to promote food diversity in Europe and in France (Sect. 2). However, can we switch the terms in this question? Is it possible to invert the expression (hyperbaton), asking whether rural development is ensured by food diversity? It would again seem necessary to answer "yes" here, since food diversity, in the terms of the law of 13 October 2014, called the "law for the future of agriculture", must for instance promote employment, protection of the environment, or the overall environment of the producer. This is then no longer a vertical policy of rural development but horizontal implementation of this policy through food diversity, in which producers are the primary creators (Sect. 3). To sum up, rural development and food diversity, if we again consider poetic writing, are "broken verses", inviting the reader to find the meaning of the stanza both in the words that precede the caesura in the hemistich and in those that follow it. Conversely, there is rural development today without food diversity, just as there is no food diversity without rural development. It then remains for us to give meaning to the stanza dedicated to rural development and to food diversity in France that, although it can be seen as a palindrome, changes in meaning depending on which way round you look at it.

2 Food Diversity Guaranteed by Rural Development

Rural development and food diversity are today major objectives both at the EU and national levels. As a result, it is not surprising that rural development policy encourages food diversity (Sect. 2.1). However, it is also clear that this policy, through its regulations, is contradictory because it sometimes creates obstacles for food diversity (Sect. 2.2).

---

[9]Council Regulation (EC) No 1698/2005 of 20 September 2005 on support for rural development by the European Agricultural Fund for Rural Development (EAFRD).

[10]Law No 2014-1170 of 13 October 2014 on the future of agriculture, food and forestry (1), *Official Journal of the Republic of France*. No 0238 of 14 October 2014.

2.1 Promotion of Food Diversity

Although for a long time, both at the EU and national levels, the focus was solely on food quantity, it is now clear that food quality has taken on significant importance and participates in, or is even equivalent to, food diversity. Regulations in France and at the EU level concerning rural development seem to be consistent with this idea. There are indeed many texts that point out this inseparable bond between quality and diversity. Firstly, community law, which points out in the Regulation of 21 November 2012 on quality schemes for agricultural products and foodstuffs[11] that "Citizens and consumers in the Union increasingly demand quality as well as traditional products. They are also concerned to maintain the diversity of the agricultural production in the Union",[12] and secondly French law that states in the first article of Title IV, Book VI of the Rural and Maritime Fishing Code entitled "Optimised use of agricultural, forestry, or food products and seafood" that "policies in the area of quality and origins of agricultural, forestry, or food products and seafood must meet the following objectives: promote diversity of products and identification of their characteristics, as well as of their production method and origin, to ensure that consumers are better informed and to meet their expectations (...)".[13] We can therefore see that food diversity, through food quality, is encouraged since quality is both a way of achieving rural development and an objective of rural development policies.[14] This statement raises several questions: why is food diversity equivalent to food quality? And, if this question is valid, how does food quality contribute to food diversity? Lastly, what impact does food quality have on food diversity and thereby on rural development?

Now that food production has been standardised, too often to meet the aims of food safety, we could forget that France, like Europe generally, has wide diversity across its territory. One of the largest regions in France, Midi-Pyrénées (since united with the Languedoc-Roussillon Region), is an excellent example of this, as can be seen in the Regional Rural Development Programme of Midi-Pyrénées[15] when it indicates that "the geographic diversity of the territory with its plains, plateaux, hillsides, massif and mountain areas, enables great diversity in the types of

[11]Regulation (EU) No 1151/2012 of the European Parliament and of the Council of 21 November 2012 on quality schemes for agricultural products and foodstuffs.

[12]In consideration of 2 in this Regulation.

[13]Article L. 640-1 of the Rural and Maritime Fishing Code.

[14]In this sense, for example, in Community law: "Support for investment (...) shall contribute to one or more of the following objectives: (...) to improve and monitor quality" (Article 25 of Council Regulation (EC) No 1257/1999 of 17 May 1999, *op. cit.*). In this sense, for example, in French law: "Policies in favour of agriculture and food, in their international, European, national and territorial aspects, are aimed at: 1 As part of food policies defined by the Government, ensuring that the population has access to safe, healthy, and diversified food, of good quality, and in sufficient quantity" (Article L. 1, -I, 1° of the Rural and Maritime Fishing Code).

[15]Regional rural development programme for Midi-Pyrénées, http://www.midipyrenees.fr/IMG/pdf/pdrr_complet.pdf.

production and agricultural methods". Because of this, optimised use of this diversity ensures food diversity. Since certain foodstuffs are unique—in that they may for instance have a strong regional identity that is an expression of a given territory and thus of a tradition, expertise, or even a specific climate—their value should be enhanced because they participate in food diversity and contribute to rural development of France and thereby of its regions. However, it is not only regional identity that justifies enhancing the value of these products, but also their intrinsic properties, including gastronomic or nutritional quality. So, how can we increase the value of these food products and thus enhance food diversity?

The answer in France and the EU is simply to make use of quality identification labels. To give just a few figures, the Midi-Pyrénées region today has 120 products identified using these procedures, illustrating the diversity of territories and expertise, and more specifically as an example, 14 products with protected designations of origin (PDOs) and 10 protected geographic indications (PGIs) in the wine production sector.[16] Regulations in this area are practically the same at the EU and national levels although some specificities should be pointed out at the national level. These regulations are organised in a pyramidal hierarchy. At the top, one finds quality and origin labels, which hierarchically cover: protected designations of origin (PDOs),[17] which correspond to higher quality than protected geographic indications (PGIs)[18] in that PDOs reflect the typicality of the food product resulting from a combination of the local territory and specific expertise, while PGIs have a less direct link between product and territory. Lastly, traditional specialities guaranteed (TSGs)[19] are more focused on a specific composition or method of production inherited from traditional practice than on the geographic location. Fundamentally, optional quality terms[20] such as labels like "product of mountain farming" and "product of island farming" simply indicate that all the production steps of the food product took place either in a mountain or island area. Although all these distinctive quality labels are collective markers that enhance the value of food products and ensure food diversity, French law has given the field a Gallic touch by adding other signs of quality or origin to these distinctive labels. An example would be the "*Label rouge*" that, considered better than a protected designation of origin and in routine use since 1960, "certifies that the food and products have specific characteristics that establish a higher level of quality, specifically resulting from their particular conditions of production or manufacture and that comply with specification requirements distinguishing them

[16]Data from the Regional Rural Development Programme for Midi-Pyrénées, *op. cit.*

[17]Articles L. 641-5 of the Rural and Maritime Fishing Code, L. 115-1 of the Consumer Code and 5, 1 of Regulation (EU) No 1151/2012 of 21 November 2012, *op. cit.*

[18]Articles L. 641-11 of the Rural and Maritime Fishing Code, L. 115-21 of the Consumer Code and 5, 2 of Regulation (EU) No 1151/2012 of 21 November 2012, *op. cit.*

[19]Articles L. 641-12 of the Rural and Maritime Fishing Code, L. 115-21 d of the Consumer Code and 17 of Regulation (EU) No 1151/2012 of 21 November 2012, *op. cit.*

[20]Articles L. 641-1 *et seq.* of the Rural and Maritime Fishing Code and 27 of Regulation (EU) No 1151/2012 of 21 November 2012, *op. cit.*

from similar marketed products".[21] Similarly, the "farm produce" quality label for cheese makes it possible to indicate that "the product was produced using traditional methods by producers who processed milk from their own farms on the actual site of production".[22] Basically, a product's value is related to its quality, its link to a territory and/or specific know-how, helping to ensure food diversity. We now need to ask what the impact of this quality is on food diversity and rural development.

It is important to remember that policies for rural development are aimed not only at producers, but also at consumers. On the consumer side, we must look beyond the fact that the aim is to better inform them, as indicated in Regulation (EU) No 1151/2012 of 21 November 2012 on quality schemes for agricultural products and foodstuffs, which states that producers can only continue to produce if they are "able to communicate to buyers and consumers the characteristics of their product under conditions of fair competition". They must also "be able to correctly identify their products on the marketplace"[23]. Indeed, studies in France have shown that one of the criteria for purchase of a food item, after price and proximity, is none other than quality.[24] Therefore, "it should be no surprise to find a link between the Consumer Code and the Rural Code. The first organises legal protection of quality labels that are determined and attributed by virtue of the second".[25] On the producer side, the quality scheme ensures the diversity of products and must help, as a priority, and in the terms of the regulations "to improve the competitiveness [of the agricultural sector]",[26] to provide "a competitive advantage to the Union's producers",[27] and to "promote fair distribution of valorisation efforts".[28] Furthermore "Producers can only continue to produce a diverse range of quality products if they are rewarded fairly for their effort".[29] As a result, justification of the quality scheme relies on the fact that "Operating quality schemes for producers which reward them for their efforts to produce a diverse range of quality products can benefit the rural economy. (. . .) In this way, quality schemes are able to contribute to and complement rural development policy. . .".[30] In addition to exporting quality food products outside their area of production, distinctive quality labels should also

[21] Article L. 641-1 of the Rural and Maritime Fishing Code.

[22] Article 9-1, paragraph 1 of Decree No 2007-628 of 27 April 2007 on cheese and cheese products.

[23] In consideration of 3, Regulation (EU) No 1151/2012 of 21 November 2012, *op. cit.*

[24] Research Centre for the Study and Observation of Living Conditions (CREDOC) (2012), Research report, Market survey 2012, Behaviours and attitudes of consumers concerning food retail. No 301.

[25] Krajeski (2009), No 525.

[26] In consideration of 13, Regulation (EC) No 1698/2005 of 20 September 2005, *op. cit.*

[27] In consideration of 1, Regulation (EU) No 1151/2012 of 21 November 2012, *op. cit.*

[28] Article L. 640-1 of the Rural and Maritime Fishing Code.

[29] In consideration of 3, Regulation (EU) No 1151/2012 of 21 November 2012, *op. cit.*

[30] *Ibid.*, In consideration of 4.

make it possible to promote employment[31] strengthen the local roots[32] of production, transformation, and marketing, thus highlighting the value of local territories, specifically those that are the most disadvantaged. Although one can only welcome this type of system, on initial analysis, it is not certain that quality always rhymes with food diversity and rural development. Firstly, we must remember that studies show that the proximity criterion is more important than quality in a consumer's decision to buy a food product.[33] It is rather rare to see producers who are engaged in quality labels also using short distribution channels for instance, since proximity serves as a guarantee of origin and freshness for the consumer. As a result, quality schemes ensure better export of food products outside the area of production and transformation, but do not necessarily ensure a significant increase in sales within the given area, unless this is encouraged by public policies for rural development that encourage "supplying of collective catering, whether public or private, with seasonal agricultural products or products with quality and origin labels",[34] for instance through food programmes. Also, although top-down classification of quality labels was adopted to this end, it remains true that increasing the number of quality signs leads to a real risk that the consumer will no longer understand the system, since the specificity of a food product ultimately ends up describing a generic product. Lastly, while it is clear that the use of quality labels participates in food diversity it is nonetheless first necessary to meet the acceptance criteria of the various specifications. However, food diversity is about more than products that can, in a certain way, be considered exceptional. Therefore, food diversity is clearly encouraged by rural development policy through quality labels, but these labels are of course not sufficient to entirely reflect food diversity. That said, this is not the most important concern. It does happen, paradoxically, that food diversity is hindered by rural development policy, although it is one of its main objectives.

2.2 *Hindering Food Diversity*

This stanza cannot be limited to praise for rural development as a tool for food diversity. Indeed, oxymoron haunts this stanza, as it haunts rural development that sometimes hinders food diversity. To be reminded of this, we need look no further than seed. Do we need to point out that most of our food originally comes from seed provided by nature that has incredible diversity? Let us use the tomato as an example—"the apple of love" when the May sun offers the promise of a ripening tomato for our pleasure in the summer. But where is the pleasure? Surely not in the tomato's standard sizing, or in its bright red colour that means that we are assured of

[31]Article L. 1, -I, 1° of the Rural and Maritime Fishing Code.

[32]Article L. 640-1 of the Rural and Maritime Fishing Code.

[33]Research Centre for the Study and Observation of Living Conditions (CREDOC) (2012).

[34]Article L. 1, -III, of the Rural and Maritime Fishing Code.

finding the same item on our plates whether we are in Lecce or Toulouse. What figure of speech gave the tomato such a claim to ubiquity? Precisely by the regulation of seeds and seedlings. It is therefore useful to examine these regulations that ensure some food diversity, but that are insufficient to sum up our overall food diversity.

French regulations in this area mirror those that were adopted at the European level. If we sum up, only seeds and seedlings that comply with criteria of distinctness, uniformity, and stability, can be included in official catalogues, whether national or EU, and can thus be marketed. Article 4, 1 of the Directive of 13 June 2002[35] stipulates that "Member States shall ensure that a variety is accepted only if it is distinct,[36] stable[37] and sufficiently uniform.[38] The variety must be of satisfactory value for cultivation and use". Similarly, Article 5, sub-paragraph 1 of the Decree dated 18 May 1981[39] stipulates that "registration in the catalogue is subject to three conditions: the variety must be distinct, stable, and sufficiently uniform". Conversely, it is only if these certification requirements are met, as evaluated by public institutions,[40] that seed can be registered in the catalogue, which is defined as "the exclusive list [our underlining] of the varieties or types of varieties for which seed and seedlings can be placed on the national market on the national territory".[41] Once registered on this exclusive list of which our food diversity is composed, marketing finally becomes possible and is defined as "the sale, storage for sale, offer of sale or any disposal, supply or transfer, with a commercial purpose, of seed or seedlings, whether for payment or not".[42] Although the apparent contradiction between food

[35]Council Directive 2002/53/EC of 13 June 2002 on the common catalogue of varieties of agricultural plant species.

[36]Article 5, 1 of this Directive indicates that "A variety shall be regarded as distinct if, whatever the origin, artificial or natural, of the initial variation from which it has resulted, it is clearly distinguishable on one or more important characteristics from any other variety known in the Community."

[37]Article 5, 2 of this Directive indicates that "A variety shall be regarded as stable if, after successive propagation or multiplications or at the end of each cycle (where the breeder has defined a particular cycle of propagation or multiplications) it remains true to the description of its essential characteristics".

[38]Article 5, 3 of this Directive indicates that "A variety shall be regarded as sufficiently uniform if, apart from a very few aberrations, the plants of which it is composed are, account being taken of the distinctive features of the reproductive systems of the plants, similar or genetically identical as regards the characteristics, taken as a whole, which are considered for this purpose".

[39]Decree No 81-605 of 18 May 1981 in application of Law dated **1 August 1905 on fraud control concerning trade in seed and seedlings.**

[40]In France, we should particularly mention the Technical Committee for Plant Breeding (CTPS), which is responsible for preparing and implementing policies on varieties, seed and seedlings. It plays a role in monitoring and registration in the French catalogue. Also see the Variety and Seed Study and Control Group (GEVES), the Official Service for Seed Control and Certification (SOC) for field crop species and vegetable species, the Inter-branch Technical Centre for Fruit and Vegetables (CTIFL), or FranceAgriMer.

[41]Vialle (2011), p. 21.

[42]Article 1, paragraph 2 of Decree No 81-605 of 18 May 1981, *op. cit.*

diversity and an exclusive list is clear from the outset, can we at least understand the stakes behind this type of certification and registration? Based on a ministerial report, "registration of varieties and certification of seed aims to guarantee, on the one hand, the authenticity and performance of new varieties placed on the market, and on the other, to guarantee the identity and quality (safety, physiological and technological quality, and purity) of seed and seedlings sold to users, both professionals, amateurs, and the general public".[43] Closer reading of this report should be cause for satisfaction, or even lead us to believe that this type of regulation should be praised. Indeed, "The process of registration in the catalogue by the public authorities, by imposing a clear distinction between varieties and official classification, makes competition between registering parties transparent and objective, and contributes significantly to competition between the parties. This leads to development of a large number of new varieties adapted to most of the constraints and production requirements for major agricultural species".[44] Nonetheless, the enthusiasm of the registering party is like that of the poet who finds that his poem is made up only of poorly rhyming verses although he had dreamed of much richer rhymes. Conversely, if a catalogue is a restrictive list, we must accept that limiting does not sit well with food diversity. However, we cannot ignore the progress that has been made in this area although it is, in the final analysis, completely insufficient, meaning that if we go back to our tomato example, the only pleasure is its visual appearance, with a sour after-taste for both consumers and producers, since the producers could only apply to register certain seed types in catalogues because others did not meet registration criteria.

A few steps forward in food diversity do however need to be highlighted. European and French regulations[45] have introduced a few exceptions by breaking the mould of required criteria for registration in catalogues of storage seed[46] and varieties without intrinsic value for marketing. Registration of storage seeds is

[43]Vialle (2011), p. 4.

[44]Vialle (2011), p. 6.

[45]In French law, see Article 3-1 of Decree No 81-605 of 18 May 1981 *op. cit.* as well as the Order dated 26 December 2010 providing a list of varieties for storage and a list of varieties without intrinsic value for commercial production and intended for specific crop conditions in the Official French Catalogue of plant species and varieties, and amending a technical regulation for registration in this catalogue. In Community law, see Commission Directive (EC) No 2002/53 of 13 June 2002 *op. cit.* as well as Commission Directive (EC) No 2008/62 of 20 June 2008 providing for certain derogations for acceptance of agricultural landraces and varieties which are naturally adapted to the local and regional conditions and threatened by genetic erosion and for marketing of seed and seed potatoes of those landraces and varieties. In addition, see Commission Directive (EC) No 2009/145 of 26 November 2009 providing for certain derogations, for acceptance of vegetable landraces and varieties which have been traditionally grown in particular localities and regions and are threatened by genetic erosion and of vegetable varieties with no intrinsic value for commercial crop production but developed for growing under particular conditions and for marketing of seed of those landraces and varieties.

[46]These are field crops, potatoes, and vegetables naturally adapted to local and regional conditions and threatened by genetic erosion.

justified to avoid erosion of diversity in seed varieties, while varieties without intrinsic value are justified because they are intended primarily for consumption by the producer. Also, the French Law of 8 December 2011 on the plant variety certificate[47] allows for 21 farm seed varieties, defined as seed with prior certification that the producer has bought from a supplier and that, after harvest, will be selected and developed by the producer to sow fields for the next crop. However, two issues arise here. Firstly, use of the seed is subject to payment of financial compensation called a "voluntary contribution [our underlining]" to selectors to finance the creation of new varieties. Secondly, only 21 farm seed varieties are authorised, meaning that all the others are excluded, for example, vegetables. Overall, these few positive steps must be praised but do not make it possible to say that regulations concerning seed and seedlings encourage food diversity. To use another image, this time from biological systematics, regulation of seed is like that of genera and not species, that of diversity without variety, although it is precisely the different varieties of seed that ensure food diversity.

Regulations on seed and seedlings hinder food diversity for at least two reasons. Firstly, registration of a variety in a catalogue is valid for ten years, renewable for 5-year periods.[48] This means that, even if they have once been registered in the catalogue, the question arises as to what happens to older varieties that are removed from the list at the end of this period, either at the request of the registering party or when marketed seed is no longer compliant with the initial description of the variety, or even when there is no longer an identified maintainer. The answer is quite simple, the varieties are removed from the list and can therefore no longer be marketed, although they may still be of interest. To ensure food diversity, it would be beneficial for regulations to be updated to make room for re-registration procedures or maintenance procedures for these varieties. Secondly, the same ministerial report that praised the merits of the regulation concedes nonetheless that among the areas for improvement, it would be necessary to broaden the possibilities for recognising "population varieties" also called "farm varieties". This is seed that the farmer samples from the harvest with the aim of future sowing but that, unlike farm seed, are not sourced from certified seed bought from a seed supplier. They are also clearly differentiated from standardised industrial seed that is selected at a research site or developed in the laboratory since these varieties are grown in controlled conditions, often non-reproducible[49] by the end grower. "Farm seed" can currently not be

[47]**Law No 2011-1843 of 8 December 2011 on plant variety certificates (1).**

[48]Article 12 of Council Directive (EC) No 2002/53 of 13 June 2002 *op. cit.* and Article 7 of Decree No 81-605 of 18 May 1981, *op. cit.*

[49]Either because of male sterility, or for example because we are dealing with "F1 hybrids".

registered or protected by a plant variety certificate[50] because it does not comply with the registration and certification conditions because of lack of uniformity and stability. Rejecting population varieties is a limitation on food diversity. Moreover, "their characteristics make them essential to adapt to the diversity and variability of territories, climates, farming practices, and human needs. Through their hardiness and their adaptability, they help to reduce chemical inputs and to address the need to feed humanity while protecting the environment".[51] Lastly, "they contribute to the self-sufficiency of the farm, unlike crops from varieties that are not reproducible by farmers, who need to purchase new seed each year".[52] The author often has this familiar dream that legislation will change but it is clear that here in fact lies Verlaine's Nightmare, that of the knight whose "fur coat out-raised distend, Beat back the wind"[53]—of hope. On 17 December last year, the European Commission announced the withdrawal of its proposal concerning new much-awaited regulations on seed because there was no political consensus! Seven years of preparation wasted and it appears that the work done will quite simply be shelved.

It can therefore be seen that rural development both encourages and hinders food diversity, in almost equal measure. To ensure implementation of rural development policy, those involved in food diversity must use this diversity policy to achieve rural development, or at least, maintain it.

3 Rural Development Through Food Diversity

Rural development through food diversity requires that those involved in food diversity use this diversity specifically to ensure their own development and thereby participate more generally in rural development. And the principal stakeholders in food diversity are of course the producers. To ensure their own development, or at least their own maintenance, they must benefit from the food diversity that the wide range of land types in France offers. To this end, two actions are often implemented in parallel: on the one hand, producers try to create

[50]Article 6 of Council Regulation (EC) No 2100/94 of 27 July 1994 on Community plant variety rights indicate that "Community plant variety rights shall be granted for varieties that are: (a) distinct; (b) uniform; (c) stable; and (d) new." Likewise, Article L. 623-2 of the Intellectual Property Code stipulates that "for application of this chapter, 'plant varieties' designates any new variety created that: 1: is clearly distinct from any existing variety that is well-known on the date of submission of the application; 2: is uniform, i.e. sufficiently homogeneous in its main characteristics, except for expected variation given the specificities of its sexual or vegetative reproduction; 3: remains stable, i.e. identical to its initial definition following its successive reproduction or, in the event of a specific reproduction cycle, at the end of each cycle".

[51]Les maisons des semences paysannes (2014), *Regards sur la gestion collective de la biodiversité cultivée en France*, Réseau Semences Paysannes, p. 4.

[52]Les maisons des semences paysannes (2014).

[53]Paul Verlaine, *Cauchemar*.

structured groups, strength in numbers so to speak (Sect. 3.1), and on the other, not forgetting that their activity involves feeding the population, producers try to become involved in all the steps of marketing their produce, in addition to their efforts to boost their own production. In sum, earning their daily bread by feeding others. This idea in the end is very simple: make every effort to get rid of intermediaries in distribution by moving even closer to the end consumer. This is known as distribution via short supply channels (Sect. 3.2).

3.1 Cooperation Among Producers to Ensure Rural Development

In the context of an open question to the Government,[54] a Senator asked the Minister what he intended to do to put a stop to the constant decrease in the number of small and medium-sized agricultural businesses. The question is particularly important since agriculture in France involves primarily small and medium-sized farms. For example, in the Midi-Pyrénées region, there are no less than 47,900 farms.[55] The minister's response was to point out the various subsidies, co-financed by the European Union, that are part of policies for rural development. These subsidies particularly aim to improve the structure of agricultural holdings,[56] to modernise them or even to diversify their activities[57] to meet several objectives of rural development policy. One of the aims is thus to preserve family-held businesses—encouraged by legislation in France—that "do not individually have the economic power needed to undertake real negotiations with companies that, for the most part, control the market in which they are active".[58] The best example of this is probably the integration contract.[59] In addition, the aim according to the Rural and Maritime Fishing code is to "provide the population with safe, healthy, diversified food of high quality in sufficient quantity, produced in economically and socially acceptable conditions".[60] Conversely, it is not possible to ask a producer to participate in food diversity if the activity is not economically viable. As a result, stakeholders in food diversity, and producers primarily, cannot hope to develop, and only barely to continue to operate, by relying solely on public subsidies. Producers, therefore,

[54]Ministerial response No 10344, *JO Sénat* 3 April 2014, p. 858.

[55]See Regional rural development programme for Midi-Pyrénées, *op. cit.*

[56]Article 2 of Council Regulation (EC) No 1257/1999 of 17 May 1999 *op. cit.*

[57]In consideration of 21, Council regulation (EC) No 1698/2005 of 20 September 2005 *op. cit.*

[58]Aubin-Brouté (2013).

[59]An integration contract is a "contract whereby an agro-food company integrates an agricultural holding into its production network so that it becomes part of its business strategy" Aubin-Brouté (2013), No 60. It is governed by Articles L. 326-1 *et seq.* of the Rural and Maritime Fishing Code.

[60]Article L. 1.-I, 1° of the Rural and Maritime Fishing Code.

with the support of public policy for rural development, have tended to join and cooperate by drawing on the food diversity that they create.

"Cooperation in agricultural activities, to reduce production or consumption costs, is a standard feature in the recent history of agriculture (...). In France (...) [and] in agriculture, the cooperative movement did not begin until the phylloxera crisis at the end of the 19th century"[61] since cooperation held very little interest in a context of subsistence farming before this time.[62] Grouping producers makes it possible not only to offer a wide range of food products that are in demand,[63] but also to "reduce production costs, whether the aim is to facilitate production by mutualised use of equipment or services or selling off of production as-is or after transformation (...) [so that cooperation] is fully integrated into agricultural policy[64] to contribute [in particular] to the viability of family-owned agricultural businesses". Cooperation, if we set aside the hypothesis of mutual aid[65] regulated by French law, requires structural grouping of producers that may take two main forms: jointly run farms and cooperative agricultural organisations.

Jointly run farms are businesses that are defined in Article L. 323-2, paragraph 1 of the Rural and Maritime Fishing code, which indicates that "Jointly run farms are considered to be fully integrated when the aim of the associates is to mutualise all their agricultural production activities corresponding to control and management of a biological cycle involving plants or animals and constituting one or more necessary steps of this cycle, including marine production activities. In the event of partial mutualisation of activities, the jointly run farm is considered partially integrated. A single jointly run farm cannot be fully integrated for certain associates and partially integrated for others". Conversely, this type of group has a cooperative approach since it aims to produce savings for its members but also productivity gains through balanced contributions by workers. The balanced contributions come from two sources: firstly, there is a legal obligation to work jointly for all associates, and secondly, the members must make a "contribution in kind" (with their labour). Although initially this type of grouping was established to create common agricultural output generated by several producers, it has had limited success. However, this

[61]Aubin-Brouté (2013), No 318.

[62]Couturier (1994).

[63]Couturier (1994), No 381.

[64]Article 20 of Council Regulation (EC) 1698/2005 of 20 September 2005 above indicates that "Support targeting the competitiveness of the agricultural and forestry sector shall concern: (...) c) measures aimed at improving the quality of agricultural production and products by: (...) iii) supporting producer groups for information and promotion activities for products under food quality schemes".

[65]Article L. 325-1 of the Rural and Maritime Fishing Code stipulates that "mutual assistance is performed between producers through exchange of services for work and operating means, including those that extend from production activities. It can be occasional, temporary or occur on a regular basis. Mutual assistance is a cost-free contract, even when the beneficiary reimburses the supplier for all or part of the costs incurred by said supplier. When it is carried out in a holding subject to rules on authorisation of marine production facilities, mutual assistance requires a written contract."

grouping has become a favoured instrument to ensure future generations remain in the activity—meaning that the reference to the familial nature of this company should be no surprise—and thus anchor the foodstuffs produced firmly in their territory. Consequently, this type of grouping clearly participates in rural development but does not necessarily draw on food diversity because it often has a familial structure. Can the same be said for cooperative companies?

Article L. 521-1, paragraph 1 of the Rural and Maritime Fishing Code stipulates that "agricultural cooperatives ensure collective use by producers of all equipment and means that facilitate or develop their economic activity, and improve or increase the output of this activity". In France, as in Europe generally, agricultural cooperatives have become very popular, as two-thirds of French farmers belong to one. This popularity, which would deserve a sonnet, can be explained by several factors. One of the main reasons is that members are the owners of their marketing and/or transformation tool which means that they can benefit from greater added value from their products and participate themselves in the development of their market prospects. Moreover, and this is probably the most important factor, they are ""undelocalisable" (...) since they are attached to specific territories that they manage by virtue of their statutes and within which they create value. Their activities and initiatives to revitalise certain rural areas (...) very often make them the main employers in the area or the region".[66] For all these reasons, it would appear that agricultural cooperatives participate in rural development. However, we still need to examine whether cooperatives draw on food diversity to do this. A mixed answer is probably closest to the truth. On the one side, if we follow the arguments of Couturier, "agricultural cooperatives favour (...) specialised uniform production, avoiding sources of reduced volumes produced by farms aiming to broaden the range of varieties or qualities; this selection of farmers leaves little room for local initiative".[67] As a result, part of the diversity produced will be set aside for reasons related to specialised and uniform production, leading to the conclusion that agricultural cooperatives do not truly draw on food diversity to ensure their development and thereby rural development as well. From another perspective, it is difficult to deny today that agricultural cooperatives are looking to reach "critical mass". To do this, they create alliances to reduce their costs, they assemble new teams of experts, and they diversify their activities by increasing the number of product sectors. As a result, we can well imagine that achieving "critical mass" leads cooperative companies to diversify their production and therefore to draw on food diversity to develop, and thus contribute to the rural development of the territories where they are rooted. Ultimately, although cooperation among producers participates inevitably in rural development, it seems difficult to claim that producers always use the food diversity they generate to reach this goal. Do we come to the same conclusion when producers and consumers aim to form closer bonds?

[66]Mangin (2012).

[67]Couturier (1994), No 508.

3.2 *Rural Development in Cooperation with Consumers*

"Irrespective of the period in history, regulations on rural law (production of plants and animals for food, energy or industrial purposes) cannot be limited to interventions targeting farming structures. Regulations extend to the production itself and also to its distribution".[68] As we have seen, rural development requires policies that aim both to enhance the value of food products and to improve production structures. At the end of the day, however, once produced the foodstuff is intended for the market. It therefore comes as no surprise that regulations, whether French or European, aim to act on the marketing of food products. An example underlining this point would be European legislation which indicates that investment aid must contribute to the streamlining of marketing channels[69] and must particularly help to enhance market opportunities.[70] But what are these marketing channels specifically?

The answer is once again clear if we look at national and EU legislation. Regulation (EU) No 1305/2013 of 17 December 2013 targets, as part of rural development programmes, short supply channels,[71] and nationally Article L. 1.-I, 9° of the Rural and Maritime Fishing Code, derived from the Law of 13 October 2014, indicates that it is important to "strengthen the local roots of production, transformation and marketing of agricultural products, including through promotion of short supply channels". Why is there so much interest in these short supply channels? Studies and programmes on rural development and food clearly explain the reasons behind this type of support from policies for rural development.

From the consumer side, the Regional rural development programme for Midi-Pyrénées indicates that short supply channels meet the expectations of consumers since they are interested in more direct links to producers.[72] This consumer approach, highlighted in various studies including the CREDOC study, and most often guided by the consumer's desire to consume local products, has repercussions on distribution channels. In France, large supermarket chains still dominate the food retail sector with slightly more than 60% of sales in 2011, but studies show that consumers are diversifying their buying patterns, with greater interest in short supply channels.[73] From the producer perspective, Couturier explains that "in their efforts to increase revenue, producers try to avoid giving up value to multiple intermediaries

[68]Krajeski (2009), No 501.

[69]Article 25 of Regulation (EC) No 1257/1999 of 17 May 1999, *op. cit.*

[70]In consideration of 28 in Regulation (EC) No 1698/2005 of 20 September 2005, *op. cit.*

[71]Article 7 of Regulation (EU) No 1305/2013 of 17 December 2013 of the European Parliament and of the Council of 17 December 2013 on support for rural development by the European Agricultural Fund for Rural Development (EAFRD) and repealing Council Regulation (EC) No 1698/2005.

[72]Regional rural development programme for Midi-Pyrénées, *op. cit.* Likewise, Report "Short supply channels and marketing" March 2009, http://agriculture.gouv.fr/IMG/pdf/rapport_du_gt_circuits_courts0409.pdf.

[73]In this sense, see French Research Centre for the Study and Observation of Living Conditions (CREDOC) (2012).

that could be added to the price of their production. To do this, they try to place their production directly in the hands of the end consumer".[74] Moreover, we must remember that there are not always industrial channels capable of exploiting the food diversity generated by producers, since some cooperatives work essentially on specialised, standardised production. This means that food diversity excluded from other distribution networks can fit very well into short supply channels and thus participate in the development of the agricultural holdings that produce it, and thereby, contribute to rural development. However, concerning these short supply channels, public policy on rural development goes beyond the interests of producers and consumers. The link between producers and consumers "results from (...) an inevitable strengthening of links to the land (...). This local approach also favours the development of local synergies between the various players involved: producers, intermediaries, artisans, retailers, local authorities, and consumers. The development of short supply channels thus goes hand in hand with the development of "food governance" at the local level",[75] which enables local anchoring of food diversity production and this production, as mentioned for example in the Regional rural development programme for Aquitaine, has "significant potential for the development of local short supply channels".[76] In short, food diversity can, through short supply channels, be a means of contributing positively to rural development. We must however still define what we understand by "short supply channels".

Short supply channels can be compared to the *enjambement* technique in poetry, whereby at the start of a line we continue a verse that had already started at the end of the preceding verse. It was in fact not that long ago that wheat was sold directly to the baker, and milk directly to neighbours. However, in French law, short supply channels are defined precisely as "any distribution network that has no or a single intermediary between the producer and the consumer".[77] In France, there is a wide range of short supply channels: sales via the internet, direct sales to restaurants, supply to local authorities through canteens, consignment systems, catering groups, fairs and farm sales, small-scale farmer's markets, collective sales points, and associations for the preservation of smallholdings (AMAP). These associations are worth mentioning because they are original and have become very popular.

The AMAP Charter defines an association for the preservation of smallholdings as "an association that aims to preserve the existence and continuity of local farms with a focus on sustainable agriculture, i.e. small-scale farming that is socially fair and ecologically sound, and that lets consumers buy quality food products of their choice at a fair price and to be informed of the origin of the products as well as the way they were produced. This should also enable consumers to participate actively in maintaining and developing local agricultural activity meeting the requirements of

---

[74]Couturier (1994), No 79.

[75]Report "Short supply channels and marketing", March 2009.

[76]Regional rural development programme for Aquitaine, p. 3.

[77]See Law No **2010-874 of 27 July 2010 on modernisation of agriculture and fishing.**

sustainable development".[78] AMAPs are an interface between consumers and producers and the producers must all follow the main principles set out in the AMAP Charter: small-scale production, solidarity with local players, support of producers towards self-sufficiency, proximity between producer and consumer, absence of intermediaries, fair price setting, and solidarity of the consumer with the producer in the event of a contingency. The relationship between the associated consumers and the producers is governed by a contract. The contract has several features that should be pointed out. First, the contract has a term that is related to the production cycle. Second, the contract sets a list of diversified products that the producers commit to providing, making it possible for them to extend the seasons and propose a varied range of food but also limit climatic and safety risks. Third, the consumers set the cost of the food products with the producer and this cost considers all the overheads of the holding as well as the revenue that must be generated annually to ensure the producer's self-sufficiency. This cost is paid up front to the producers so that they have working capital and sufficient disposable funds to invest and pay certain expenses. Fourth, this contract is above all non-binding in that if the producer cannot supply the associated consumers because of a contingency, the consumers carry this risk, enabling the producer to always have the same level of revenue. AMAPs thus participate in ensuring rural development, with relocalisation of production and maintenance or creation of employment, by drawing widely on the food diversity of a given territory. AMAPs have today become victims of their own success because it is common to wait more than a year before being able to join the association, since supply is far below the level of demand.

In conclusion, all these developments form a type of *mise en abyme,* with rural development and food diversity in France as endless reflections of one another. Rural development feeds food diversity, much as food diversity feeds rural development, because behind each expression, there is diversity among the men and women involved, whether producers, distributors, or consumers, as well as diversity among the territories where these stakeholders are located. The fruits of this labour, both good and bad, serve literally to feed some, and figuratively to feed others. The poet has the same relationship with his poem, if he uses "broken verses", we should still remember that there is hidden meaning, that remains to be revealed. And we must not forget this about the poet's world: "Because we want the nuance again, Not the colour, nothing but the nuance!/Oh, the nuance alone betroths/Dream to dream and flute to horn!"[79]

[78]AMAP Charter, p. 3, http://www.reseau-amap.org/docs/chartedesamap.PDF.

[79]Paul Verlaine, *Art Poétique*.

References

AMAP Charter. (associations for the preservation of smallholdings), http://www.reseau-amap.org/docs/chartedesamap.PDF

Aubin-Brouté RJ (2013) Le contrat en agriculture, Contribution à l'étude du contrat comme instrument de l'action publique. LGDJ, Paris

Couturier I (1994) La diversification en agriculture. L'Harmattan, Paris

Regional rural development programme for Midi-Pyrénées, http://www.midipyrenees.fr/IMG/pdf/pdrr_complet.pdf

Krajeski D (2009) Droit rural. Defresnois, Paris

Les maisons des semences paysannes (2014) Regards sur la gestion collective de la biodiversité cultivée en France. Paysannes, Réseau Semences

Mangin P (2012) L'entreprise coopérative: un modèle anti-crise. Le Monde, Paris

Morize J (1992) Manuel pratique de vulgarisation agricole. Maisonneuve et Larose, Paris

Report "Short supply channels and marketing" (2009) http://agriculture.gouv.fr/IMG/pdf/rapport_du_gt_circuits_courts0409.pdf

Research Centre for the Study and Observation of Living Conditions (CREDOC) (2012) Research report, Market survey 2012, Behaviours and attitudes of consumers concerning food retail

Vialle P (2011) Semences et agriculture durable. Report for the Ministry of Agriculture, Food and Fisheries, Rural life and Local Affairs

Regional Quality Labels for Agro-Food Products

Clelia Losavio

1 Protection of Food Quality and Multilevel Governance

The European legislator is heavily present in food law with many acts that, as is well-known, prevail over the law of Member States. Often, among other things, these interventions take the form of regulations that, being directly applicable, do not envisage transposition acts by the Member States. Consequently, apart from the competence to legislate on the food sector—arising from the division of competences between European Union and Member States—the actual regulatory capacity of State depends on the areas not covered by the European legislator.[1] Therefore, also in matters related to agriculture and food[2] as for the other matters coming under concurrent competence[3]—for which, by virtue of the subsidiarity principle, there is an equal theoretical possibility of intervention by the European Union and Member States—one must look at the size of Community intervention in practice to understand exactly how much room is left for the national legislator. While primary health protection needs are the reason for choosing to ensure a high level of uniformity in

[1]Mangiameli (2014).

[2]The Treaty does not mention the subject "Foodstuffs". However, whether it is included in the subject *Agriculture*—Art. 43 TEU (ex 37 TEC) on the Common Agricultural Policy is indicated by the EU legislator as the first juridical base in Regulation 1151/2012 on the quality of agricultural and food products and in Regulation 178/2002 on food safety—or in the *Health Protection*—Art. 168 TEU (ex 152 TEC) is the juridical base of Regulation 852/2004 on the Hygiene of foodstuffs and, in addition to the Art. 43, of the abovementioned Regulation 178—it would be a subject for concurrent powers. On the powers in food matters refer to Germanò et al. (2014), pp. 13–16.

[3]Germanò et al. (2014).

C. Losavio (✉)
Institute for the Study of Regionalism, Federalism and Self-Government "Massimo Severo Giannini", National Research Council, Rome, Italy
e-mail: clelia.losavio@cnr.it

A. Isoni et al. (eds.), *Food Diversity Between Rights, Duties and Autonomies*, LITES – Legal Issues in Transdisciplinary Environmental Studies 2,
https://doi.org/10.1007/978-3-319-75196-2_9

food safety,[4] hence reduced scope of action for the State and Regional levels,[5] the same cannot be said for the task of protecting and promoting food quality.[6]

Indeed, in this connection, in pursuing the rural development agricultural policy,[7] the European legislator actually encourages Member States to create quality schemes that are alternative and complementary with those of the EU[8]: geographical indication schemes (Protected Designations of Origin—PDOs and Protected Geographical Indications—PGIs) and the Traditional Specialities Guaranteed (TSGs) scheme, governed by Regulation 1151/2012[9] and, albeit in a different way, the organic production system.[10]

With the expression *quality schemes* or *quality system* reference is made to the possibility for public bodies as well to set up optional certification schemes that, by promoting the "added value" of a product, ensure the presence of given characteristics or attributes of a category of products or compliance with specific methods of production or processing.[11]

In reality, also Article 28 of mentioned Regulation 1151 on food quality, by referring to "optional quality terms" (that are signs whereby consumers are informed that a given product belongs to a given quality scheme), specifies that European law does not prevent national indications from being maintained, provided that they are governed by the regulation and, as is obvious, are consistent with European law.

As we read in the Commission Communication "EU best practice guidelines for voluntary certification schemes for agricultural products and foodstuffs", in the case in which the certification systems are supported by public bodies (such as regional or national authorities, for instance), conformity with EU law is ensured not only, as for other schemes, through consistency with the consumer information and labelling requirements and with competition rules, but also by making sure that the actions taken "may not lead to restrictions based on the national origin of producers or

[4]According to Germanò et al. (2014, p. 16) "For the safety and health related aspects, food is a subject that the Union took over when on 28 January 2002 it issued Regulation 178/2002 on Food safety".

[5]Germanò et al. (2014) and Losavio (2007a).

[6]The quality being referred to is the "objective", observable and measurable quality, and hence a quality that can also be certified and controlled. Cfr. Zanon (1997), p. 496; Albisinni (2000), pp. 127–128; Germanò (2010a).

[7]Regulation (EU) 1305/2013 of the European Parliament and the of the Council of 17 December 2013 on support for rural development by the European Agricultural Fund for Rural Development (EAFRD) and repealing Council Regulation (EC) 1698/2005.

[8]See Recital 14 of Regulation (EU) 1305/2013 and Art. 16 ("Quality schemes for agricultural products, and foodstuffs") that envisages support for farmers who participate also in quality schemes recognised by Member States (par. 1, lett. b).

[9]Regulation (EU) 1151/2012 of the European Parliament and the of the Council of 21 November 2012 on quality schemes for agricultural products and foodstuffs.

[10]Council Regulation (EC) 834/2007 of 28 June 2007 on organic production and labelling of organic products and repealing Regulation (EEC) 2092/91.

[11]Recital 44, Regulation 1151/2012.

otherwise impede the single market".[12] "Any support for certification schemes granted by a Member State or through State resources (...) must comply with State aid rules".[13]

In practice, such schemes, even if promoted by a Member State, should be open under transparent and non-discriminatory criteria to all operators present in the EU market who are willing and able to comply with the specifications.

Therefore, these are the boundaries within which national and regional authorities can act to adopt instruments that may enhance and promote agricultural products and foodstuffs. And the issue of enhancing the territorial specificities of agricultural products and foodstuffs through national or regional public trademarks must come to terms with these very limits.

2 Public Geographic Trademarks and Compatibility with EU and Competition Law

Geographic public trademarks, created to enhance and promote national or regional products by their respective local public bodies that are the owners of the trademark, were institutions that are commonly found in the tradition of the various EU Member States. In Italy, for instance, the "national trademark for the export of fruit and vegetables" was established as early as 1927. The owner of the trademark was the National Institute for Foreign Trade,[14] while in 1966 the Sicily Region gave life to the "Quality assurance and promotion of Sicilian products trademark",[15] and the Apulia Region, in 1988 gave life to an "origin and quality trademark for Apulian agricultural and animal products".[16] And also the Veneto Region in 1988[17] created the "Paniere Veneto" trademark.[18]

In the early nineties however, in 1992 to be precise, which was also the year when the first regulation on DOPs and IGPs[19] was approved—the Commission launched a large survey in Member States aimed at "drawing up inventories of trademarks and other national quality assurance denominations in the area of agricultural products

[12]OJ C 341, 16 December 2010, p. 5.

[13]Article 107 of the TFEU.

[14]Established with Act 1272/1927.

[15]Sicily, Regional Law 14/1966.

[16]Apulia, Regional Law 8/1988 "Rules on the use of the trademark of origin and quality for actions promoting provate consortia for the economic enhancement of Apulian agricultural and animal products".

[17]Veneto, Regional law 11/1988 "Initiatives for the enhancement of agricultural products and foodstuffs of Veneto".

[18]See, among others, Masini (1999), p. 657; Losavio (2007b), p. 174; Trapè (2012), p. 78.

[19]Council Regulation (EEC) 2081/92 of 14 July 1992 on the protection of geographical indications and designations of origin for agricultural products and foodstuffs.

and foodstuffs set up and managed by public bodies".[20] This overview was extended, in subsequent years, also to the States that joined the Union. After examining the quality trademarks that emerged from the survey to ensure compatibility with Community regulations, the Commission started an infringement procedure for 120 of them for violation of the European competition rules: such national measures were considered to be "measures whose effect was equivalent to the quantitative restrictions on imports" and, particularly, as measures "that favour national products, or that attach preference to them other than aid" (Directive 70/50), prohibited by Article 34 of the Treaty on the Functioning of the European Union (TFEU),[21] at the time Article 28 of the ECT. To this we must add that, with approval of Regulation 2081/1992, the DOPs and the IGPs are considered to be the only trademarks having the nature of "public law" suited to enhancing the geographic origin of an agrifood product at the European level.

The infringement procedure turned into an appeal to the Court of Justice only for a few of them since, in most cases, the States promptly amended their domestic provisions.[22] Among the trademarks brought before the Court of Justice, mention can be made of the German countryside quality assurance trademark ("CMA") introduced to promote the marketing of national agricultural products and foodstuffs, restricted to products produced in Germany with German or imported raw materials, was deemed by the Court to have "at least potentially restrictive effects on the free movement of goods among Member States", since it "could induce consumers to purchase the products bearing the CMA trademark, to the detriment of imported products".[23] The same happened with ten French regional quality trademarks, among which, for instance, "*Label régional Normandie*" and the "*Label régional corse*", applied only to the products of the respective Regions of France, favoured, according to the Court, "the marketing of goods of national origin to the detriment of imported goods" according to the Court, and ultimately created and maintained "a disparity in treatment between these two categories" of products.[24] And with the same motivations the Belgian "*label de qualité wallon*" trademark, exclusively applied to the products produced or processed in Wallonia, was deemed to be incompatible with Community law.[25]

[20]This is stated in point 6 of the conclusions of General Attorney Mischo of 5 December 2002, case 6/02, *Commission vs. French Republic*.

[21]Art. 34: "Quantitative restrictions on imports and all measures having equivalent effect shall be prohibited between Member States".

[22]Losavio (2007b), pp. 173 et seqq.

[23]CJEC, Case C-325/00 *Commission of the European Communities v Federal Republic of Germany* [2002] *ECR* I-09977.

[24]CJEC, Case C-6/02 *Commission of the European Communities v French Republic* [2003] ECR I-02389.

[25]CJEC, Case C-255/03 *Commission of the European Communities v Kingdom of Belgium* [2004].

Among other things, in a judgment issued a few years earlier,[26] the Court had recalled some remarks already made in the well-known Eggers judgment of 1978.[27] In this judgment, the Court stated that "the Member States are empowered to lay down quality standards for products marketed on their territory and may make the use of designations of quality subject to compliance with such standard, but only *on the condition that such standards and designations*—unlike the position in the case of registered designations of origin and indication of origin—*are not linked to a requirement that the production process for the products in question be carried on within the Country,* but are dependent solely on the existence of the intrinsic objective characteristics which give the products the quality required by law". Conversely, in these schemes for enhancing agricultural products, what should be emphasised are the *qualitative characteristics* of the product, irrespective of their geographic origin.[28]

This orientation has been confirmed and further spelled out in many documents of the Commission. As an example, suffice it to read a note by the European Commissioner for Agriculture of February 1998 in which, in urging Member States to adapt their legislation on national or regional trademarks on quality and origin of products to Community law, the Commissioner laid down precise rules: "only the inherent characteristics of a product may constitute a requirement for the granting of trademarks of quality and origin; the qualitative characteristics required to grant a trademark cannot merely be the compulsory requirements imposed by Community and National food legislation, but should represent a *quid pluris* that sets them apart from other products of the same category; the origin of the product cannot be a condition for obtaining a trademark; any Community entrepreneur shall have access to a trademark without territorial limitations".[29]

Coming back to the present day, the 2014 to 2020 European Union *Guidelines* for State aid in the agricultural and forestry sectors and in rural areas[30] state that the quality schemes of agricultural products recognised by Member States are compatible with the competition rules if they are, among other things,[31] "open to all

---

[26]The *Pistre* judgment of May 1997 on the use of the word 'mountain' (CJEC, Joined Cases C-321/94, C-322/94, C-323/94, C-324/94 [1997] *ECR* I-02343). The French legislation governing the use of the word "mountain" for products coming from the mountain areas across the national territory was declared to be incompatible with Community law and particularly with Article 28 of the EC Treaty.

[27]Judgment of the Court of 12 October 1978—Joh. Eggers Sohn & Co. v Freie Hansestadt Bremen, Case 13/78.

[28]Germanò (2010b), pp. 267–270; Dani (2000), p. 1057; Gencarelli (2005), p. 75; Carmignani (2005), p. 135; Losavio (2007b), p. 172.

[29]Document SG(98) D/1618 of 24.02.1998.

[30]OJ C 204 1 July 2014, p. 1.

[31]... consistent with the following criteria: i) the specificity of the end product protected by such schemes shall derive from strict obligations that ensure: specific characteristics of the product, or specific production methods, or a quality of the product that is significantly above the current commercial standards in terms of public health, animal health and environmental protection; (ii) the

producers".[32] As to the indication of the origin of products, in referring to promotion campaigns and activities, the Communication states that "where the activity concerns products covered by quality schemes other than schemes for Union-recognised denominations, the origin of the products may be mentioned provided it is *secondary* in the message. To determine whether the reference to origin is secondary, the Commission will consider the overall amount of text and/or the size of the symbol including images, as well as the general presentation referring to the origin as compared with the text and/or symbol referring to the key sales pitch, meaning, the part of the promotion not focused on the origin of the product".[33]

3 Public Regional Trademarks of Quality and Origin: The Italian Case

Since regional bodies had to give up the role of enhancers and promoters of the products from their territories—an aspect that is "inherent" in these institutions that are the expression of the territories they represent—because of the potential restrictive effects on the free circulation of goods of national measures like those establishing public geographic trademarks, the regional bodies "reinvented" themselves as certifiers of the objective characteristics or properties of the products, giving rise to quality schemes that at least in theory were open to all entrepreneurs operating in the European agrofood market.

And so, in Italy for instance, the various Regions established "collective quality marks" that focus on the qualitative value of their agrofood products by certifying the presence in them of given "objective" qualitative characteristics or of specific production modalities. In most cases the production systems are more environment friendly, as, for instance, integrated production.[34]

In the Italian case two different types of collective regional quality marks can be identified. With the trademarks that we might call "first generation" trademarks, the Regions substantially "renounced" indicating the geographic origin of the product, and they kept the reference to the territory only in the name of the Region, that is present, but exclusively as the body issuing the trademark or certifying that the

quality scheme shall be accessible to all producers; (iii) the quality scheme shall comprise binding production rules, whose compliance shall be checked by public authorities or by an independent control body; (iv) the quality scheme shall be transparent and ensure full traceability of the agricultural products.

[32]Section 1.1.9. "Aid for the participation of producers of agricultural products in quality schemes". The Communication refers, for certification schemes, to the already mentioned EU best practice guidelines for voluntary certification schemes.

[33]1.3.2 Aid for promotion measures in favour of agricultural products, p. 66.

[34]Integrated production, aimed at reducing to a minimum the use of chemical substances (products for crops and fertilizers) but also reduced consumption of water and energy. As in the case of the trademarks of Emilia-Romagna, Tuscany, Veneto.

product is compliant with the rules. This is the case, for example, of the trademark of the Emilia-Romagna Region, established[35] in accordance with Regional Law 28/1999,[36] after extensive exchanges with Brussels. The main indication on this trademark indicates "Controlled quality"—written in large bold letters; the secondary indication is: "Environmentally friendly and health sensitive integrated production", and provides information about the subject of the certification and finally, in smaller print compared to the other two indications, a regulatory reference with the name of the Region "Regional Act no 28/1999 of Emilia Romagna".

As already stated, this trademark was approved by the European Commission because it was "not aimed at qualifying only the regional production, but that of any Community entrepreneur whose products are consistent with those particular quality criteria". To be clear, any EU product, regardless of where it is produced, can claim the Emilia-Romagna quality trademark provided that it complies with the relevant production conditions.

Based on similar remarks, the Commission deemed the "Agriqualità" trademark established by the Tuscan Region with Regional Law 12/1999,[37] to be compatible with European law whose secondary indication states: "Produced from integrated agriculture. Trademark issued by the Region of Tuscany". Also, in this case, therefore, the sole geographic reference is the name of the Region, that however appears only as name of the body that is the owner of the trademark.[38] Soon thereafter, with the same criteria, also the "Verified Quality (VQ)" of the Veneto Region[39] and "QM – Guaranteed Quality from Marche"[40] were also registered.

The "second generation" regional trademarks are different by nature and function from the trademarks described thus far. Besides being registered as "collective Community trademarks",[41] they reintroduce the indication of origin that appears in the trademark as "secondary indication" compared to the main message that remain the quality. The requirement of granting access to all European Union producers is complied with envisaging the possibility of "replacing" the indication of origin, in the trademark and in the logo, depending on the real place of production. So, for

---

[35]Cfr. Emilia-Romagna, Deliberation of the Regional Government 1436/2001.

[36]"Enhancement of agricultural products and foodstuffs obtained using techniques that respect the environment and consumers' health".

[37]"Rules for the enhancement of the agricultural products and foodstuffs obtained using integrated production techniques and protection against misleading advertising".

[38]See "Regulation on the Collective trademark Agriquality, Product obtained from integrated agriculture" in accordance with Art. 3, Regional Law 25/1999.

[39]Veneto, Regional Law 12/2001 "Protection and enhancement of quality agricultural products and quality agrifood products".

[40]Marche, Regional Law 23/2003 "Interventions in support of schemes certifying the quality and traceability of agricultural and agrifood products". Cfr., also, Deliberations of the Regional Government 257/2006 and 536/2006 approving the final texts of the rules on the use of the "QM – Quality guaranteed by the Marche Region" trademark and of the guidelines for drafting the rules on branded products and services.

[41]Council Regulation (EC) 207/2009 of 26 February 2009 on the Community trade mark.

example, in the case of quality trademark indicating with indication of origin "Quality Products of Apulia"[42]—it too the result of consultation with the European Commission[43]—the elements of origin indicated in the lower part of the quality trademark ("Apulia") can be in turn replaced by the true area of origin. In practice, the descriptive formula will state "Quality Products *Apulia*" if they are Apulian products, "Quality Products Tuscany" if they are from Tuscany, or, again, "Quality Products *Navarra*" or "Quality Products *Baden-Württemberg*" if the products come respectively from the Spanish or German Regions.[44] The trademark "Alto Adige Quality"[45] works in the same way where "the elements of origin of the quality assurance trademark are replaced depending on the Region of origin" (Article 3, LP no 12/2005).

The "Trentino Quality" trademark (DGP no 986/2011) and the "Quality Controlled by the Abruzzi Region" trademark (RL 6/2012 and DGR 902/2013) appear to be somewhat different.[46] Indeed, both these trademarks, though being quality marks with a secondary function in providing the indication of origin, do not envisage any other Region of origin being indicated in the trademark and the logo. While in the case of the Trentino trademark, there could be doubts as to whether it was intentional or not, the same cannot be said for the Abruzzi trademark. Indeed, in this case it is precisely the name of the trademark "Quality Controlled *by the* Abruzzi Region" that indicates that, just as in the case of the so-called "first-generation" trademarks, the reference to the Region of origin disappears in favour of the Regional body issuing the trademark. One might wonder, then, how is this secondary function of the indication of origin expressed and above all how is it guaranteed to a producer who is not based in Abruzzi.

4 Doubts on the Distinctive Function of Regional Trademarks of Quality and Origin

As said earlier, these regional trademarks are to be deemed compatible with European Union law because they have the approval of the Commission. In most cases, among other things, the "final" version of these trademarks is the outcome of the indications provided by the Commission in its notification and of the invitation

[42]Deliberation of the Regional Government 1100/2012 "Approval of the logo and of the regulation on the use of the Apulia Quality Products trademark".

[43]In 2011, the Commission invited the regional authorities to adopt amendments to ensure compliance of the use of the trade mark with Article 34 TFEU.

[44]See Article 2, Deliberation of the Regional Government 1100/2012.

[45]Bolzano, Provincial law 12/2005 "Measures for ensuring quality of foodstuffs and adoption of the 'quality trademark with indication of origin'".

[46]"Interventions in support of quality and traceability of the agricultural products and foodstuffs of the Abruzzi Region".

addressed to the regional authorities "to emphasize the prevalence of the quality assurance indication over the reference to the origin of the product".

The fact that they are "formally" compatible, however, does not eliminate doubts about their validity or appropriateness as distinctive signs in the marketplace. Conversely, it appears to us—as likewise pointed out in the literature—that these quality trademarks run the risk of creating confusion rather than clarity in the marketplace, in that they present information that is not easy to interpret by consumers and in some cases, it is even misleading.[47]

With respect to the first type of regional trademarks—namely those where the indication of origin appears only in the reference to the certifying regional body—we cannot omit to mention that the presence in the trademark of the name of the Region presents to the consumer a reference to the territory in a manner that is not "openly manifest but in any case suggestive", a reference to a territory that runs the risk of "overshadowing the true function of the trademark which is that of ensuring that the product is the outcome of a given production process", but instead, it attracts the consumer's attention to a given geographic origin of the product that may not be true. Indeed, as pointed out, the Tuscan Region could certify products coming from any territory of the European Union that comply with the relevant production conditions.[48]

But also the second type of regional trademarks—the quality trademarks with a secondary indication of origin—albeit perhaps not misleading regarding the indication of the true origin of the product, that as said earlier, may change depending on where the company applying for the trademark is based, is not really compatible precisely with the characteristics of the legal instrument used, in that it "misrepresents the potential and the functions" of a distinctive trademark—the geographic collective trademark—that is acceptable only if attached to its being used by the producers in a given geographic area.[49] Among other things the use of a "collective trademark with variable *content*, depending on the origin of the products" runs the risk of being in contrast with the underlying goal pursued by the Union, namely fair competition in the market.[50] To this day, indeed, each Region can, at least in theory, issue a quality certification to producers from different geographic areas provided that the trademark is adapted, from case to case, to indicate the right origin. A producer in Apulia, for instance, could obtain the "Apulia Quality Products" trademark from the Apulia Region and at the same time also the "Apulia Quality" trademark from the Region of Bolzano, which as envisaged in the implementation rules, is obtained by modifying the "Alto Adige Quality" trademark to reflect the correct origin of the product. To this we may add that the two trademarks, that are so similar as to be likely confused, could instead certify the presence in the product

[47]Di Lauro (2005), pp. 163 ss; Carmignani (2005), p. 151.

[48]Di Lauro (2005), p. 163; Carmignani (2005), p. 151.

[49]Canfora (2013), pp. 160–161.

[50]Canfora (2013), p. 163.

of characteristics that are qualitatively different, since this depends on the regional production rules whose compliance is required to have access to the trademark.

5 Regional Instruments for Enhancing the Value of Local Products

One may wonder whether there might be more convincing solutions, whether there are other avenues that the public body wishing to enhance the geographic origin of its products may take. Input for reflection on the issue is offered by two different European documents.

The first is a Resolution of the European Parliament of January 2014 on "regional branding: towards best practice in rural economies".[51] In this document, to tell the truth, is not particularly clear in its formulation and contents, the EU Parliament appears to have the intention of highlighting the yet unexpressed potential of "territorial regional branding". In pursuing the goal of promoting a territory in all its aspects the trademark should distinguish in the marketplace not only the agricultural products but all the resources of a given territory—tourist, cultural, craftsmanship, entrepreneurial—that contribute to shaping its identity. For this purpose, the Member States and the regional administrations are invited to adopt participatory governance processes to enable the development of territorial projects that enhance the potential of existing resources. The territorial quality trademark applied transversally to all the products, goods and services of a given territory, would be the instrument for promoting this enhancement of the quality of the territory in the marketplace. With specific emphasis on agricultural products, the regional trademark should promote the products that are known at the European level to have a strong link with the territory, and hence the DOP, IGP and STG products.

Now, undoubtedly, the invitation by the European Parliament has its *raison d'etre* because it embodies a territorial and participatory development logic that is designed, on the one hand, to involve the local communities, and on the other, to enhance the specific resources of a territory as the strengths and drivers of development. This is definitely in line, amongst other things, not only with the rural development policy but also with the new cohesion policy that in the 2014–2020 Programme shifts its focus away from being an "economic and social" programme to being an "economic, social and territorial" programme in the sense of being "place-based". There are, however, many questions that need to be discussed such as the added value that such a trademark would attach to agricultural products and foodstuffs since, as already pointed out, it would come to add to a geographic indication (DOP, IGP) that inherently "communicates the territory of origin". On the other

[51] P7_TA-PROV(2014)0017. The Resolution follows the report of the Committee on Agriculture and Rural Development and the opinions of the Committee on Transport and Tourism and the Committee on Regional Development (A7-0456/2013).

hand, if this were not so, it is not clear why, being of public ownership, it should not come up against the same problems of compatibility with the European competition law mentioned above.

The positions expressed by the European Commission on the importance of *local agriculture* would appear to be more interesting, as well as the debate that is growing among the European institutions on the possibility of establishing a relevant labelling scheme. In the 2011 Green Paper on promotion measures and information provision for agricultural products,[52] indeed, the Commission pointed out that "regional and local farming has hidden potential that is not currently being fully exploited" and that regional and local markets are "an essential meeting place for producers and consumers" because "they enable the former to receive the rewards for their labours more efficiently and the latter to contribute to the development of their local areas, reduce the environmental impact of their consumption habits and access a wide variety of products rooted in their traditions and ways of life".[53] And moreover the Commission points out how these circuits "can help improve environmental performance, in terms, for example, of limiting CO2 emissions or packaging", and also from a cultural and social point of view since the local supply chains "encourage collaborative decision-making and local governance that is more sensitive to the specific needs of the areas concerned and preserve and support local traditions, while at the same time linking the product to a geographical area shared by producers and consumers".

"In a territorial approach"—the Commission concludes—"the involvement of producers in *quality systems* and the promotion of these products plays a key role in meeting growing expectations in relation to local production, tradition and authenticity, reaffirming the social link between consumers and producers and capitalising on the freshness, innovation and nutritional qualities of produce and product awareness".

As to the case for establishing a labelling scheme for local farming and direct sales, in its December 2014 Report submitted to Parliament and the Council,[54] the Commission, albeit without a final response on this issue,[55] seems to set aside the idea of establishing an EU certification and labelling scheme in favour of a simpler and more streamlined system—perhaps in the form of an optional quality

[52]"Promotion measures and information provision for agricultural products: a reinforced value-added European strategy for promoting the tastes of Europe" of 14 July 2011—COM(2011) 436 def.

[53]"Short supply chains are a means to improve the income of producers, ensure survival of a large number of companies thanks above all to higher margins, decrease transportation costs and ensure greater autonomy in the agro-industrial sector".

[54]COM(2013) 866, final Report of 6 December 2013, *on the case for a local farming and direct sales labelling scheme.* The *Report* follows up on Art. 55 of Regulation 1151/2012.

[55]Instead, it paved the way for a deeper debate. Already in the Annex to the Report, indeed, the Commission identified "*Proposed issues to be addressed in the framework of discussions on the Report from the Commission to the European Parliament and the Council on the case for a local farming and direct sales labelling scheme*".

indication—but it does not appear to oppose national or regional initiatives supporting the "local" origin of products and, hence, still in principle, not even any distinctive trademarks that communicate such specific conditions. What appears to be possible is the admissibility of a form of labelling certifying the origin from a given restricted geographic area ["local" to be precise], over and above envisaging specifically given quality characteristics associated with the product.[56]

And this opening by the European institutions may perhaps allow to rethink regional legislation on quality trademarks to find more effective solutions. In doing this, however, we need to bear in mind that the "local" quality of a product concerns not the characteristics of the product per se, but the way in which the product is marketed.[57] Hence, a regional trademark or any other form of labelling issued by a regional public authority on local agricultural produce should convey in clear language the message embodied in this expression, namely the closeness between production and consumption. And this characteristic—"being local"—is not suited to distinguishing the product in the marketplace, namely it is what it is because of the place where it is marketed. Conversely, a product is a local product if the area where it is produced coincides with the area where it is marketed, but this characteristic becomes evident only at the time of trading. And so a trademark that tells the consumer about the geographic proximity between place of production and place of purchase cannot take on the form of a collective trademark; rather some other distinctive sign that does not circulate with the product needs to be devised. Some Italian Regions have started to move in this very direction. The Veneto Region, for instance, in its Regional Law 7/2008 ("Rules for guidance and support in the consumption of zero kilometre agricultural products"), as amended in 2010,[58] envisages the issuing of an ad hoc logo to catering, hospitality and retail companies operating in the regional territory that during the year can prove that 30% of their agricultural purchases are 'zero kilometre' produce.[59] This sign can be placed outside their premises and can be used in their promotional activity. A similar provision was adopted also by Sicily in its Regional Law25/2011.[60] More recently

[56]Canfora (2013), p. 151.

[57]Ultimately, the multiplying effects on the local communities, in terms of environmental friendliness and socio-economic benefits, are the elements that attribute to this category of products "that extra something" (cfr. Report by the Commission to Parliament and to the Council *Council on the case for a local farming and direct sales labelling scheme,* cit., p. 5). It is obvious that other characteristics linked to the special properties of a given product may be added (namely that the local products are also DOPs, IGPs or biological etc.), but these aspects should remain distinct.

[58]Veneto, Regional Law 3/2010.

[59]A similar rule was introduced also by Basilicata in Regional Law 12/2012 ("Rules for supporting the consumption of regional zero kilometre agricultural products"), which was declared to be unconstitutional by the Constitutional Court, Judgment 209 of 3–18 July 2013, on which cfr. Masini (2013), pp. 670–672.

[60]"Interventions for the development of agriculture and fishing. Rule on crafts, cooperation and trade. Amendments to the budget".

also Lombardy with its Regional Law14/2013,[61] promoted "the establishment of an ad hoc sign, that does not have the value of a quality trademark, to identify the companies that indicate the restaurants and hotels that operate in the Region ensuring that most of the agricultural products they use are seasonal produce and comply with the principle of coming from nearby farms".[62]

Now, in light of the above, and considering the European competition rules on the one hand, and the poor "transparency" of regional trademarks with the secondary function of origin, on the other, it seems that the enhancement of local produce and the short supply chain may be a good instrument that the Regions can use to enhance regional products as well.

Even though the expressions "local products" and "regional products" do not perfectly coincide, they do identify sectors that are to a large extent comparable.[63] Over and above the issue of the trademark, therefore, the regional institutions may "promote" local products in different ways by creating ad hoc places for direct sales, by facilitating supply in public procurement for collective catering, etc.[64] In conclusion it might be said that the enhancement of regional products may occur through instruments other than collective trademarks of origin. This is valid at least for the national market.

Of course, the same remarks cannot apply to the international market where a trademark identifying the products from a given territory—and the symbolic value attached to the latter—plays a much more important function, since it can be a driver that has a strong commercial impact. But this function is played, at least to some extent, by the indications of origin (PDOs and PGIs) that go well beyond evoking a territory, because they certify the presence in the product of characteristics that are an inherent expression of the territory.[65]

---

[61]The law supplements the Consolidated Act of regional laws on agriculture, forestry, hunting, fishing and rural development (Lombardy, Regional Law 31/2008) with rules on rural markets and the promotion of local products.

[62]It is further recalled that at the national level the bill "Rules for enhancing agricultural products and foodstuffs from high quality "zero kilometre" short supply chains", envisages, amongst other things, the establishment of the "zero kilometre" supply chain trademark "that identifies agricultural products and foodstuffs with proven environmental sustainability and reduced amounts of pollutants in their production and distribution processes". This statement "can be included .... in the sales receipt in markets and commercial facilities that sell such products" by the companies registered in an ad hoc list.

[63]On local products refer to Sirsi (2014), pp. 499–505.

[64]Refer to other Italian regional laws on the enhancement and promotion of local products and "zero kilometre" produce, cfr. Losavio (2011), pp. 93–107.

[65]On this issue see Germanò et al. (2014), pp. 149 et seqq.; Trapè (2012), pp. 167 et seqq.

References

Albisinni F (2000) Azienda multifunzionale, mercato, territorio. Giuffrè, Milano

Canfora I (2013) I marchi regionali di qualità e la correttezza dell'informazione dei consumatori: libera circolazione delle merci vs. tutela dell'agricoltura locale? Riv dir agr 1:149–165

Carmignani S (2005) La tutela del consumatore tra comunicazione e informazione. In: Germanò A, Rook Basile E (eds) Il diritto alimentare tra comunicazione e sicurezza dei prodotti. Giappichelli, Torino, pp 135 et seqq.

Dani M (2000) Gli strumenti di politica e economica regionale e i loro limiti comunitari: il caso dei marchi collettivi territoriali. Le Reg 6:1057–1086

Di Lauro A (2005) Comunicazione pubblicitaria e informazione nel settore agro-alimentare. Giuffrè, Milano

Gencarelli F (2005) I segni distintivi di qualità nel settore agroalimentare e le esigenze del diritto comunitario (Commento ai Reg. 2081/92/CE e 2082/92/CE). Dir Un Eur 1:75–97

Germanò A (2010a) Le politiche europee della qualità alimentare. In: Adornato F, Albisinni F, Germanò A (eds) Agricoltura e alimentazione. Principi e regole della qualità. Disciplina internazionale, comunitaria, nazionale. Giuffrè, Milano, pp 189 et seqq.

Germanò A (2010b) Manuale di diritto agrario. Giappichelli, Torino

Germanò A, Ragionieri MP, Rook Basile E (2014) Diritto agroalimentare. Le regole del mercato degli alimenti e dell'informazione alimentare. Giappichelli, Torino

Losavio C (2007a) Alimenti non sicuri e tutela della salute: il potere residuale degli Stati membri tra libera circolazione delle merci (artt. 28 e 30)e armonizzazione (art. 95). Dir giur agr alim amb 6:369–379

Losavio C (2007b) Il consumatore di alimenti nell'Unione europea e il suo diritto ad essere informato. Giuffrè, Milano

Losavio C (2011) I prodotti agricoli a "chilometri zero" nelle leggi regionali. Agr Ist Merc 3:93–107

Mangiameli S (2014) Concurrent powers of the European Union and their impact on the powers of Member States. Paper presented at the IACFS Conference 2014, Cape Town, 5–8 November 2014

Masini S (1999) Funzioni e limiti del marchio regionale a connotazione geografica. Prime osservazioni alla l.r. Toscana n. 12/1999. Dir giur agr amb 12:657–661

Masini S (2013) Il sacrificio del chilometro zero sul terreno del libero scambio e il ruolo debole della Corte costituzionale. Dir giur agr, alim amb 11:669–672

Sirsi E (2014) Regole e implicazioni giuridiche della produzione e del consumo di cibo locale. In: Studi in onore di Luigi Costato. Volume Terzo: I multiformi profili del pensiero giuridico. Jovene, Napoli, pp 499–518

Trapè I (2012) I segni del territorio. Profili giuridici delle indicazioni di origine dei prodotti agroalimentari tra competitività, interessi dei consumatori e sviluppo rurale. Giuffrè, Milano

Zanon E (1997) La qualità dei prodotti alimentari con particolare attenzione alla disciplina delle denominazioni di origine e delle indicazioni di provenienza. Riv dir agr 1:495 et seqq.

Agro-Food Typicality and Cultural Heritage: The Case of the Mediterranean Diet

Aida Giulia Arabia

1 Introduction

There are no doubts that typical foodstuffs are part of the cultural heritage of a place and of a people. As Massimo Montanari[1] says, food, above all the food prepared using traditional and quality products is culture. Now, a question that has been discussed by several authors in the literature is: can typical foodstuffs be considered cultural heritage—and as such be subjected to the same forms of protection in place for the other items of cultural heritage? Why is it necessary to answer this question before going any further?[2]

Because defining the subject matter means delimiting the field of competence and hence identify the institutional subject that is empowered to act through tasks and functions, whether it be to provide protection, or promotion and enhancement. And for a subject area like typical foodstuffs—protected or not by PDOs, PGIs and TSGs[3]—that besides being culture, as said earlier, also reflect agriculture, territory, diet, health, tourism, crafts and trade. Given the multisectoral aspect of foodstuffs, reducing the subject matter only to cultural assets would be a limiting factor rather

[1]Montanari (2006).

[2]Montelione (2000), Mirri (2004) and Alabrese (2005).

[3]To encourage the protection and differentiation of these products from their "imitations", the European Union has laid down precise provisions, shared by all Member States. Particularly, it has established three certification schemes: Protected Designation of Origin (PDO), Protected Geographical Indication (PGI) and Traditional Specialty Guaranteed (TSG). For further reading on the legal profiles of the designation of origin of agricultural products and foodstuffs, see Trapè (2012), pp. 167–278.

A. G. Arabia (✉)
Institute for the Study of Regionalism, Federalism and Self-Government "Massimo Severo Giannini", National Research Council, Rome, Italy
e-mail: aidagiulia.arabia@cnr.it

A. Isoni et al. (eds.), *Food Diversity Between Rights, Duties and Autonomies*,
LITES – Legal Issues in Transdisciplinary Environmental Studies 2,
https://doi.org/10.1007/978-3-319-75196-2_10

than being a strength. Indeed, especially from the standpoint of the Regions, while on the one hand typical foodstuffs are one of the most suited instruments for influencing, elaborating and reviving traditions, on the other hand they constitute the instrument for conveying and promoting quality agriculture, healthy diets, tourist itineraries and commercial initiatives for the products of original craftsmanship that can help boost the economy that for years has been struggling and is still stagnating.

After some general remarks on the tangible and intangible cultural heritage protected by UNESCO, this chapter focus on the roles of the Regions in upgrading the intangible cultural heritage and then providing instruments for promoting and enhancing the traditional products and the Mediterranean diet that uses such products.

It is "in the local context that the forms and expressions of tradition and cultural identity are created and nurtured",[4] and it is the local context that we need to look at (not only the institutions but also the third sector, the local associations that organise cultural events and other similar bodies) to revitalise ancient traditions, including culinary traditions.

2 Cultural Heritage and the Code

Before answering our question, some general remarks need to be made.

Firstly, does cultural assets consist only of things or is it their inherent cultural immanence that make them such? It is clear that only by leaving aside their material aspects that we can include also popular traditions and folklore (and therefore also food, music, dances, rites and so on), thus opting for a broad definition of cultural assets as witnessed by many international but also regional and local activities, and initiatives that are dealt with below. However, while according to the leading thinkers[5] cultural assets are not so much the things but rather the intangible value they represent, the Italian Government has preferred—after a parenthesis during which he seemed to value the intangible aspect of the cultural heritage, defining it as a "record having the value of civilization" (Art. 148, Legislative Decree 112/1998)[6]—the material concept according to which the item and its cultural

[4]Gualdani (2014), p. 4.

[5]Giannini (1976), p. 24.

[6]The appreciation for the elimination of the limit to the material nature of an element of cultural heritage—present both in Act 1089/1939 that spoke expressly of "things of artistic and historic interest", and in the Consolidated Text of legislative provisions on matters related to cultural and environmental heritage (Legislative Decree 490/1999) that did not codify a comprehensive notion of cultural heritage but continued to refer to the individual kinds of cultural assets recognised by the legislation in force (the "things" governed by Act 1089)—would, according to the literature, ensure "the necessary evolution of the general category of cultural heritage". Find this expressly dealt with in Chiti (1998), p. 3.

value cannot be separated since the two elements live in a "symbiotic relationship".[7]

Indeed, in the Cultural Heritage Code (Leg. Decree 42/2004), the legislator again linked the intangible value to the *res,*[8] defining cultural heritage as the "tangible and intangible things that, according to Articles 10 and 11, have an artistic, historic, archaeological, etnoanthropological, archival and bibliographic interest and the other items identified by the law, or on the basis of the law, as records having the value of civilization" (Art. 2(2)). The legislator also emphasises as the asset belonging to private citizen will be elevated to cultural assets by a declaration assessing the existence of the cultural interest required by law in the item in question (Art. 13(1)).

Given this background, the legislator finally repealed the mentioned provision of Leg. Decree 112/1998 that, instead, could have broadened "excessively the list of cultural items that would be part of the heritage".[9] In a nutshell, the Code has linked its scope of efficacy to the material records of civilization thus limiting the effect of "an uncontrolled expansion of the area of protection, beyond the regulatory limits that the latter is currently acknowledged".[10]

This choice eventually led to the exclusion from the scope of the Code of the intangible cultural items protected by the 2003 UNESCO Convention (for the Safeguarding of the Intangible Cultural Heritage) and the 2005 UNESCO Convention (on the Protection and Promotion of the Diversity of Cultural Expressions). Indeed, Art. 7 bis[11] lays down that "the expressions of collective cultural identity" contemplated by the UNESCO Conventions "can come under the provisions" of the Code "if they are represented by tangible items and there are preconditions and conditions for the application of Article 10". If according to some scholars, the new provision of the Code seems to pretend it adopts the international provisions, confirming "the superiority of the provisions of the Code",[12] according to other authors the provision should be interpreted not as an absolute denial of the existence of intangible heritage by our legal order, rather as the need to protect such heritage through "the development of adequate instruments and institutes that are different from those pertaining to tangible heritage".[13]

[7]Morbidelli (2014), p. 4.

[8]Disapproval of the reconstruction of an item of cultural heritage where attention is paid exclusively to the "thing" was already expressed by Cassese (1976), p. 177.

[9]Zucchelli (2012), p. 156.

[10]Such position was already expressed by Cammelli (2000), p. 21, in his comment to the 1999 Consolidated Text of the provisions on cultural heritage (dlgs. no 490).

[11]Article introduced through an amendment to the Code by Leg. D. 62/2008.

[12]Tarasco (2008), p. 2263.

[13]Severini (2014), p. 2.

3 UNESCO'S Commitment to Safeguard the Intangible Cultural Heritage

The attention devoted by UNESCO to the intangible cultural heritage dates back to long before the approval of the 2003 Convention. Indeed, prior to that date, the United Nations Organization had already promoted many programmes, reports and actions focused on culture. Suffice it to mention the foundation back in 1959 of the International Centre for the Conservation and Restoration of Cultural Property (ICCROM), present in 100 Countries across the world, that has given life to projects for the protection of collections and places linked to the cultural heritage, and also the Third Report on Intangible Culture drawn up just before the Convention and published following the International Istanbul Forum of 2002 that saw the participation of 110 Countries and 72 Ministers of Culture.[14]

The 2003 UNESCO Convention for the Safeguarding of the Intangible Cultural Heritage (that came to add to the 1972 Convention Concerning the Protection of the World Cultural and Natural Heritage) ratified by Italian Parliament with Act 167/2007, encompasses, in the concept of intangible cultural heritage "the practices, representations, expressions, knowledge, skills – as well as the instruments, objects, artefacts and cultural spaces associated therewith – that communities, groups and in some cases, individuals, recognize as part of their cultural heritage" (Art. 2.1 of the Convention).

The intangible heritage, therefore, comprises traditions, mostly expressed orally, language, performing arts, social practices, rituals and festive events, knowledge and practices concerning nature and the universe, traditional craftsmanship (Art. 2.2 of the Convention). Conversely, we are talking about the so-called intangible cultural heritage whose identifying substrate is exclusively its being intangible, distinguished from the intangibility-value that instead is inherent in the cultural heritage made up of things. According to the Convention (Art. 2.1) the feature that characterises "intangible" cultural heritage is its being transmitted from generation to generation orally and hence it is constantly recreated by communities and groups in response to their environment, their interaction with nature and their history, and provides them with "a sense of identity and continuity" thus "promoting respect for cultural diversity and human creativity".

Intangible cultural heritage, therefore, typical of the cultural traditions of communities, make up, as a whole, the intangible cultural heritage safeguarded by UNESCO. To be part of such heritage the item needs to be entered and described in the appropriate national and international lists. At UNESCO, the Intergovernmental Committee has in place two lists respectively for the "Intangible Cultural Heritage of Humanity" (Art. 16 of the Convention), and the "List of Intangible Cultural Heritage in Need of Urgent Safeguarding" (Art. 17). For Italy, two entries of

[14]Many other examples can be found in Tarasco (2008), cit., pp. 2264–2265.

interest here were made in the first list, namely the "Mediterranean diet"[15] in 2010, and the traditional agricultural practice of cultivating the 'vite ad alberello' (head-trained bush vines) of the community of Pantelleria (zibibbo).[16] In 2015 the Italian UNESCO Committee selected the "Art of Neapolitan Pizza Makers" as candidate for inscription. This is the first step in the international negotiations that involves 163 States. From the 1st April 2015 to 15 November 2016, independent UNESCO evaluators will be invited to examine Italy's candidacy (or rather, candidacies, because there is also falconry), and by November 2016 they will decide whether the art of pizza-making (and falconry) is to be recognised as an intangible heritage of humanity.

The inscription of an item of intangible cultural heritage on the lists determines only that the International Community and individual States have the "duty to safeguard"[17] such item. The safeguarding is inclusive, in pursuance of Art. 2(3) of the 2003 UNESCO Convention, of all measures aimed at "ensuring the viability of the intangible cultural heritage, including the identification, documentation, research, preservation, protection, promotion, enhancement, transmission, particularly through formal and non-formal education, as well as the revitalisation of the various aspects of such heritage". In other terms, the true goal of the proclamation is not the mere recognition of the value of given items of intangible heritage, but the concrete commitment of the States to promote their knowledge and enhance their value.

Although it does not have the sort of constitutive efficacy that would impact subjective juridical situations (capable of modifying or extinguishing them), the inscription of an item on the lists has exclusively declarative and programmatic purposes that in any case can determine positive effects on the tourist-cultural economy of the State and, particularly, on the place where the intangible item originated and continues to be upheld. In a nutshell, the inscription on the lists, paves the way to actions for salvaging the intangible heritage and it prompts civil society to rediscover traditional and identity values. And finally, especially the traditions involving food may play a critical role in prompting greater attention to dietary issues in general and to nutrition education in particular.[18]

---

[15]The intangible cultural heritage of humanity already includes the Mexican cuisine (2010), the Japanese cuisine (2013), and the gastronomic meal of the French (2010).

[16]The other Italian items inscribed on the list are the Opera dei pupi, Sicilian puppet theatre (2008), the Canto a tenore, Sardinian polyphonic singing by a group of four men (2008), Traditional violin craftsmanship in Cremona (2012), shoulder-borne structures for processions and popular rites (2013).

[17]Bartolini (2013), p. 111.

[18]The knowledge of local certified typical products (PDO, PGI, Biological) and of products obtained by using certified integrated production methods is, amongst other things, an educational awareness raising strategy—required of Member States (that apply for funds by cofinancing the Programme)—present in the European Programme "Fruit in the schools" introduced by Council Regulation (EC) No 1234 of 22 October 2007 and by Commission Regulation (EC) of 7 April 2009. The programme envisages the distribution of fruit and vegetables to children at school with the aim of encouraging young people to adopt good and healthy food habits.

4 Commitment of the Regions in Promoting and Enhancing Intangible Heritage

Before moving on to analyse the regional initiatives in the safeguarding of intangible heritage and, particularly of the "Mediterranean diet", it is worth pointing out that even under the original Title V of the Italian Constitution, before the UNESCO Convention, the local governments had passed several laws on some aspects involving their intangible cultural heritage identity (especially concerning the protection of some dialects, some linguistic minorities and folk traditions, including some historic celebrations).[19]

Regarding the actual enhancement of typical and traditional productions, special mention needs to be made first and foremost about the many regional legislative actions supporting the promotion of quality agricultural products, local (short supply chain) and "Zero-kilometre" products with the aim of encouraging the production and consumption of such products, as well as dietary education initiatives on such products[20] obtained by using techniques that safeguard the environment and consumer health.[21] Secondly, the regional initiatives for the creation of the so-called "wine routes" or "olive oil routes", "truffle routes", "milk routes", the "route of tastes", and "of typical products".[22] These are tourist itineraries with special destinations often linked to "sagre" (celebrations at the time of harvesting a given products) and fairs, having an environmental-naturalistic, historic, artistic and gastronomic value that, for these very characteristics are well suited to making known traditional foods.[23]

---

[19]See Liguria, Regional Law 32/1990, containing provisions for the study, safeguarding, enhancement and social use of some categories of cultural items, particularly dialects and popular traditions; Piedmont, regional Law 26/1990 concerning the safeguarding, enhancement and promotion of the knowledge of the original language heritage; Emilia-Romagna, regional Law 45/1994, law on the protection and enhancement of dialects; Basilicata, Regional Law 16/1996, on the promotion and safeguarding of ethnic and language minorities of Greek-Albanian origin; Sardinia, Regional Law 26/1997, concerning the promotion and enhancement of the Sardinian language and culture. These laws were recently totally or partly repealed and replaced by other legislative measures that have maintained the same title as the original law. See, for instance, Basilicata, Regional Law 17/2004, Piedmont, Regional Law 11/2009, Emilia-Romagna, Regional Law 14/2014. In some cases, the provisions of the original law were included in the Consolidated Texts. See, for instance, Liguria, Regional Law 33/2006 (Consolidated Text of cultural matters, Article 17 "Conservation of dialects and of popular traditions"). On these issues, see Bartole (1999) and Imparato (2010).

[20]See Bolzano, Province Law 11/2000, Friuli Venezia Giulia, Regional Law 21/2002, Marche, Regional Law 5/2004, Trento, Province Law 13/2009, Basilicata, Regional Law 5/2010, Puglia, Regional Law 43/2012.

[21]See Emilia-Romagna, regional Law 28/1999.

[22]See, for instance, Umbria, Regional Law 38/1999, included in the Consolidated Text on agriculture (Regional Law 12/2015, Articles 29–38), Veneto, Regional Law 17/2000 (amended with Regional Law 23/2004), Molise, Regional Law 50/2005, Tuscany Regional Law 45/2003 (amended with Regional Law. 4/2013).

[23]Attili and Losavio (2010), p. 662. See, furthermore, Strambi (2006), p. 204.

The renewed interest in this subject field, with an expansion of the elements that need to be protected, including the tradition linked to the typical features of foods, instead originates from the UNESCO Convention that some Regions transposed into their legal orders using different sources. The first to take action was the Sicily Region that with Council Decrees 77 and 78 of 26 July 2005, set up, at the Department for Cultural Heritage and Sicilian Identity, the Register of Intangible Heritage (REI) with the goal of identifying, protecting and enhancing the oral and intangible cultural heritage of the Region, and the Board for the Registration of Intangible Heritage, that examines the applications for registration, and confirms and/or deletes items from the Register. On 5 March 2014, with Decree 571 a new Register was established whose name is Register of the Intangible Heritage of Sicily (REIS).[24] In 2009 Sicilian cuisine was entered into the Register and in 2014 the red tuna of Favignana was entered, a product of high gastronomic quality, at risk of extinction and processed by a handful of companies that operate throughout the island, particularly in the territories of Trapani and Palermo. In the wake of the Regional Register, many Municipalities or Unions of Sicilian Municipalities established their REIL (Register of Intangible Heritage of local interest) with the precise aim of identifying, drawing up inventories and protecting from the risk of extinction the intangible cultural heritage of local interest.[25] The Lombardy Region was the second Region to take action, with a law at first (Regional Law 27/2008) and then by setting up its own REIL, coordinated and managed by the Archive of Ethnography and Social History—for valorising the intangible cultural heritage.[26] Among other things, Lombardy is the leading partner of the ECHI Project (Italian-Swiss Ethnographies for the enhancement of intangible heritage) that includes Piedmont, Valle d'Aosta, the Province of Bolzano and the Wallis, Ticino and Grigioni Cantons.[27]

---

[24]The Register includes six books: Book of Crafts, Knowledge and Techniques; Book of Celebrations Feasts and Ritual Practices; Book of Dialects, Parlances and Jargons; Book of Expressive Practices and of Oral Repertoires; Book of Human Living Treasures; Book of Symbolic Spaces.

[25]By way of example, the REIL was established by the Union of the Elmo Ericini Municipalities and by the Municipality of Noto.

[26]The Register comprises five sectors: the Register of Oral works, dedicated to the oral traditions (both formal and non formal), to expressive modes and language registers, and to communal values and memories; the Register of the Performing Arts and of Entertainment, dedicated to artistic-expressive performances like music, dances, plays, figurative arts including also the traditional performance of street shows and street artists; the Register of Rituals, dedicated to religious and lay rites, to feasts, to social communal practices celebrating life and annual cycles, entertainment and other significant moments in the cultural life of individuals and communities; the Register of Naturalistic knowledge, dedicated to the practices and knowledge of nature and the universe; the Register of Technical knowledge, dedicated to craftsmanship, farming techniques, agricultural and crafts knowledge about the production of lasting and short-lived objects, diets, husbandry, hunting and fishing, etc.

[27]The Project is based on a shared strategy in support of Communities in defending and maintaining cultural and language specificities and pluralities and in preserving some traditional activities, particularly in border areas. For further details refer to www.echi-interreg.eu.

Finally, it is worth mentioning also the other Regions that have acted to protect their intangible heritage with legislative initiatives although they have remained proposals and have not been turned into laws.[28]

5 The Case of the "Mediterranean Diet" as Protection of Tradition and as Model Diet Plan

Going back to the issue of the "Mediterranean diet", it is worth emphasising, primarily, that it is "a nutritional model that has remained constant over time and space, consisting mainly of olive oil, cereals, fresh or dried fruit and vegetables, a moderate amount of fish, dairy and meat, and many condiments and spices, all accompanied by wine or infusions, always respecting beliefs of each community".

Conversely, we are talking about the use of a series of typical products produced in the places where the communities live. According to UNESCO "the Mediterranean diet (from the Greek «diaita», or way of life) encompasses more than just food. It promotes social interaction, since communal meals are the cornerstone of social customs and festive events. It has given rise to a considerable body of knowledge, songs, maxims, tales and legends". And this is apparently "the added value"[29] that accompanies the choices UNESCO makes where the safeguarding of the diet combines with respect for the territory and biodiversity, thus ensuring also the conservation and development of traditional activities and crafts linked to fishing and farming[30] of some Mediterranean communities (Soria in Spain, Koroni in Greece, Cilento in Italy and Cefchaouen in Morocco). In November 2013, this recognition was extended also to Cyprus, Croatia and Portugal.

What can the Regions do to enhance the "Mediterranean diet" as an intangible cultural asset—and not only—and all its related activities?

Indeed, while on the one hand, by stating that only the "expressions of collective cultural identity", represented by tangible items" can be considered cultural assets, Article 7 bis of the Code has reduced the scope of international regulations because it was incapable of coordinating the concept of cultural heritage adopted by Article 2 of the Code with that of cultural heritage envisaged by the 2003 UNESCO Convention[31]; on the other hand it is also true that the UNESCO Conventions exist and have attributed a "value of civilization" also to intangible heritage. And this is where the local bodies become the main players in the promotion and enhancement of the "diet", precisely because they are the places where the traditions

---

[28]See, by way of example, the bill of the Piedmont Region no 628 submitted on 25 June 2009 and the bill of the Calabria Region no 100 submitted on 5 November 2010.

[29]Colella (2013), p. 585.

[30]For some reflections on the relationship between agriculture and food cultures, see Adornato (2009).

[31]Casini (2008), p. 1062.

(food traditions in this case) were born. The "diet" being common heritage may also be "transferred" and spread out to the rest of the territory and perhaps even "exported" to support and promote its use.

So perhaps, as pointed out initially, going beyond the idea of considering typical and traditional products, and particularly the "Mediterranean diet", as cultural assets would not be so harmful.[32] Indeed, it is of fundamental importance to safeguard and protect, not so much the products, but the traditional production procedures as cultural activities that are typical or peoples and their territories.

It is precisely the cultural component that, alongside the farming, dietary, environmental and landscape component, inherent in traditional quality products, should play a decisive role in the conservation of ancient and traditional methods and procedures in an increasingly industrialised market.[33] Evidence of the foregoing is Regional Law 59/2012 of the Calabria Region bearing the title "Recognition of the historic method «Moscato at the Government of Saracena» as cultural heritage". Apart from the title of the law, the contents emphasise that it is an intangible cultural asset and specify that the category comprises also the "artisanal winemaking methods documented also by accredited historic sources and/or handed down, even only orally, characterised by uniqueness and exclusiveness and that run the risk of being forever lost", precisely like the "Method" of boiling the grapes in the process of making the Saracena moscato. In the classification by subject, used to draw up the Report on the 2013 legislation of the Chamber of Deputies,[34] this law was entered (by the Region) under the heading "Cultural heritage and activities". Now why should we emphasise this aspect?

Because these items-procedures can find protection, promotion and enhancement by attributing them to several tangible sectors, which are influenced above all by the regional policies, including the cultural activities as pointed out by the literature referred to. Also, in the case of the Mediterranean diet, the local bodies could promote it and enhance it because it refers to a way of life that embraces farming, health protection, healthy diet, protection of the environment and of the landscape, healthy territory, and cultural activities. This involves partially policies that are traditionally regional, and partially policies where the Regions, despite the Constitutional Court judgments curtailing regional autonomy, have succeeded in carving out room for their intervention that in some cases has been very significant and original.

---

[32] According to Mirri (2004, p. 291) the consideration of the typical and traditional products as cultural assets might also be seen as a "contradiction", since these items are for consumption.

[33] Mirri (2004), p. 291.

[34] The chapter "Tendencies and problems of Regional legislation" of the 2013 Report is edited (as are the relevant chapters in all the editions of the Report) by ISSiRFA-CNR and can be read in www.issirfa.cnr.it>Rapporti Issirfa>Rapporti sulla legislazione delle regioni.

Campania and Calabria[35] are two Regions that have passed laws to enhance and diffuse the "diet". The former motivated its intervention as a way of enhancing a "development model based on values of this type of diet and way of life from the cultural, social, historic, gastronomic, diet, environmental, landscape and customs standpoints" (Campania, Regional Law 6/2012, art. 1), while the latter states that the Mediterranean diet is a way of life and a "health-improving nutritional discipline" (Calabria, Regional Law 45/2013, art. 1).

These two different purposes of the mentioned laws take us back to the statement made earlier about the fact that this subject is linked to several fields like agriculture, historic and cultural heritage, nature conservation for Campania, food and health protection for Calabria. Not only do the purposes expressed show the disciplines that prevail in each case and the institutes envisaged for consulting and monitoring purposes, but they are also proof of the different concrete fields affected by the measures concerning the "diet". The institutes are: the Regional Observatory for the Mediterranean Diet, in Campania, that has set up an operational centre at the General Directorate for agricultural, food and forestry policies and has advisory, proactive and monitoring functions; the Mediterranean Diet Foundation in Calabria, based in the municipality of Nicotera, chosen by the Region as being representative of the Mediterranean diet because it is the place where the International cooperative study on cardiovascular epidemiology was carried out. Again, we see the different intentions of the legislators embodied in the purposes of the two laws.

Moreover, also the goals that the two legislators intended to pursue follow different strategies.

The legislator from Campania intended to promote the diet throughout the "UNESCO sites and areas characterised by typical products with known brands, as an integrated and synergistic strategy for enhancing the tangible and intangible cultural heritage of humanity" and for this purpose he envisages the development of "innovative models of economic and tourist attraction for enjoying the products of the diet within specific and historic-cultural landscapes of the territories of provenance"; he then intended to promote "healthy lifestyles based on the Mediterranean diet as the best diet for preventing social disorders linked to nutrition", a result that can be achieved by using and spreading the use of the products of the diet in canteens and in the networks of collective catering. Finally, the strategic goals also comprise a series of training and educational activities on the diet, cultures and landscapes associated with it and the promotion of relationships and cultural, scientific and economic exchanges among the Mediterranean communities that share that "diet".

[35]Other Regions (Apulia and Sicily) have instead, developed bills that are still being discussed. Among other things, for several years, Apulia has been organising the Mediterranean Diet Festival, hosted each year by a different municipality. Besides promoting the typical products that are part of their diet (oil, wine and bread, to mention just a few), it is an opportunity for rediscovering villages, natural and architectural beauties and ancient traditions (including outdoor games). Meetings are held with experts and nutritionists who provide indications and suggestions about food "habits". Similar initiatives are organised also in Campania and Calabria.

The legislator from Calabria, instead, focused on the enhancement of the products used in the diet by popularising the diet through training courses for the operators in the catering field to equip them with the skills, practices and traditions linked to this nutritional model, with special emphasis on the farming, harvesting, preservation, processing and consumption of the food, and the diffusion of certified typical products in canteens and networks of collective catering. He then focused on the medical-scientific-statistical studies—in collaboration with the Institute for Scientific Research on the Mediterranean Diet and Nutrigenomics, based in Reggio Calabria—on the degree of compliance of the Calabrian population with the Mediterranean diet, methods of preparation and combination of the foods comprised in the diet. The aims of the Calabrian law are as a consequence consistent with another Regional Law (Regional Law 29/2004, supplemented by Regional Law 13/2005) that established the Regional Healthy Diet Week to be celebrated every year in February—"in order to raise awareness and promote education and information in the area of healthy foods produced according to traditional methods, across the Calabrian territory with a view to ensuring the wellbeing of our consumers" (Article 1). With the same Act, the legislator acknowledged the importance of the activities carried out by the Institute especially its data base on the quality of the foodstuffs of the Mediterranean diet and the studies and research on disorders deriving from unhealthy dietary habits whose results are shared with the training institutions and the food production sector.

6 Conclusion

All the things said above respond to the initial question and confirm the remarks made in the introduction.

Although typical products, including the Mediterranean diet that exclusively uses the typical products, are a part of the cultural heritage[36] they are not "assets", but cultural activities. Moreover, they are also the vehicle for enhancing regional policies ranging from promoting multifunctionality in agriculture, taking care of the landscape, of the natural, historic and cultural heritage, promoting sustainable tourism, healthy diets, health protection and cultural activities. In other terms,

[36]The decree issued by the Minister for Agricultural, Food and Forestry Policies and the Minister for Cultural Heritage of 9 April 2008 expressly states that the traditional agrifood products contained on the lists in Ministerial Decree of 18 July 2000 and subsequent amendments "are an expression of Italy's cultural heritage". This decree, published after Leg. Decree no 62 of 2008 amending the Cultural Heritage Code (issued in March), appears to be a declaration after the failure to extend the discipline of the Code also to intangible cultural heritage not supported by a 'thing', which is a true "declaration" of cultural asset. After all, as already pointed out, in the case of typical and traditional products it is not a question of promoting and enhancing items for consumption, but rather the methods and the production and processing procedures of a product, which in themselves are culture (Mirri 2004 *cit.*, p. 291).

recovering old products and old methods of production means recovering an environmental and natural balance that had hitherto been destroyed; it means recovering supply chains for food security and safeguarding important intangible items so that we may revitalise cultures inherited from the past.

In intangible heritage, and specifically in the procedures and techniques involved in food processing, which are the outcome of specialised and traditional knowledge, promotional and enhancement activities play a significant role. Instead, other measures that are regularly adopted to safeguard "material" cultural heritage do not appear to be applicable; reference is being made, for instance, to the prohibition to make changes or restrictions on circulation that, in any case, are exclusively a field of State competence.

If there is still any sense in continuing to keep the Regions alive, the power to make laws on some "subjects" needs to be left to the local governments. One such subject is without question that of intangible cultural items that originate in the territories and therefore should be promoted and enhanced at the territorial level, also to facilitate and encourage their fruition.[37]

The same holds for typical products and particularly for the "Mediterranean diet". The latter, among other things, besides being shared with other six communities in the Mediterranean, and besides having an impact on health protection and on food intake, can be easily "exported" beyond its place of birth, according to a bottom-up approach going from the region, and particularly the precise place of origin, to the rest of the national territory and even beyond.

Evidence of this is the fact that also the State legislator, perhaps prompted by the regional actions, started a legislative initiative in 2003 aimed at enhancing and promoting the diet "as a model of cultural and social development based on specific skills, knowledge, practices and traditions handed down from past generations, as well as a nutritional regime recognized by the scientific community as being beneficial for health".[38] Regarding the promotion of healthy lifestyles, the bill encourages "the diffusion of the use of products and specialties of the Mediterranean diet in the networks of collective catering and, in particular, in school canteens". In addition, schools are called upon to provide educational and information activities on the Mediterranean diet and on the cultures and landscapes associated with it (Article 2 (2), letter b and d), thus envisaging a sort of link between intangible cultural heritage and the tangible cultural heritage.

---

[37]According to the literature (Morbidelli 2014, p. 5), it is the multiplicity and variety of the intangible cultural heritage that requires that the promotion and enhancement measures of such heritage be "«naturally» a competence of the community in which they are born and of which they constitute a value and an element of identity".

[38]DDL S. no 313 addressed to the Presidency on 26 March 2013 was assigned to the 9th Standing Committee (agriculture and agrofood production) on 15 May 2013. The DDL is currently being examined by the Committee.

References

Adornato F (2009) Cultura alimentare e politiche agricole. Agric Istit Merc 3:5–10

Alabrese M (2005) Evoluzione della nozione giuridica di "bene culturale" e i prodotti agro-alimentari tradizionali. In: Strambi G, Alabrese M (eds) I prodotti agro-alimentari tipici e tradizionali come "beni culturali". Edizioni Campano, Pisa, pp 9–13

Attili P, Losavio C (2010) I prodotti agroalimentari tradizionali: frammenti di una disciplina incompleta. Riv di dir agr 4:637–688

Bartole S (ed) (1999) Le regioni alla ricerca della loro identità culturale e storica. Giuffrè, Milano

Bartolini A (2013) Beni culturali. Enciclopedia del diritto (Annali), pp 93–132

Cammelli M (2000) La semplificazione normativa alla prova:il testo unico dei beni culturali e ambientali. In: Cammelli M (ed) La nuova disciplina dei beni culturali e ambientali. Il Mulino, Bologna, pp 7–27

Casini L (2008) La disciplina dei beni culturali dopo il d.lgs. no 62/2008. "Erra l'uomo finché cerca". Gior di dir amm 10:1060–1068

Cassese S (1976) I beni culturali da Bottai a Spadolini. In: Cassese S (ed) L'amministrazione dello Stato. Giuffrè, Milano, pp 153–183

Chiti MP (1998) La nuova nozione di "beni culturali" nel d.lgs. 112/1998: prime note esegetiche. Aedon 1:1–7

Colella L (2013) La "dieta mediterranea" come patrimonio dell'umanità: dalla tutela dell'Unesco alla legge regionale della Campania no 6 del 2012. Dir e giur agr, alim e dell'amb 10:583–587

Giannini MS (1976) I beni culturali. Riv trim di dir pub 1:3–38

Gualdani A (2014) I beni culturali immateriali: ancora senza ali? Aedon 1:1–7

Imparato EA (2010) Identità culturale e territorio tra Costituzione e politiche regionali. Giuffè, Milano

Mirri MB (2004) I prodotti agricoli "tradizionali" come beni culturali? In: Germanò A (ed) L'agricoltura dell'area mediterranea: qualità e tradizione tra mercato e nuove regole dei prodotti alimentari. Profili giuridici ed economici. Giuffrè, Milano, pp 283–291

Montanari M (2006) Il cibo come cultura. Laterza, Roma-Bari

Montelione E (2000) La produzione agroalimentare di qualità come bene culturale. Riv di dir agr 3:462–486

Morbidelli G (2014) Il valore immateriale dei beni culturali. Aedon 1:1–8

Severini G (2014) Immaterialità dei beni culturali? Aedon 1:1–4

Strambi G (2006) Le strade del vino, dell'olio e dei sapori: il quadro giuridico di riferimento. Riv di dir agr 2:204–238

Tarasco AL (2008) Diversità e immaterialità del patrimonio culturale nel diritto internazionale e comparato: analisi di una lacuna (sempre più solo) italiana. For amm: CdS 7–8:2261–2287

Trapè I (2012) I segni del territorio. Giuffrè, Milano

Zucchelli C (2012) Art. 13. In: Sandulli MA (ed) Codice dei beni culturali e del paesaggio, II ed., pp 147–184

The Role of Gastronomy and Typical Foods in the Tourism Experience

Luca Zamparini

1 Introduction

The recent decades have witnessed the passage from traditional tourism, where the main motivation for a holiday was the need to rest and to recover from the stress of everyday life,[1] to modern tourism, where tourists consider the culture and environment that characterize their destinations as key elements of their holiday periods. In this new context, gastronomy and typical food have emerged as determining factors in the choice of tourists' destinations and in the satisfaction for the consequent tourism experiences. Several studies have attempted to estimate the relevance of food in the overall tourists' expenditures.[2] It has emerged that in most contexts, the expenditures on food and eating represent roughly one quarter of the overall tourism expenditure and that they rank second, following those on transportation. Moreover, the development of an important food sector can represent a key element that may contribute to the economic development of the tourist destinations.

Consequently, the role of typical food and gastronomy has attracted the attention of scholars aiming to highlight the determinants of consumption of beverages and of food in tourists' destinations. The present chapter critically considers the reasons that determine why some tourists choose typical foods while others rely on food that is very similar to the one consumed in everyday contexts. The chapter is structured as follows. Section 2 discusses the three main approaches to describe the role of food at tourist destinations; namely, food as a tourist attraction and as an impediment, food's role in the overall tourism experience and its linkages to overall tourism

[1] See Krippendorf (1987).

[2] See Telfer and Wall (2000), Minister of Industry of Canada (2003) and Kivela and Crotts (2006).

L. Zamparini (✉)
Department of Legal Studies, University of Salento, Lecce, Italy
e-mail: luca.zamparini@unisalento.it

A. Isoni et al. (eds.), *Food Diversity Between Rights, Duties and Autonomies*, LITES – Legal Issues in Transdisciplinary Environmental Studies 2,
https://doi.org/10.1007/978-3-319-75196-2_11

supply. Section 3 proposes an analysis of the evolution of the literature that has aimed at explaining the determining factors for the consumption of local food at tourists' destinations. Section 4 considers the role of globalization on local food consumption by tourists and critically considers the interactions between globalization and localization and the consequent dichotomy between convergence and divergence in food consumption. Section 5 offers some concluding remarks.

2 The Three Main Approaches to Describe the Role of Food at Tourist Destinations

A study by Hegarty and O'Mahony has described food as a form of art that is part of the local destination culture.[3] According to the authors, one of the characteristic traits of each culture lies in the preparation and consumption of food in a sophisticated manner. For Levi-Strauss this can be considered as one of the proofs of civilization.[4] Consequently, there are remarkable differences in "the basic ingredients from which food is prepared, the ways in which it is preserved, prepared and cooked, the amount and variety available at each meal, the tastes that are liked and disliked, the customs and traditions of serving food, the implements and utensils which are used, and inherent beliefs about the properties of particular foods".[5] This implies that visiting a non-familiar destination exposes the tourists to a different gastronomy than the one that they are accustomed to in their everyday life. This section proposes a description of the three main approaches that have been proposed in the tourism literature to describe the role that food may have at tourist destinations: (a) food as a tourist attraction and as an impediment; (b) food as an experience in tourism and (c) food and its linkages to overall tourism production.

2.1 *Food as a Tourist Attraction and as an Impediment*

A seminal paper by Cohen has proposed a taxonomy of possible tourist roles.[6] The main issue of the paper was related to the description of tourists as subjects that, when travelling to a distant or uncommon destination, look for novelty, strangeness and peculiar experiences. However, only a limited number of tourists are really eager to give up completely their everyday life tenets. For most tourists, it is necessary to have a certain degree of familiarity, also in food consumption, to be satisfied by the tourism experience. This is particularly relevant when international tourist activities

[3]See Hegarty and O'Mahony (2001).
[4]See Levi-Strauss (1978).
[5]See Hegarty and O'Mahony (2001), p. 5.
[6]See Cohen (1972).

are considered. On the contrary, Hjalager and Richards have emphasized the role of food as a tourist attraction.[7] However, their study was related to Western tourists visiting renowned Western destinations. Cohen and Avieli[8] have reconsidered the taxonomy by Cohen[9] and the distinction proposed by Fischler[10] between neophobic and neophilic tendencies in taste to underline the twofold role of food at tourist destinations as both an attraction and an impediment. It appears that destinations' cuisines are appreciated by most tourists only when they resemble typical dishes of everyday life or when they have already been assimilated by going to ethnic restaurants in their own countries.[11] Generally, these ethnic restaurants fuse their peculiar recipes and dishes with the local ones to increase their demand potential among residents. According to a study by Warde and Martens, the familiarity with ethnic cuisines is restricted to a limited number of middle to high class perspective tourists.[12] Lower and lower-middle classes tourists have not generally been exposed to foreign foods before their trips and they tend to avoid them also during their holidays.

Another component that constitute a possible impediment of local food is the fear of illness or sickness that derive from consuming local dishes that may have ingredients that are not fully tolerated by visitors. Given that tourists consider their holiday time as "non-ordinary" and qualitatively more important than their everyday ordinary time they fear that any health related circumstances may threaten the possibility to enjoy several coming experiences during the trip (see Sect. 2.2).[13] Typical behaviors of neophobic tourists is the avoidance to drink local water or to eat at street hawkers' shops although the dishes are full of flavor and appear to be appreciated by local people. These considerations are among the causes of the impediment to taste local food. Another issue that is important to consider for tourists travelling to remote countries is the fact that the restaurants or parlors that offer genuine local food do not normally have menus that are translated in foreign languages. It is thus very difficult for tourists to attend these venues and they prefer establishments that provide translated menus or pictures of the dishes and, generally, offer a fusion of local and other countries' food and ingredients.[14] In many cases this is what characterizes restaurants in hotels and in popular touristic neighbors.

[7]See Hjalager and Richards (2002).

[8]See Cohen and Avieli (2004).

[9]See Cohen (1972).

[10]See Fischler (1988).

[11]See Mintz (1996).

[12]See Warde and Martens (2000).

[13]See Graburn (1977).

[14]See Cohen and Avieli (2004).

2.2 Food's Role in the Overall Tourism Experience

To analyze the role of food in the tourism experience, it is important to elicit the differences between experiencial tourists and recreational (or traditional) ones. According to a seminal study by MacCannell, the former ones seek to experience the authentic life of others and are characterized by neophilic tendencies (see Sect. 2.1) with respect to local foods and to local eating habits.[15] The author distinguishes between a front and a back region in tourist destinations. "The front is the meeting place of hosts and guests or customers and service persons, and the back is the place where members of the home team retire between performances to relax and to prepare".[16] Experiencial tourists aim at living the back region of their tourist destinations. This requires both their willingness and the availability of locals that may be reluctant to let visitors enter their back regions. In few cases, visitors can enter the back regions of the natives, also in terms of local foods and eating habits. In most of the cases, local destinations create tourist settings where some form of staged authenticity is offered to visitors that are not allowed to enter the back regions of locals.

On the other hand, recreational tourists normally aim at relaxing themselves in areas where they feel safe and to some extent recognize certain patterns and neighborhoods that are not very different from those that characterize their everyday life. They normally do not attach a relevant value to authenticity. Moreover, they have the tendency to be more self-indulging during their holidays and to spend more money than at home to enjoy their holiday periods. They tend to be neophobic (see Sect. 2.1) with respect to foods and drinks and tend to prefer familiar food (possibly of a high quality) rather than seeking for authenticity.

The predominance of the former category of tourists (experiencial tourists) characterizes the early developments of tourist destinations. As destinations become popular and enter the mature stage of their life cycles, recreational tourists tend to predominate. In terms of foods and eating establishments, this normally generates two specific and parallel tendencies. In local restaurants there will be a growing number of foreign dishes aiming at accommodating the needs of the recreational tourists. Moreover, the quantity and variety of local preparations will be reduced and, in most cases, homogenized to the taste of the prevailing category of visitors.

Quan and Wang have proposed a structural model of the food tourist experience that embodies the marketing/management approach and the social sciences one.[17] Consequently, the tourist experience is constituted by the attractions that are the main motivations to travel and the basic consumer activities (among them eating) that are necessary at the tourist destinations. The disappointment in the main attractions will normally determine a rapid decay of the tourism destination. However, the availability of efficient transport means, of comfortable accommodation

[15]See MacCannell (1973).

[16]See MacCannell (1973), p. 590.

[17]See Quan and Wang (2004).

facilities and of good quality food and restaurants are important supporting activities that are necessary for the exploitation of the competitive potential of a destination. In some cases, parts of the supporting activities, and particularly food and restaurants, may turn to be peak experiences. This is particularly the case for food festivals, wine tourism and gastronomic tourism where food and drinks become the main motivation for travelling.

2.3 *Food and Its Linkages to Overall Tourism Supply*

Food, jointly with other activities, attractions, transport and accommodation, is one of basic and relevant elements that constitute the tourism supply of a destination as stressed, among others, by Reynolds.[18] This is because tourism is a composite product made up of several complementary goods and services that are bought by visitors to a determined area. Consequently, food is also important as it contributes to build the image of a tourist region. Hjalager and Antonioli Corigliano have provided a taxonomy of the ways in which food and the destination may be connected.[19] The first possible way is related to the idea of complementarity. In this case, the core activities in the area are linked to specific attractions (i.e. a theme park, a monument, a museum) but food is included in the marketing strategies of the firm or of the local administration to attract more tourists and to differentiate the firm or area from other competing firms or destinations in a logic of monopolistic competition. The second dimension is related to the inventory issue. In this case, the entrepreneurial and administrative efforts aim at creating new tourist products and/or experiences. A pivotal role is played by the food heritage sites and by peculiar foods and recipes that characterize the area. The targeted demand is the one that attaches a high value to the culture of food and eating. The third dimension is the superficial one. In this case, quality plays a minor role while the environment is considered as the most important element in the food/eating dimension of the holiday. Tasting local products and dishes is anyway a means to get in touch with the local population and culture. Lastly, the relationship between food and tourism destinations may be a disconnected one. In this case, the products and brands are globalized and the act of eating has almost nothing to do with encountering local culture, people and values.

It is important to stress that not all tourist destinations are equally suited to promote and market peculiar foods and dishes. Local entrepreneurs and administrations should thus carefully consider the possibility to introduce local food promotion strategies and marketing. On the other hand, it is important to consider that large leakages and loss of local identity can occur when the tourism industry relies heavily on imported products and that the relationships between food production and

[18]See Reynolds (1994).

[19]See Hjalager and Antonioli Corigliano (2000).

tourism can either imply conflict or coexistence or symbiosis.[20] However, it should be noticed that there may be many obstacles to increase the amount of local food in the tourism industry. This may be because of behavioral reasons, that are normally related to the demand side of the tourism market, but also to physical, economic, technological and marketing issues originating in the supply side.[21] Physical attributes may be related to the scarce amount of precipitations, to the low quality of land or to soil types. Economic reasons may be determined by the distribution of land ownership, competing productions in other regions with lower labor costs and/or higher quality lands. In a seminal paper, Latimer argued that efficient planning and assistance are important prerequisites to increase the number of local products used in the tourism industry.[22] Considering the case of Jamaica, Latimer suggested that infrastructural and transport related investments may be less useful than legal assistance with contracts, backup supply arrangements and insurance against inefficient or lack of delivery. Regular meetings among the most important local stakeholders would also play an important role. The efforts towards a larger use of local food products may also lead to more regional employment, to a higher tourism multiplier and to a more sustainable development of tourist activities.

3 Determining Factors for the Consumption of Local Food at Tourists' Destinations

An analysis proposed by Fields suggested that the determinants of food and beverages consumption can theoretically be included in four categories: (a) physical motivators; (b) interpersonal reasons; (c) status and prestige purposes and (d) cultural determinants.[23] Such categories can be applied to consumption at home and in everyday life but they are also useful to explain the role of gastronomy and of typical food at tourists' destinations The first set of motivators are, of course, the most basic ones that imply the necessity of food to provide a positive experience through sensory perceptions, such as taste, smell and sight. The second set of reasons are related to the necessity to consume food in a friendly environment or jointly with relatives. In the case of holidays, there is also the necessity to share ideas and impressions on food that, in most cases, is not commonly eaten at home. The status and prestige purposes characterize the need to eat good food in a renowned restaurant as a possible means to be distinguished from other people in terms of social status. Lastly, and most interestingly in terms of the analysis proposed in this chapter, the cultural determinants can have an important impact on the gastronomic decisions of tourists. As it was stated in the introductory section, modern day tourism

[20] See, among others, Taylor et al. (1991).

[21] See Telfer and Wall (1996).

[22] See Latimer (1985).

[23] See Fields (2002).

is determined by the need to have a (new) experience. Tasting typical food is thus among the factors that allow to have a feeling of a different culture (see Sect. 2.2).

A study by Tikkanen[24] has reconsidered the Maslow's hierarchy of basic needs and motivations.[25] It has distinguished them into physiological, safety, social, esteem and self-actualization. These categories can be applied in the context of tourism behavior with respect to food consumption. According to Tikkanen, in the cases in which physiological needs are the main motivators, food is considered as an important attraction that can help to characterize the overall tourist product. The safety needs may constitute an impediment to local food consumption (see Sect. 2.1) but they also motivate internal and international research that may give rise to international health and safety conferences. If social needs are the main component, food becomes the focal point of special events and festivals that may gather local residents and tourists. This may lead to social interactions and experiences. Food can in these cases either complement wider trip motivation or constitute the main focus of a determined trip or short holiday. Mason and Paggiaro have argued that the satisfaction of tourists in the consumption of local food products can have important beneficial effects to the local development of the tourist sector and of the overall economic system by means of increased awareness, loyalty, emotional connection, increased involvement and brand differentiation.[26] In this context Mason and Paggiaro use the notion of "festivalscape" to describe the physical environment that puts together the event atmosphere and tangible factors, such as food quality, souvenirs, availability of auxiliary facilities. Esteem needs lead to culinary food tourism where visitors are satisfied in the cases in which they can experience new tastes. Lastly, self-actualization needs are fulfilled through the participation to workshops, conferences and fairs that increase the tourist knowledge and increase her/his self-respect.

Kim et al.[27] have reconsidered the theoretical discussions by Fields[28] and Tikkanen[29] and have proposed a model that tries to account for food consumption by tourists at holiday destinations based on three different categories of factors; demographic, psychological and motivational. Demographic factors included gender, age and education while psychological factors considered food neophilia and neophobia (see Sect. 2.1). The motivational factors were composed of: (a) exciting experience; (b) escape from routine; (c) health concern; (d) authentic experience; (e) learning knowledge; (f) togetherness; (g) prestige; (h) sensory appeal; and (i) physical environment.

Neophilic tourists tend to consider local food as an exciting experience and as an escape from routine meals, although they are happy with the food that they normally

---

[24]See Tikkanen (2007).

[25]See the discussion in Stephens (2000).

[26]See Mason and Paggiaro (2012).

[27]See Kim et al. (2009).

[28]See Fields (2002).

[29]See Tikkanen (2007).

consume at home. Another category of tourists may prefer to consume local food as it is considered healthier given that it is prepared with fresher ingredients. On the other hand, some tourists may be reluctant to consume meals that are prepared locally if they do not deem that the necessary hygienic conditions are met by the owners of the eating establishments. The cultural dimension was considered with the authentic experience and the learning knowledge factors. In particular, the latter is related to the possibility not only to experience local cultures but also to enrich one's intellectual tastes. Togetherness implies the possibility to strengthen social and family bonds and to build personal relationships once on holiday. For some tourists, sharing food with others may be more important than the quality of food in itself. Coherently with the theory proposed by Tikkanen, the prestige factor can be related to the possibility to enhance tourists' ego by showing to friends and relatives that they have been able to taste local food or local drinks (i.e. wine).[30] An important motivational factor to consider is the sensory appeal of the dishes. A good flavor is especially deemed as a prerequisite for the consumption of local food. However, other senses are also involved by means of the smell and visual image of meals. Lastly, the physical environment plays an important role within the motivational factors. It is related to the degree of cleanness of local restaurants and food stalls. Moreover, a comfortable atmosphere and the idea that the venue is a traditional one can have an important role in nudging tourists to consume local food at a certain location. A study by Meiselman et al. has shown that elements such as interiors, sound and lighting of a restaurant can highly influence the choice and that identical food can be sensed and appreciated differently in heterogeneous surroundings.[31]

A further model was proposed by Mak et al. by adding the religious influence and the exposure effect/past experience to the factors that had already been considered by previous literature.[32] Religious influence is important for a series of reasons. Certain foods are prohibited (i.e. Islam or Judaism), some fasting practices are mandatory (i.e. *Ramadan*) and some particular preparation methods must be observed in order not to mix specific ingredients according to specific religious creeds. The authors propose the example related to the Muslim religion where food is divided into *halal* (allowed) and *haram* (forbidden). Muslim people must then abide to these prescriptions also while on holiday. The authors also discuss the relevance of past experience especially in a globalized context in which a large share of tourists have travelled more than once to distant destinations and have thus had the opportunity to taste exotic food and have acquired familiarity with those dishes/meals. This familiarity can then enhance their propensity to consume local food during further holiday periods.

[30]See Tikkanen (2007).

[31]See Meiselman et al. (2000).

[32]See Mak et al. (2012a).

4 The Role of Globalization on Local Food Consumption by Tourists

The previous sections have provided an analysis of the role of food at tourist destinations and of the determining factors for its consumption. The present section will present and discuss the role of globalization on local food consumption by tourists. Part of the literature has considered globalization merely as a threat to local gastronomic image and identity with the ensuing loss of authenticity of local tourist regions.[33] This impact would be particularly momentous for destinations that consider food and gastronomy as one of the main attractors of tourism demand (see Sect. 2.3). Another line of research has emphasized the possibility that globalization may provide an incentive to improve and, to a certain extent, reinvent typical gastronomic products of specific tourist destinations. According to this vision, food cultures and traditions should not be considered as static but as dynamic in response to internal and external stimuli.[34] One example in this respect has been provided by Ram that has analyzed the effects of the arrival of the McDonald's chain in Israel.[35] It has emerged that such occurrence has led to the revival of a local dish (*falafel*) rather than its disappearance. More specifically, this dish has then been marketed in two distinct forms. The gourmet version has started to be served in elegant restaurants and hotels while the more standardized version has appeared more consistently in food stalls and has also been branded and sold internationally by franchise chains.

Mak et al. have proposed a conceptual model of the influence of globalization on local culinary supply and on tourist food consumption.[36] The main relationships of the model are represented in Fig. 1. The first relationship that emerges from Fig. 1 is related to the convergence in food consumption that originates from globalization. Such convergence, in turn, reinforces globalization. On the other hand, localization forces aim at increasing the divergence in food consumption. These different trends generate three contrasts; homogenization v. heterogenization, global consciousness v. local consciousness and global culture v. local culture. These contrasts can create a thrust to produce "glocalized" products that merge some international trends with local ingredients and recipes. Mak et al. propose the example of Hong Kong milk tea, which has substituted fresh milk with evaporated milk and has become very popular among both residents and tourists.

Previous literature (see Sect. 2) had stressed that local restaurant and food parlor may modify their recipes to meet coming from tourists.[37] The attendance of these establishments by residents may then modify also the gastronomic culture of the tourist destination.

[33]See, among others, Harrington (2005) and Fox (2007).

[34]See, among others, Mennell (1985) and Molz (2004).

[35]See Ram (2004).

[36]See Mak et al. (2012b).

[37]See, among others, Reynolds (1994) and Telfer and Wall (1996).

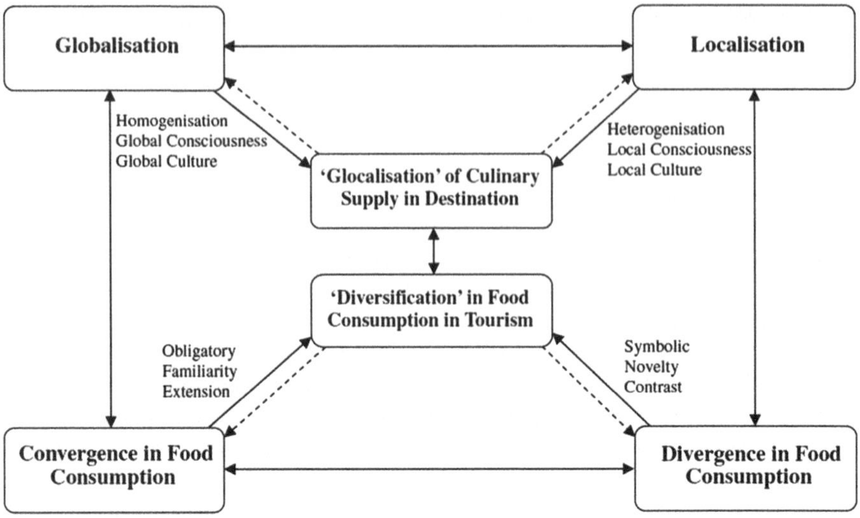

Fig. 1 Influence of globalization on culinary supply and food consumption in tourism. Source: Mak et al. (2012b)

The relationships between convergence and divergence in food consumption generate three dichotomous dimensions (symbolic v. obligatory, novelty v. familiarity and contrast v. extension)[38] that exemplify the differences between neophobic and neophilic tourists.[39]

5 Conclusions

This chapter has considered the role of gastronomy and typical food in the tourism experience. It has emerged that there is a clear distinction between neophobic and neophilic tourists. The former ones are not eager to taste unfamiliar food while the latter look for novelty. Consequently, it is impossible to state that there is a clear role of food at tourist destinations. Firstly, it may be either an attraction or an impediment. However, most tourists appreciate local food only when it is similar to the typical dishes of everyday life. On the other hand, it is important to stress that not all tourist destinations are equally suited to promote and market peculiar foods and dishes. Private and public local stakeholders need to carefully plan the most appropriate strategies to introduce local food promotion and marketing. The chapter has also stressed that there may be many impeding factors to the increase of the amount of local food in the tourism industry. This may be because of behavioral reasons, that

[38]See Mak et al. (2012b).

[39]See Sect. 2.1 and Fischler (1988).

are not only related to the demand side of the tourism market, but also to physical, economic, technological and marketing issues that characterize local tourist destinations. However, the implementation of a larger use of local food products may also lead to more regional employment, to a higher tourism multiplier and to a more sustainable development of tourist activities. It has then been highlighted that there are several categories of determining factors that can be usefully applied to explain the consumption of local food by tourists. Contrarily to part of the literature, it has emerged that globalization is not only a threat to local gastronomic image and identity but it can also provide incentives to improve and, to a certain extent, reinvent typical gastronomic products of specific tourist destinations. A dialectic, and in most cases fruitful, relationship emerges between local cultures and *stimuli* coming from tourists and globalization. It is then important that local stakeholders adopt cooperative strategies to turn the threats into opportunities to improve the food culture of the region and to generate growth and sustainable development.

References

Cohen E (1972) Toward a sociology of international tourism. Soc Res 39:163–182

Cohen E, Avieli N (2004) Food in tourism. Attraction and impediment. Ann Tour Res 31 (4):755–778

Fields K (2002) Demand for the gastronomy tourism product: motivational factors. In: Hjalager A, Richards G (eds) Tourism and gastronomy. Routledge, London, pp 37–50

Fischler C (1988) Food, self and identity. Soc Sci Inf 27:275–292

Fox R (2007) Reinventing the gastronomic identity of Croatian tourist destinations. Int J Hosp Manag 26:546–559

Graburn N (1977) Tourism: the sacred journey. In: Smith V (ed) Hosts and guests: the anthropology of tourism. University of Pennsylvania Press, Philadelphia, pp 17–31

Harrington RJ (2005) Defining gastronomic identity: the impact of environment and culture on prevailing components, texture and flavors in wine and food. J Culinary Sci Technol 4 (2–3):129–152

Hegarty JA, O'Mahony GB (2001) Gastronomy: a phenomenon of cultural expressionism and an aesthetic for living. Int J Hosp Manag 20:3–13

Hjalager A, Antonioli Corigliano M (2000) Food for tourists – determinants of an image. Int J Tour Res 2:281–293

Hjalager A, Richards G (eds) (2002) Tourism and gastronomy. Routledge, London

Kim YG, Eves A, Scarles C (2009) Building a model of local food consumption on trips and holidays: a grounded theory approach. Int J Hosp Manag 28:423–431

Kivela J, Crotts J (2006) Tourism and gastronomy: gastronomy's influence on how tourists experience a destination. J Hosp Tour Res 30(3):354–377

Krippendorf J (1987) The holidaymakers: understanding the impact of leisure and travel. Butterworth-Heinemann, London

Latimer H (1985) Developing island economies: tourism v. agriculture. Tour Manag 6:32–42

Levi-Strauss C (1978) The origin of table manners. Harper & Rowe, London

MacCannell D (1973) Staged authenticity: arrangements of social space in tourist settings. Am J Sociol 79(3):589–603

Mak AHN, Lumbers M, Eves A, Chang RCY (2012a) Factors influencing tourist food consumption. Int J Hosp Manag 31:928–936

Mak AHN, Lumbers M, Eves A (2012b) Globalisation and food consumption in tourism. Ann Tour Res 39(19):171–196
Mason MC, Paggiaro A (2012) Investigating the role of festivalscape in culinary tourism: the case of food and wine events. Tour Manag 33:1329–1336
Meiselman HL, Johnson JL, Reeve W, Crouch JE (2000) Demonstrations of the influence of the eating environment on food acceptance. Appetite 35:231–237
Mennell S (1985) All manners of food: eating and taste in England and France from the middle ages. Blackwell, Oxford
Minister of Industry of Canada (2003) National tourism indications. Quarterly Estimates Fourths Quarter, System of National Accounts, Catalogue No. 13-009-XPB
Mintz S (1996) Swallowing modernity. In: Watson J (ed) Golden arches east: McDonald's in East Asia. Stanford University Press, Stanford, pp 174–211
Molz JG (2004) Tasting an imagined Thailand: authenticity and culinary tourism in Thai restaurants. In: Long LM (ed) Culinary tourism. University Press of Kentucky, Lexington, pp 77–93
Quan S, Wang N (2004) Towards a structural model of the tourism experience: an illustration from food experiences in tourism. Tour Manag 25:297–305
Ram U (2004) Glocommodification: how the global consumes the local – McDonald's in Israel. Curr Sociol 52(1):11–31
Reynolds P (1994) Culinary heritage in the face of tourism. Prog Tour Recreat Hosp Manag 6:189–194
Stephens DC (ed) (2000) The Maslow business reader. Abraham H. Maslow. Wiley, New York
Taylor BE, Morison JB, Fleming EM (1991) The economic impact of food import substitution in the Bahamas. Soc Econ Stud 40:45–62
Telfer DJ, Wall G (1996) Linkages between tourism and food production. Ann Tour Res 23 (3):635–653
Telfer DJ, Wall G (2000) Strengthening backward economic linkages: local food purchasing by three Indonesian hotels. Tour Geogr 2(4):421–447
Tikkanen I (2007) Maslow's hierarchy and food tourism in Finland: five cases. Br Food J 109 (9):721–734
Warde A, Martens L (2000) Eating out: social differentiation, consumption and pleasure. Cambridge University Press, Cambridge

Part II
Food Diversity in a Multidimensional Perspective: Food Traditions, Food Security, Right to Food, Market's Evolutionary Processes

Diversity of Food Traditions: A Historical Perspective on Invention and Transformation

Alessandro Isoni

1 Introduction

"Novelties attract us only when they are dead, incapable of arousing vital currents; that is what gives rise to the extraordinary phenomenon of the constant formation of myths which would be venerable if they were really ancient, but which are really nothing but sinister attempts to plunge us back into a past that attracts us only because it is dead"—thus the Prince of Salina described the Sicilian people to Chevalley, special envoy of the Italian government. Chevalley had the task of proposing Salina as a candidate for the Senate of the Kingdom of Italy following the annexation of the Kingdom of the Two Sicilies by the rest of the Italian peninsula. Meanwhile, Salina, one of the most important exponents of the Sicilian aristocracy, refused this nomination while making another bitter observation about Sicily: "Do you really think, Chevalley, that you are the first who has hoped to canalize Sicily into the flow of universal history? I wonder how many Moslem imams, how many of King Roger's knights, how many Swabian scribes, how many Angevin barons, how many jurists of the Most Catholic King have conceived the same fine folly; and how many Spanish viceroys too, how many of Charles III's reforming functionaries! And who knows now what happened to them all! Sicily wanted to sleep in spite of their invocations; for why should she listen to them if she herself is rich, if she's wise, if she's civilized, if she's honest, if she's admired and envied by all, if, in a word, she is perfect?".[1]

[1]See Tomasi di Lampedusa (1958), part Four.

A. Isoni (✉)
Department of History, Society and Human Studies, University of Salento, Lecce, Italy
e-mail: alessandro.isoni@unisalento.it

A. Isoni et al. (eds.), *Food Diversity Between Rights, Duties and Autonomies*,
LITES – Legal Issues in Transdisciplinary Environmental Studies 2,
https://doi.org/10.1007/978-3-319-75196-2_12

This contribution sets out with some extracts from one of the most beautiful, twentieth century, Italian history novels, as these bring into focus at one glance some key issues to be addressed here.

A first issue concerns the authenticity—or, at least, the likely authenticity—of traditions with respect to historical reality. Conversely, to what extent can folkloric reconstructions be considered historically genuine? Or, on the contrary, when is the supposed authenticity of such so-called traditions invalidated, unveiling them at times as veritable inventions.

A second issue closely relates to the first and concerns the ahistorical use of history, implying a kind of theoretical short-circuit able to invalidate any serious attempt at reconstructing and interpreting complex realities, such as those concerning the manifold cultural identities of peoples around the world.

A reference to Lampedusa's famous novel, *The Leopard*, immediately resolves these thorny issues, offering a valuable prelude to the analysis of food traditions that constitutes the main objective of this paper. This analysis focuses on the issues related to the 2010 declaration of the "Mediterranean diet" as Intangible Cultural Heritage of Humanity by UNESCO. It examines when and how the concept of a Mediterranean diet spread, as well as what factors led up to the UNESCO nomination. Above all, this discussion considers the reasons for an affirmative decision by UNESCO after a previous refusal in 2008, and identifies the resulting declaration as a clear indication of the lack of vitality of invented traditions.

2 Between History and Myth: The Invention of Tradition

Prior to proceeding, it is worth returning to a previous citation from *The Leopard*. This quote represents an extraordinary opportunity to fully appreciate the sensitivity of history as a subject that needs to be handled with care and is likely to lead to misunderstandings in inexperienced hands.[2] In the second, above-quoted excerpt the Prince of Salina describes a kind of ancestral apathy characterising Sicily—importantly, not the Sicilians—regarding the policies of different foreign rulers superseding one another on the island after the end of the Ancient World.

This extract follows on from another famous passage, in which the Prince of Salina refers to the peculiarities of Sicilians from being ruled by different civilizations over the centuries. The specific features characterising the Sicilian people are seen to be both the outcome of climatic and geographical factors, as well as because of cultural facts, which may be attributed to a kind of "colonial" *longue durée*.[3] This interpretation, however, suggests a great aporia that may be summarized by the following questions: were Sicilians at the time of the Arab domination similar to

[2]Gilmour (2003), Bottino (1973), Masi (1996), Lanza Tomasi (2001), Vitello (2008) and Nigro (2012).

[3]Braudel (1958), pp. 725–753.

those who viewed the enlightened reformism of Charles III of Bourbon with interest? Were eighteenth century Sicilians the outcome of Arab supremacy, as well as a result of previous superseding dominations? Conversely, are Sicilians a manifestation of historicity or, instead, out of step with their times and collocated in an ahistorical dimension aspiring to eternity?[4] These questions highlight Lampedusa's paradoxical use of history: in the pages of his novel the supposed historical depth, both of facts and problems concerning Sicily, is posited as the outcome of cultural influences. These influences locate a sequence of actual facts, with roots in historically verifiable episodes, in a timeless dimension.[5]

This paradox—implying the use of historical events to adopt mythical dimensions—directly reflects a typical phenomenon of human existence, i.e. the invention of tradition. This concept clearly refers to the famous book edited by Eric J. Hobsbawm and Terence Ranger, *The Invention of Tradition*, with various contributions analysing how traditions have, at times and for very different reasons, been totally invented.[6] A brief overview of what is meant by the "invention of tradition" and which goals motivate this ideological activity, is of relevance here. According to Eric Hobsbawm "invented traditions" are taken "to mean a set of practices, normally governed by overtly or tacitly accepted rules and of a ritual or symbolic nature, which seek to inculcate certain values and norms of behaviour by repetition, which automatically implies continuity with the past".[7] In this context, it is important not to confuse traditions with customs, which usually dominate so-called "traditional" societies. Eric Hobsbawm clears the field of potential misunderstandings by underlining that the goal of traditions, including invented ones, is "invariance": in effect, "the past, real or invented, to which they refer imposes fixed (normally formalized) practices, such as repetition". In some ways, "inventing traditions is essentially a process of formalization and ritualization, characterized by reference to the past, if only by imposing repetition".

Concisely, the key elements of invented traditions can be identified as continuity with the past, repetition and immutability. Each of these appear in the pages of *The Leopard* sign-posting misrepresentations of history. Moreoever, they are often seen as essential requisites to justify historical interpretations that validate political and cultural designs. Similarly, according to Eric Hobsbawm, invented traditions can be organized in three ideal types, often overlapping and aimed at achieving various targets. Traditions of the first type establish or symbolize social cohesion or the membership of groups, be these real or artificial communities; the second type of traditions establishes or legitimizes institutions, status or relations of authority; meanwhile, the last type pools together those invented traditions whose main purpose regards socialization—that is the inculcation of beliefs, value systems and conventions of behavior.

---

[4]Martinoni (2011) and Orlando (1998).

[5]Bloch (1949), Febvre (1952) and Furet (1982).

[6]Hobsbawm and Ranger (1983).

[7]Hobsbawm (1983), pp. 1–14.

3 Food Traditions: The "Mediterranean Diet"

It is important to underline how invented traditions are often the most evident sign of moments of transition. They offer real clues to crisis situations, when rapid transformations within societies lead to the weakening, or even destruction of traditional cultural patterns. The dissolution of previous traditions inevitably leads to the creation of new cultural patterns, able to both reinforce social cohesiveness and generate new values tightly binding groups together.

As such, invented traditions provide a significant service in limiting upheaval during critical phases of change, when the demise of old patterns may endanger an entire society. In response to such threats, people tend to resort to old "raw material" to invent new traditions, thereby avoiding a total loss of meaning. However, since such processes of transition usually last a long time, it is generally very difficult to locate a specific instant when the invention of such traditions starts. Additional problems mark the study of food traditions, as these have their roots mostly in oral traditions, making it highly challenging to identify any sources of information related to them.[8]

In this sense, a study of the tradition of homemade tomato sauce provides an insightful example: widespread in Southern Italy, this is not only a way of economically supporting households to face the winter months, but also represents a time-honoured practice passed down over the centuries, able to transmit key values, such as resourcefulness and conviviality.[9] However, research reveals that the homemaking of tomato sauce is a relatively new practice—the main condiment previously being bell pepper sauce—thus unmasking this invented tradition as a mythological element of Italian food traditions. In fact, tomatoes are a key element of the so-called Mediterranean diet and, since the mid-twentieth century, have become a culinary and health-conscious benchmark.[10]

Interestingly, any reference to the term "Mediterranean diet" generally refers to an invented tradition, which—unlike other such traditions—not only has a father and mother but also a date of birth. The American spouses, Ancel and Margaret Keys may be seen to be the parents of the "Mediterranean diet" and it is worth considering what led them to formulate this innovative interpretation of Mediterranean food habits.[11] Ancel Keys may be confidently defined as one of the most influential twentieth-century food scientists. He was able to combine his scientific competence and historical and cultural sensitivity, with a great sense of culinary taste and a love of nature.[12] It is worth remembering, equally, that Ancel Keys helped to create the so-called K-ration, an individual daily combat food portion, which was introduced by the US Army during World War II to feed American soldiers all along the front

[8]Harris (1998).

[9]Giancristofaro (2012).

[10]Moro (2014).

[11]Truswell (2010).

[12]Mancini and Stiemler (2004), pp. 52–57.

lines. Following the war, Ancel Keys continued to study food issues and, in 1950, published his research on the effects of malnutrition on human beings: this work aroused the interest of the FAO and he was subsequently invited to attend the First World Workshop on Famine in 1951.[13] During this workshop, held in Rome, Ancel Keys stressed the close relationship between diet and heart disease, stimulating the interest of Italian scientists, who declared that this problem was relatively absent in the South of Italy. In 1952, Ancel and Margaret Keys left for Naples to verify the incidence of heart disease in southern Italy and its relationship to food habits.[14] With this journey from Oxford to Naples the adventure of the "Mediterranean diet" began, giving rise to a concept which was to be widely adopted thanks to the decision of Ancel and Margaret Keys to take residence in Pioppi, a coastal village in the Cilento district.

Empirical records collected by the two American scientists confirmed that among the lower-classes of southern Italy cholesterol rates and heart disease numbers were low, while the upper class in Naples and people from the north of Italy showed a high cholesterol rate and heart disease statistics like those of US citizens. Conversely, a high protein diet, typical of rich countries, proved to be a high-risk diet for heart disease, whereas the Mediterranean diet, based on carbohydrates, vegetable proteins, vegetables, fruit and olive oil, was clearly able to ensure both physical and psychological well-being.[15] It was not by chance that the Keys' study was entitled "*Eat well and stay well*", confirming an indissoluble link between diet and well-being.[16]

The complex framework hypothesized by Ancel Keys and his research team, devoted to the so-called Seven Countries Study, enjoyed the favour of the UN World Health Organization, which in 1994 adopted a large part of this study to establish its nutritional guidelines; similarly, in recent years the UN Food and Agriculture Organization declared the Mediterranean diet as one of the most sustainable in the world, especially when considered from an agroecological point of view.[17]

Until now this discussion has shown how the diet apparently adopted in southern Italy was literally "discovered" by an American scientist. This does not, however, explain the birth of the "Mediterranean diet" as a concept. To begin, the definition of this concept was set out during a time span between 1959—when the aforementioned book "*Eat well and stay well*" was published—and 1975, when a new edition of the book was published with the revised title "*How to eat well and stay well. The Mediterranean way*". This second edition marked a transition from description to prescriptive measures reflecting the typical path of invented traditions.[18] Distinctive patterns were introduced, based on supposedly shared diets that acknowledged food habits and the traditions of local communities.

[13]Keys (1950).

[14]Keys et al. (1954), pp. 328–335.

[15]Rubba et al. (2007) and Stamler (2013).

[16]Keys and Keys (1959).

[17]Toshima et al. (1994) and Kromhout et al. (2002).

[18]Keys and Keys (1975) and Nestle (1995).

However, what was the meaning of a Mediterranean diet in the Keys' opinion? While this question could be dismissed as an idle remark, it actually offers a valuable starting point to examine what anthropological and cultural patterns significantly contributed to defining the original features of this so-called Mediterranean diet.[19] According to Ancel and Margaret Keys, the Mediterranean world could be seen to correspond with Pioppi, a small village situated near the ancient Greek city of Elea, once home to the Eleatics, a pre-socratic school of philosophy founded by Parmenides and including other members, such as Zeno of Elea and Melissus of Samos.[20] In the Keys' eyes, the historical stratification and cultural wealth of Cilento represented a great cultural wealth, allowing them to align and compare civilizations lying 2500 years apart.[21] To a certain degree, this supposed historical depth presumed by the Keys revealed itself to be a myth that was essential to justify cultural and gastronomic choices blending ancient and modern practices.[22] In the Keys' perspective, the Mediterranean world was a mythological place. Even the relatively novel name of this sea basin—as a metaphor of a complex reality, residual concerning the different civilizations along its shores—did not manage to distract their attention from a unifying conception of this area, a view which was apparently natural and confirmative of the typical *Grand Tour* clichés—sunlight, sea and violent passions. The sobriety of habits identifying people from Cilento seemed to be the main characteristic of the Mediterranean diet. This was marked by seasonality, the absence of red meat and, above all, pervaded by conviviality and a passion for good flavors. These latter elements inevitably transform the process of food consumption into a tool of symbolic—and potentially mythological—production.

4 The Mediterranean Diet as UNESCO Intangible Cultural Heritage

In 1995, 40 years after the "discovery" of the Mediterranean diet, Ancel Keys returned to the key question motivating his entire life's work: what is the Mediterranean diet? His own response is noteworthy here, as it variously confirms our hypothesis: according to Ancel Keys, the Mediterranean diet "is what the Mediterranean natives eat. But as we know and think of it now, it is a relatively new invention. Tomatoes, potatoes, and beans, for example, came from America long after Christopher Columbus discovered the New World. Today, we are interested in

[19]Mackenback (2007).

[20]Pugliese Carratelli (1980), pp. 269–280.

[21]Interestingly, the Keys' home in Pioppi was named "Minnelea", syneresis of Minneapolis—the American town where Ancel Keys had his class—and Elea, the old toponym of Pioppi, to indicate the historical continuity between Magna Graecia and Cilento.

[22]Braudel (1977–1978).

the Mediterranean diet as it is now and has been in recent times".[23] In these few lines, Ancel Keys contradicts some key assumptions on which his research in the 1950s was based. First, he denies the genuineness of the Mediterranean diet: prior to his research, nobody made any reference to or imagined such a definition. Conversely, it was clearly an invention, an abstract concept imposed onto a complex reality. At nearly 90 years of age, Ancel Keys himself dismissed the ahistorical and mythological features assumed to be inherent to a Mediterranean diet: his reference to vegetables of American origin inevitably broke the perceived line of historical continuity within a supposedly unchanging and everlasting world, seen to be characterized by the same distinctive tastes and courses of the Ancient World.[24]

Ancel Keys was led to give a definitive answer to his pivotal question defining a Mediterranean diet by the evident change of food habits occurring in the Mediterranean region, with people's lives ever more resembling those of others living in the wealthy western hemisphere. This change provoked considerable bitterness in Keys. As part of the so-called Italian "economic miracle", he witnessed how Italy was transformed from a poor agricultural country into a great industrial power within a very short time. This was marked by the mass migration of nine million farmers from southern Italy towards the northern cities of the Po valley between 1955 and 1971.[25]

In reality, the discovery of the Mediterranean diet coincided with a critical threat to its existence, thereby confirming a typical paradox of invented traditions. Meanwhile, this concept became widely diffused and it has, over time, increased its sphere of reference, moving well beyond food traditions to encompass a set of features nowadays commonly summarized as the "Mediterranean way of life".[26] In this broader sense, the term "Mediterranean diet" relates to the ancient Greek concept of *dìaita*, in the context of which diet referred to rules of conduct characterized by moderation, frugality and the observance of life cycles, thereby becoming an actual philosophy or lifestyle.[27]

In this cultural atmosphere, a Mediterranean way of life was seen to be characterized by low stress levels and, as such, viewed as highly desirable. This was enhanced by the advent of economic globalization, provoking localistic effects in response to the increasing loss of cultural and social identities.[28] One of the key expressions of this phenomenon was the 2008 nomination of the Mediterranean diet to become part of the UNESCO list for Intangible Cultural Heritage. Article 2 of the UNESCO Convention for the Safeguarding of the Intangible Cultural Heritage, signed in Paris in 2003, provided that "the "intangible cultural heritage" means the practices, representations, expressions, knowledge, skills – as well as the instruments, objects, artefacts and cultural spaces associated therewith – that communities,

---

[23]Keys (1995).

[24]Cantarelli (2005).

[25]Ginsborg (1990).

[26]Cassano (2011), Cassano and Consolo (2000) and Cassano and Zolo (2007).

[27]Latouche (1999, 2011).

[28]Appadurai (1996) and Bauman (1999).

groups and, in some cases, individuals recognize as part of their cultural heritage".[29] In 2008 UNESCO rejected the nomination of the Mediterranean diet, supported by Italy, Spain, Greece and Morocco, with a final decision that is worth examining.[30] Significantly, the Mediterranean diet does not wholly characterize these countries: in some of them—such as Italy and France—food traditions are divided along a "celtic line" dividing northern and southern regions.[31] Moreover, this nomination was managed by national cabinets, according to a *top-down* approach, which overlooked local communities, thus revealing an artificial imposition, other than applying a definition of the Mediterranean diet that was limited to a list of typical foods. These limitations came under close scrutiny at a 2009 conference held in France, at which a clear understanding of food traditions as complex phenomena emerged. Culinary habits were seen to embody wider traditions, including oral transmissions, the arts, dance, folk song and poetry, as well as regional and patronal festivals, religious rites and more.[32] Above all, this new definition of food traditions had a twofold aim: it was an attempt to view food traditions within a long-term perspective and it posited a clear invite not to embalm food traditions within stereotypical formulas.

Consequently, the renewed 2008 nomination focused on specific local communities—Pollica in Italy, Soria in Spain, Koroni in Greece and Chefchaouen in Morocco—thereby taking a *bottom-up* approach that was able to highlight both commonalities, as well as unique features characterising specific communities, all of which belonged to the more ample Mediterranean *koinè*.[33] This twofold identity was brought to light by UNESCO, which defined the Mediterranean diet as "a social practice based on the set of skills, knowledge, practices and traditions ranging from the landscape to the cuisine, which in the Mediterranean basin concern the crops, harvesting, fishing, conservation, processing, preparation and, particularly, consumption. This set, recreated within and by the communities identified in the territories of the four State Parties, is unavoidably linked to a seasonal calendar marked by nature and religious or ritual meanings. The Mediterranean Diet as a unique lifestyle, determined by the climate and by the Mediterranean area, is also manifested through related festival and celebrations. These events become the receptacle of gestures of mutual recognition, hospitality, neighbourliness, conviviality, intergenerational transmission and intercultural dialogue. This is how among these communities in particular, and the Mediterranean people in general, there is a feeling of rebuilding identity, of belonging, and of continuity, allowing them to recognise this element as an essential part of their shared intangible cultural heritage".[34]

---

[29]Clemente and Mugnaini (2001) and Mariotti (2013).

[30]Skounti and Tebbaa (2011).

[31]Montanari (1997).

[32]Grimaldi (1990).

[33]Scovazzi et al. (2012).

[34]UNESCO (2010) Nomination file 00394. http://www.unesco.org/culture/ich/index.php. Accessed 10 Jan 2016.

Within this description three main features of intangible cultural heritage stand out: a kind of epiphany of this heritage (objective part); the community that is characterised by it (subjective or social part) and, finally, the wider cultural field (spatial part). Most significantly, according to this understanding, food practices represent a type of cultural experience, reaching well beyond food itself and involving the actual identities of various groups of people.[35]

5 The Damnation of King Midas

Before continuing, it is valuable to underline another important feature: namely, that the UNESCO description of the Mediterranean diet is quite ambiguous. It consists of both material and immaterial issues denoting its elusive complexity.[36] Ultimately, some of the key words found in the UNESCO declaration are difficult to understand, reflecting the typically shaky ground on which invented traditions stand. In reality, manifold aporias are the clearest confirmation that the UNESCO declaration certifies the end of the Mediterranean diet as a social practice. Numerous factors corroborate this hypothesis, among which the most important concerns the totemic use of the Mediterranean diet. The aim of Ancel Keys' research was of particular significance here. It was devoted to demonstrating the existence of a healthy diet, typical of Mediterranean people, to devise a dietary pyramid able to replace the fatty diets of the Western world and to reduce high heart disease rates.[37] In line with this, the Mediterranean diet was considered a cure-all able to ensure longevity thanks to the healthy lifestyle it guaranteed.

To make this diet more attractive, Ancel Keys referred to heterogenous food specific elements, thereby creating a cultural patchwork reminiscent of a food Arcadia, in which some historical features relating to the Ancient World were adopted to erase features of other times viewed as less functional to the plan of creating a narrative specific to the Mediterranean.

Ancel Keys succeeded in this idealization of food traditions, producing a completely ahistorical identity of Mediterranean culinary traditions, reflecting a model common to romantic novels, where the exotic prevails over reality.[38] The UNESCO description of a Mediterranean diet actively contributed to this by stating "that one group deserves special consideration: women, who always play a very important, often essential, role in the transmission of expertise, in the recreation of rituals, traditional gestures and celebration, in the safeguarding of techniques, in the respect of seasonal rhythms and the inclusion of all cultural, social and environmental values in the education of the new generations", thus presenting a male

[35]Scovazzi (2012).

[36]Scepi and Petrillo (2012).

[37]Buckland et al. (2008), La Vecchia and Bosetti (2006) and Trichopoulou and Vasolopoulou (2000).

[38]Moro (2013).

chauvinist, traditionalist and grotesque portrait, largely cut off from the actual role of women in Mediterranean societies, considering even the differences to be expected between the northern and southern coast.

A clear liking for ancient world heritage and the creation of ahistorical identities are two elements that commonly guide local development planning. Through the construction of brands that can strongly publicise these preferences, such development frequently results in the omission of any attempts to understand local communities.[39] In parallel, the UNESCO Committee was also aware of perspectives exclusively viewing the Mediterranean diet as a business, thus circumventing any issues of heritage transmission and going so far as to recreate false heritage replicas. Conversely, it is vital to discern whether traditions are genuine phenomena or, instead, follow pre-arranged patterns. If is it true that traditions are characterized both by supposed temporal continuity and significant inalterability, then any novel interpretation of such traditions breaks the invisible line of discernment between the genuine and pre-arranged. This equally reveals to what degree traditions are not respected, possibly because they are often seen to be behind the times.

In practical terms, the inscription of the Mediterranean diet in the UNESCO list for Intangible Cultural Heritage has two key outcomes: first, it reveals how protected traditions usually "live on" only because they are safe-guarded, while having lost any life-breath of their own; secondly, the exploitation of tourism has condemned the Mediterranean region to "the damnation of King Midas". King Midas, the mythological king who could transform everything he touched into gold, quickly realized that this perceived gift was actually a calamity, as everyone he met lost their lives as they were transfixed into gold. This is precisely what is happening to the Mediterranean people nowadays, when they are condemned to a traditionalist conception of their identity, deprived of their cultural complexity, enchained to an imaginary past and therefore unable to imagine a different future.

References

Appadurai A (1996) Modernity at large: cultural dimensions of globalization. University of Minnesota Press, Minneapolis

Bauman Z (1999) In search of politics. Polity Press, Cambridge

Bellini N (2000) Il marketing territoriale. Sfide per l'Italia nella nuova economia. Franco Angeli, Milano

Bloch M (1949) Apologie pour l'histoire ou métier d'historien. Armand Colin, Paris

Bottino G (1973) Saggio su "Il Gattopardo" di Giuseppe Tomasi di Lampedusa. Realizzazioni grafiche artigiana, Genova

Braudel F (1958) Histoire et sciences sociales. La longue durée. Ann ESC 4:725–753

Braudel F (1977–1978) La Mediterranée. L'espace et l'histoire, Les hommes et l'héritage. Arts et Métiers Graphiques, Paris

[39] Bellini (2000) and Cercola et al. (2010).

Buckland G, Bach A, Serra-Majem L (2008) Obesity and the Mediterranean diet: a systematic review of observational and intervention studies. Obes Rev 9(6):582–593
Cantarelli F (2005) I tempi alimentari del Mediterraneo. Cultura ed economia nella storia alimentare dell'uomo. Franco Angeli, Milano
Cassano F (2011) Modernizzare stanca. Perdere tempo, guadagnare tempo. Il Mulino, Bologna
Cassano F, Consolo V (2000) Lo sguardo italiano. Rappresentare il Mediterraneo. Mesogea, Messina
Cassano F, Zolo D (2007) L'alternativa mediterranea. Feltrinelli, Milano
Cercola R, Izzo F, Bonetti E (2010) Eventi e strategie di marketing territoriale. I network, gli attori e le dinamiche relazionali. Franco Angeli, Milano
Clemente P, Mugnaini F (eds) (2001) Oltre il folklore. Tradizioni popolari e antropologia nella società contemporanea. Carocci, Roma
Febvre L (1952) Combats pour l'histoire. Armand Colin, Paris
Furet F (1982) L'Atelier de l'histoire. Flammarion, Paris
Giancristofaro L (2012) Tomato Day. Il rituale della conserva di pomodoro. Franco Angeli, Milano
Gilmour D (2003) L'ultimo gattopardo. Vita di Giuseppe Tomasi di Lampedusa. Feltrinelli, Milano
Ginsborg P (1990) A history of contemporary Italy: society and politics: 1943–1980. Penguin Books, London
Grimaldi P (1990) Il calendario rituale contadino. Il tempo della festa e del lavoro fra tradizione e complessità sociale. Franco Angeli, Milano
Harris M (1998) Good to eat: riddles of food and culture. Waveland Press, Long Grove
Hobsbawm EJ, Ranger T (eds) (1983) The invention of tradition. Cambridge University Press, Cambridge
Keys A (1950) The biology of human starvation. University of Minnesota Press, Minneapolis
Keys A (1995) Mediterranean diet and public health: personal reflections. Am J Clin Nutr 61:1321–1323
Keys A, Keys M (1959) Eat well and stay well. Doubleday, Garden City
Keys A, Keys M (1975) How to eat well and stay well: the Mediterranean way. Doubleday, Garden City
Keys A, Fidanza F, Scardi V et al (1954) Studies on serum cholesterol and other characteristics on clinically healthy men in Naples. Arch Int Med 93:328–335
Kromhout D, Menotti A, Blackburn H (2002) Prevention of coronary heart disease: diet, lifestyle and risk factors in the seven countries study. Kluwer, Norwell
La Vecchia C, Bosetti C (2006) Diet and cancer risk in Mediterranean countries: open issues. Public Health Nutr 9(8A):1077–1082
Lanza Tomasi G (2001) I luoghi del Gattopardo. Sellerio, Palermo
Latouche S (1999) Le Défi de Minerve. Rationalité occidentale et raison méditerranéenne. La Découverte, Paris
Latouche S (2011) Vers une societé d'abondance frugale. Contresens et controverses sur la décroissance. Mille et une nuits, Paris
Mackenback JP (2007) The Mediterranean diet story illustrates the "why" questions are as important as the "how" questions in disease explanation. J Clin Epidemiol 60:105–109
Mancini M, Stiemler J (2004) Diet for preventing cardiovascular diseases: light from Ancel keys, distinguished centenarian scientist. Nutr Metab Cardiovasc Dis 14:52–57
Mariotti L (2013) La convenzione sul patrimonio intangibile e i suoi criteri di valorizzazione, tutela e protezione. Voci Ann sci um 10:88–97
Martinoni R (2011) Troppo poco pazzi. Leonardo Sciascia nella libera e laica Svizzera. Olschki, Firenze
Masi G (1996) Come leggere Il Gattopardo di Giuseppe Tomasi di Lampedusa. Mursia, Milano
Montanari M (1997) L'Europa a tavola. Storia dell'alimentazione dal Medioevo a oggi. Laterza, Roma-Bari
Moro E (2013) La dieta mediterranea tra i presocratici e l'UNESCO. Retoriche di ancestralizzazione e politiche di patrimonializzazione Voci Ann sci um 10:111–122

Moro E (2014) La dieta mediterranea. Mito e storia di uno stile di vita. Il Mulino, Bologna 2014
Nestle M (1995) Mediterranean diets: historical and research overview. Am J Clin Nutr 61:1313–1320
Nigro SS (2012) Il Principe fulvo. Sellerio, Palermo
Orlando F (1998) L'intimità e la storia. Lettura del "Gattopardo". Einaudi, Torino
Pugliese Carratelli G (1980) Tra Cadmo e Orfeo. Contributi alla storia civile e religiosa dei Greci d'Occidente. Il Mulino, Bologna
Rubba R, Mancini FP, Gentile M, Mancini M (2007) The Mediterranean diet in Italy: an update. World Rev Nutr Diet 97:85–113
Scepi G, Petrillo PL (2012) La dimensione culturale della dieta mediterranea patrimonio immateriale dell'umanità. In: Golinelli GM (ed) Patrimonio culturale e creazione di valore. Verso nuovi percorsi. Giuffré, Milano, pp 247–273
Scovazzi T (2012) La definizione di patrimonio culturale intangibile. In: Golinelli GM (ed) Patrimonio culturale e creazione di valore. Verso nuovi percorsi. Giuffré, Milano, pp 151–186
Scovazzi T, Ubertazzi B, Zagato L (eds) (2012) Il patrimonio culturale intangibile nelle sue diverse dimensioni. Giuffré, Milano
Skounti A, Tebbaa O (2011) De l'immatérialité du patrimoine culturel. UNESCO/Université Cadi Ayyad Marrakech, Marrakech
Stamler J (2013) Toward a modern Mediterranean diet for the 21st century. Nutr Metab Cardiovasc Dis 23:1159–1162
Tomasi di Lampedusa G (1958) Il Gattopardo. Feltrinelli, Milano (English Translation by Colquhoun A (1960) The Leopard, Collins, London)
Toshima H, Koga Y, Blackburn H (eds) (1994) Lessons for science from the seven countries study: a 35-year collaborative experience in cardiovascular disease epidemiology. Springer, Tokyo
Trichopoulou A, Vasolopoulou E (2000) Mediterranean diet and longevity. Br J Nutr 84:205–209
Truswell AS (2010) Cholesterol and beyond. The research on the diet and coronary heart disease 1900–2000. Springer, Dordrecht
UNESCO (2010) Nomination file 00394. http://www.unesco.org/culture/ich/index.php. Accessed 10 Jan 2016
Vitello A (2008) Giuseppe Tomasi di Lampedusa: il Gattopardo segreto. Sellerio, Palermo

Food Security: A Challenge for a Global Governance

Fabio Pollice, Valentina Albanese, and Giulia Urso

1 "Nourish the Planet" Beyond the Anthropocentric Vision

The description of the rationale underpinning the theme of the World Exposition Milan 2015 takes off with the question: "Is it possible to secure nutrition that is good, healthy, sufficient and sustainable for all humanity?".[1] The vision—despite of the reference to the "planet" that the very name of the event incorporates—is very much anthropocentric. It would be more appropriate to ask "Given that humanity has increasing nutritional needs, does the planet have a future?".

The question, as discussed in this chapter, is not at all rhetorical, since increasing population together with a (more than proportional) increase in food consumption bring along considerable sustainability issues. The potential increase in food production causes irreversible risks to the planet's ecosystem. On the other hand, as many analyses show, if the planet's capabilities have not theoretically been exceeded yet, certainly the consumption of its resources has exceeded the planet's capacity to replace such resources. Therefore, the non-sustainability of human development is already a reality, regardless of the further increase in human kind's consumption.

[1]Cfr. Expo 2015 (2015), *Guida al tema*, Milano (http://www.expo2015.org/archive/it/cos-e/il-tema.html).

F. Pollice (✉)
Department of History, Society and Human Studies, University of Salento, Lecce, Italy
e-mail: fabio.pollice@unisalento.it

V. Albanese
Department of History and Cultures, University of Bologna, Bologna, Italy
e-mail: valentina.albanese@unibo.it

G. Urso
Social Sciences Unit, Gran Sasso Science Institute, L'Aquila, Italy
e-mail: giulia.urso@gssi.it

A. Isoni et al. (eds.), *Food Diversity Between Rights, Duties and Autonomies*,
LITES – Legal Issues in Transdisciplinary Environmental Studies 2,
https://doi.org/10.1007/978-3-319-75196-2_13

From this side, the question that the promoters of Expo 2015 bring to the fore cannot but be considered clearly rhetorical.

The current consumption trend, and the direction it is taking on a planetary scale, makes food security non-sustainable.[2] Humankind's future lies in the adoption of a nutrition policy on a global scale that maintains consumption within the energetic need threshold and that directs such consumption towards products with lower impact on the environment, adapting at the same time cultivation plants to pedoclimate conditions (meaning, adapting cultivations to sites rather than sites to cultivations), reducing also food exchanges to the minimum levels to avoid extra damage to the environment because of products transportation.[3]

It is important to understand that these are objectives that must not be left in the market's hands, given the fact that capitalism is moving in a totally different direction, as it will be discussed further. Nor can they be dealt with by single countries that, when free to decide their own nutrition policies, generally demonstrate inadequate sensibility towards the environment prioritizing instead national interests to those of the entire humankind. The only solution is the set-up of a global, nourishment governance model that guarantees equal access to available resources within the limits that the ecosystem safeguard imposes.[4]

2 The "Food Security" Objective

"Food security" is a much complex and articulated concept; it has to do with both geo-economics and geo-politics. To properly understand these interdependent fields, it is worth focusing on the very definition of food security, recalling the one of the 1996 World Food Summit: food security is related to the insuring at any time to every single human being [. . .] physical, social, economic access to sufficient, safe and nutrient foods, able to cover nutritional needs and preferences to lead a safe and active life.

The word "access" explicitly refers to a crucial issue: the one of food resources property, most of which is already under private control and out of the collective availability. It is the same process that is happening to water availability that, in most parts of the globe, is undergoing privatisation processes,[5] a fact aggravated by the increasing water shortage that jeopardises large areas of our planet. The guarantee to the access to food resources cannot be connected to intergenerational equity: one of the two pivotal principles of sustainable growth. Everyone must have equal

[2]Pollice (2013a), Lovejoy and Hannah (2005) and Maes (2013).

[3]Guha-Khasnobis et al. (2007).

[4]Solbrig et al. (1994).

[5]Pollice (2013b).

possibilities to access food resources, with no economic, social, racial, political forms of discrimination.[6]

It is pointless to underline that capitalism not only did not prove to be capable of securing such a right. It even proved to be the reason for an unwise consumption of food resources and for an uneven distribution of opportunities to access food resources. Corrective actions—by means of re-distribution policies or interventions to help weakest categories—not only were not diffusely applied; recently they were even cancelled or significantly reduced, as welfare policies' evolutions of the last years in most of western countries demonstrate.

Nevertheless, unequal opportunities to access food resources are never only of an economic nature. Most of them come from a political matrix and, in some areas of the planet, are used to make pressure on specific ethnical or socio-geographic minorities to leave. Restricting the access to water or making use of the land can force migrations that facilitate the path to capitalistic appropriations. Here geo-economic and geo-political dimensions already show themselves with their reciprocal dependency. It is when the scale becomes bigger that they take on an even more problematic configuration.

Countries must guarantee food security to their populations, but if the territory on which those countries exercise their political control can't satisfy their populations' nutritional needs, then resources must be found abroad, taking them from other populations. Large part of neo-colonialism, from a strictly economic point of view, can be read as a form of capitalistic exploitation of the "periphery" by dominant economies; on the other side—and cases like *land grabbing* tend to confirm it—it can also be read as a strategy of a geo-political nature aiming to the control of food resources. Analogies between old and new colonialism forms have also to do with resources exploitation and with the consequences that they can produce on the environment.

The exploitation of production potentials of the territories under colonial control used to take place directing such potentials towards cultivations that satisfied the needs of the dominating countries by imposing mono-cultivations: a form of exploitation that is still practised although the political control (political and military specifically) has come to an end, with serious effects on the environment going from the destruction of entire ecosystems characterized by high biodiversity—it suffices to recall the deforestation of the equatorial stripe—to desertification processes caused by the excessive exploitation of the soil.[7] In these cases it is quite evident how any country's food security can be guaranteed at the expense of others and, not rarely, by diminishing those countries' possibilities to positively respond to their own nutritional needs in the future, with irreversible consequences on a planetary scale.

The iniquity of neo-colonialism forms encompasses the merging effects of unequal exchanges and cultural colonization. In fact, while the former tends to

[6]AA VV (2006).

[7]Cervigni (2001) and Dinelli (2012).

increase competition among dominant economies, reducing possibilities of economical redemption of dominated countries, cultural colonization leads dominated populations to adopt nutritional habits of the dominating ones that, as discussed above, is not sustainable on a quantitative level, as well as on a typological one, since they make use of products whose production has a high impact on the environment, as largely demonstrated by the increasing incidence that consumption of food of animal origin is having within the nutritional habits of those countries that are undergoing a process of economical expansion.

The "food security" objective when looked at a national scale produces distortions of this kind. It endangers not only the pursuit of the objective on a planetary scale but on the ecosystemic one, too, since they tend to charge other territories with their problems of inadequate food resources, without considering the consequences that exploitation's intensity and modes can have on those areas.

3 Unsustainable Nourishment

The right to nourishment translates into the presence, still unlikely in many areas of the planet, of two conditions: *food security* and *food safety*, by now included, and rightly, in every world policy against malnutrition. They were firstly mentioned in The Universal Declaration of Human Rights[8] but were defined later within the 2000 Resolution[9] by means of a few but fundamental words: "the right of everyone to have access to safe and nutritious food, consistent with the right to adequate food and the fundamental right of everyone to be free from hunger to be able to fully develop and maintain their physical and mental capacity". To fulfil the right to nourishment, in those very terms, it is necessary that three prerequisites are all together present: availability, access, use. Let us examine them separately. The first prerequisite, availability, stands for food quantity, quality and diversity. In this sense, it is worth mentioning that the relation between food availability and food demand is very much influenced by environmental factors acting on a planetary scale like, for example, climate change, problems linked to water shortage, waist of food resources, the conversion and loss of soil, the injection of polluting substances into the environment and, most of all, the decrease in biodiversity.[10] "Access" is the second prerequisite; it refers to all those aspects—social, economic, political—for long time underestimated in the framework of actions against food uncertainty. It is clear now that unequal food access often stems from environmental factors (like climate change) but also from social factors, like those taking place inside the families and that are typical of developing countries where hunger is more concentrated. Within

[8]The Universal Declaration of Human Rights (1948).

[9]ONU Resolution 2000/10, UN (2000).

[10]Porter et al. (2014).

certain social contexts, for example, the unequal food distribution in favour of the family's male adult is systematic and generally accepted.[11]

Again, in relation to food access and among distortion factors, political ones must be considered. Food uncertainty can be both and at the same time reason for and consequence of conflicts. If, on the one side, hunger causes wars, conflicts cause the destruction of subsistence means in urban and rural areas, undermining agricultural production on the small scale. This way, political instability translates into one of the main causes for hunger, threatening food security in various ways. All food uncertainty circumstances are characterized by some forms of conflict and transform food security crises into devastating famine.

Food uncertainty can represent moreover a catalysing factor for civil and political disorders. Death rate caused by conflicts related to food uncertainty and hunger can largely exceed deaths directly caused by violence.[12]

The third prerequisite that guarantees *food safety* and *food security* is employment: risks related to this factor are of varied nature such as, among others, clean water unavailability, deficient hygiene conditions, inadequate food manipulations.

Obviously, all risks and factors considered so far are not equally distributed on the globe, and geographical analyses of food security are a preliminary step towards the evaluation of possible solutions to such equilibrium. Evaluating new nourishment geographies can give evidence to the location and amount (namely the quantity) of food demands globally.

In its more recent report, FAO shows how currently in the world one out of nine human beings can have food enough to lead a healthy and active life.[13] The largest majority of undernourished people is concentrated in developing regions, and this concerns more than 800 million people (44.4% less than in 1990–1992 data). Nevertheless, these figures, at a first sight encouraging, must be confronted with the planet population growth.

The *World Population Prospects* referred to 2014 (United Nations 2015)[14] essentially confirms the previous estimates of demographic trends (United Nations 2013).[15] The report shows that the world population reached 7.2 billion in 2014, and will increase of over 2 billion before 2050, therefore reaching 9.6 billion. FAO figures not only make clear that the population is growing, they also show location aspects: the very crucial issue. The world population growth is not uniform, entirely concentrated in the less developed regions and, more generally, in African and Asian continents, while in developed countries trends are opposite.[16]

---

[11] According to FAO evaluations, this would be one of the main reasons why bad nutrition and even hunger concerns in a higher percentage women and children.

[12] Tisdell (1999).

[13] FAO (2015).

[14] World Population Prospects, United Nations (2015).

[15] World Population Prospects, United Nations (2013).

[16] The population of 48 countries or areas in Europe should decrease between 2015 and 2050. In some cases, we will see a decrease in the population equal to or even higher than 15% before 2050.

Taking for well-founded United Nations' estimates that in 35 years time the world population will be 2.4 billion, we are projected to an unavoidable consumption increase (UNITED NATIONS 2011). Such increase will be more than proportional compared to the population growth because it will be concentrated, as already said, in those less developed countries where food demand is currently much lower than the one recorded elsewhere.

It is reasonable to expect that in countries with demographic growth, as incomes increase food consumption rises, at least until incomes get to the same levels as those in developed countries.

FAO figures related to hunger decrease at the global level confirm such estimate. Between 1992 and 2012 they show much better results than expected. The most recent FAO estimates show that about 805 million people are constantly undernourished.[17]

Still it must be considered that in 2050 people living in countries with an average daily consumption per person of 3000 Kcal will be 52% of the world population (4.7 billions), while people living in countries with an average lower than 2500 Kcal will be 2.6% of the world population (240 million).

Based on these considerations, FAO considers that between now and 2050 agricultural products demand will increase of an average annual rate of 1.1%: about half the percentage of the average increase recorded within the last 40 years on a planetary scale. This estimate is grounded essentially on two separate long period trends: on the one hand, the progressive decrease in the average increase of the world population; on the other hand, the rise, in some of the most populated countries of the planet, in the levels of consumption per person.[18]

The prevision,[19] fully acceptable, is that an increase in the demand of agricultural products will be concentrated where population increases, that is, where most part of

---

Among these countries are Bosnia and Herzegovina, Bulgaria, Croatia, Hungary, Japan, Latvia, Lithuania, Republic of Moldova, Romania, Serbia and Ukraine.

[17]FAO (2014).

[18]Pollice and Albanese (2015).

[19]About to the increase rates' decrease trend of the world population, we already underlined that it showed a slowing down in the last years, to the point that the United Nations Department of Economic and Social Affairs was obliged in 2010 to rise previous estimates. Same adjustments were made regarding consumption levels. Consequently, FAO's conclusions are not fully convincing. First, it must be considered that if in developed countries the level of daily consumption per person is currently about 3400 Kcal—and that it is meant to increase further by 2050—it is highly possible that in developing countries, because of the rise in income levels, there will be a process of convergence towards developed countries' consumption levels. FAO moreover underlines (FAO 2010) that a missed convergence would be because of the persistence of large—social and/or geographic—poverty fractions within single countries, surely not because of championing behaviours on the consumption ground. Conversely, it is stated that worldwide demand will not increase beyond estimates because poverty will still be widespread.

undernourished people reside[20]: Sub-Saharan region, Middle East, North Africa, Latin America and Caribbean, Southern Asia and Eastern Asia.

Among them, Africa is the fastest growing geographical area. According to the 2015 *World Population Prospects* more than half of the increase of the world population estimated by 2050 will concern the African region. Between 2010 and 2015 Africa already showed a demographic increase of the yearly rate of 2.55% and therefore, 1.3 of the 2.4 billion people that will sum to the current world population will be located in Africa. The second fastest growing area is Asia. Between 2015 and 2050 its population will increase of 0.9 billion. North America, Latin America, Caribbean and Oceania follow in decreasing order.

In conclusion, the growth of the population remains particularly high in those 48 countries defined by the United Nations as the least developed countries (LDC), 27 of which are African countries. Even if the increase rate of least developed countries slows down in the next years, their populations should double from 954 million in 2015 to 1.9 billion in 2050, with a further increase to 3.2 billion in 2100. Among the countries that, by 2100, will have a population five times larger than the current one there are Angola, Burundi, The Democratic Republic of Congo, Malawi, Mali, Niger, Somalia, Uganda, Tanzania and Zambia. Naturally, the high increase of the population in these countries will make even more difficult for local governments to eradicate illiteracy, inadequate hygiene and inequality, together with undernourishment.

4 Consumption Guidance, Markets Governance. The Need for a Global Governance

The world population increase trend together with the increase in consumption per person will lead to an increase in food demand that the world current agricultural system is not capable to sustain.

Consequently, humanity food security can be guaranteed only through a significant increase of agricultural productions, but this would have such negative impacts on biodiversity and environment equilibriums that not only mankind would be at risk; the whole planet's ecosystem would be.

Food security is a global question, as climate change is, and therefore it must be dealt with on a global scale, coordinating strategies and investments.

[20]Some data of a national scale are particularly interesting. In 7 years India's population will surpass China's, currently of 1.380 million compared to 1.310 million of India. By 2022, both countries will probably reach 1.4 billion people. After that, India's population should increase to reach 1.7 billion in 2050. The population of Nigeria, at the present the seventh largest country of the world, is growing very fast to the point that it is supposed that by 2050 it will be the third largest country of the world. Also, by 2050 six out of the ten largest countries of the world could exceed 300 million: China, India, Indonesia, Nigeria, Pakistan and United States of America.

Nevertheless some countries, because of "lack of attention" by the international community, deal with food emergency getting hold of other countries' productive capacities, as it happens with land grabbing—a new form of economic colonialism interesting mostly the African continent but already expanding in Latin America and Asia –. It is an illegal practice strictly linked to the geo-economic and geo-financial relevance that cultivated lands, a limited and rare resource, have taken up in the last decades as a consequence of the progressive expansion of food demand.[21]

It is precisely those countries' predatory behaviours that make clear how dramatic the situation is, and how urgent the need to solicit the international community about the adoption of a strategic, global plan for the management of food resources.

A plan that aims to guarantee a sustainable progress for agriculture worldwide, capable to cope with increases in food demand while preserving biodiversity and the ecosystem and, last but not least, ensuring a more equal resources distribution, reflecting the intergenerational equity principle, which is fundamental not only to sustainable development but also to world peace.[22]

Kevin Morgan supports "alternative" nutrition as an ethical, sustainable way to overcome world hunger.[23] Humanity faces environmental problems that seem to worsen day by day. We are spectators of a constant flux of news regarding climate change, natural disasters, water shortage and deforestation. They all do not certainly work in favour of biodiversity and food security.

We are aware of the increasing toxic chemical substances that are tolerated in the food we eat, and we therefore ask together with Paul Steinberg why such problems still remain, and what can be done to have society taking a more sustainable path.[24] How can we aim to growth, solve the alarming undernourishment and bad nourishment problems following the way to sustainability?

Agricultural productivity increase certainly contributes to determine (or improve, in less critical cases) the access to food resources on a local as well global level; nevertheless, it has to be considered that the increase in food prices and the global financial crises have furthermore reduced access to food, and have contributed to the worsening of nourishment and health conditions since food quality and quantity have been reducing.

Therefore, the increase of agricultural productivity cannot be the only process undertaken to solve crises that are more composite and faceted, as nourishment crisis is. In fact, the increase in food prices and unemployment rates have caused a purchasing power decrease for large part of the people, causing a reduction of food consumption or a shift towards cheaper food, with high concentration of fat and sugar, violating the Food safety & Food security principles discussed above.

International agreements regulating the agro-industrial system (mainly issued by the World Trade Organisation) consider food as a mere market product. With food

---

[21]Sellari (2015).

[22]Lockie and Carpenter (2010).

[23]Morgan (2010).

[24]Steinberg (2015).

been considered as a consumer good rather than a primary resource as it is, current and future food crises can be solved only by means of an increase of the production as a whole, with clear disadvantages for biodiversity and sustainability.

Food, instead, is necessary to survival. For this reason, it is a human right. This means that food cannot be dealt with by market actors.

The right to food is implicitly based on the improvement to food access for the most vulnerable social segments of the population. Efforts must be concentrated on the improvement of food distribution rather than food production. Undernourishment cannot be interpreted based on its availability. The problem must be considered in terms of its accessibility.

Hunger is not because of planetary shortage of food. It is because of lack of accessibility. Food resources are not rare, they are unevenly distributed, not accessible to specific sections of the population or geographical areas.[25] Local governments (the so called *public accountability*) must ensure that their policies are constantly guided by the urgency to mitigate undernourishment and bad nourishment via food distribution improvements and higher resilience for the weakest segments of the population when facing environmental or political pressures.

If the total amount of foodstuffs does not seem to interest industrialised northern countries, where food waste reaches 40% of edible food and where obesity must be fought, the access to healthy, nourishing, varied food that is compatible with local ecosystems and with low impact on the climate and the environment, is an issue that concerns Western as LDC consumers.

Food as a right is therefore a global issue that must be dealt with locally. With such a new cultural paradigm, the principle of food security focused on the increase of agricultural productivity is replaced by a food supremacy principle, that implies the reactivation of the deep and direct bond—of a cultural, social, political, environmental and of course economic nature—between human settlements and their ecosystems.

The localisation of both production and distribution systems actually allows for a decrease of transportation needs and therefore lower the impact on climate, while the so called short supply chains ensure more democratic access, better food quality and freshness, and higher cultural diversity.

Chemical agriculture—derived by the green revolution, strictly linked to pesticide, fertilisers, hormones and to its biotechnological advancements—through a global revision of agricultural policies, should be replaced by local agriculture and livestock, short supply chains centered on local pedoclimates specificities and, lastly, founded on cultivation practices with low environment impact.[26]

[25] Sen (1982).

[26] Shiva (2006, 2009).

5 Security Limits

Based on the above considerations, it seems clear that "food solidarity" should support "food security", so that the entire planet's food resources can be distributed according to world population's needs, ensuring everyone, regardless of their birthplace, an equal access to food resources.

Consequentially, on the one side, the international community must ensure that adequate quantities of food resources—both in type and quality terms—are produced to satisfy humanity's needs, and on the other side that such quantities are distributed among the world population not according to mere market logics with the aim of favouring populations with higher purchasing power, but according to equity principles, taking account of food needs and activating compensation policies that sustain consumption where there is real need.

What seems to be utopia, in an economic system that remains solidly anchored to the capitalistic model, turns out to be the only possible way to ensure those sustainable growing conditions that the capitalistic interests themselves cannot function without, making utopia an objective actually acceptable by both political and economic powers.

Some considerations can help substantiate the validity of this assertion. The actual food resources distribution, the usual appropriation of it through the market—suffice to mention land grabbing or unequal exchange—and/or forms of political and military control over other populations' food resources, rise tensions that produce conflicts and migration pressures that end up with negative effects on the evolution of the world economy, causing uncertainty and instability.

An increasingly significant part of migration flows stems right from the unequal distribution of food resources and, consequently, to limit them it is fundamental to address the international community's efforts towards the improvement of the production capacities of those areas where such flows originate.

An improvement that can be obtained via both infrastructure and technological investments, as well as through interventions on human capital, a pivotal prerequisite to make territories able to untap productive potentials.

Nevertheless, even the adoption of a world agricultural policy will not be able to guarantee "global food security" if two precise actions, which must be considered priorities and cannot be delayed, will not take place. On the one hand, pacification (and the current geo-political framework does not seem capable to guarantee it, with severe repercussions on certain regions' food production) and, on the other hand, a fight against climate change that today is the most serious threat to food security. This is a utopia we are forced to believe in, if we wish to give new generations and the entire planet a future.

References

AA VV (2006) Biodiversità ambiente agricoltura alimentazione. Accademia Nazionale delle Scienze detta dei 40. Roma
Cervigni R (2001) Biodiversity in the balance: land use, national development and global welfare. Edward Elgar, Cheltenham
Dinelli G (2012) La biodiversità agricola nella strategia nazionale per la biodiversità. Pàtron, Bologna
FAO (2010) The state of food and agriculture report. FAO, Rome
FAO (2014) The State of Food and Agriculture Report. Rome
FAO (2015) The state of food insecurity in the world. Nomination file. http://www.fao.org/3/a-i4646e.pdf. Accessed 20 July 2015
Guha-Khasnobis B, Acharya Shabd S, Davis B (eds) (2007) Food insecurity, vulnerability and human rights failure. Basingstoke, Palgrave Macmillan, New York
Lockie S, Carpenter D (eds) (2010) Agriculture, biodiversity and markets: livelihoods and agro-ecology in comparative perspective. Earthscan, London
Lovejoy TE, Hannah L (2005) Climate change and biodiversity. Yale University Press, New Haven
Maes F (ed) (2013) Biodiversity and climate change: linkages at International, National and local levels. The IUCN Academy of Environmental Law Series. Cheltenham, Northampton
Morgan K (2010) Local and green, global and fair: the ethical foodscape and the politics of care. Environ Plan 42:1852–1867
Pollice F (2013a) Geopolitica dell'acqua. Il ruolo della cooperazione. INEA, Atti del Workshop internazionale #ShareWaterSaveWater. Cooperate for a New Water Culture. INEA, Rome, pp.11–20
Pollice F (2013b) La sicurezza insostenibile. Boll. Soc Geog It, VII(1): 159–173
Pollice F, Albanese V (2015) Biodiversità e sicurezza alimentare. Gnosis, XXI(1):39–45
Porter JR, Xie L, Challinor AJ, Cochrane K, Howden SM, Iqbal MM, Lobell DB, Travasso MI (2014) Food security and food production systems. In: Contribution of Working Group II to the Fifth Assessment Report of the Intergovernmental Panel on Climate Change. Field CB, Barros VR, Dokken DJ, Mach KJ, Mastrandrea MD, Bilir TE, Chatterjee M, Ebi KL, Estrada YO, Genova RC, Girma B, Kissel ES, Levy AN, MacCracken S, Mastrandrea PR, White LL (eds) Climate change 2014: impacts, adaptation, and vulnerability. Part A: global and sectoral aspects. Cambridge University Press, Cambridge, United Kingdom and New York, pp 485–533
Sellari P (2015) Land Grabbing. Geopolitica e il Global-colonialismoagricolo. Gnosis, XXI (1):155–163
Sen A (1982) Rights and agency. Philos Public Aff 11:3–39
Shiva V (2006) Il bene comune della Terra. Feltrinelli, Milan
Shiva V (2009) Ritorno alla terra. Fazi Editore, Rome
Solbrig OT, van Emden HM, van Oordt PGWJ (1994) Biodiversity and global change. CAB International, Wallingford
Steinberg P (2015) Who rules the hearth? OUP, Oxford
Tisdell C (1999) Biodiversity, conservation and sustainable development: principles and practices with Asian examples. Edward Elgar, Cheltenham
United Nations (1948) Universal Declaration of Human Rights. New York, USA Nomination file. http://www.un.org/en/documents/udhr/index.shtml. Accessed 20 July 2016
United Nations (2000) The right to food. Commission on Human Rights resolution 2000/10.Office of the United Nations High Commissioner for Human Rights. Geneve, Switzerland
United Nations (2011) World population prospects: the 2010 revision. Population division department of economic social affairs. United Nations, New York
United Nations (2013) World Population Prospects: The 2012 Revision. Population Division Department of Economic and Social Affairs, United Nations, New York
United Nations (2015) Concise Report on the World Population Situation in 2014. Population Division Department of Economic and Social Affairs. United Nations, New York

Considerations on the Subject of Food Security and Food Safety

Francesco Fabrizio Tuccari

1 Introduction

Connected to the problem-theme of food diversity, considered both in general terms and with reference to the crucial points of agro-bio-diversity and of sustainable rural systems, is the subject of food security.

This connection is considered in the more general context of risk[1] and therefore of uncertainty and insecurity,[2] which has become endemic in the contemporary world, and is particularly evident in the present crisis.[3] The connection appears so close as to make us think of food security as the so-called last mile of food diversity. Their very delicate balance is endangered by natural calamities, warfare and by territorial, climatic, demographic, socio-economic, institutional and political, technological and energy factors. All these have a negative effect on the condition of rural areas and of small farmers, both guaranteeing such diversity.[4]

This is what can be understood already on the logical level of simple association of ideas, since food diversity and all that revolves about it are mutually implicated with security. In fact, they refer automatically and immediately to security when they do not even presuppose it, from which a relationship of real one-to-one correspondence results.

Such a connection takes on more defined outlines to the point of becoming the possible object of empirical observation, where one pays attention to the title and

---

[1] Albisinni (2011).

[2] Amorosino (2010).

[3] Tallacchini (2012).

[4] Certomà (2010).

F. F. Tuccari (✉)
Department of Legal Studies, University of Salento, Lecce, Italy
e-mail: francesco.tuccari@unisalento.it

A. Isoni et al. (eds.), *Food Diversity Between Rights, Duties and Autonomies*,
LITES – Legal Issues in Transdisciplinary Environmental Studies 2,
https://doi.org/10.1007/978-3-319-75196-2_14

particularly to the subtitle[5] of this particular group research, to which the knowledge and experience of the components have contributed. Its declared aim is to help in the creation of a legal system capable of effectively recognising, promoting and protecting the various diversities of food; a system whose guarantee to put into action and maintain in action, lies exactly in the security of its assets, its interests and reference values.

To make this connection still closer are the circumstances of time and place in which this group work developed. It began in 2015, that is in the year of Expo, the Milan Exhibition, whose logo is "Feed the planet, energy for life",[6] the year also indicated for realising the ambitious objective of halving world hunger. This work, also in striking conjunction with the inauguration of Expo, has seen the divulgation of its first results in Italy,[7] where the FAO and the EFSA are based, the two most important organizations that deal with food in general and with food security in particular at a world and European level. Here, as common in countries belonging to the Mediterranean basin, an ideal frontier between the North and the South of the globe, questions of food security, because of demographic and economic reasons, assume peculiar connotations which it is difficult to find elsewhere.[8]

The limited space available makes it possible to outline only the most characteristic and distinguishing features of the articulated and complex subject of food security.

2 The Notion of Food Security

It is accepted today that, in the absence of an adequate normative definition, the correct treatment of the subject of food security, going back to the 1970s, should start with an approach that distinguishes the two concepts of food security and of food safety.

Food security, which concerns the quantitative aspect of the food problem, was first understood as availability of food and then as physical and economic access to food, conceived as a fundamental resource to be guaranteed to the whole of humanity as a necessary but not yet sufficient condition of life.

Food safety, which instead concerns the quality of food and drink, is defined in a previous, narrower version and in a more recent, wider version. The first is based on the defence of mainly hygienic-sanitary needs concerning the protection of the health of populations and the prevention and repression of commercial frauds.[9] The second pays attention also to a wide range of assets, interests and values. These are capable of combining variously with each other, such as landscape,

[5]"«Food diversity» between rights, duties and autonomies. Legal perspectives for a Scientific, Cultural and Social Debate for the Right to Food and Agroecology".

[6]Ramajoli (2015a), p. 271.

[7]3rd L.A.I.R. Group International Workshop ("Law and Agroecology/Ius et Rus"/University of Salento) "Food diversity" between rights, duties and autonomies. Legal perspectives for a Scientific, Cultural and Social Debate for the Right to Food and Agroecology, Lecce, May 6–7, 2015.

[8]Greco (2009).

[9]Cammeo (1905).

environment, territory, sustainable development, the non-debasement of agricultural and food production as an answer to the risks of economic globalization and of progress.[10] It also respects the safety of workers, free circulation of goods,[11] as well as "the observance of religious dictates, traditional social and cultural models, requirements of loyalty in commercial traffic and relations (business to consumer and business to business)",[12] which have further prospects compared with those of the original strictly criminological area.[13]

These two aspects of the notion of food security/safety have remained distinct and differentiated for a long time and have therefore been considered as two realities, each one substantially separated from the other.

Although from a logical point of view security precedes safety, from the historical-legal point of view the opposite phenomenon has been recorded. In fact, while the problem of access to food, typical but not exclusive of developing countries (so called "convergent societies") was faced above all in facts and technically, for example by preparing an adequate system of transport of "surpluses where there had been shortages",[14] and then juridically too. On the contrary, the various questions connected to the quality of food, typical but even they not exclusive of the industrialized and in any case more advanced countries (so-called "competitive or affluent societies"), have since ancient times been faced directly at an organizational level with instruments peculiar to law in book and law in action.[15]

Today no-one doubts the reciprocal, physiological convergence of safety and security. They have become so interpenetrated and complementary as to create an integrated, homogeneous and legally significant notion of food security/safety.

This must not, however, lead us to consider food security/safety a monotonous and undiversified entity for at least two reasons. First, because of the intrinsic, ontological diversity between the two dimensions of "security" and of "safety". If considered from the sole aspect of economic law, the former is the object of a "public good" and the latter of a "merit good", with all its consequences from the point of view of the respective legal treatment.[16] Secondly, because of the vast and variegated production of norms on the matter, which makes of food safety a "macro-concept needing to be divided and itemized according to the particular case", in such a way that it becomes almost obligatory to make a fine distinction between "food safety" (in general) and "food safety" (in particular).[17]

In this way "food safety" (in general), understood as the minimum common denominator of every and any kind of food, operates on the more general hygienic-

---

[10]Malagoli (2004), Costato (2007), Vadalà (2011) and Venturi (2012).

[11]Giuffrida (2015).

[12]Ricci (2014), p. 20.

[13]Distaso (2009).

[14]Costato (2010).

[15]Cammeo (1905).

[16]Jannarelli (2009) and Amorosino (2010).

[17]Russo (2010).

sanitary level with the purpose of preventing or eliminating outside risks or the compensation of consequent damage. Moreover, it is based on the ban to commercialize unsafe products, on the principles of the separation between management and evaluation of risk and traceability and re-traceability of products, as well as on the discipline of the responsibility of both the producers and the distributors.[18]

"Food safety" (in particular), on the contrary, referring to particular categories of foods, such as the so-called "novel foods" and the GMO, require added control as a protection against inside risks, that is those risks deriving from the composition of the food (so-called toxicological safety), from incorrect information given to consumers (so-called informative safety) and from the inaccurate use of products (so-called nutritional safety).[19] They are based on the obligation to have preventive permission respecting principle of substantial equivalence with already existing foods, tested when tasted. They most also respect a more specific and complete discipline in the question of labelling, also concerning the particular and not secondary aspect of nutritional information.

The situation described, which almost closes the course of the food chain, drawing a link between food security and food safety, is completed by an increasing tendency today to involve consumers actively. In this way they themselves vouch for and become responsible for their habits concerning the choice and the assumption of food and drink, the aim of which is to avoid incorrect practices in purchasing and in the household management of food.[20]

To confirm the importance of the distinction between "food safety" (in general) and "food safety" (in particular) is the narrowing of the gap in GMO matters between the North American and the European model. The former is founded on later controls and on compensatory protection, whereas the latter, although using these instruments too, considers them as the last resort of the system and usually prefers interventions of a preventive kind.[21]

3 The Objective of Food Security/Safety

Apart from its notional aspect, food security/safety has also a practical objective as a valid and basically stable solution to the more general food problem.

This objective can be reached directly based on the homonymous doctrine (food security/safety), which accepts the prospects of international organizations for the general good of all people. According to this doctrine, the more general food problem derives not so much from the lack or quantitative or qualitative insufficiency of food of which there is even an abundance but rather from obstacles to

[18]Pisanello (2008), Canfora (2009) and Russo (2014).

[19]Costato and Rizzioli (2010).

[20]Belliggiano (2009) and Biglia and Gabbai (2014).

[21]Sirsi (2008) and Visani (2015).

gaining access to food. These obstacles must be removed by means of targeted interventions on food policies and on the operative instruments, to avoid clear imbalance in the distribution of food stuffs, and to recover the correct ratio between the lower and lower earnings of producers and the higher and higher purchase costs of products,[22] that may lead to so-called agflation.

Nonetheless, the objective of food security/safety can be reached in other ways, which present significant points of contact with the above doctrine, but also not secondary differences with respect to it.

One of these alternative ways is that expressed by the well-known formula of the right "to adequate food" or "to adequate diet",[23] in which the dimensions of security and of safety are examined and reconstructed in more open and fluids terms,[24] thanks also to the diversity of meanings given to the adjective "adequate".[25] First of all, there is the literal meaning, which gives the idea of the assumption of the correct quantity of good quality food, such as to ensure a person's nourishment, without recourse to chemical treatment (particularly pesticides and similar substances), bearing in mind his dignity,[26] culture, religion, sex and work. Then there is the meaning of "available", referring both to production in view of an easier economic access, and to the food in view of preparing a valid system of distribution. Then there is "accessible", which means food for all (particularly for the weaker and more exposed sections of the population,[27] unfortunately today continually on the increase),[28] the cost of which should be kept stable and should not damage or prevent the respect of other primary interests and value such as health, housing and instruction.

---

[22]Costato (2011).

[23]Art. 25.1 Universal Declaration of Human Rights (1948); Art. 11.2 International Covenant on Economic, Social and Cultural Rights, adopted and opened for signature, ratification and accession by the General Assembly (1966); General Comment N° 12. The right to adequate food (Article 11 of the International Covenant on Economic, Social and Cultural Rights), adopted by the Committee on Economic, Social and Cultural Rights (1999); General Comment N° 14. The right to the highest attainable standard of health (Article 12 of the International Covenant on Economic, Social and Cultural Rights), adopted by the Committee on Economic, Social and Cultural Rights (2000); General Assembly Resolution 63/187 on the Right to Food, Sixty-third session, agenda item 64(b) (2008).

[24]Ricci (2012).

[25]Bottiglieri (2014).

[26]Ziegler (2003).

[27]Including sectors of society in which fragility and exposition derive from physical or psychic reasons, as for example in the case of the newly-born, of children, of the elderly, of the disable, of chronic or terminal invalids, of expectant mothers or women during puerperium, of mothers, of victims of natural calamities, of prisoners, of indigenous minorities for whom access to their land of origin may be dangerous, of refugees, etc.

[28]This is the case of the so-called refugees for hunger, who currently represent the most widespread sub-category (see previous note), for whom there are no specific protective regulations. Moreover, there are serious doubts about the possibility of extending and applying analogically the protective regulations intended for other refugees in general: Ziegler et al. (2011), p. 66.

However, there are various ways to obtain food security/safety, as the doctrine of adequacy apart from implying a lasting lack or insufficiency of food, holds that the availability of food must find valid recognition and an effective guarantee at a legal level more than a political one. This principle was recently underlined in the opening statement of the Milan Charter,[29] the availability of food constitutes object of a fundamental human right, indeed one of the "most fundamental of fundamental rights".[30]

A further alternative way, which is intermediate between the two methods outlined, is that of food sovereignty. This assumes as a starting point the view of the non-government organizations and institutions of civil society that pay particular attention to the conditions and needs of rural populations, a prospect which is not new to the European culture and system,[31] and maintains that the objective of food security/safety should first pass through political and then legal intervention. With reference to the latter this doctrine must in any case be based on fundamental rights, although tailored to the specific needs of the populations in question.[32]

Neither must another doctrine be ignored towards food security/safety, that of neo-liberal origin, which, by means of agrarian techniques the access to information and the rules and instruments of the market, also connected among themselves, aims at increasing the profitability of agricultural production and the growth of trust in producers with the intention of inducing them to lavish greater energy in terms of work and investments.

4 The System of Food Security/Safety

Understood in this way, food security/safety is the pivot around that the food sector moves, which, although it has only recently been fully "juridicalized", at least as far as the European system is concerned,[33] already presents the features of a system in continuous and rapid evolution.

[29]"We man and women, citizens of this planet, sign this document, named Milan Paper, to take on precise commitment in relation to the right to food which we maintain should be considered a fundamental human right. Indeed, we consider a violation of human dignity the failure to provide access to healthy, sufficient and nourishing food, clean water and energy".

[30]Rodotà (2011).

[31]Vaccari (2015).

[32]Food Sovereignty: A Future without Hunger, Rome, November 11–17, 1996, in http://www.viacampesina.org/imprimer.php3?id_article38. Accessed 28 Dec 2015; Food Sovereignty: A Right for All, Rome, June 13, 2002, in http://croceviaterra.it/FORUM/DOCUMENTI%20DEL%20FORUM/political%20statement.pdf. Accessed 29 Dec 2015.

[33]Ramajoli (2015b) connects the full "juridification" of the food sector to the Regulation 2002//178/CE of the European Parliament and Council of 28th January 2002 "which fixes the general requisites of food legislation, founds the European Food Safety Authority (EFSA) and establishes procedures in the field of Food Safety". The Authoress points out that "Before this date, as the European Commission has recognized in its Green Book and in its White Book on food security/safety a fragmentary view, without organicity and at times incoherent has been dominant in the matter" (p. 3).

As already mentioned, the original dimension of food safety has the object of intervention on the part of national States for the hygienic-sanitary protection of food and drink according to a mixed model of administrative control and penal repression,[34] has been followed by a significant widening of the horizon. Today it also includes food security/safety and enlarges its field of action to super-national and intra-national areas, developing fundamental aspects such as that of "administrative prevention"[35] which consists not only of prohibitions and control (both by public authorities and private organisms) but also of instruments for controlling and programming.[36] It also develops "a sense of responsibility" in producers,[37] certifiers, distributors and consumers,[38] by means of the preparation of adequate mechanisms of accountability and transparency.[39]

The above-mentioned complexity of this system, which recent studies qualify as a real sectional legal system,[40] can be found in all its essential points; meaning, at the level of subjects, organization, sources, principles, legal situations and of their protection.

As far as the subjects are concerned, there is a changeover from traditional, closed model of government to the more innovative open ones of governance in which side by side with national States are various figures.[41] Of global importance are public organizations, such as the World Trade organization and those of mixed composition, such as the Codex Alimentarius Commission.[42] Of European significance are the Union with its various institutions,[43] including independent ones like EFSA,[44] and at a sub-state and local level (Regional or equivalent structures, and other public administrations, of that territory or not, also including independent ones).[45] Nor on this subject can the transformation of the role and the physiognomy of figures, such as that of consumers, be ignored, who differently from the past are now also involved in the circuit of chain responsibility, which according to a realistic expression goes "from field to fork".

The first result of this is the co-existence of various organizer models like that of direct administration (sometimes, and this is the novelty, even on the part of supernational institutes)[46] and of indirect administration (in the forms of co-administration

---

[34]Papaldo (1958) and Pericu (2009).

[35]Ramajoli (2015a), p. 274.

[36]Amorosino (2010).

[37]Ramajoli (2015a), p. 274.

[38]Biglia and Gabbai (2014).

[39]Albisinni (2015) and Amorosino (2015).

[40]Bevilacqua (2012).

[41]Jannarelli (2009).

[42]Viti (2009).

[43]Maltoni (2006) and Savino (2007).

[44]Gabbi (2009).

[45]Cassese (2002a, b) and Longobardi (2009).

[46]See Arts. 53 and 54 of (already mentioned) Regulation 2002/178/CE about the power of direct intervention of European Commission in case of food risk regarding one or more Member States: Albisinni (2009), pp. 10–14.

and of localized integration). Along with these there are also private organizer models, whose common characteristic is the fact that they were all founded, although with different interests, on the associative idea.

Another result is the co-existence of numerous sources of different origin, nature and position—international, transnational and national of every kind, of law in book and law in action, of hard law and soft law, of civil law and common law, of public and private law. This implies every consequent imaginable necessity for agreement, which, however, presently is either completely lacking or in any case is still far from being said to be effectively reached.

Referring to general principles, two macro-models of food security/safety can be seen and distinguished. One particularly popular in the WTO circuit is characterized by putting into practice the principle of prevention, which foresees the certainty or at least the probability of scientifically demonstrated risk. The other model, that is the European one, is founded on the principle of precaution, which implies the uncertainty or in any case the possibility in the abstract of risk, which obviously requires even more previous protective intervention.

With the principles of prevention and precaution, which on food security/safety are the most important of all, are all the others which have been mentioned when referring to the conceptual evolution of security/safety, both further principles, as those of self-administration, and administrative simplification.

These principles either come immediately and directly from the food sector or in any case find in it new life that re-moulds them and puts them back into the circuit of the general system or of the sectorial or sectional system.

In the light of all this, it is possible to look at food security/safety as an important "vehicle of institutional innovation"[47] and, more in general, to consider food law not only as a "workshop of the new European law"[48] but also as a fertile soil for the development and experimentation of the increasingly emerging global law.

5 Food Security/Safety, Subjective Juridical Situations: Instruments, Mechanisms of Protection and Remedies

Subjective juridical situations and their protection pass once again through the formula of the "right to adequate food". On its possible relation with other conterminous rights and with relative instruments and mechanisms of protections and remedies there does not seem to be unanimity of views.

On this matter, in fact, there is first of all the traditional position of those who maintain that the "right to adequate food" should be placed within other and vaster rights like the classical rights to life and to health.[49] It should also be placed within

[47] Albisinni (2009).

[48] AIDA-IFLA (2015).

[49] Golay (2009) and Cordini (2014).

rights that have emerged more recently, the right to work, to well-being, to development and to the minimum acceptable standards of existence and dignity.[50]

An innovate current of thought maintains, on the contrary, that the "right to adequate food" has the dignity and autonomous dignity of an undeniable basic right, recognized and guaranteed formally and directly towards all or towards particular categories of people (children, needy schoolchildren, prisoners and imprisoned refugees, etc.). From this right, they maintain, come the conditions and pre-requisites for the exercise of all the other rights usually included in the Constitutions of the different States.[51]

Others, placing themselves, in an intermediate prospective consider the "right to adequate food" as a juridical situation, sometimes derived sometimes autonomous. This depends on the conformation either abstract or concrete, which has been given to it from time to time by regulations that are written and based on the principles of the various systems regarding the subjects and/or assets, interests and values to be protected.[52]

The analysis of the different sources, considered at their different levels and in their different ranks and compared or even only juxtaposed, reveals the existence of a multi-dimensional system, which it would be difficult to reduce to units[53] and, indeed, even to quality as a real system, as it is in reality a hybrid mixture of microcosms that can be connected in the direction of the common final objective of food security/safety.

A further question is that of the effectiveness of the protection of the "right to adequate food".

At an international level, in fact, despite the noteworthy effort made by the organizations in charge, this right is practically non-existent because of the absence of links or of reduced links of most of the sources of international law, both hard and soft law. In any case it hardly exists because of State sovereignty, which can impede the satisfaction of violated human rights, particularly if activated individually.

In a different way, at the State level and above all in legal systems peculiar to developing, so-called convergent societies, protection of the "right to adequate food" is recognized both collectively and individually[54] and is put into practice in the most varied ways.[55] This can take place in a direct or indirect way, as well as in immediately perceptive terms or merely programmatic ones, with reference to sources of inferior level regarding what is necessary to carry out this right.

However, there are also organizations that consider the right examined as a subjective legal situation that is "financially conditioned", in this way considerably limiting both the opportunities and the potentialities of protection. There are also

[50]Sicardi (2014).

[51]Bottiglieri (2014).

[52]Santini (2012).

[53]Ricci (2012) and Lupone et al. (2013).

[54]Borghi (2012).

[55]Bottiglieri (2014).

organizations that consider the "right to adequate food" as a simpler constitutional objective or as a mere directive line of action for future State policies. Consequently, this right, far from being able to be conceived as absolute or in any case above all the others, must from time to time be connected to and balanced with various public and private interests which are concretely important and in the end, it may be recessive or even yielding.

On the contrary, protection prepared competitively, even if in a sectorial perspective, by the World Health Organization (WHO) appears to be much more effective in coping with every possible risk of food derivation. It is founded on rigorous measures, which bind the Member States directly, committing them to carry out the measures immediately, with few exceptions, moreover very strictly interpreted.[56]

Therefore, the idea has become accepted that the "right to adequate food", understood as a subjective juridical situation, must be opened to a dimension of duty able to complete and put this right into effect from the substantial point of view, as well as that of protection. This is a further dimension working on the ethical,[57] legal level,[58] which regards public and private subjects and is founded on principles of solidarity, of vertical and horizontal subsidiarity, of differentiation and adequateness.

This right/duty,[59] whose protean evolutive dynamics seem to suggest that it should obtain necessary satisfaction, is becoming an integral and substantial part of the structure of food security/safety, which as has been stated above, appears to be a system "with variable geometry". Or, to render the concept even clearer, because of its continuous progress, it is presented as a system "with variable valve timing". Consequently, the treatment of this right/duty is as difficult as it is stimulating, also considering successive developments in the enquiry.

References

Acconci P (2011) Tutela della salute e diritto internazionale. CEDAM, Padova

Albisinni F (2009) La sicurezza alimentare veicolo di innovazione istituzionale. http://www.rivistadirittoalimentare.it/rivista/2009-04/ALBISINNI.pdf. Accessed 19 Jan 2016

Albisinni F (2011) Sicurezze e controlli: chi garantisce cosa? http://www.rivistadirittoalimentare.it/rivista/2011-04/ALBISINNI.pdf. Accessed 13 Jan 2016

Albisinni F (2015) Funzioni pubbliche e competenze dei privati fra accountability e trasparenza: verso un diverso ordine nella Food Law. http://www.rivistadirittoalimentare.it/rivista/2015-01/ALBISINNI.pdf. Accessed 26 Jan 2016

Amorosino S (2010) Sicurezze ed insicurezze in campo alimentare tra regolazioni e programmazioni. http://www.rivistadirittoalimentare.it/rivista/2010-03/AMOROSINO.pdf. Accessed 18 Jan 2016

---

[56] Acconci (2011) and Golay (2011).

[57] Visani (2015).

[58] Biglia and Gabbai (2014) and Ramajoli (2015a).

[59] Visani (2015).

Amorosino S (2015) Trasparenze, certezze e sicurezze dei prodotti e dei mercati agroalimentari: correlazioni e funzioni. http://www.rivistadirittoalimentare.it/rivista/2015-01/AMOROSINO.pdf. Accessed 26 Jan 2016

Associazione italiana diritto alimentare (AIDA) – Italian Food Law Association (IFLA) (2015) Il diritto alimentare, laboratorio del nuovo diritto europeo. Un bilancio a 10 anni dalla costituzione dell'AIDA. http://www.rivistadirittoalimentare.it/rivista/2015-01/ROVIGO.pdf. Accessed 28 Jan 2016

Belliggiano A (2009) Percezione della sicurezza alimentare e nuovi modelli di organizzazione della produzione. http://www.rivistadirittoalimentare.it/rivista/2009-04/BELLIGGIANO.pdf. Accessed 07 Jan 2016

Bevilacqua D (2012) La sicurezza alimentare negli ordinamenti giuridici ultrastatali. Giuffrè, Milano

Biglia C, Gabbai S (2014) Consumare alimenti sicuri è un diritto? In: Macchia P (ed) La persona e l'alimentazione: valutazione clinica e diritto alla salute. Profili clinici, giuridici, culturali ed etico-religiosi (Atti del Convegno, Asti, 30 novembre 2012). Aracne, Roma, pp 119–124

Borghi P (2012) Le azioni di classe nel settore alimentare. http://www.rivistadirittoalimentare.it/rivista/2012-01/BORGHI.pdf. Accessed 28 Jan 2016

Bottiglieri M (2014) Il diritto ad un cibo adeguato: profili comparati di tutela costituzionale e questioni di giustiziabilità. In: Macchia P (ed) La persona e l'alimentazione: valutazione clinica e diritto alla salute. Profili clinici, giuridici, culturali ed etico-religiosi (Atti del Convegno, Asti, 30 novembre 2012). Aracne, Roma, pp 124–153

Cammeo F (1905) Sanità pubblica: fonti e organizzazione. In: Orlando VE (ed) Primo trattato completo di diritto amministrativo italiano, part II, vol 4. Società editrice libraria, Milano, pp 213–394

Canfora I (2009) Sicurezza alimentare e nuovi assetti delle responsabilità di filiera. Rivista di diritto alimentare. http://www.rivistadirittoalimentare.it/rivista/2009-04/CANFORA.pdf. Accessed 12 Jan 2016

Cassese S (2002a) La nuova disciplina alimentare. In: Cassese S (ed) Per un'autorità nazionale della sicurezza alimentare. Giuffrè, Milano, pp 13–22

Cassese S (2002b) Proposte per un'autorità nazionale per la sicurezza alimentare. Ragiusan: 206–213

Certomà C (2010) Diritto al Cibo, Sicurezza Alimentare, Sovranità alimentare. http://www.rivistadirittoalimentare.it/rivista/2010-02/CERTOMA.pdf. Accessed 13 Jan 2016

Cordini G (2014) Salute, ambiente e alimentazione: i profili costituzionali. In: Macchia P (ed) La persona e l'alimentazione: valutazione clinica e diritto alla salute. Profili clinici, giuridici, culturali ed etico-religiosi (Atti del Convegno di Asti, 30 novembre 2012). Aracne, Roma, pp 9–18

Costato L (2007) Perché una rivista di diritto alimentare. http://www.rivistadirittoalimentare.it/rivista/2007-01/COSTATO.pdf. Accessed 22 Jan 2016

Costato L (2010) Non solo "allarmi" per la sicurezza alimentare. http://www.rivistadirittoalimentare.it/rivista/2010-03/COSTATO.pdf. Accessed 10 Jan 2016

Costato L (2011) Sicurezza alimentare e Rivolte in Nord Africa. http://www.rivistadirittoalimentare.it/rivista/2011-02/COSTATO.pdf. Accessed 13 Jan 2016

Costato L, Rizzioli S (2010) Sicurezza alimentare. In: Digesto discipline privatistiche, sezione civile, Aggiornamento. UTET, Torino, pp 929–937

Distaso M (2009) La sicurezza e la qualità degli alimenti di fronte alle nuove incertezze. http://www.rivistadirittoalimentare.it/rivista/2009-04/DISTASO.pdf. Accessed 11 Jan 2016

Gabbi S (2009) L'Autorità europea per la sicurezza alimentare. Giuffrè, Milano

Giuffrida M (2015) Il diritto fondamentale alla sicurezza alimentare tra esigenze di tutela della salute umana e promozione della libera circolazione delle merci. http://www.rivistadirittoalimentare.it/rivista/2015-03/GIUFFRIDA.pdf. Accessed 20 Jan 2016

Golay C (2009) The right to food and access to justice. http://www.fao.org/righttofood/en. Accessed 12 Jan 2016

Golay C (2011) Droit á l'alimèntation et accés á la Justice. Bruylant, Bruxelles
Greco S (2009) La sfida alla sicurezza alimentare nel Bacino mediterraneo. http://www.rivistadirittoalimentare.it/rivista/2009-04/GRECO.pdf. Accessed 04 Jan 2016
Jannarelli A (2009) La governance della sicurezza alimentare: l'esigenza di un monitoraggio costante del modello. http://www.rivistadirittoalimentare.it/rivista/2009-04/JANNARELLI.pdf. Accessed 13 Jan 2016
Longobardi N (2009) A proposito di autorità italiana per la sicurezza alimentare fra disciplina europea e interventi normativi nazionali. http://www.rivistadirittoalimentare.it/rivista/2009-04/LONGOBARDI.pdf. Accessed 21 Jan 2016
Lupone A, Ricci C, Santini A (2013) The right to food towards a global governance. Giappichelli, Torino
Malagoli C (2004) Sicurezza alimentare. Agg soc (5):441–444
Maltoni A (2006) Sicurezza alimentare. In: Cassese S (ed) Dizionario di Diritto pubblico, vol II. Giuffrè, Milano, pp 5545–5549
Papaldo A (1958) Alimenti e bevande (igiene degli). In: Enciclopedia del diritto, vol II. Giuffrè, Milano, pp 50–61
Pericu G (2009) La disciplina pubblicistica dei prodotti alimentari. In: Pericu G (ed) Scritti scelti. Giuffrè, Milano, pp 351–394
Pisanello D (2008) Obblighi di sicurezza alimentare nel mercato unico europeo tra gestione del rischio e responsabilità d'impresa. Dir comunit sca intern (4):695–734
Ramajoli M (2015a) Dalla "food safety" alla "food security" e ritorno. Amm (2–3):272–292
Ramajoli M (2015b) La giuridificazione del settore alimentare. Paper presented at the symposium "1865–2015: a 150 anni dall'unificazione amministrativa italiana". Firenze, 15–16 ottobre 2015
Ricci C (2012) Contenuti normativi del diritto a un cibo "adeguato" a livello internazionale. In: Ricci C (ed) La tutela multilivello del diritto alla sicurezza e qualità degli alimenti. Giuffrè, Milano, pp 33–60
Ricci C (2014) Salute e alimentazione: profili di diritto internazionale. In: Macchia P (ed) La persona e l'alimentazione: valutazione clinica e diritto alla salute. Profili clinici, giuridici, culturali ed etico-religiosi (Atti del Convegno, Asti, 30 novembre 2012). Aracne, Roma, pp 19–29
Rodotà S (2011) Il diritto al cibo. http://saperedemocratico.it/stefano-rodota-il-diritto-al-cibo. Accessed 19 Jan 2016
Russo L (2010) La sicurezza delle produzioni "tecnologiche". http://www.rivistadirittoalimentare.it/rivista/2010-02/RUSSO.pdf. Accessed 14 Jan 2016
Russo L (2014) La responsabilità del produttore e del distributore. http://www.rivistadirittoalimentare.it/rivista/2014-01/RUSSO.pdf. Accessed 15 Jan 2016
Santini A (2012) Brevi note sulla complessità del diritto alimentare europeo. In: Ricci C (ed) La tutela multilivello del diritto alla sicurezza e qualità degli alimenti. Giuffrè, Milano, pp 231–240
Savino M (2007) Autorità e libertà nell'Unione europea: la sicurezza alimentare. Riv trim di dir pub (2):413–442
Sicardi S (2014) Introduzione. In: Macchia P (ed) La persona e l'alimentazione: valutazione clinica e diritto alla salute. Profili clinici, giuridici, culturali ed etico-religiosi (Atti del Convegno, Asti, 30 novembre 2012). Aracne, Roma, pp 6–8
Sirsi E (2008) Il problema della coesistenza tra colture transgeniche, convenzionali e biologiche negli USA: il caso alfalfa, commento a US District Court of the Northern District of California 13 Feb 2007. Rivista di diritto agrario (2):86–101
Tallacchini M (2012) Sicurezza e responsabilità in tempi di crisi. http://www.rivistadirittoalimentare.it/rivista/2012-01/TALLACCHINI.pdf. Accessed 17 Jan 2016
Vaccari S (2015) Food right and food sovereignty: the legacy of Milan Charter. http://www.rivistadirittoalimentare.it/rivista/2015-03/VACCARI.pdf. Accessed 22 Jan 2016
Vadalà G (2011) La sicurezza alimentare. Economia e ambiente (4–5):25–32
Venturi (2012) Sicurezza alimentare tra scienza e diritto. In: Cordini G, D'Ancona PI (eds) Alimentazione, salute e ambiente. Qualità delle produzioni alimentari, tutela del consumatore, salute, sviluppo rurale e sostenibilità ambientale (Atti del Convegno di Studi, Martano, 1–2 ottobre 2011). Grifo, Lecce, pp 157–171

Visani S (2015) Modelli normativi a confronto: regolamentazione degli Ogm tra UE e USA. Giurisprudenza in materia di brevettabilità degli organismi viventi. http://www.rivistadiritto alimentare.it/rivista/2015-03/VISANI.pdf. Accessed 25 Jan 2016

Viti D (2009) Il governo della sicurezza alimentare nella globalizzazione dei mercati: il caso Cina. http://www.rivistadirittoalimentare.it/rivista/2009-04/VITI.pdf. Accessed 11 Jan 2016

Ziegler J (2003) Le droit á l'alimèntation. Mille et une nuits. Librairie Arthème Fayard, Paris

Ziegler J, Golay C, Mahon C, Way SA (2011) The fight for the right to food. Palgrave Macmillan, London

Right to Food and "Tragedy" of the Commons

Antonio Gusmai

1 Introductory Notes

The theme of "common goods" has lately reached a relevant position among the studies of jurists, especially since it was formally introduced by the Rodotà Committee's report.[1]

The term "common goods" is often used with respect to goods that everyone should be able to afford. Should we use the words of the abovementioned committee, common goods "express functional utility to the exercise of fundamental rights as well as the free development of a person" and therefore "must be protected and safeguarded by the legal system, also for the benefit of future generations". Still according to the Committee, common goods are, *inter alia*, "rivers, streams and their springs; lakes and other waters; the air; parks as defined by law, forests and wooded areas; mountainous areas at a high altitude, glaciers and perpetual snow; shores and coastlines declared environmental reserves; protected wild flora and fauna; archaeological, cultural and environmental goods and other protected areas of the landscape".[2]

[1]The committee for amending the regulations of the Civil Code on Public Goods (called "Rodotà" from the name of its chairman), was formed in 2007 by the Prodi government and the Minister of Justice Mastella, with the purpose of defining a systematic regulatory framework for measures of privatization aimed at lightening the burden of the public debt. Particularly, the Committee was set up at the Ministry of Justice by decree of the Minister in June 21st 2007 and accomplished its work by drawing up a draft law that delegated to the government the power of amending the Chapter II, the Title I of the Book III of the Civil Code, as well as other parts of the same book linked to it. Unfortunately, as it is known, no discussion has been held about this draft law in the institutional framework.

[2]See Article 1, section 3, letter *c)* of the Proposal of Regulation.

A. Gusmai (✉)
Department of Legal Studies, University of Salento, Lecce, Italy

A. Isoni et al. (eds.), *Food Diversity Between Rights, Duties and Autonomies*,
LITES – Legal Issues in Transdisciplinary Environmental Studies 2,
https://doi.org/10.1007/978-3-319-75196-2_15

They are mostly environmental goods, even if artificial resources are included: historic and cultural heritage, physical infrastructures (aqueducts, routes of communication), computer resources (the Internet or other networks) or immaterial goods such as knowledge, if they are qualified, as we will see shortly, by a certain modality of *governance* and *management*.

Therefore, despite the complex heterogeneity[3] of the commons, it seems that their juridical importance can be expressed as follows: the conservation and the reproduction of these goods, be they global or local, material or immaterial, is a necessary condition to maintain a democratic social order.[4]

Because of its nature, the theme dealt with arouses interest and is subject to a multidisciplinary approach: many speculations are focused on theoretical and philosophical questions (either as scientific analysis or with purposes that are merely political or educational), but many other analyze the economical and juridical aspects more specifically.

To narrow the scope of the analysis for a theme that is so complex and boundless, the following pages will briefly deal with three specific questions from a constitutional perspective, without claiming of exhaustivity. First, the juridical significance of the economic theory of the commons; second, the constitutional consequences of this doctrine; ultimately, a possible explanation why food should not be considered to be a common good, but rather a *social good*.

2 The Juridical Significance of the Economic Theories of the Commons

The economic theory of the commons has a mother and a father.

The father, Garrett Hardin, had hypothesized, in a successful article about the "tragedy of the commons", that a coercive regulation could be a useful solution to control the individual use of scarce open-access resources. In a world with scarce resources, his concern was that open-access goods, of common use, would end up running out if there were not a socially-accepted regulation for limiting the use of and the access to them.

According to the thesis supported by Hardin, if men do not restrict the individual freedom, they risk ending up like the inhabitants of Easter Island, because they would destroy the common goods on which the life of the human species depends.

The American ecologist analyzes the goods by considering the combinations of two couples of characteristics relating to their modalities of consumption: *a)* "rivalry" *versus* "non-rivalry" in the consumption; *b)* "excludability" *versus* "non-excludability" from the consumption. Briefly, the "common goods" fall into the case

[3]See, at least, the attempt of classification made by Cerulli Irelli and De Lucia (2014), pp. 6 et seqq.

[4]The studies carried out by Rodotà seem to be particularly focused on this aspect (2013), pp. 459 et seqq.

in which the good is *rival* (e.g. the cubic meters that John Doe draws from a lagoon cannot be used from other people), but *non-excludable* (it may be impossible to keep a person from drawing water from the lagoon, even if he/she does not pay anything). Because of rivalry, and fearing the depletion, everyone tends to forestall the consumption of goods by other people and maximize his/her own consumption. Moreover, even if the amount available (supply) decreases and the price increases, the demand does not decline, because no one can be excluded. So, everyone is urged to consume goods until the moment in which the cumulative effect of individual consumption will correspond to the depletion of resources. The result is the "tragedy of the commons", i.e. the unavoidable tendency towards congestion and hyper-consumption and the consequent depletion of natural resources, if they are managed as "open-access" common goods.

It is interesting to notice that Hardin does not use the word "tragedy" to mean "ruin", "irreparable damage", "unfortunate outcome", but to describe from a social point of view a dilemmatic situation that recalls the "prisoner's dilemma", i.e. a situation for which no optimal solution exists as long as men accept the anthropological assumption based on which all the individuals behave according to the standard of the *homo oeconomicus*. To perpetuate the dilemma, people must therefore act in a rational, instrumental and selfish way and must not be able to reduce their consumption based on voluntary agreements nor to respect the juridical regulations issued for this specific purpose.[5]

The mother, Elinor Ostrom, developed instead a prescriptive model—in some respects more specific—and acknowledged that the commons can be subjected to community-based management by users, proposing a *model of self-governance as an alternative to the public model*. Conversely, between public and private management common goods should be subjected to *common management*.

According to Ostrom, who was awarded the Nobel Prize for economics, the solution for the "tragedy of the commons" based on the self-organization of the community is perfectly possible. Of course, this perspective modifies in a substantial way the range of the economic institutions that the economists consider to be worthy of attention and makes it possible to downsize the rather dominant role that the free market has today. In fact, starting from the study of some cases of community-based management of irrigation systems in some communities of developing countries—but also cases of multilevel management of reservoirs in the USA, where common possession is effective—Ostrom has shown that it is possible to establish social rules and institutions for access to natural resources and avoid the "tragedy" through *self-governance*. The basic concept is to channel the energies of the organized civil society into the invention of new ways to manage common goods that comply with models of *participative democracy*.[6] In practical terms, the management model must

[5]See Hardin (1968), pp. 1243–1248.

[6]See Lucarelli (2013), pp. 45 et seqq.—with respect to the forms that the participative democracy in the management of common goods can assume. See also Marella (2012), p. 24, who recognizes the

be congruent with the nature of the good in question: if it is common, then also its management must be so.[7]

In fact, the "tragedy of the commons" suggested that privatization strategies should be adopted by the mainstream ideology, i.e. strategies for introducing private property rights in the place of common goods to manage the effective excess demand—compared to supply—through the variation of the price.

In this way, since people are compelled to pay prices that are higher than the marginal utility of the good when the resource is running out, the neoliberal policies have obtained an automatic way to discriminate men depending on their willingness to pay a price on the market.[8] But what is neglected is that the price does not express the value of a good and that the discrepancy that arises between the price and the value of goods is so great that even the economists cannot find a solution. It has been noticed, in fact, that this problem is inherent in the very idea of the prices determined by the logic of profit, which is bound to produce—through the free market—a huge social inequality.[9]

In the juridical sphere, these arguments led some scholars to seriously question the categories on which the social and political thinking of the modern Occident is based. Ugo Mattei expresses an interesting proposition that contemplates the demolishment of the traditional (juridical and non-juridical) categories: according to him, common goods cannot be seized if we do not free our minds from the most rooted conceptual frameworks through which we view reality. This proposition goes beyond the traditional juridical and political categorizations and questions classical dichotomies such as "to be" and "to have" or "subject" and "object". A common good must not be considered to be a mere object, a tangible part of the external world. It cannot be seized through the mechanistic and reductionistic logic typical of Illuminism, which sharply separates the subject from the object. In short, it cannot be attributed to the modern idea of goods. The common good exists in a "qualitative relationship". We do not "possess" a common good (an ecosystem, for example), but we "are" somewhat a part of the common good (we are a part of an urban and rural ecosystem). So, the separability between "to be" and "to have", subject and object, is called into question by a political theory that places common goods in the foreground. These goods can be defined as such thanks to the material contexts in which they become relevant and not because of their ontological, objective or mechanical characteristics.[10]

As a result, from a juridical and functional perspective, recognizing an entity as a common good means declaring its incompatibility with both the logic of public systems (characterized by the delegation to the government and its bureaucracy) and

"collective and/or participative management" as a decisive element of a possible juridical statute of common goods.

[7]For more detailed studies in this regard, see Ostrom (1990); Ostrom (1998), pp. 1–22.

[8]In this regard, see Sacconi (2015), pp. 175 et seqq.

[9]See Patel (2009).

[10]In this sense Mattei (2012), pp. 52–53.

the logic of private systems typical of possessive individualism, whose main goal is to accumulate wealth and resources.

In a perspective of radical overturn of the traditional principles of the bourgeois constitutionalism, it is necessary to find juridical solution "beyond" the traditional dichotomies State/market, public/private property. Common goods are the basis of an authentic participative democracy based on the individual responsibility and commitment to get long-term benefits for everyone.

While these doctrines have obtained no legislative validation so far, they have been partially accepted by Italian law. In fact, with respect to the juridical nature of the *Valli da pesca della laguna veneta*[11] [fishing valleys of the Venetian lagoon], the Court of Cassation invited to view common goods "beyond a merely patrimonial and proprietary perspective", underlining a public and collectivistic functionality and purpose. After identifying the constitutional foundations of the collective use of public goods in Articles 2, 9, 42 of the Italian Constitution, the judges have asserted that discussing in terms of exclusive dichotomy between *public and private goods* means identifying in part the mere ownership of goods, neglecting the fundamental aspect that they are categorized according to the functions and the interests linked to them.[12]

In short, from a *benicomunista*[13] perspective, this means studying and devising structures of participative and authentically democratic government which can attract the more motivated administrators, incentivize the pursuit of a transnational and transgenerational logic (e.g. the ecological one) and verify their actions through the universal exercise of the fundamental right to access.[14] The identification of common goods is reached "from the bottom", meaning through activities, social conflicts and—above all—through the widespread perception that a certain good must meet collective needs.[15]

3 Some Considerations of Constitutional Order

Within the Italian doctrinal debate, severe criticism[16] has been expressed against the above-mentioned theories.

[11]The dispute was started because some private farmers claimed the right to own the bodies of water situated within the lagoon, where intensive fish farming is carried out. The State, which was the other party, laid claim to the ownership of these areas. The Supreme Court established that the lagoon was an area of public property.

[12]Civil Cassation, United Sections, no. 3665 of the 14th February 2011.

[13]I.e. the perspective of the Italian advocates for the juridical theses on common goods.

[14]See Mattei (2012), pp. 60–61.

[15]Lucarelli (2013), p. 66.

[16]See the caustic counterarguments of Vitale (2013).

For constitutional lawyers—it has been said with poetical irony—the scientific value of the juridical construction of common goods "ha la resistenza di un fiocco di neve fluttuato in casa dalla porta semiaperta".[17] It has been said—conversely—that it is scientifically tenuous.

In fact, the juridical theories about common goods tend to overstep not only the proprietary egotism as an elemental paradigm for private law, but also the State sovereignty as a necessary filter for the community to manage and enjoy resources.[18] On the one hand, it is thought that—in the course of time—the citizens have been dispossessed of the popular sovereignty by the institutions, which are stuck in a representative circuit all in all self-referential. On the other hand, heavy criticism is levelled at the big "Leviathan" of these days—the capitalist economy—that aims, through the accumulation of material wealth, at taking the possession of the State sovereignty and putting it in the "invisible" hands of the economic subjects that dominate the market.[19]

So, if on the one hand the *benicomuniste* theories do not appear to be lacking foundation and discernment, because they have brought to light the "human idiocies" related to the plunder of nature[20] that the biological ecologists denounced time ago (besides the juridical artifices that the capitalistic production system has used to create a plutocratic society characterized by inequality), on the other hand they seem to be willing—maybe too easily—to get rid of both the constitutional function of the "public" dimension and the responsibility coming from the egotism of "private" subjects.

In fact, from a juridical point of view, it is not easy to place within the constitutional system the thesis according to which the *distinction* between public and private—rather than their contraposition—should be overstepped, so that the dominant role can be played by the "common", i.e. a dimension that directly refers to the civil community as such and not to a public institution that represents the citizens. In fact, if one intends to adopt such perspective, one will encounter an enormous difficulty: the necessity to identify the reference community (a neighborhood?, a municipality? a region or the State? the whole humanity?), to establish its boundaries, its overall structure and the internal organization, besides the rules of inclusion and exclusion. One aspect is almost ineluctable: the contradiction between the tendency to limit the extension of the community that participates in the management of common goods according to horizontal rather than top-down modules—in compliance with the model of participative democracy—and the necessity to effectively guarantee the universal and equal right to enjoy the goods that—precisely as common goods—afford the exercise of the fundamental rights and the free development of human individuals.[21]

---

[17]di Plinio (2015), p. 189.

[18]On the problems related to food sovereignty, Jannarelli (2011), pp. 299–325.

[19]See Nivarra (2012), especially pp. 85 et seqq.

[20]See De Lucia (2013), pp. 93–112; Sanchez (2010), pp. 61–75.

[21]D'Andrea (2015), pp. 9–10.

Since both the public and the private dimensions are fundamental for democracy, the constitutional system must rather have the possibility to limit both, because it is aware that if no confidence is placed in any of them, they both can have the power to violate human rights.[22]

If so, common goods risk being a magic flute towards the nothing, if they are abstractly thought to play the role of almighty and all-inclusive *salvatores mundi*. They can have a juridical importance only if they are firmly built on the "subversive" bases that exist but have not yet been implemented by the republican Constitution: particularly by carrying out the project of social democracy contemplated by Articles 2 and 3, paragraph 2 of the Italian constitution, whose explicit purpose is to reach the personal and social emancipation of the whole community.

So that the "common" good can be *beyond*, in the sense that it can develop the vision of a new world. But, according to the Constitution[23] this can realistically happen *from* and *in* the public sphere.

It is not a coincidence that the "new" constitutionalisms focused on the "common" sphere—like the ones that have been historicized in the constitutions of Ecuador and Bolivia—have assimilated (i.e. positivized), a vision of the world that considerably differs from the anthropocentric vision of the European and North American constitutionalisms.

In fact, the Andean *nuevo constitucionalismo* deals with the "dilemmas" related to the "externalities" caused by the collective choices from a constitutional point of view and does not consider the problems related to nature and sustainable development to be mere questions of political "discretion" within an unchanged framework of rules regarding authority-freedom. In opposition to the liberal-democratic perspective of the Occident, which focuses on the individual, the perspective of *nuevo constitucionalismo* is focused on *ecology*, so that society and nature are given constitutional roles that are subjective and active. As a result, a citizen can for example apply for a *recurso de amparo* against the administration to defend the "rights of nature", since its "common" resources are fundamental for the life and the wellbeing of everybody.[24]

Conversely, vis-à-vis an ecosystem that is being destroyed (by the conflict between limitless economic growth and limited world) and within such constitutional regulations, the famous "*trust dilemma*" described by Rousseau is not focused on the prey, but rather on the survival/extinction of the hunter himself. With the awareness that the "preference" (the hunter appropriates the natural goods) has become "tragic" compared with the "choice" (the hunter decides the good he

[22]In this sense Fioravanti (2014), pp. 7 and 44 et seqq.

[23]This is the thought of Algostino (2013), pp. 28–29.

[24]In this respect, see Baldin and Zago (2014). See also Carducci (2014), pp. 138–151. About the same topic and from his "benicomunista" perspective, Mattei (2012), p. 22, notices that the new Bolivian Constitution, together with the Ecuadorian one, is the juridical model that deals with the concept of common goods in the most advanced way available, should the humanity intend—after putting aside the Occidental arrogance—to revise its development model.

wants to hunt), men are explicitly considered to be a part of nature by the Constitution, rather than its almighty manipulators.

It has already been said[25] that this became the *Constituent Dilemma* of the twenty-first century, *how* and *what* to decide about this "tragedy".

That is why, compared to the "classical" categorizations of the European and North American democratic systems, these models do not fulfil the anthropomorphic requirements of representativeness and freedom that define democracy as exclusively human responsiveness. Between democracy and nature, between negotiable or deliberative choices and coercion in the name of the "*pro natura*", the latter must prevail, not only to meet "non-balanceable" needs, but above all to pull the constitutional regulations in a different, non-negotiable direction.

Otherwise—this seems to be the main threat that the Andean constitutionalism intends to remedy—the economic benefits and rights will always (or almost) prevail over all the others. Therefore—it also seems to express—in the "name" of nature, the power cannot be fully democratic, because it would fall into the self-destructive "human stupidity" above denounced from biologists and ecologists.

Anyway, this is the issue that needs to be highlighted the most: as long as these questions are not perceived and metabolized by the juridical culture of the Occident through a brand-new constituent process, theses and theories like the ones about common goods are unlikely to become concretely functional and juridically valid.

One thing is certain: in that South of the World that is being "decolonized" from the representations of the colonial reason of the Occident, these themes acquire new postulates and paradigms about the future that must be known and compared to ours to really identify a "new statute" of citizens.[26]

4 Overview on the Right to Adequate Nutrition

At the international level, a first juridical formulation aimed at safeguarding the right to adequate nutrition can be found in Article 25 of the Universal Declaration of Human Rights—proclaimed by the United Nations General Assembly in Paris on 10 December 1948. The purpose of this article is to positivize the wider right to an adequate standard of living,[27] stating in the first paragraph that: "everyone has the right to a standard of living adequate for the health and well-being of himself and of his family, including food, clothing, housing and medical care and necessary social services, and the right to security in the event of unemployment, sickness, disability, widowhood, old age or other lack of livelihood in circumstances beyond his control".

[25]See Linebaugh (2008).

[26]In this respect, see the reflections of Cassano (2004).

[27]In this regard, see Eide (2001), pp. 133 et seqq.

The same universalistic safeguard of the right to adequate nutrition is offered by Article 11 of the International Covenant on Economic, Social and Cultural rights, a multilateral treaty adopted on 16 December 1966 within the framework of the UN and in force from the 23 March 1976.

Just like Article 25 of the Declaration, the first paragraph of Article 11 demands that the Nations recognize "the right of everyone to an adequate standard of living for himself and his family" and explicitly defines "adequate nutrition" in terms of a material condition that must be guaranteed to everybody, since it is essential for life.

Within national law, on the contrary, the general situation seems to be the following: no fundamental right to adequate nutrition is expressly provided by the regulations of the Italian Constitution. For now, it is therefore a fundamental right with an external origin, which—as we will see—has led the national jurisprudence to implicitly declare it in its verdicts.

In fact, unlike the Bolivian Constitution—whose Article 20 establishes that "each individual has the universal and equal right to access to the service of drinking water [...]. Access to water is a human right and *cannot be subjected to concession and privatization*"—the management of water services in Italy has risked being entrusted to private citizens.[28] Moreover—this is the whole point—access to food is almost entirely[29] at the mercy of the social forces that are freely deployed on the free market, even when it comes to the bare minimum of subsistence.

Since—at the constitutional level—the regulations of the Italian Republic do not have (unlike other modern forms of constitutionalism) the *explicit* "obligation to guarantee [...] healthy and adequate nutrition, sufficient for the whole population" (still Article 16 of the Bolivian Constitution), we could say that food has easily become accessible on the markets for people who—thanks to a sufficient salary to buy it—have the possibility to own food *iure privatorum*.[30]

This means that our Constitution lacks a public service of food assistance aimed at ensuring the poor the possibility to be freed from the inconvenience of "hunger".[31]

---

[28]As the Italians know, *ex* Article 75 of the Constitution, that is to say through the abrogative referendum of 12 and 13 June 2011, they have been in favor of the abrogation of Article 23 *bis* of the Decree-Law no. 112/2008, i.e. the abrogation of the therein contained regulations concerning the privatization of public services of economic relevance, including the management of water. In this regard see Staiano (2011), p. 21; and, ultimately, Chirulli (2015).

[29]"Almost", because even if the problem of hunger has not been resolved, a few local services are taking hold in several municipalities of the Republic, including *urban vegetal gardens* (in different varieties, such as didactic, social and associative gardens), different types of *soup kitchens,* local markets, etc.

[30]As it can be inferred, the attention is shifted—*ex* Article 42 of the Constitution—to the problems concerning the "social function" of ownership, especially the ones relating to its "accessibility for everyone". With respect to the "constitutional system of ownership", see Rodotà (2013), pp. 273–421.

[31]Unfortunately, law often tends to distance this word from its real meaning, using instead expressions such as "undernourishment", "malnutrition", "food uncertainty" to talk about a problem—the vital lack of food—that still afflicts millions of people. In this regard, Caparròs (2015), pp. 15 et seqq.

These conditions have made it possible for the right to access to nutrition to be recognized as fundamental by the judges through *inter-constitutional interpretations*.[32] This means that they have had—in the interpretative/applicative phase and formally referring only to national law—to give more sense to the principles of our Constitution by drawing further regulatory meanings—to attribute to the pre-existing[33] constitutional legality—from the outside (i.e. from the rights and the principles contained in supranational documents).

5 Food and Market

From the assumptions crystallized in the agreements of the World Trade Organization[34] (WTO) to the constitutional rights of the single nations of the European and North American Occident, food is considered to be a mere "good". Any other dimension is neglected, especially the one according to which food, together with water, is a basic resource for the populations' life.

To be convinced of this, it is sufficient to read Article 5 of Regulation (EC) no. 178/2002. When it comes to defining the purposes of food law, the article explicitly highlights how important the "safeguard of the *consumers' benefits*" and the "*free movement* of food and animal feed produced and put on the *market* within the community" are. Moreover, if we read this article together with the previous Article 2 (which provides a definition of "food"), we come to the conclusion that EU food law mainly—if not exclusively—deals with the "right *of nutrition*" (i.e. with the rules of production, processing, distribution, commercialization and movement of food, addressed to "consumers"), rather than with the "right *to the nutrition*" of the population, i.e. the individuals' right to feed themselves in an adequate way to live in a respectable way.

It seems, then, that EU law does not take care of regulating the vital "access" to food in the European territory. Instead, it only deals with the aspects related to its "commodification", thus influencing the evolutive interpretation of the Member States' Constitutions and ultimately causing—through its tendential *primauté*—slow and silent transformations from social into neoliberal democracies.

The dominant principles of the free market and "privante"[35] property—which deprives people also of consumer goods—have a considerable impact on the

[32] As I have better defined and explained in other works of mine, of which you can see, if you wish, Gusmai (2015a), pp. 47 et seqq.

[33] With respect to the inter-systemic and inter-constitutional character of Italian law, determined by its openness towards external regulations, see the recent considerations of Ruggeri (2015a).

[34] See the Marrakesh Agreement establishing the WTO, signed on 15 April 1994, which includes the Agreement on Agriculture.

[35] That is, in Italian, private property that "excludes", "deprives" someone of something. This suggestive expression is used by Mattei (2014).

possible meanings that the regulations of the republican Constitution can or cannot take on in the practical application.[36]

For this work, it is sufficient to consider the blows that EU law has dealt to Article 42, paragraph 2 of the Constitution, in the point in which it is stated that the recognition and the guarantee of private property given by law must have "the aim of ensuring its social function and of making it accessible for everybody".

In fact, the competition in the free market—that is the fair of the so-called "best"—cannot ensure the "food-goods" to everybody alone, since the word "everybody" itself does not make it possible to deny access to these goods to certain categories of individuals.

The article can be given another correct interpretation, in line with the whole system of social democracy described by the Constitution, according to which the legislator—to ensure the collective enjoyment at least of the goods essential for life (such as food)—must restrict the freedom "of trade", even by preventing private citizens[37] to claim any right to property, so that the Republic can effectively guarantee "essential levels for the performances regarding civil and social rights".[38]

It has been said, in this respect, that—if the laws regarding property are given an interpretation inspired by Article 42 of the Constitution seen as a foundation for a right to property aimed at ensuring the *social function* of goods, rather than an absolute subjective right—it would be possible to assert that if the social function that allows law to guarantee the right to property is lacking, then also the proprietary title of which goods are the object[39] will be lacking.

However, it is clear that within the current structure of the European juridical/economic system, this type of guarantee seems to be inexorably condemned—if not disregarded—to be slowly buried in the juridic cemetery of desuetude.

The right to adequate food[40] is then no longer perceived as a question that exclusively regards the poor and needy people of the world and is increasingly seen as an entitlement that concerns the human condition as a whole, at any latitude.

In a global world ruled by markets that are increasingly independent of the sovereignty of the single Nations,[41] it is unavoidable that also the phenomenon of poverty increases its dynamism, affecting entire segments of population and

[36]The Italian Constitution is the product of different and opposing political forces. That is why—regardless of the influence of EU law—it is full of "antinomies and conflicts" that can be also found in its second part, i.e. the so-called "organizational" part. This is what has been noticed by Mortati (1962), pp. 222 et seqq.

[37]See Rodotà (2013), p. 387.

[38]As provided by Article 117, paragraph 2, letter m), of the Italian Constitution. In this regard, see Ruggeri (2015b), pp. 541 et seqq.

[39]Lucarelli (2015), p. 117.

[40]In this respect, please allow for references to Gusmai (2015b).

[41]. . .to the point that the doctrine has recently talked about "absence of legislation" and about "law with no author" within the global dimension of law. Ferrarese (2012), pp. 54 et seqq. Other authors have instead a contrary opinion. Among them and above all, see Luciani (2012), p. 3, who asserts instead that "Nations still play a central role in the supranational scenario".

reducing them to misery with incredible speed compared to the past. It has been noticed that the presence of the poor in the common spaces of the everyday life of rich countries and cities has increased again: they beg, sleep on the streets, rummage through rubbish and queue up at the soup kitchens.[42]

It is not a coincidence that, in 2011, even the big global food companies supported the PVOs (private voluntary organizations) and the international organizations[43] in demanding that G20 did something for the "*global food security*".

Although food is a complex sociocultural[44] element, a "basic element for citizens" and, as such, a "precondition of the democracy itself",[45] today one must necessarily turn to the market to adequately feed oneself, even with no salary. Which clearly represents a paralogism, a nonsense that collides with the Constitution, which states that the legislator must establish suitable plans and controls, so that the free economic activity "can be aimed at and coordinated for social purposes" to not "cause damage to safety, freedom and human dignity" (Article 41 of the Constitution).

As it has been said at the beginning, Italian law does not include effective juridical remedies that ensure the needy food assistance (which is provided by Articles 3—paragraph 2—and 38 of the Constitution). In this regard, an effort has been made by introducing the social card that, however, has turned out to be, all in all, an unsuccessful measure,[46] because of the political and administrative inefficiency and to the lack of funds of the local institutions caused by the European Austerity measures.

Without considering that the "state of poverty" is not ascribable to a mere economic question: it is often neglected that it regards the juridical possibility of "access to the means of subsistence".[47]

Anyway, in Italy over 3 million families do not have adequate food. The income of the poor population—unlike the rich one—has further decreased because of the

[42]Saraceno (2015), p. 9.

[43]In this regard, see De Schutter (2013).

[44]This aspect has been stressed by the President of the Republic himself, during his speech at the closing ceremony of Expo 2015 (31st October). In fact, he underlined that "the social and cultural importance of food policies is crucial and their value must be passed down through a serious educational commitment". See Mattarella (2015). Ultimately, Ghiazza (2015) curated an interesting collection of articles about the relationship of food with literature, cinema, linguistic structures, semiotic and psychological implications.

[45]To use the words of Rodotà (2013), p. 486.

[46]However, considerable criticism has been made about the disappointing empirical data showing that the social card (a card to purchase essential goods, aimed at boosting the income of the needy) introduced by the Labour Law Decree of 10 January 2013 (published in the O.J. no. 102 of 3 May 2013), not only has been refinanced, but also reinforced and strengthened by some regulations that are currently being examined by the government. Therefore, it is not possible, now, to make further evaluations.

[47]In this regard, see Cacciari (2014), p. 69.

economic crisis: 18% of the Italians aged 15–64 years skip meals and 40% have reduced the overall consumption.[48]

Indeed, precisely with respect to the social card introduced in 2008, the Constitutional Court—through a systematic and evolutive interpretation of some regulations (Articles 2, 3—second paragraph—and 38 of the Constitution), has removed food from the sovereignty of the free market and has recognized the legitimacy of "*a State intervention that includes also the prevision of the proper and prompt provision of a certain providence*"[49] in favor of the needy.

In this way, at least, the minimum juridical conditions of our system have been legitimized, so that a fundamental right[50]—namely the right to adequate nutrition—can be safeguarded and put into effect through innovative public interventions.[51]

After all, in an interventionist nation oriented toward society—such as Italy—forms of public intervention in the economy, aimed at directing and coordinating the public and private economic activity for social purposes are constitutionally imposed by both the social emancipation policy described in the 2nd paragraph of Article 3 of the Constitution (and recalled in the 2nd paragraph of Article 4, in the point that refers to the "material progress" of society), and the constitutional recognition of social rights.[52] If these regulations should turn out to be retrogressive compared to other sources, the balance underlying the social—albeit not easy—democracy of the Founding Fathers[53] could be upset.

6 Food and Common Goods

At this point, after providing the outlines of the doctrines relating to common goods and the right to food, a question arises: is it possible to see food as a common good to rescue from the "tragedy"?

The answer is not easy, but I will try hereafter—albeit concisely—to illustrate some arguments against this possibility.

At the economic level, the fact that the Occidental contemporary societies are based on the *Grundnorm* of the free market makes food a good that is "rival", but

[48]Saraceno (2015), p. 98.

[49]This is what has been decided by the Italian Constitutional Court with the sentence no. 10 of 2010 ("In law" section, point 6.4). Through the debate about the regulation that introduced the "social card" in Italy for the first time, (Article 81 of Decree Law no. 112 of 2008), the Court seems—through this sentence—to have explicitly recognized the right to nutrition as both a "fundamental right" and a "social right", based on Articles 2, 3—second paragraph—and 38 of the Constitution.

[50]The matter concerning the actualization of fundamental rights is dealt with in detail by Baldini (2015).

[51]This is what many economists hope, including Mazzucato (2014).

[52]Colapietro and Ruotolo (2015), p. 634.

[53]With respect to the concept of "social democracy", see d'Albergo (2004), pp. 161 et seqq.

mostly "excludable" (unlike what is stated by Hardin) from the logic of the prices set by the capitalist production processes.

Except the marginal cases of self-sustenance of property owners, who—as such—have the possibility to consume and find some food in the plots of land they own (of course not all, which means that nutrition is inadequate), food is available on the markets and is moreover processed by other economic actors acting in the same economical and speculative context. Since food is subjected to the rules of demand and supply, people with an insufficient income cannot have access to it, because they cannot pay its price.

A different argument applies to another fundamental good: drinking water.

Since the economic system cannot prevent people from acquiring water to feed themselves freely (from a public fountain, for example)—although this good has been the object of attempts of commodification and privatization—in this case the theory of common goods has had the possibility to take root and find expression. In fact, as it has been said above, the political slogan "water common good" led the Italians, in 2011, to side against the privatization of "public services of economic relevance" (including water) through the referendum that took place on the 12th and the 13th June of the same year. From an economic point of view, unlike food, water is partially "non-excludable", besides being "rival", which is perfectly in line with the economic theory of common goods.

At the juridical level, things do not change that much. As it has been underlined, going "beyond" the public and the private dimensions not only means disregarding the sociocultural principles of the Occident, which religiously worships the free market in a fundamentalistic way, but also collides with the republican Constitution and its principles.

In fact, beyond the dichotomy public/private, there is no offshoot of the republican Constitution that can give firm guarantees to citizens and residents, which juridically risk falling into the nothing and being deprived of a positive protection. Moreover, the underestimation of the political and cultural context (which legitimizes the system[54]) and the positive data (the formal Constitution that does not know common goods) may turn out to be a lethal mix for the structure of constitutional democracy. In fact, the Constitution has been calibrated to create balance between the public and the private spheres and hold them together, rather than to go beyond them. It is not about taking a reactionary position towards the doctrines of common goods or being traditionalists, obtuse and closed to new ways of seeing the juridical relationships. On the contrary, it is about knowing the historic and material conditions to fully understand the good that lies in common goods are not ripe. The biggest risk is to lose sight of the unexpressed potentialities of the "public" dimension in favor of a "common" dimension, which—since it is not clearly defined in its

[54]Mattei (2012), pp. XIV–XV, acknowledges in his introduction that, since common goods are loaded with political values, their compatibility with the common ideal must not be taken for granted.

juridical traits—risks strengthening even more the benefits of the few "private" citizens that already possess half of the global wealth.[55]

Let us focus again on the right to food, which is undoubtedly a fundamental right. The positive aspect does not lie in the fact that it is "common", but in the fact that everybody can have access to its enjoyment, so that the whole community has the possibility to live a respectable life, free from hunger. Therefore, to be constitutionally valid, the adjective "common" must be read as synonym with "everybody", according to what is said by Article 42 of the Constitution with respect to the individuals' material accessibility to those that, like it or not, are *de iure condito* the "terrible" rights of public or private owners.

Therefore, a higher scientific value can be attributed to the doctrine that, rather than a common good, considers food to be an "*asset*" when it goes beyond one's possibility to pay for the minimum amount of food needed for survival and a "*social good*" when it can guarantee the fundamental right to human subsistence.[56] Conversely, food can be considered to be a good subjected to free exchange on the market (thus becoming subjected to property rights) when only the individuals who have a certain income can have access to that good. On the contrary, if the economic condition cannot even guarantee the subsistence, food unavoidably turns into a *social good*, i.e. a good that embodies a *social right* that, as such, compels the institutions of the Republic to remove any obstacle that hinders *de facto* the enjoyment of the fundamental right to access to this good (Article 3, paragraph 2 of the Constitution).

This leads to the conclusion that if we really want to definitively resolve the issue of access to food, we must deal with it from a different perspective: we must no longer place it exclusively within the context of the economic principles underlying the process of production and consumption of food.

As Karl Marx taught us, food is a product and an object of consumption at the same time.[57] Agriculture as the cultivation of the *ager* and farms as functional instruments[58] have enclosed the agricultural industry in a frame of economic factors and food law in a frame of rules concerning relationships that do not relate to it. It is not a coincidence that food policies are mostly subjected to agro-economic regulations and that—in jurisprudence—they are a matter of agricultural law. The clearest evidence is represented by the regulations issued by European law so far. The "economic" figures of "producer" and "consumer" are the most relevant "legal entities" for the law of the Old Continent. As seen, all European law concerns the

[55]According to the data provided by Oxfam (updated to January 2016), the situation is nothing short of embarrassing: only 62 people possess the half of the global wealth. The data can be consulted at this link: *http://www.oxfamitalia.org/eventi/sfida-lingiustizia-di-basta-ai-paradisi-fiscali*.

[56]See Ferrajoli (2012), pp. 599 et seqq.

[57]Marx (2012), p. 44.

[58]See Segré (2011).

series of production and consumption processes aimed at meeting the needs of trade.[59]

For the European food policies and for the resulting law, food safety is a priority, in the sense that the "healthiness of the agricultural produce" and the consumers' safeguard represent the basic goals of the European common market.[60] The principle of the Lisbon Treaty regarding the "social market economy" is still an empty formula; just as food policies do not consider the *civis europaeus* to be a "human being" with his vital needs; the political and juridical attention towards food and nutrition as a source of individual rights or as social goods aimed at ensuring people their fundamental rights is today marginal.

7 What Can Be Done to Avoid the "Tragedy"? Paradoxical Conclusion

Once food is deservedly placed among the fundamental rights, another strange legal regulation must be considered, which can be implicitly inferred from the system: a law that invites the fast, just like Occidental doctors often write in their prescriptions.

In fact, if one simply takes note of what is provided by the formal juridical system, it is unavoidable to realize that Italian law is clearly unreasonable, because it provides a public service that guarantees access *only* to water, while food, albeit essential for life, is left at the mercy of the forces of the free market.

The problem arises especially when these disparities of safeguard affect indigent people—not protected by the State—who claim the right *to* nutrition because it is fundamental right, and not because they are mere "consumers" who have access to the rights *of* nutrition. The issue is far from being secondary and shifts the attention from the level of "*lex mercatoria*" to the level of "*lex fundamentalis*"—i.e. a "supra-constitutional" level—because ultimately it is all about dealing with rules and principles that must not only affect the State incomes, but also the dignity of human beings.[61]

Nowadays, the challenge to food seems more than ever to be a political challenge.[62] And, like all the political questions that are closely related to the fundamental rights and principles, it is a challenge that—because of its "ecumenical" value—has above all a constitutional significance.[63] Somehow, the challenge to the right to "adequate" nutrition seems to envisage a different worldview according to

[59]Ferrari and Izzo (2012), p. 9.

[60]See Gabdin (2011), pp. 243 et seqq.

[61]Human dignity is defined a "supra-constitutional" value by Ruggeri and Spadaro (1991), pp. 343 et seqq.

[62]Although on different reasons, the same opinion is expressed by De Castro (2015), p. 81.

[63]The closeness of constitutional law to the political dimension has been recently underlined by the Italian Constitutional lawyers Zagrebelsky and Marcenò (2012), p. 65.

which the forward-looking intelligence of men and the institutional activities (at any level, but above all "from the bottom") cooperate to build new organizational models different from the current ones.

In the light of what has been said, it seems necessary to clarify another point: there is the need to experiment political and institutional models able to upset the current individualist and anthropocentric worldview and in favor of social policies of solidarity that can coexist in harmony with every thing around us or, to be more precise, with the "environment" seen as natural heritage that includes *also* men,[64] in front of whom "terrible" rights and limitless economic powers must surrender, if necessary.

It is not about sharing peaceable perspectives nor about believing in a world where it is suddenly possible to live in a new phantomatic state of nature characterized by the total absence of social conflict. Such perspective would be unhistorical and naive.

More simply, it is about thinking that in a period of economical, financial, environmental and structural crisis like the one that is affecting the current society—in which poverty is threatening a higher and higher number of citizens and families of the so-called middle class, which is about to die out—dealing with "food availability", with its "adequacy" and—above all and from a constitutional perspective—with its "economic accessibility" can only turn out to be useful to reconsider the traditional competences and the main functions of the public administrations that are close to the population. Therefore, "inside" the public sphere, in support of the social democracy contemplated by the Constitution, rather than "beyond", as the *benicomuniste* doctrines want.

As I have explained in other works of mine,[65] it has been explained that, to reduce the macroscopic disparity of approach towards two rights (to water and to food) that are equally fundamental, it would be appropriate to introduce into the system a *permanent public service of food assistance* aimed at ensuring adequate food to all the indigents legally residing in the territory of the Republic. In a nutshell, starting "from the bottom", through a minimum contribution from the residing population with an income above a certain threshold, it is possible to create in the municipalities a local service: a public food service (through food stamps, family allowances to buy food, agreements with restaurants, supermarkets, etc.), to not weigh on the already precarious economic situation of the local institutions.

Consider, for example, a municipality with 50,000 taxpayers with an income above a certain threshold. If each taxpayer should donate *only 50 cents per day* to support the local service of food assistance, a monthly amount of 750,000 Euros would be available for the residents with an income that does not guarantee access to adequate food. In this way, the municipality would have the possibility to always give 5000 residents 150 Euros per month to be spent exclusively for nutrition.

---

[64] In this respect, Ostrom (2009), pp. 419 et seqq.

[65] See Gusmai (2015b), pp. 13 et seqq.

Regardless of the numerical data, which require anyway a detailed and constant study on the poverty rates of each municipality, this possible perspective "of solidarity" oriented towards the Constitution (Articles 2 and 53)[66] could represent—better if combined with innovative measures against food waste[67]—an effective political action to publicly implement values and principles that, so far, the ruling class has all in all left in the orbit of what is juridically irrelevant.[68]

Ultimately, at the constitutional level, "reasonableness"—that, as it has been effectively said, represents a true "architectural principle of the system"[69]—gives way to positive "irrationality" if the legislator does nothing to balance the juridical safeguard of two goods that—it has been said repeatedly—both embody the same fundamental right to nutrition.[70]

Of course, since these goods have different qualitative and quantitative features by nature, the economy can diversify their value (of exchange) depending on the production methods that a certain State allows the economic entities to adopt.

However, from the perspective of constitutional law and, particularly of the safeguard of fundamental rights, no doubt that no distinction can be made between water and food. Otherwise, as it has been asserted, law would be completely subordinated to the economy, which would be the only reasonable "structure"[71] on

[66]It is well-known that Article 2 of the Italian Constitution—in a dialectical relationship with the "*inviolable rights* of man"—expects that "the *fundamental duties* of political, economic and social solidarity be fulfilled". By fulfilling the duty to contribute (Article 53 of the Constitution) to public expenditure for a public local service easy to manage and control (e.g. through the online publication of both the balance sheet of the institution that provides the service and the incomes of users), the population could allow local institutions to withdraw a derisory amount of money from the medium-high incomes and redistribute wealth in favor of the weakest subjects who have not access to adequate nutrition.

[67]Think of the recent French regulation against food waist that, *inter alia*, compels supermarkets to give unsold food to the needy. In fact, according to FAO the amount of food that ends up in the garbage accounts for over 35% of the global production, for a cost of about 1 trillion dollars per year. While these words are being written, also the Italian legislator is discussing a draft law against food waste (presented by the MPs Chiara Gadda and Massimo Fiorio).

[68]It is not a coincidence that in the section about the "*Right to food*" of the most prestigious encyclopedia of international law (*Max Planck Encyclopedia of Public International Law*), it can be read that the right to food is one of the less respected human rights, as shown by the data about global malnutrition and hunger. See Mechlem (2008).

[69]D'Andrea (2005), pp. 203 et seqq.

[70]With respect to the possible "arbitrariness of legislation" and the terminological distinction between the "irrationality", "unreasonableness" and "injustice" of juridical choices, see once again Zagrebelsky and Marcenò (2012), pp. 195 et seqq.

[71]This is a reference to Marx, who considers law to be a "super-structural" manifestation of the mode of production, since its character is not autonomous and historically determined. According to his theses, law is not something beyond the classes that are struggling, something that impartially judges the actions of social actors: on the contrary, it is the reflection of certain social relationships (which it ratifies and legitimizes) as well a way through which the ruling class dominates the other classes. In this respect, see the considerations of Bobbio (1997), pp. 153–166.

which every other social phenomenon depends (not only law, but also politics, religion, art, philosophy, etc.).

The same thing would happen if an attempt were done to resolve the problem by searching the fine-mesh net of the formal juridical logic to find a possible "scientific value freedom".[72] Which, if used to give sense to the fundamental right to adequate nutrition, leads to an aristotelian syllogism: all men must eat to live (*major premise*); food and water can be ingested (*minor premise*); food and water are vital for men (*conclusion*).

So, it is clear that the constitutional system cannot permit indiscriminate "exclusions" from access to food, which—on par with access to water—must be "universal", since also food is fundamental for life.

Since no one has the right to behave like he was the last human being on the Earth and no one has the right to appropriate and deny the others access to collective goods,[73] once food is (re)considered by law to be a fundamental sociocultural good, the "rivalry" for the acquisition of food on the market needs to be downsized proportionally to its guarantee for those who cannot afford it.[74]

If no service of public assistance ensures the poor *also* "adequate food", the juridical paradox is an unavoidable conclusion: according to national law, one could legally and legitimately assert that it is possible to live only on water—with no food—in an "adequate" way.

References

Algostino A (2013) Riflessioni sui beni comuni tra il "pubblico" e la Costituzione. Costituzionalismo.it: 3, pp 28–29. http://www.costituzionalismo.it. Accessed 25 Jan 2016

Baldin S, Zago M (2014) Le sfide della sostenibilità. Il buen vivir andino dalla prospettiva europea. Filodiritto, Bologna

Baldini V (2015) La concretizzazione dei diritti fondamentali. Tra limiti del diritto positivo e sviluppi della giurisprudenza costituzionale e di merito. Editoriale Scientifica, Napoli

Bobbio N (1997) Né con Marx né contro Marx. Editori Riuniti, Roma, pp 153–166

Cacciari P (2014) Vie di fuga. Crisi, beni comuni, lavoro e democrazia nella prospettiva della decrescita. Marotta & Cafiero editori, Napoli

Caparròs M (2015) La fame. Einaudi, Torino

Carducci M (2014) Per un costituzionalismo "Handy" nell'epoca dei mutamenti costituzionali incostituzionali. Alternative per il socialismo 32:138–151

Cassano F (2004) Homo civicus. La ragionevole follia dei beni comuni. Dedalo, Bari

Cerulli Irelli V, De Lucia L (2014) Beni comuni e diritti collettivi. Politica del diritto 1:3–36

[72]This is a clear reference to the famous "methodological" essay by Weber (1917).

[73]Cacciari (2014), p. 81.

[74]The fact that the laws produced by the State (and by the consequent activities of public institutions) often create a high level of social inequality has been recently underlined by Stiglitz (2012).

Chirulli P (2015) I servizi pubblici locali e il servizio idrico integrato tra legislatore, Corte costituzionale e (presunta) "volontà" referendaria. Federalismi.it: Rivista di Diritto Pubblico italiano, comparato, europeo 7:1–39. http://www.federalismi.it. Accessed 25 Jan 2016
Colapietro C, Ruotolo M (2015) Diritti e libertà. In: Modugno F (ed) Diritto Pubblico, Torino
d'Albergo S (2004) Diritto e Stato tra scienza giuridica e marxismo. Sandro Teti Editore, Roma, pp 161 et seqq
D'Andrea L (2005) Ragionevolezza e legittimazione del sistema. Giuffrè, Milano
D'Andrea L (2015) I beni comuni nella prospettiva costituzionale: note introduttive. Rivista AIC 3:9–10
De Castro P (2015) Cibo. La sfida globale. Donzelli Editore, Roma
De Lucia V (2013) Towards an ecological philosophy of law: a comparative discussion. J Hum Rights Environ 2:93–112
De Schutter O (2013) Rapport intérimaire du Rapporteur spécial sur le droit à l'alimentation. A/68/ 288, 7 August 2013, pp 7–8
di Plinio G (2015) Ambiente. In: Somaini E (ed) I beni comuni oltre il comune. IBL Libri, Torino
Eide A (2001) The right to an adequate standard of living including the right to food. In: Eide A, Krause C, Rosas A (eds) Economic, social and cultural rights, 2nd edn. Brill, Dordrecht, pp 133 et seqq
Ferrajoli L (2012) Principia iuris. Teoria del diritto e della democrazia. Teoria della democrazia, vol II. Laterza, Roma-Bari, pp 599 et seqq
Ferrarese MR (2012) Prima lezione di diritto globale. Laterza, Roma-Bari, pp 54 et seqq
Ferrari M, Izzo U (2012) Diritto alimentare comparato. Il Mulino, Bologna
Fioravanti M (2014) Pubblico e privato. I principi fondamentali della Costituzione. Editoriale Scientifica, Napoli, pp 7 et 44 et seqq
Gabdin D (2011) Droit de l'alimentation et droit agricole européens: quelles articolation? Revue européenne de droit de la consomation 2:243 et seqq
Ghiazza S (2015) Cibo e/è Cultura. Quaderni di Ateneo, vol XV. Bari, Università degli Studi di Bari Aldo Moro
Gusmai A (2015a) Giurisdizione, interpretazione e co-produzione normativa. Cacucci, Bari, pp 47 et seqq
Gusmai A (2015b) Il diritto fondamentale al cibo adeguato tra illusioni e realtà. Dirittifondamentali.it: Dottrina. http://www.dirittifondamentali.it/unicas_df/. Accessed 25 Jan 2016
Hardin G (1968) The tragedy of the commons. Science 162:1243–1248
Jannarelli A (2011) Profili giuridici del sistema agro-alimentare tra ascesa e crisi della globalizzazione. Cacucci, Bari, pp 299–325
Linebaugh P (2008) The Magna Charta Manifesto. Liberties and commons for all. UCLA Press, Berkeley
Lucarelli A (2013) La democrazia dei beni comuni. Nuove frontiere del diritto pubblico. Laterza, Roma-Bari
Lucarelli A (2015) Beni comuni e funzioni sociali della proprietà. Il ruolo del Comune, in Beni comuni e cooperazione. In: Sacconi L, Ottone S (eds) Beni comuni e cooperazione. Il Mulino, Bologna
Luciani M (2012) Funzioni e responsabilità della giurisdizione. Una vicenda italiana (e non solo). Rivista AIC 3:3
Marella MR (2012) Oltre il pubblico e il privato. Per un diritto dei beni comuni. Ombre Corte, Verona
Marx K (2012) Lineamenti fondamentali di critica all'economia politica. Manifestolibri, Roma
Mattarella S (2015) Intervento del Presidente della Repubblica Sergio Mattarella alla cerimonia conclusiva di EXPO 2015. http://www.quirinale.it/elementi/Continua.aspx?tipo=Discorso&key=143. Accessed 25 Jan 2016
Mattei U (2012) Beni comuni. Un manifesto. Laterza, Roma-Bari
Mattei U (2014) Senza proprietà non c'è libertà (Falso!). Laterza, Roma-Bari

Mazzucato M (2014) Lo Stato innovatore. Sfatare il mito del pubblico contro il privato. Laterza, Roma-Bari

Mechlem K (2008) Food, right to, international protection. Max Planck Encyclopedia of Public International Law

Mortati C (1962) Costituzione dello Stato (dottrine generali e Costituzione della Repubblica italiana). Enc dir, vol XI. Giuffrè, Milano, pp 222 et seqq

Nivarra L (2012) Alcune riflessioni sul rapporto fra pubblico e privato. In: Marella MR (ed) Oltre il pubblico e il privato. Per un diritto dei beni comuni. Ombre Corte, Verona, pp 85 et seqq

Ostrom E (1990) Governing the commons, the evolution of institutions for collective action. Cambridge University Press, Cambridge

Ostrom E (1998) A behavioral approach to the rational choice theory of collective action. Am Polit Sci Rev 91:1–22

Ostrom E (2009) A general framework for analyzing sustainability systems. Science 325:419 et seqq

Patel R (2009) The value of nothing. How to reshape market society and redefine democracy. Portobello Books, London

Rodotà S (2013) Il terribile diritto. Studi sulla proprietà privata e i beni comuni. Il Mulino, Bologna

Ruggeri A (2015a) Come costruire e preservare la tradizioni dei diritti in Europa? II seminario annuale di Diritti Comparati "Costruendo le tradizioni dei diritti in Europa", Macerata 11 dicembre 2015. http://www.dirittifondamentali.it/unicas_df/. Accessed 25 Jan 2016

Ruggeri A (2015b) Per uno studio sui diritti sociali e sulla Costituzione come sistema (notazioni di metodo). Consulta OnLine, Fasc. II, pp 541 et seqq

Ruggeri A, Spadaro A (1991) Dignità dell'uomo e giurisprudenza costituzionale (prime notazioni). Politica del diritto 3:343–377

Sacconi L (2015) Beni comuni, contratto sociale e governance cooperativa dei servizi pubblici locali. In: Sacconi L, Ottone S (eds) Beni comuni e cooperazione. Il Mulino, Bologna, pp 175 et seqq

Sanchez L (2010) Darwin, artificial selection, and poverty. Contemporary implications of a forgotten argument. Polit Life Sci 29(1):61–75

Saraceno C (2015) Il lavoro non basta. La povertà in Europa negli anni della crisi. Feltrinelli, Milano

Segré A (2011) Politiche per lo sviluppo agricolo e la sicurezza alimentare. Carrocci Editore, Roma

Staiano S (2011) Note sul diritto fondamentale all'acqua. Proprietà del bene, gestione del servizio, ideologie della privatizzazione. Federalismi.it: Rivista di Diritto Pubblico italiano, comparato, europeo 5:21. http://www.federalismi.it. Accessed 25 Jan 2016

Stiglitz JE (2012) The price of inequality. How today's divided society endangers our future. W.W. Norton & Company, New York-London

Vitale E (2013) Contro i beni comuni. Una critica illuministica. Laterza, Roma-Bari

Weber M (1917) Der Sinn der "Wertfreiheit" der soziologischen und ökonomischen Wissenschaften. Logos. Internationale Zeitschrift für philosophie der kultur 7(1)

Zagrebelsky G, Marcenò V (2012) Giustizia costituzionale. Il Mulino, Bologna

The Constitutional Protection of the Right to Food in Bolivia and Ecuador

Anjeza Sina

1 A Global Overview of Right to Food

This section provides a general overview of right to food. We want to emphasize that the right to food is enshrined in the Constitution mostly in poor countries. Rodotà claimed that "the right to nutrition – healthy, safe, sufficient – is considered as one of the "most fundamental of fundamental rights".[1] This suggests that there is a relationship between the right to food and the Constitution. In fact, a good protection system requires not only the ratification of the relevant human rights treaties but, arguably, also their constitutional protection and further implementation, as necessary, through the enactment of appropriate legislation.

1.1 International and Constitutional Protection of the Right to Food

Although, the "most fundamental of fundamental rights" such as the human right to adequate food has been recognized in different international instruments, it is essential a legal protection at national level through constitutional provisions and national legislation. Most notably normative, first of all, Art. 25 of the Universal Declaration of Human Rights (UDHR) recognizes the right to an adequate standard of living, including food; Art. 11 of the International Covenant on Economic, Social and Cultural Rights (ICESCR) recognizes the right to an adequate standard of living,

[1]Rodotà (2011).

A. Sina (✉)
Department of Legal Studies, University of Salento, Lecce, Italy
e-mail: anjeza.sina@studenti.unisalento.it

A. Isoni et al. (eds.), *Food Diversity Between Rights, Duties and Autonomies*,
LITES – Legal Issues in Transdisciplinary Environmental Studies 2,
https://doi.org/10.1007/978-3-319-75196-2_16

including adequate food, and the fundamental right to be free from hunger as a separate right (160 States Parties); the Convention on the Elimination of All Forms of Discrimination Against Women (CEDAW) that recognizes in Art. 12 the right of pregnant and lactating women to special protection concerning adequate nutrition and in Art. 14 the right of rural women to equal access to land, water, credit and other services, social security and adequate living conditions (186 States Parties); Art. 25 of the Convention on the Rights of the Child (CRC) recognizes the right to the highest attainable standard of health, and Art. 27 states the right to an adequate standard of living that, in both articles, includes food and nutrition (193 States Parties).

In examining the right to food it is important to consider the definition of the UN Special Rapporteur on the Right to Food as the "right to have regular, permanent and unrestricted access, either directly or by means of financial purchases, to quantitatively and qualitatively adequate and sufficient food corresponding to the cultural traditions of the people to which the consumer belongs, and which ensures a physical and mental, individual and collective, fulfilling and dignified life free of fear".[2]

The International Covenant on Economic, Social and Cultural Rights indicates that the right to adequate food entails a "fundamental right of everyone to be free from hunger".[3] The Committee has further defined the scope of this right, clarifying that it implies: the availability of food in a quantity and quality sufficient to satisfy the dietary needs of individuals, free from adverse substances, and acceptable within a given culture; [and] the accessibility of such food in ways that are sustainable and that do not interfere with the enjoyment of other human rights.[4] For example, the General Assembly has referred to the complex character of the global food crisis, in which the right to adequate food is threatened to be violated on a substantial scale, is a combination of several major factors, such as the global financial and economic crisis, environmental degradation, desertification, and the impacts of global climate change, as well as natural disasters.[5]

From a study conducted by FAO in 2011 has emerged as the right to food is present in more than one hundred Constitutions.[6] Sometimes it is explicitly acknowledged as a human right (in 23 cards), in some cases, by implication through the right to an adequate standard of living, welfare, development or objective of public policy. The study briefly identifies those countries in which the right to food is directly applicable under international human rights treaties being incorporated into the

[2]OHCHR (2010).

[3]ICESCR (1966), 993 U.N.T.S. 3, 16 December 1966 (entered into force 3 January 1976), Art. 11 (2).

[4]UN Committee on Economic, Social and Cultural Rights CESCR (1999), General Comment No. 12: The Right to Adequate Food (Art. 11 of the Covenant). http://www.refworld.org/docid/4538838c11.html. Accessed 14 December 2015.

[5]OHCHR (2013).

[6]Cf. Knuth and Vidar (2011).

domestic legal system; others countries the right to food takes place through its incorporation into the constitution and through framework laws and sectoral laws.

A total of 23 constitutions recognize—as mentioned—the right to food explicitly as a human right, of these nine countries recognize the right as a separate and stand-alone right: Bolivia (Art. 16), Brazil (Art. 6), Ecuador (Art. 13), Guyana (Art. 40), Haiti (Art. 22), Kenya (Art. 43) and South Africa (Art. 27.1). The Interim Constitution of Nepal recognizes an individual right to food sovereignty (Art. 18.3) and Nicaragua (Art. 63) provides for the right of every person to be free from hunger.

Therefore, a legal workshop extremely interesting is represented by constitutional and infra constitutional experience in Andean countries, especially for their "bio-centric and ecological shift".[7] The so-called "Andean nuevo constitucionalismo" identified by the Constitutions of Ecuador (2008) and Bolivia (2009), mark a sharp change. Indeed, beyond providing the legal substantive scope of the right to food, they ensure some tools to guarantee quantitatively and qualitatively adequate and sufficient food.[8]

In Bolivia, Art. 16 of the current Constitution reads as follows: "Every person has the right to water and food. The State has the obligation to guarantee food security for all through a healthy, adequate and sufficient food. Meanwhile, Art. 13 of Ecuadorian Constitution reads: "Persons and community groups have the right to safe and permanent access to healthy, sufficient and nutritional food, preferably produced locally and in keeping with their various identities and cultural traditions. The Ecuadorian State shall promote food sovereignty."

1.2 Food Security

Recent food price increases and volatility have created a sense of urgency to better understand and address food security across developed and developing countries.

Food security is a complex phenomenon that was originally measured by the ability of a country to access enough food to meet the dietary energy requirements of its population, it is also a flexible concept as reflected in the many attempts at definition in research and policy usage.[9] However, since that original definition, the term "food security" has been used in different ways, until a definition was widely accepted at the 1996 World Food Summit (FAO 1996): "Food security, at the individual, household, national, regional and global levels [is achieved] when all people, at all times, have physical and economic access to sufficient, safe and nutritious food to meet their dietary needs and food preferences for an active and

[7]See Pierri (2015), pp. 451–469.

[8]In Venezuela (1999), Ecuador (2008), and Bolivia (2009) new constitutions were adopted as a 'Turn to the Left' emerged and a "new Andean Constitutionalism" was born, see Schilling-Vacaflor (2011).

[9]See Pinstrup-Andersen (2009).

healthy life". This definition is again refined in The State of Food Insecurity 2001: "Food security [is] a situation that exists when all people, at all times, have physical, social and economic access to sufficient, safe and nutritious food that meets their dietary needs and food preferences for an active and healthy life". It is known that there are multiple factors affecting an individual's ability to easily access food that adequately satisfies his or her needs. Overall, food security interventions in practice span several sectors, including agriculture, rural development and infrastructure, health, education, and social protection.

Traditional knowledge systems are being harnessed and enriched with findings from modern systems of science and technology to protect communities and their ecosystems throughout Latin America, but it is not quite. Although agriculture interventions typically improve the production of agriculture goods promoted—as a recent review shows—they fail to increase households' total incomes, improve their dietary diversity, increase absorption of macronutrients, or reduce malnutrition prevalence.[10] We have considered that, *inter alia*, food security is just one of the factors contributing to good nutritional status at the individual, household and community levels.

2 The Right to Food in the Andean "Nuevo constitucionalismo"

The purpose of this section is to learn from the very encouraging developments concerning the realization of the right to food that are currently taking place in Bolivia and Ecuador, also to gather information relevant to the protection and promotion of this right.

2.1 Main Characteristics of the Andean Region

The Andean Region of Latin America has experienced governance and economic difficulties that had compounded its structural problems of poverty. The Andean Region, characterized by the high Andes, the Amazon and coastal climates, is exposed to recurrent floods, tropical storms, cold waves, seismic and volcanic activity. Socioeconomic conditions within these countries constitute a very high level of risk. The Andes are a mountain range with the most varied climates and microclimates on Earth, presenting a vast biodiversity. Human survival may seem easy in such favorable conditions, but in reality, it appears difficult and complex because of the frequent climate changes. The Andean region has during recent decades witnessed a marked expansion of nature conservation initiatives. From the

[10]Masset et al. (2012).

preservation of the region's rich biodiversity, to the maintenance of watershed integrity, to the promotion of ecotourism.[11] Nevertheless, the population below the poverty line in rural areas reaches 52.8%, while in the urban areas is 27.8% (Martinez et al. 2009).

To be noted that in Latin American the "Derecho Humano a la Alimentación Adecuada" is explicitly recognized at the constitutional level in: Bolivia, Brazil, Costa Rica, Cuba, Ecuador, El Salvador, Guatemala, Honduras, Mexico, Nicaragua, Panama and Paraguay. Those mentioned in the implicitly: Colombia, Peru and Venezuela. It is advisable to improve spaces for dialogue between Municipal governments (because at this level the vulnerability of rural populations becomes visible) and civil society as a strategy for the integration of obligations and the principles related to the "Derecho Humano a la Alimentación Adecuada", food security and nutrition in rural communities in Latin America.[12]

Constitutional reforms in the Andean countries since the 1990s have been considered among the most radical because of the introduction of mechanisms to enhance popular participation at the local level and to narrow the gap between state and civil society. Andean countries have also pioneered what has been called multicultural constitutionalism, whereby traditionally excluded ethnic groups of the population are constitutionally recognized.[13]

Bolivia and Ecuador are, among others, receiving technical and financial support from the FAO to conduct consultations with civil society and develop the framework law.[14]

In the relationship between man and nature protection of nature it can only be realized through the recognition of the constitutional nature as subject with the man recovering well cosmology of indigenous peoples who see nature as Pachamama, "Mother Earth". Furthermore, the two Andean Constitutions, ecologically oriented, state the primacy of the "Mother Earth": the human being is not only the company; it is above nature; he lives in nature and with nature; so, relations with it can no longer be left to the indifference of law or regulation only to serve the interests and property of individuals and communities. The experiences of the "new Andean constitutionalism" as the Constitution of Ecuador in 2008 and Bolivia in 2009 trace its inspiration by the need to secure future generations access to food and the sustainability of development and the reproducibility of resources (Carducci 2014).

Therefore, it is overcome the idea of accumulation and unlimited growth as ultimate goal of development, saying a new ordering principle: sumak kawsay (in Quichua), Andean principle of 'good life' (buen vivir, in Spanish). The inclusion of the concept of the "good life" in the new constitutions of both Bolivia and Ecuador reflects a new regional commitment to implement an alternative development strategy, which embodies the variety of cosmologies of mountain peoples

[11] Himley (2009), pp. 832–842.

[12] See Kopp (2015).

[13] See Lalanders (2012), pp. 163–182.

[14] De Schutter (2010).

throughout the Americas; similar expressions of the "good life" are a central part of the philosophical heritage of other indigenous groups, many of whom live in the highlands.[15] Here we find a different conception of food and lifestyle from the current Western model. In this sense arise new constitutional principles such as food sovereignty, the right to food, the qualification of natural resources as "common goods" not subject to any public or private property, the subordination of the primacy of the economy to the preservation of biodiversity.

2.2 *Bolivia's Legislative Framework*

As mentioned above, in 2009, Bolivia has taken a major step towards the realization of the right to food by including the latter in its Constitution.

Landlocked in the middle of South America, Bolivia ranges across the high Altiplano and the Andes Mountains, to the tropical hills of the Yungas, the expansive forests of the Amazon and the lowland grassy plains of Santa Cruz. Bolivia is the second poorest country in Latin America and has one of the highest levels of child malnutrition and under-nourishment in the continent. Although Bolivia is a country rich in mineral resources—natural gas, oil, and metals including silver, gold, iron, zinc and tin—the vast majority of its people have not benefited from this natural wealth and remain poor and malnourished. Chronic malnutrition affects more than one in four Bolivian children. The highest levels of malnutrition are amongst Bolivians living in rural areas, especially in the high plains of the Altiplano regions of Potosi and Chuquisaca, but also in the valleys and tropical lowland departments of Beni and Pando.[16] Malnutrition levels are much higher amongst the poorest families. The harsh agricultural conditions, and past failures to invest in small-scale agriculture, account for the rapid urbanization of Bolivia, as millions have migrated to the cities, or abroad. The difficult conditions of subsistence agriculture stand in strong contrast to the modern agro-industrial plantations and cattle ranches that dominate Bolivia's eastern lowlands. Although highland families grow fava beans, quinoa, potatoes, oca (a tuber) and many keep animals such as sheep or llamas, many of their products are sold rather than eaten, because of the need to generate income. Unfortunately, if landless families in rural areas cannot afford to rent land they are even poorer and work as sharecroppers or wage workers. We must consider what most agricultural work is done by hand with little access to machinery even to plough the fields, and there has been little investment in irrigation and other infrastructure that would allow increased production. Consequently, this has resulted in very high levels of malnutrition of families, especially micronutrient malnutrition, amongst altiplano families because their diet is inadequate.[17]

---

[15]Barkin (2012), pp. 12–22.

[16]See UN Human Rights Council (2008).

[17]See Ziegler (2008) and De Schutter (2010).

The new constitution recognizes the right to food and the right to water as fundamental human rights of all Bolivians, with the correlative obligation of the State to guarantee food security, through safe, sufficient and adequate food for the whole population (Art. 16). It also states that public expenses and national investments will be allocated, through participatory mechanisms, on a priority basis to social issues including food policies (Art. 321); that genetically modified organisms are prohibited (Art. 256); and that international human rights instruments proclaiming more favorable rights than the constitution shall take precedence (Art. 257).

Bolivia had already brought food security and sovereignty issues to the centre stage of its developmental agenda: "Bolivia's National Development Plan 2006–2010" already referenced food security and sovereignty as components of both its developmental and its sovereignty axes (Government of Bolivia, 2006). Those policy guidelines materialized in the Plan de Apoyo a Seguridad Alimentaria, which has become a key pillar of the Morales administration's Revolucion Agraria (Government of Bolivia, 2006). Recently, food security and sovereignty also became visible elements of the Government Plan 2010–2015 (Government of Bolivia, 2010).

At national level, the National Council of Food and Nutrition (CONAN)[18] has been key to the inclusion of the right to food into the Multisectoral program Zero Malnutrition and in many other programs and projects. FAO has supported the decentralization and institutional building process of CONAN by promoting the incorporation of the right to food into sub national and municipality development plans.[19] Because of the implementation of policies led by CONAN—an interministerial body aimed at promoting the right to adequate and sufficient food—the malnutrition rate dropped from 38% in 1990–1992 to 19.5% in 2012–2014. The multisectoral program "Zero malnutrition", aimed at improving the overall health of children under 5 years of age, pregnant women and nursing mothers brought the chronic malnutrition rate down to 16% in 2014.[20]

2.3 *Food Sovereignty in Ecuador*

Ecuador, located in South America, has a population of 13 million inhabitants. Although an oil exporter, it is still one of the poorest countries in the region, because

[18]In Bolivia the National Council for Food and Nutrition (CONAN) was reformed in 2006 while under the coordination of the Ministry of Health. It is supported by the leadership of the President in carrying out its mandate, which encompasses the coordination of various stakeholders involved in interventions aimed at the realization of the right to adequate food.

[19]The FAO Project GCP/GLO/324/NOR - Integrating the Right to Adequate Food and Good Governance in National Policies, Legislation and Institution. http://www.fao.org/fileadmin/templates/righttofood/documents/project_b/GCP-GLO-324-NOR.pdf. Accessed 14 December 2015.

[20]See OHCHR (2015).

of inequity in the country's distribution of resources. The result of failed government policies to implement equitable food distribution it is linked with undernourished of a significant proportion of poor population such as indigenous rural population and urban poor.

In 2008, Ecuador includes the right to food in its Constitution, approved on September 28 by 64% of the population in a referendum. For the first time, in 1996 Via Campesina introduced Food sovereignty at the World Food Summit.[21] The political opening for institutionalizing food sovereignty in Ecuador was the making of the new Constitution where the process was decentralized and participatory. This process opened the opportunity for social movements to incorporate the indigenous tradition and notion of Sumak Kawsay.[22] Social movements were influential in incorporating food sovereignty into Ecuadorian Constitution as an entitlement for the attainment of Sumak Kawsay (Art. 13), that later developed into a food sovereignty legal framework with the approval of the Food Sovereignty Law (LORSA) in 2009. It is within the framework of Sumak Kawsay and for the attainment of a 'good living' that food sovereignty emerges as a right, a strategy, and an obligation from the State to ensure everyone can achieve self-sufficiency with respect to healthy and culturally appropriate food on a permanent and continual basis (Art. 281 Constitution). Sumak Kawsay (Buen vivir) is proposed not as a declination of development but as an alternative. In this framework, Nature becomes subject of rights, so considering it not as a mere factor of production, but rather as "an inherent part of social being".[23] Ecuador's framework law includes several remarkable provisions. One of the strengths of the Ecuadorian framework law is the emphasis placed in several provisions on small-scale farmers, who in many underdeveloped countries constitute most people affected by hunger and food insecurity. People can only realize their right to food when they can participate in the decisions relevant to them. By way of the organic law on the food sovereignty regime (LORSA), the Republic of Ecuador has provided a clear provision on social participation within its Food Sovereignty and Nutrition System for all matters related to food sovereignty laws and policies (Ecuador, 2009: Art. 31).

Another clear provision touches upon the representation of non-governmental actors in its Plurinational and Intersectoral Conference on Food Sovereignty (Ecuador, 2009: Art. 33). The Conferencia Plurinacional e Intercultural de Soberania Alimentaria (COPISA) was created in 2010 as a participatory organization responsible for drafting nine-supplementary laws that support the LORSA as a participatory organization responsible for providing participatory spaces for the deliberation and formulation of nine supplementary laws (Art. 32 of the LORSA) relating to: use

[21]Via Campesina is a transnational movement that represents peasants, small and medium-scale farmers, rural women, farm workers, and indigenous organizations that promotes food sovereignty as an alternative model for agriculture based on ethics and values in which culture and social justice ensure a future without hunger. See Desmarais (2007), p. 38.

[22]Peña (2013).

[23]Dávalos (2008).

and access to land, seeds and agro-biodiversity, agricultural development, agro-industry and agricultural employment, animal and plant health, public access to credit, agricultural insurance and food subsidies.

Moreover, to be noted that it is recognized as a responsibility of the State and society the protection of intellectual property rights on national seeds and agro-biodiversity and their free use and conservation; in fact, LORSA adds, however, the specification of "native origin" that would exclude non-native varieties, although locally improved.

Inside this innovative way of understanding the world and the social reproduction, food sovereignty as a constitutional right presents a significant opening to promote a citizen's revolution of the food system. Therefore, a leading role is assigned to the State as a guarantor of food sovereignty, most of the inherent texts consist of the description of the duties it must assume.

Despite incorporating indigenous moral economies into the Constitution, the government has been unable to fulfill promises of resource distribution. Lastly, high levels of food insecurity resulting from an unequal distribution of resources and an extremely vulnerable and impoverished population and chronic malnutrition affects 25% of children under five and can often reach 35% in some northern border provinces. Perhaps the inadequacy of forming an effective state-society relationship based on citizen rights to nature it does not help effective protection to the right to food.

3 Final Remarks

As we have seen, the empirical analysis concentrates in Bolivia and Ecuador, two of the poorest countries in the western hemisphere, with dismal rates of extreme poverty in the rural sector and an ambitious strategy of food sovereignty, above all with constitutional provisions in right to food. Legal recognitions do not translate into implementation of food sovereignty laws and policies, in fact the process can be slow and uncertain, as the experiences of Bolivia and Ecuador demonstrate.[24] Laws alone are not sufficient to realize the right to food in a country. Through the adoption of a national strategy for the realization of the right to food and the implementation of the programs placed under such a national strategy, the right to food is operationalized and put into action at the local level. Indeed, these two countries have implemented a host of economic and social policies to mitigate the effects of hunger and malnutrition, including reduction in consumption taxes, price controls, import tariffs, export controls the issue of hunger is not yet resolved.[25]

[24]Lambek et al. (2014).

[25]Barret (2013).

References

Barkin D (2012) Communities constructing their own alternatives in the face of crisis. Mt Res Dev 32:S12–S22

Barret BC (ed) (2013) Food security and sociopolitical stability. Oxford University Press, Oxford

Carducci M (2014) Costituzionalismo e sopravvivenza umana, 9 maggio 2014, in www.diritticomparati.it

Dávalos P (2008) Reflexiones sobre el sumak akwsay (buen vivir) y las teorías del desarrollo. ALAI (Agencia Latino Americana Informaciones) http://alainet.org/active/2561. Accessed 14 Dec 2015

De Schutter O (2010) Countries tackling hunger with a right to food approach. Significant progress in implementing the right to food at national scale in Africa, Latin America and South Asia. http://www2.ohchr.org/english/issues/food/docs/Briefing_Note_01_May_2010_EN.pdf. Accessed 26 Dec 2015

Desmarais AA (2007) La Vía Campesina: globalization and the power of peasants. Fernwood Publishing and Pluto Press, Halifax and London

FAO (1996) Declaration on world food security. In: World Food Summit. FAO, Rome

FAO Project GCP/GLO/324/NOR - Integrating the Right to Adequate Food and Good Governance in National Policies, Legislation and Institution. http://www.fao.org/fileadmin/templates/righttofood/documents/project_b/GCP-GLO-324-NOR.pdf. Accessed 14 Dec 2015

Himley M (2009) Nature conservation, rural livelihoods, and territorial control in Andean Ecuador. Geoforum 40(5):832–842

Knuth L, Vidar M (2011) Constitutional and legal protection of the right to food around the world. Food and Agriculture Organization of the United Nations, Rome

Kopp Valdivia C (2015) Espacios de diálogo como estrategia para integrar el derecho a la alimentación en municipios rurales de América Latina, Organización de las Naciones Unidas para la Alimentación y la. Agricultura, Rome

Lalander R (2012) Neo-constitutionalism in twenty-first century Venezuela: participatory democracy, deconcentrated decentralization or centralized populism? In: Nolte D, Schilling-Vacaflor A (eds) New constitutionalism in Latin America: promises and practices. Ashgsate, London, pp 163–182

Lambek N, Claeys P, Wong BL (eds) (2014) Rethinking food systems: structural challenges, new strategies and the law. Springer, Netherlands

Martinez R, Palma A, Atalah A, Pinheiro AC (2009) Food and nutrition insecurity in Latin America and the Caribbean, ECLAC/WFP, Santiago

Masset E, Haddad L, Cornelius A, Isaza-Castro J (2012) Effectiveness of agricultural interventions that aim to improve nutritional status of children: systematic review. https://doi.org/10.1136/bmj.d8222

OHCHR (2010) UN Special Rapporteur on the right to food. http://www.ohchr.org/EN/Issues/Food/Pages/FoodIndex.aspx. Accessed 13 Dec 2015

OHCHR (2013) Mapping human rights obligations relating to the enjoyment of a safe, clean, healthy and sustainable environment, Individual Report on the UN General Assembly and the Human Rights Council. Report No. 6. http://www.ohchr.org/Documents/Issues/Environment/MappingReport/6.HRC-UPR-25-Feb.docx. Accessed 15 Dec 2015

OHCHR (2015) Report of the United Nations High Commissioner for Human Rights on the activities of his office in the Plurinational State of Boliva. http://www.ohchr.org/EN/HRBodies/HRC/RegularSessions/Session28/Pages/ListReports.aspx. Accessed 13 Dec 2015

Peña K (2013) Institutionalizing food sovereignty in Ecuador. Paper presented at food sovereignty: a critical dialogue. International Conference, Yale University, September 14–15, 2013

Pierri M (2015) Agrobiodiversity, intellectual property rights and right to food: the case of Andean countries. In: Monteduro M, Buongiorno P, Di Benedetto S, Isoni A (eds) Law and agroecology: a transdisciplinary dialogue. Springer, pp 451–469

Pinstrup-Andersen P (2009) Food security: definition and measurement. Food Sec 1(1):5–7

Rodotà S (2011) Il diritto al cibo, in www.alfabeta2.it/2011/07/26/il-diritto-al-cibo

Schilling-Vacaflor A (2011) Bolivia's new constitution: towards participatory democracy and political pluralism? Eur Rev Lat Am Caribb Stud 90:3–22

UN Human Rights Council (2008) Report of the Special Rapporteur on the Right to Food, Ziegler J: addendum: mission to Bolivia, 30 January 2008, A/HRC/7/5/Add.2, http://www.refworld.org/docid/47c6c3332.html. Accessed 14 Dec 2015

Ziegler J (2008) Report of the Special Rapporteur on the right to food, Addendum: Mission to Bolivia, UN Human Rights Council. http://www.refworld.org/docid/47c6c3332.html. Accessed 14 Dec 2015

The Impact of Biofuels on the Realization of the Human Right to Food

Martina Cutazzo

1 The Right to Food in International Law

From a biological point of view, food represents, along with water, the most essential element for human survival: an insufficient and poor nutrition can limit the physical and mental development of a human being, can cause diseases, suffering, loss of productivity and earning power, and may also lead to poverty and death. A regular access to a healthy, safe, nutritionally and culturally adequate diet ensures a life free from anxiety, satisfying and dignified.

At international level, the adequate and permanent access to food is an inalienable and fundamental human right. It is not simply a right to a minimum ration of calories, proteins or other specific nutrients. It is a right to all nutritional elements that a person needs to decently live a healthy and active life, free from hunger, food insecurity and malnutrition.[1]

According to international human rights law, the right to food is inherent in all human beings by virtue of their humanity alone.[2] This right is of crucial importance for the enjoyment of all rights. One might well say that it is even the "solid and inviolable rock upon which the whole edifice of human rights is founded",[3] among them, above all, the right to life and to health: "From a biological point of view, the priority given to the right to food seems to be justified [. . .] No right has meaning or value if the hunger affects [. . .] Rights have value just for living people [. . .]".[4]

[1]United Nations Human Rights – Office of the High Commissioner for Human Rights (2010), p. 2.

[2]Cfr FAO (2007).

[3]Benedict (2008).

[4]Bensalah-Alaoui (1989), p. 30.

M. Cutazzo (✉)
University of Salento, Lecce, Italy
e-mail: studiolegalecutazzo@alice.it

A. Isoni et al. (eds.), *Food Diversity Between Rights, Duties and Autonomies*,
LITES – Legal Issues in Transdisciplinary Environmental Studies 2,
https://doi.org/10.1007/978-3-319-75196-2_17

From a structural point of view, the right to food—belonging to the category of economic, social and cultural rights—must be available, accessible and adequate.

Availability refers to the possibility of feeding oneself and one's family directly from productive land or animal husbandry, or through other activities such as fishing, hunting or gathering or even through distribution systems (markets and shops).

The concept of *accessibility* encompasses both economic and physical access to food. Firstly, accessibility means that food must be inexpensive: it implies, i.e., according to the UN Committee on Economic, Social and Cultural Rights, that "personal or household financial costs, associated with the acquisition of food for an adequate diet, should be at a level such that the attainment and the satisfaction of other basic needs are not threatened or compromised".[5] Physical accessibility means that food should be accessible to everyone, "including physically vulnerable individuals, such as infants and young children, elderly people, the physically disabled, the terminally ill and persons with persistent medical problems, including the mentally ill", for whom it may be difficult to get food.[6] Access to food must also be guaranteed "to people in remote areas and to victims of armed conflicts or natural disasters" and to other socially disadvantaged group, such as prisoners.[7]

"*Adequacy* means that the food must satisfy dietary needs, taking into account the individual's age, living conditions, health, occupation, sex, etc.".[8] Food should also be "safe for human consumption and free from adverse substances, such as contaminants from industrial or agricultural processes".[9] Lastly, adequate food must be culturally acceptable. "For example – according to the United Nations Human Rights – Office of the High Commissioner for Human Rights – aid containing food that is religious or cultural taboo for the recipients or inconsistent with their eating habits would not be culturally acceptable".[10]

It is worth noting that the right to food is not a right to be fed, but primarily the right to feed oneself in dignity. In this regard, the concept expressed by the UN High Commissioner for Human Rights is lighting: "Individuals are expected to meet their own needs, through their own efforts and using their own resources. To be able to do this, a person must live in conditions that allow him or her either to produce food or to buy it. To produce his or her own food, a person needs land, seeds, water and other resources, and to buy it, one needs money and access to the market. The right to food requires States to provide an enabling environment in which people can use their full potential to produce or procure adequate food for themselves and their families. However, when people are not able to feed themselves with their own means, for

[5]UN Committee on Economic, Social and Cultural Rights (1999), p. 5.

[6]Ibidem.

[7]United Nations Human Rights – Office of the High Commissioner for Human Rights (2010), p. 3.

[8]Ibidem.

[9]Ibidem.

[10]Ibidem.

instance because of an armed conflict, natural disaster or because they are in detention, the State must provide food directly".[11]

A precondition for the full enjoyment of the right to food is *food security*, a concept variously defined, interpreted and evolved over the years.[12] Earlier concept focused on *physical availability* of food: in the 1974 World Food Conference, food security was defined, in fact, only as "availability at all times of adequate world food supplies of basic foodstuffs to sustain a steady expansion of food consumption and to offset fluctuations in production and prices".[13] In 1983, FAO—influenced by Amarthya Sen's 1981 seminal study *Poverty and Famines: An Essay on Entitlement and Deprivation*—revised and broadened the concept of food security to include *economic access* by poor and vulnerable people to available supplies: the goal of food security was to ensure that "... all people at all times have both physical and economic access to the basic food that they need".[14] In 1986, the World Bank report *Poverty and Hunger* introduced, for the first time, the distinction between chronic food insecurity, associated with problems of continuing or structural poverty and low incomes, and transitory food insecurity, which involved periods of intensified pressure caused by natural disasters, economic collapse or conflict.[15] The 1996 World Food Summit finally adopted a still more complex definition: "Food security, at the individual, household, national, regional and global levels [is achieved] when all people, at all times, have physical and economic access to sufficient, safe and nutritious food to meet their dietary needs and food preferences for an active and healthy life".[16] The 1996 World Food Summit definition of food security is still widely used today, with the addition of the word "social" to the expression "physical, social and economic access".

2 The International Legal Basis of the Human Right to Food

The right to food is recognized in several instruments under international law. Many initiatives and organizations have also been created in defense of this right.

It has its contemporary origin within the foundational document of international human rights law: the main reference point is located, in fact, within the Universal

---

[11]Cfr United Nations Human Rights – Office of the High Commissioner for Human Rights (2010), pp. 3–4.

[12]For an historical analysis of the concept of food security see FAO – Economic and Social Development Department (2003).

[13]United Nations (1975).

[14]FAO (1983).

[15]World Bank (1986). See also FAO (2003).

[16]FAO (1996).

Declaration of Human Rights, Article 25, which recognized solemnly, for the first time, the human right to food as part of the right to a standard of living.[17]

Twenty years later, the UN Committee on Economic, Social and Cultural Rights enshrined the right to adequate food and provided the main direction on States obligation at national level. In its Article 11, the International Covenant on Economic, Social and Cultural Rights acknowledged "the right of everyone to an adequate standard of living for himself and his family, including food, clothing and housing..." and, in addition, "the fundamental right of everyone to be free of hunger".[18] Particularly, by signing the Covenant, States agreed to take steps to the maximum of their available resources to achieve progressively the full realization of the right to food and recognized the essential role of international cooperation and assistance in this context.

The first World Food Summit, organized by FAO in Rome in 1996, represented a decisive step for the institutionalization of the right to food. On that occasion, the representatives of 185 countries and the European Community renewed global commitment at the highest political level to eliminate hunger and malnutrition and to achieve sustainable food security for all people. On its first day, the Summit formally adopted two key documents fully negotiated and prepared upon prior to the meeting itself: the *Rome Declaration on World Food Security* and the *World Food Summit Plan of Action*. Particularly, the Declaration contained seven commitments that placed the basis for achieving sustainable food security for all (among them, the important commitment to halve the number of people living in hunger and extreme poverty by 2015), and the Plan of Action spelled out the objectives and actions relevant for practical implementation of these commitments. For the first time, the relationship between people and food—historically considered to be only a biological or a scientific question—became a political question.

In reaction to the World Food Summit, on 12 May 1999, the UN Committee on Economic, Social and Cultural Rights issued its *General Comment No. 12*, widely considered as the official interpretation of Article 11 of the 1966 International Covenant on Economic, Social and Cultural Rights. The Committee affirmed that "the right to adequate food is indivisibly linked to the inherent dignity of the human person and is indispensable for the fulfilment of other human rights enshrined in the International Bill of Human Rights".[19] It also stated that "the right to adequate food is realized when every man, woman and child, alone or in community with others,

[17]UN General Assembly (1948) The Universal Declaration of Human Rights, Paris, 10 December 1948, Article 25: "Everyone has the right to a standard of living adequate for the health and well-being of himself and of his family, including food, clothing, housing and medical care and necessary social services, and the right to security in the event of unemployment, sickness, disability, widowhood, old age or other lack of livelihood in circumstances beyond his control".

[18]UN General Assembly (1966) International Covenant on Economic, Social and Cultural Right, 16 December 1966. http://www.ohchr.org/EN/ProfessionalInterest/Pages/CESCR.aspx.

[19]UN Committee on Economic, Social and Cultural Rights (1999), p. 2.

has physical and economic access at all times to adequate food or means for its procurement".[20]

In April 2000, the appointment of a *Special Rapporteur on the Right to Food* by the UN Commission on Human Rights marked the final consecration of this right. The Special Rapporteur is an independent expert appointed to examine and report back on the right to food in the world, to analyse ways and means of overcoming existing and emerging obstacles for the full realization of this right, to promote the adoption of measures at the national, regional and international levels, to work in close cooperation with all States, intergovernmental and non-governmental organizations (on 4 September 2000, the UN Commission on Human Rights appointed Jean Ziegler as the first Special Rapporteur on the Right to Food. In 2008 the second Rapporteur, Olivier De Schutter, was appointed by the UN Human Rights Council, which replaced the Human Right Council in June 2006. The current Special Rapporteur is Hilal Helver). On its first report, presented at the 57th ordinary session of the Commission on Human Rights, Ziegler provided a new definition of the right to food. It was the evident result of a different exegetical approach to the issue. He explained that the right to food is defined as "the right to have regular, permanent and unrestricted access, either directly or by means of financial purchases, to quantitatively and qualitatively adequate and sufficient food corresponding to the cultural traditions of the people to which the consumer belongs, and which ensures a physical and mental, individual and collective, fulfilling and dignified life free of fear".[21]

In recognition of the persistent problem of hungry and poverty (in 2000, 826 million people were still hungry), world leaders participating in the 2000 Millennium Summit adopted the *United Nations Millennium Declaration*, which became the basis for the formulation of the *Millennium Development Goals*. The overarching goal was—once again—to halve world poverty by 2015. The Millennium Development Goals have been elaborated as global timebound and quantified targets for addressing several key dimensions of extreme poverty, including income poverty, hunger, disease, lack of shelter, and exclusion.

In June 2002, FAO held a new world meeting to examine the progress made in eradicating hunger. The purpose of the meeting, which was named *World Food Summit: five years later*, was to track progress made since the summit of 1996 and consider ways to accelerate these efforts. Recognizing the obvious failure of previous policies in defense of the right to food, the Head of States and Government renewed the commitments of the 1996 World Food Summit and advocated the creation of a worldwide voluntary instrument on the right to food. In reaction to that call, on September 2004, the 127th of the FAO Council adopted the *Voluntary Guidelines to Support the Progressive Realisation of the Right to Adequate Food in the Context of Nation Food Security*, providing practical guidance to States in their implementation of the right to adequate food.[22]

[20]Ibidem.

[21]UN Commission on Human Rights (2001), p. 7.

[22]FAO (2005).

In June 2008, FAO convened a *High-Level Conference on World Food Security* that was attended by Heads of State and Government, Ministers, and Representatives of 180 countries and the European Union. Those attending issued a declaration, the *Declaration of the High-Level Conference on World Food Security: the Challenges of Climate Change and Bioenergy,* which outlined both immediate/short term measures and more intermediate/long-term measures that its signatories committed to undertake to overcome the crisis and ensure the realization of the goals of the 1996 World Food Summit.

On 10 December 2008, the UN General Assembly adopted the *Optional Protocol to the International Covenant on Economic, Social and Cultural Rights*, making the right to food (and the others economic, social and cultural rights) justiciable at the international level.

In September 2015, at an historic UN Summit, world leaders adopted the *2030 Agenda for Sustainable Development* and its 17 *Sustainable Development Goals* (SDGs or Global Goals).[23] Universal, inclusive and indivisible, the Agenda calls for action by all countries to improve the lives of people everywhere. Unlike the Millennium Development Goals, the Sustainable Development Goals apply to all countries, but they are not legally binding. Nevertheless, countries are expected to take ownership and establish a national framework for achieving the Global Goals. Regarding the right to food, SDG 2 (*Zero Hunger*) aims to end hunger, achieve food security and improved nutrition and promote sustainable agriculture. On 1 January 2016, the 17 Sustainable Development Goals came into force.

Today, the right to food is also recognized in other international conventions protecting specific groups, such as the *Convention on the* Elimination of All Forms of Discrimination against Women (1979), the Convention on the Rights of the Child (1989) *and the Convention on the Rights of Persons with Disabilities* (2006). The right to food is also recognized in some regional instruments, such as the *Additional Protocol to the American Convention on Human Rights in the Area of Economic, Social and Cultural Rights* (1988), the *African Charter on the Rights and Welfare of the Child* (1990) and the *Protocol to the African Charter on Human and Peoples' Rights on the Rights of Women in Africa* (2003).

3 Biofuels: Is the Cure Worse Than the Disease?

Despite the solemn affirmation of its existence, the right to food is violated all the time, virtually anywhere in the world. Nowadays, in fact, the right to food could seem an expected, obvious and unquestionable right, but as the information supplied by specialized agencies shows, this right is still denied to more than a fifth of people in the world.

According to the FAO State of Food Insecurity and Nutrition in the World 2017, despite progress made in reducing chronic hunger, some 815 million people in the

[23]United Nations General Assembly (2015).

world do not have enough food to lead a healthy active life. That is about one in nine people on Earth. The vast majority of the world's hungry people live in developing countries. Sub-Saharian Africa remains the region with the highest number of undernourished people, affecting an alarming 22.7% of the population in 2016. Poor nutrition causes nearly half (45%) of deaths in children under five (3.1 million children each year) and 155 million children under five years of age across the world suffer from stunted growth.[24] It is a matter of defeat without an appeal, an unpleasant defeat, difficult to understand because the Earth can feed easily twice the current population, by guaranteeing daily rations of between 2400 and 2700 calories for every human being[25]: indeed, as Jean Ziegler often stated, "suffering hunger is not a fatality... If the distribution of food was equal everywhere in the world, all people would have something to eat".[26]

But if there is enough food for everybody and if there is no connection between world population growth and hunger, then nowadays what allows to have a "world in a world, an enclave of happiness in a world of pain", to relegate a billion of people to "nonexistence", to suffer the killing order of the world?[27]

The food or fuel issue is not easy to all and is more complex than it usually appears. Many factors have contributed to creating this situation, among them structural famine that results from social, economic, political and cultural factors. This kind of famine is neither accidental, nor exceptional, nor as result of natural factors, but caused by human action and it plays an important role in the relationship between man and food. Unlike a short-term famine (for example an earthquake, a drought, a hurricane, a flood or other unfavorable environmental situations), which brings about a temporary difficulty of supplying food, structural famine means a permanent lack of food, a persevering and lasting absence of adequate diet.

In this context, biofuels represent a big cause of hunger and a very serious threat to the concrete realization of the human right to food: their use is certainly determined by human choices. The example of the largest producer of biofuels in the world, the United States of America, is glaring. In 2010, the USA burned 138 million tons of corn and hundreds of million tons of grain to manufacture ethanol and biodiesel. If we have a car that runs on biofuel, we would have to use 352 kilograms of corn to fill up a 50-liter tank. But turning food crops into fuel, without first examining the impact on global hunger would be—as it was said—a recipe for disaster[28]: the corn is, in fact, one of the key staple food crops in the world, the primary source of calories and nutrients for nearly billion people worldwide.[29]

But what are biofuels? And how do biofuels pose a threat to the realization of the right to food?

---

[24]FAO, IFAD, UNICEF, WFP and WHO (2017).

[25]Economic and Social Council, UN Commission on Human Right (2001), p. 2. Cfr also FAO (1995), p. 16.

[26]Ziegler (2002 [1999]), p. 17.

[27]Cfr Ziegler (2002 [1999]), p. 138.

[28]Ziegler (2007).

[29]ActionAid (2012a) Biofuelling hunger..., p. 5.

Biofuels are ecological fuels, alternative to traditional fuels, produced from the transformation of food feedstocks. It can be said that biofuels were born together with man. Solid biofuel, such as wood and charcoal, and liquid biofuels, such as olive oil and whale oil, have been used, in fact, ever since man discovered fire both for domestic use and for transport.[30] During the twentieth century, great scientists and engineers, such as Nikolaus August Otto, Rudolph Diesel and Henry Ford, showed eager enjoyment and interest for biofuels.[31] Really it seemed that biofuels were the future. However, the advent of petroleum severely changed the outlook again: in terms of cost, efficiency and supply, oil seemed to have no rivals. Three oil crises since the 1970s—the 1973 OPEC oil export embargo, the 1979 Iranian Revolution oil crisis and the 1990 Gulf War—and, more recently, the fight against climate change and greenhouse gas (GHG) emissions and the rural economic development generated a renewed interest in biofuels. Many countries, such as the US and Brazil, became modern large-scale producers of biofuels. According to the recent HLPE Report on *Biofuels and Food Security*, in less than just one decade, world biofuel production has increased five times, from less than 20 billion litres/year in 2001 to over 100 billion litres/year in 2011.[32]

Since their arrival on the world economic scene, in fact, biofuels have been proposed to public opinion and markets as a valid promising solution of clean energy, an eco-friendly choice, a sustaining alternative to fossil fuels, very efficient from an economic point of view, moreover it is a new income stream for farmers and a new possibility of business in general: indeed, the use of the prefix *bio* evokes something absolutely natural, renewable and safe. On the contrary, traditional fuels are dangerous, exhaustible over time and, most frequently, from countries where there is a strong political instability and difficult situations of conflict.

But since the beginning, the debate about new biofuels has inflamed ethic, policy and right law. The title of a working paper for the Organisation for Economic Co-operation and Development reflects and summarizes strikingly the growing and undeniable concern towards this phenomenon: is the cure worse than disease?[33] While some people believe that biofuels are the future of the planet in terms of energetic solution, energetic security and meaningful reduction of pollution, more others think that talking about biofuels means, indeed, talking *tout court* about the biggest catastrophe of our days.

By analysing some numbers from specialized organizations (such as F.A.O.) and by NGO_S (as Action Aid) you can have an overview of the current situation:

- first-generation biofuels refer to fuels derived from food sources like soy, cereals, sugar cane, sugar beet for ethanol and colza oil, sunflower oil, palm oil for biodiesel;

[30]For a short history of biofuels see Webb (2013).

[31]Cfr Webb and Coates (2012), p. 69. See also Kovarick (2013), pp. 1–22.

[32]High Level Panel of Experts (HLPE) (2013).

[33]OECD (2007).

- between 2500 and 4000 liters of water are necessary to produce a liter of biofuel;
- approximately 200 kg of maize are needed to fill the tank of a car;
- the consumption of 10% of biofuels just in the transport sector, as an alternative to fossil fuels, requires the use of 26% of land that can be brought into production around the globe;
- according to World Bank, the increase in biofuels consumption has contributed about 75% to the rise in food prices and the related consequence of a recent food crisis[34];
- In 2007, liquid biofuel contributed only 0.36% of the total energy consumption in the world. To achieve this modest fraction of the total energy use, 23% of US coarse grain production was used to produce ethanol and in the EU about 47% of all vegetable oil production was used to produce biodiesel[35];
- between 2008–2010 two-thirds of large-scale land acquisition in developing countries have had as aim the production of biofuels for the consumption of rich countries[36];
- in 2008 alone, biomass used for biofuel production could have fed 127 million people, reducing world hunger by about 15%[37];
- according to Agricultural Outlook 2011–2020, by 2020, an estimated 12% of global coarse grain production, 16% of vegetable oil production and 33% of sugar cane production will be used for biofuel production;
- by 2020, the EU aims to have 10% of the transport fuel come from renewable sources, of whom no more than 7% from first generation biofuels: according to experts' estimate, in terms of additional emissions, it would be like putting up to 26 million cars more on European roads[38]; at the same time, if the target was reached an estimated 600 million extra people may be hunger by this date.[39]

4 Biofuels and EU Law

Support biofuels industry policies come from EU, which is at the forefront of energy policy and renewable energy development and is expected to be, by far, the largest producer and user of biodiesel in the world.[40] Before 2003, the policies for biofuels were mainly developed at a national level. At the EU level, there was only the Council Directive 85/536/ECC, which authorized blends of 5% ethanol and 15%

[34]Mitchel (2008).

[35]FAO (2009), p. 10.

[36]Deininger et al. (2011), p. 52.

[37]Actionaid (2013).

[38]Bowyer (2011).

[39]ActionAid (2010), p. 5.

[40]For an historical analysis of EU policies for biofuels see Henrique et al. (2013), pp. 17–36; Alabrese (2014); ActionAid (2012b).

Ethyl Tertiary Butyl Ether (ETBE) on petrol. But, at that time, some important processes were evolving that came to play a significant role in the development of biofuel policies in Europe, among them the evolution of the global environmental agenda and the commitments made by European countries at the United Nations Conference on Environment and Development (also known as the Earth Summit).

In 2003, the Directive 2003/30 on the promotion of the use of biofuels or other renewable fuels for transport was adopted. It aimed to introduce a non-binding target of a 2% market share for biofuels in 2005 and a 5.75% share in 2010. To achieve the proposed target introduced by the biofuels directive, the Council adopted a third measure in October 2003, namely the directive on energy taxation (2003/96/EC) providing Member States with regulatory discretion to use tax policy to foster the penetration of biofuels in their markets.

In 2006 the Commission's suggestion of a new strategy on biofuels was published. It contained a clear emphasis on technological innovation and promotion of energy security. The next year, a new communication of the Commission—entitled *An Energy Policy for Europe*—laid the basis for major regulatory initiatives in the subsequent years.

The year of 2009 was a turning point in EU policies on biofuels: from that moment, the European Union has set, on this issue, rigorous sustainability criteria for Member States to reduce dependence on traditional fuels, considered a serious source of concern both for the environment and for security of supply: already in 2007 the European Commission called biofuels as "the only direct substitute for fossil fuels in transport that is available on a significant scale", launching powerful programs in favor of them: biofuels are today a widespread and firmly established reality in many countries.[41]

In January 2008, the European Commission finally put forward its proposals for legislation on climate and energy, known for its 20-20-20 targets of 20% renewable energy, 20% increased energy efficiency and 20% less GHG emissions by 2020. However, during the decision-making process undertaken between the European Parliament and the Council, the term biofuels was replaced by renewable energy in transport, broadening the scope of technologies that could be used for the 10% mandatory target in 2020. The European Parliament and Council reached broad consensus in December of 2008, and the final directive text was formally adopted in April 2009 (as 2009/28/EC or RED, Renewable Energy Directive). A new fuel quality directive (2009/30/EC) was published shortly after the RED.

Today, according to the amendment added to the Directive 98/70/EC and to RED by the Directive (EU) 2015/1513 of the European Parliament and of the Council of 9 September 2015 (ILUC),[42] by 2020, the EU aims to have 10% of the transport fuel

---

[41]European Commission (2007).

[42]Directive EU 2015/1513 of the European Parliament and of the Council of 9 September 2015 amending Directive 98/70/EC relating to the quality of petrol and diesel fuels and amending Directive 2009/28/EC on the promotion of the use of energy from renewable sources. http://eur-lex.europa.eu/legal-content/EN/TXT/PDF/?uri=CELEX:32015L1513&from=IT. In these years, the majority of the EU Member States have implemented the Directive (EU) 2015/1513. With the Legislative Decree n. 51 of March 21, 2017, published on April 27, 2017 Italy adopted and transposed into its own legal system the ILUC Directive.

come from renewable sources, of whom no more than 7% from first generation biofuels.

On November 30, 2016, the European Commission published a formal proposal to the EU Council and the European Parliament to recast Renewable Energy Directive (RED).[43] The proposed new directive—called RED II—proposes a set of policy measures to achieve a 27% renewable energy share from energy consumed by the electricity, heating and cooling, and transportation sectors by 2030. Regarding renewable energy for transportation, RED II would mandate that 6.8% of transportation fuels must derive from renewable sources, specifically advanced alternative fuels and renewable electricity.

In 2026, the Commission will issue a proposal for promoting the use of renewable energy post 2030.

5 Biofuels in International Law

At international level, the debate about biofuels shows an absolute absence of common ideas and solutions to save the human right to food.

In June 2008, the *High-Level Conference on World Security* opened at FAO Headquarters in Rome to discuss about the connection between human right to food and food security, climate changes and bioenergies. The Conference concluded that a global and coherent dialogue about biofuels was required—showing, in this way, a shy inclination to them—although underlining the need to verify the total sustainability of these resources.

It is important to report also the initiatives regarding biofuels taken by UNCTAD, United Nations Conference on Trade and Development. Launched in 2006, the UNCTAD BioFuels Initiative promotes sustainable energy through international cooperation. It also provides policy guidance, ideas and examples on how to address possible shortcomings that countries might face when engaging in this new market.

It is right, however, to underline another attitude for biofuels at international level: we speak about the condemnation of this practice repeatedly expressed by the UN Special Rapporteur on the right to food. In 2007, Jean Ziegler severely inveighed against biofuels at the UN General Assembly. He said: "Rushing to turn food crops – maize, wheat, sugar, palm oil – into fuel for cars, without first examining the impact on global hunger is a recipe for disaster".[44] He also said directly that any diversion of land from food or feed production to production of energy biomass embodies "a crime against humanity". So, he called for a five-year moratorium until cellulosic biofuels processes could be developed.[45] In his first annual report presented, in 2008,

[43]European Commission (2016).

[44]United Nations General Assembly (2007), p. 8.

[45]UN independent rights expert calls for five-years freeze on biofuels production. http://www.un.org/apps/news/story.asp?NewsID=24434&Cr=food&Cr1=#.VsWUI_l6SM8.

at the United Nations Human Rights Council, the successor Olivier De Schutter also denounced the increase in food price, needed for the survival of millions of people, because of the strengthening of biofuels industry.[46]

According to the Rapporteur, no ethical justification legitimize the production of biofuels when it impedes access to adequate and sufficient alleviation and prevention of hunger.[47] The most widely used justification, that replacing fossil fuels by biofuels can reduce greenhouse gas emissions and thereby reduce global warming, is mostly not tenable. Recent studies have shown that, considering the entire production process, biofuels lead to as much and sometimes more greenhouse gas emissions than the use of fossil fuel.[48] The justification that biofuel production would in the future improve to some significant degree the energy security of developed countries, by ensuring the diversification of energy supplies, is equally unjustifiable. To do so would in fact require a vast allocation of land to their production, at the expense of many other crops. For example, to reach the target of 20% of energy from renewable sources by 2020, Europe would need an area twice as big as Belgium to be devoted only to the biofuel production.[49] This change is clearly unachievable for various and obvious reasons.

At the 2014 World Economic Forum, also the United Nation Environment Program (UNEP) denounced the negative impact of growing demand of biofuels on the realization of the human right to food. The report *Assessing Global Land Use: Balancing Consumption with Sustainable Supplie*s, produced by the International Resource Panel and presented in that occasion, outlines the need and options to balance consumption with sustainable food production. According to the study, it is not only the increase in demand for food in developing countries, but also the frantic search for biofuels to influence food prices increase, which have had dramatic impacts on the lives and livelihoods, including increased infant and child mortality, of those already undernourished or living in poverty. The rush to biofuels also enhances land use and causes the frightening increase of deforestation (calculated about 13 million hectares year). Recognizing that land is a finite resource and that humanity is at a critical juncture, the report builds on the "Safe Operating Space" (SOS) concept, which underpins the notion that once human activity has passed certain thresholds or tipping points, defined as "planetary boundaries", there is a risk of irreversible and abrupt environmental change.[50]

---

[46]United Nations General Assembly (2008).

[47]Among all, see United Nations General Assembly (2014).

[48]Crutzen et al. (2008), pp. 389–395.

[49]Maturo (2011).

[50]UNEP – International Resource Panel, Working Group on Land and Soils (2014).

6 Future Prospects of Biofuels Impact on the Human Right to Food

Ultimately, there is no ideal solution to the substantial and multiple problems posed by the advent of biofuels: energy continues to play in the delicate game of geo-political balance, made of sophisticated mediations that are often difficult to settle. However, further reflection is required.

Based on the data provided, today world biofuels production provides just 10–12% of global energy, with an estimated usage of about 30 million hectares of land for this purpose.[51] Whereas, according to estimates, in 2040 global consumption of biofuels will require more than 100 million hectares of land to be allocated to their production,[52] most of which is located in territories outside Europe (sub-Saharan Africa in the first place).

It is easy to understand that the future of the planet's food is incompatible with an estimated consumption of bioenergy. Neither the intensification of biofuels production or use of the so-called marginal or degraded land; neither the use of the so-called advanced biofuels (such as biofuels from agricultural and forestry residues or waste from woodworking or even seaweed or crops not used for human consumption) will be able to solve the problem of an unsolvable equation between the finiteness of resources and the needs of a growing and potentially more hungry population.[53]

A strong and fast turnaround in our approach to biofuels is therefore essential to change, based above all on the launch of an energy policy that supports the responsibility, social and environmental sustainability, as well as respects for human rights. Hunger was and will always remain a question of rights, to which only the responsibility of the people in government can give a serious and practical response, in the certainty that as stated by Ziegler: "No victims of hunger is an inevitable victim".[54]

References

ActionAid (2010) Meals per gallon. The impact of industrial biofuels on people and global hunger. http://www.actionaid.org.uk/sites/default/files/doc_lib/meals_per_gallon_final.pdf. Accessed 3 Dec 2015

ActionAid (2012a) Biofuelling hunger: how US corn ethanol policy drives up food prices in Mexico. http://www.actionaid.ie/sites/files/actionaid/biofueling_hunger_aausa.pdf. Accessed 9 Dec 2015

ActionAid (2012b) Il pieno che lascia a secco i poveri. La politica europea sui biocarburanti e il suo impatto sulla sicurezza alimentare e l'accesso alla terra nei Paesi poveri. https://www.actionaid.it/app/uploads/static/il_pieno_che_lascia_a_secco_i_poveri_2012.pdf. Accessed 14 Dec 2015

[51] IEA (2011).

[52] Gallagher (2008).

[53] Sensi (2015).

[54] Ziegler (2002 [1999]), p. 26.

Actionaid (2013) No food for fuel. http://www.actionaid.org/it/italia/2013/08/biocarburanti-e-ora-di-agire. Accessed 4 Dec 2015

Alabrese M (2014) Biofuel Regulation in the EU: a Failure in the path towards environmental sustainability and food security? In Monteduro M, Buongiorno P, Di Benedetto S, Isoni A (eds) (2015) Law and agroecology. A Transdisciplinary Dialogue. Springer, Berlin

Benedict XVI (2008) Address of His Holiness Benedict XVI to H.E. Mr Acisclo Valladares Molina New Ambassador of the Republic of Guatemala to the Holy See. Libreria Editrice Vaticana, Città del Vaticano

Bensalah-Alaoui A (1989) La sécurité alimentaire mondiale. Librairie générale de droit et de jurisprudence, Paris

Bowyer C (2011) Anticipated indirect land use change associated with expanded use of biofuels and bioliquids in the EU. An analysis of the national renewable energy action plans. Institute for European Environmental Policy, London. http://www.ieep.eu/assets/786/Analysis_of_ILUC_Based_on_the_National_Renewable_Energy_Action_Plans.pdf. Accessed 15 Dec 2015

Crutzen PJ et al (2008) N2O release from agro-biofuel production negates global warming reduction by replacing fossil fuels. Atmos Chem Phys 8:389–395

Deininger K et al (2011) Rising global interest in farmland. Can it yield sustainable and equitable benefits? The World Bank, Washington, DC

Economic and Social Council, UN Commission on Human Right (2001) The right to food. Report by the Special Rapporteur on the right to food, Mr. Jean Ziegler, submitted in accordance with Commission on Human Rights resolution 2000/10. UN doc. E/CN.4/2001/53, 7 February 2001. Accessed 3 Dec 2015

European Commission (2007) Communication from the Commission to the Council and the European Parliament. Biofuels Progress Report. Report on the progress made in the use of biofuels and other renewable fuels in the Member States of the European Union. http://eur-lex.europa.eu/legal-content/EN/TXT/?uri=celex:52006DC0845. Accessed 18 Dec 2015

European Commission (2016) Proposal for a directive of the European parliament and of the council on the promotion of the use of energy from renewable sources (recast). https://ec.europa.eu/energy/sites/ener/files/documents/1_en_act_part1_v7_1.pdf. Accessed 3 Mar 2018

FAO (1983) World food security: a reappraisal of the concepts and approaches. Director General's Report, Rome

FAO (1995) Dimensions of need. An atlas of food and agriculture. Rome

FAO (1996) Rome Declaration on World Food Security and World Food Summit Plan of Action. World Food Summit 13–17 November 1996. Rome

FAO (2003) Trade reforms and food security. Chapter 2: food security: concepts and measurement. http://www.fao.org/docrep/005/y4671e/y4671e00.htm#Contents. Accessed 11 Dec 2015

FAO (2005) Voluntary Guidelines to Support the Progressive Realisation of the Right to Adequate Food in the Context of Nation Food Security. http://www.fao.org/docrep/009/y7937e/y7937e00.htm. Accessed 21 Dec 2015

FAO (2007) Le droit à l'alimentation: un défi pour les droits de l'homme au XXI siècle. Journée mondiale de l'alimentation, 16 Octobre 2007. http://www.fao.org/fileadmin/templates/wfd2007/pdf/WFDLeaflet2007F.pdf. Accessed 18 Dec 2015

FAO (2009) The Right to Food and The Impact of Liquid Biofuels. http://www.fao.org/righttofood/publications/publications-detail/en/c/49268/. Accessed 6 Dec 2015

FAO - Economic and Social Development Department (2003) Trade reforms and food security. Conceptualizing the Linkages. Publishing Management Service, FAO, Rome

FAO, IFAD, UNICEF, WFP and WHO (2017) The state of food insecurity and nutrition in the world 2017 building resilience for peace and food security. http://www.fao.org/3/a-I7695e.pdf. Accessed 2 Mar 2018

Gallagher E (2008) The Gallagher review of the indirect effects of biofuels production. Renewable Fuels Agency, St Leonards-on-Sea

Henrique P et al (2013) The European biofuels policy: from where and where to? Eur Energy J 3(1): 17–36

High Level Panel of Experts (HLPE) on Food Security and Nutrition of the Committee on World Food Security (2013) Biofuels and food security http://www.fao.org/fileadmin/user_upload/hlpe/hlpe_documents/HLPE_Reports/HLPEReport5_Biofuels_and_food_security.pdf. Accessed 3 Dec 2015

IEA (2011) Technology Roadmap: Biofuel for Transport. http://www.iea.org/publications/freepublications/publication/Biofuels_Roadmap_WEB.pdf. Accessed 22 Dec 2015

Kovarick B (2013) Biofuels in history. In: Singh BP (ed) Biofuels crops: production, physiology and genetics. CABI, Cambridge

Maturo E (2011) La fame nel mondo e i biocarburanti. In: Scaglione D (ed) La bicicletta che salverà il mondo. La lotta alla fame raccontata a cavallo di un sellino. Postfazione. Infinito Edizioni, Formigine

Mitchel D (2008) Policy Working Paper 4682: a note on rising food prices. The World Bank, Washington, DC

OECD (2007) Biofuels: is the cure worse than the disease? OECD, Paris

Sensi R (2015) Biocarburanti: ultimo appello per Bruxelles. http://www.altreconomia.it/site/fr_contenuto_detail.php?intId=5018. Accessed 13 Dec 2015

UN Commission on Human Rights (2001) The right to food. Report by the Special Rapporteur on the right to food, Mr. Jean Ziegler, submitted in accordance with Commission on Human Rights resolution 2000/10. E/CN.4/2001/53, 7 February 2001. http://daccess-dds-ny.un.org/doc/UNDOC/GEN/G01/110/35/PDF/G0111035.pdf?OpenElement. Accessed 21 Dec 2015

UN Committee on Economic, Social and Cultural Rights (1999) General Comment No. 12: The Right to Adequate Food (Art. 11 of the Covenant), 12 May 1999. http://www.refworld.org/docid/4538838c11.html. Accessed 21 Dec 2015

UNEP – International Resource Panel, Working Group on Land and Soils (2014) Assessing Global Land Use: Balancing Consumption with Sustainable Supplies. http://www.unep.org/resourcepanel/Portals/50244/publications/Full_Report-Assessing_Global_Land_UseEnglish_(PDF).pdf. Accessed 11 Dec 2015

United Nations (1975) Report of the World Food Conference, Rome, 5–16 November 1974. New York

United Nations General Assembly (2007) The right to food. Report of the Special Rapporteur on the right to food, Jean Ziegler, submitted in accordance with General Assembly resolution 61/163. http://daccess-dds-ny.un.org/doc/UNDOC/GEN/N07/487/05/PDF/N0748705.pdf?OpenElement. Accessed 3 Dec 2015

United Nations General Assembly (2008) Report of the Special Rapporteur on the right to food, Olivier De Schutter. Building resilience: a human rights framework for world food and nutrition security. http://daccess-dds-ny.un.org/doc/UNDOC/GEN/G08/155/08/PDF/G0815508.pdf?OpenElement. Accessed 3 Dec 2015

United Nations General Assembly (2014) Final report: The transformative potential of the right to food. Report of the Special Rapporteur on the Right to Food Olivier De Schutter, presented to the 25th Session of the UN Human Right. http://www.srfood.org/en/documents. Accessed 3 Dec 2015

United Nations General Assembly (2015) Transforming our world: the 2030 Agenda for sustainable development, resolution adopted by the general assembly on 25 Sept 2015. http://www.un.org/ga/search/view_doc.asp?symbol=A/RES/70/1&Lang=E. Accessed 3 Mar 2018

United Nations Human Rights – Office of the High Commissioner for Human Rights (2010) The Right to Adequate Food. Fact Sheet No. 34. United Nations, Geneva

Webb A (2013) A brief history of biofuels: from ancient history to today. http://www.biofuelnet.ca/2013/07/31/a-brief-history-of-biofuels-from-ancient-history-to-today/. Accessed 11 Dec 2015

Webb A, Coates D (2012) Biofuels and Biodiversity. Secretariat of the Convention on Biological Diversity. Montreal, Technical Series, 65

World Bank (1986) Poverty and Hunger: issues and options for food security in developing countries. Washington DC

Ziegler J (1999) La faim dans le monde expliquée à mon fils. Le Seuil, Paris; Italian edition: Ziegler J (2002) La fame nel mondo spiegata a mio figlio, 1st edn (trans: Reinhart MC). Net, Milano

Ziegler J (2007) The Impact of Biofuels on the Right to Food. Report No. A/62/289. United Nations General Assembly, New York. http://www.righttofood.org/A62289.pdf. Accessed 15 Dec 2015

The Matter "Alimentation" in the Italian Constitution

Antonello Denuzzo

1 The (Present) Constitutional Frame of Reference

The first paragraph of Art. 117 of the Italian Constitution states the principle of equal orders between state law and regional law and states that both meet the same internal limits (the Constitution) and external (both international and European obligations). The second to fifth paragraphs pertaining to the legislative power, which is exclusive of the State in matters referred to in the second paragraph, competitor between state and regions regarding the matters listed in the third paragraph while in the fourth paragraph is assigned exclusively to the regions with reference "to any matters not expressly reserved to State law" thus in all matters not specified in the preceding second and third paragraphs. In the fifth paragraph of Art. 117 legislative power is attributed to the Regions for the implementation and enforcement of European Union acts "in matters within their competence", matters of the third and fourth paragraphs. Finally, the sixth paragraph has as its object the regulatory powers, that it is up to the State in matters submitted for its exclusive legislation and to the Regions "in all other matters."

Each subject listed in the second and third paragraphs has its own *nomen,* i.e. a label, expressed in an object (for example, weights and measures, currency), in relations (for e.g. between the Republic and religious denominations), in business (for e.g. foreign trade), in interest (e.g. defence, education, social security), in the legal system (for example, citizenship, marital status), in legal systems (for e.g. legal, civil, criminal, procedural), in organisational units of the public power (e.g. organisation of the state and the organs of local authorities).

A. Denuzzo (✉)
University of Salento, Lecce, Italy
e-mail: antonello.denuzzo@unisalento.it

A. Isoni et al. (eds.), *Food Diversity Between Rights, Duties and Autonomies*,
LITES – Legal Issues in Transdisciplinary Environmental Studies 2,
https://doi.org/10.1007/978-3-319-75196-2_18

For the allocation of powers, the *nomen* is not the only decisive matter, but also the objective that the law aims to achieve: thus, will identify the protection of competition or the protection of the environment, i.e. so-called purpose-matters, which are the exclusive competence of the State. The distinctive feature of the purpose-matters is that to be transversal, thus to affect the so-called subject-matters even when the latter are of Regional exclusive competence. Essentially the purpose-matters enable the State to legislate on the regional subject-matters.

Before the constitutional reform of 2001 the instruments that provide a uniform legal framework through the entire territory were represented from the limit of the national interest, that the Constitutional Court considered as a limit of legality of regional laws, and—in the administrative field—by powers of guidance and coordination of the regional functions.

Disappeared from the formulas of the new Title V the criterion of national interest and the overall function of guidance and coordination, the State recovers the role of guardian of unitary bodies through: (a) the category of purpose-matters of the second paragraph of Art. 117 of the Constitution affecting the regional matters of the third and fourth paragraphs; (b) the determination of the fundamental principles in the areas of concurrent legislation of the third paragraph.

The Italian Constitution refers to the alimentation in the third paragraph of Art. 117, qualifying it as a matter under concurrent legislative power of the State and Regions and not as a subject-matter because in the constitutional provision the food is not considered *tout court*; nor it is activity-matter because it does not apply only to the methods of food production. The alimentation coincides, more generally, with a way of being—the "feeding"—in its variations of food security (supplies) and food safety (alimentation safety).

Ultimately the discipline of alimentation concerns: (a) primary production, i.e., agriculture as a producer of food; (b) secondary production, that is, the food industry; (c) the allocation of food on the market; (d) finally, the allocation on the market of products that, having been intended to be ingested by men, must respond to the fundamental principle of respect for health. So, the alimentation is connected: (a) with matters of agriculture, industry and trade that fall within the residual competence (and therefore, as a rule, exclusive) of the Regions; (b) with matters of competition, which is the sole responsibility of the State; (c) with matters of health protection, also competing under the third paragraph of Art. 117 of the Constitution.

2 The Issues of "Alimentation" Included in the Concurrent Competition of the EU Regarding Health Protection

In the TFUE, there is no provision on alimentation: however, the food must meet essential conditions pertaining to health to be allocated on the single food market. On the other hand, the TFUE contains provisions relating, directly or indirectly, the health: the Art. 168 foresees that EU action shall complement national policies on public health, while Article. 169 requires the Union to contribute to safeguarding the consumers health. So, it can be concluded that, through the provisions on health, the TFUE implicitly contemplates the issue of alimentation.

We cannot conduct an analysis of EU sources without referring to the Regulation No. 178/2002 of 28 January 2002 establishing the general principles of food law, establishing the European Food Safety Authority and outlining procedures to ensure safety (in essence it is the discipline of the food market) and has its reason to be in the *bailamme* that the season of mad cow disease, the dioxin chicken and mercury fish had also caused in the EU institutions.

According to Regulation No. 178/2002 the EU's competence extends to the regulation of food and feed, ingredients and food additives, solvents, preservatives and sweeteners, allergenic substances, genetically modified organisms contained in food and food hygiene.

It is reasonable to believe that alimentation is instead a competitor matter for which is applicable the principle of subsidiarity, as may be deduced from Article 114 paragraph 10 TFUE, according to which all "the harmonization measures (in health, safety, environmental and consumer protection matters) include, where appropriate, a safeguard clause authorizing the Member States to take, for one or more of the non-economic reasons in Art. 36, provisional measures subject to a Union control procedure".[1]

On this point some clarifications are necessary. According to the present communitarian legal order the EU has exclusive competences and concurrent competences. The first are divided into absolute expertise and skills that require additional States interventions, while the concurrent competences are those for which there is the level of Union and state government, in that both—the EU and the Member States—have equal opportunity to intervene according to another principle very relevant, that of subsidiarity.

The term subsidiarity postulates a relationship between two subjects, one of which comes to the aid of the other as a supplementary and only to the extent that the latter is not able to fulfill the functions that normally compete. The principle of subsidiarity is expressed then according to a negative mode, for which the superior authority is required not to interfere within the activities of the inferior authority, and a positive mode, for which the superior authority has the duty to intervene.

In the EU law should prevail the negative profile because the Treaties prefer the actions of the States; yet the principle of subsidiarity has been used whenever territorial subdivisions of the single market have not been possible and has appeared incompressible the need for a unit treatment. Conversely, the EU has given the power to regulate those matters (or those "portions" of matters) that are transnational profiles, which cannot be satisfactorily regulated by action of the Member States on their own.

[1] See Tondi della Mura (2007) on the subsidiarity principle.

Thus, it can be concluded that when there is a subsistence of functions that require the unitary exercise, the European Union can allocate to itself, by applying the principle of subsidiarity, the relative legislative competence.

In fact, the Art. 5 TUE, after stating in the second paragraph that "competences not conferred upon the Union in the Treaties remain with the Member States", in the third paragraph states that "under the principle of subsidiarity, in areas that do not fall within its exclusive competence the Union act only if the objectives of the proposed action cannot be sufficiently achieved by the Member States, either at central level or regional and local level, but can be better achieved at Union level".

In the standardisation of the areas in which the European Union exercises its competence applies also another principle, which is that of proportionality. Under this principle the content and form of Union action shall not exceed what is necessary to achieve the objectives, in the sense that the EU, in the choice of means of action, must always adopt what leaves more freedom to the states and to the extent that binding is essential. If you must consider the principle of subsidiarity when it is decided whether to promote an action at EU level, the principle of proportionality is definitely in question in all actions of the Union, even when its competence is exclusive.

Returning to the Regulation No. 178/2002, its objective is to establish the discipline of a domestic food market through the adoption of security principles aimed at protecting the safety of a high level of protection of human health throughout the chain of the agricultural product, which is followed from the farm to the table.

It is the law that sets the binding rules of the production and processing of agricultural products for the food market, to ensure: the wholesomeness of the food; transparency in the steps involved along the agri-food chain; verification of compliance with standards of safety and quality for the protection of the final consumer; strict liability for damage caused by "defect" foods. Basically, it is the law and not the market to determine what is negotiable as food, having taken on the task to protect the interests, not only non-economic ones, involved in the food market and considered worthy of being subtracted from any type of transaction from private individuals.

The safety of the food market that Regulation No. 178/2002 aims to protect is guaranteed, however, only on the condition that there is a constant monitoring food risks, with a continuous exchange of information between the competent authorities in the field every time there is the presence of a danger. The risk must be scientifically assessed by the European Food Safety Authority, managed by the Commission balancing the various options in light of the evaluation results and communicated not only at the end of the analysis, but also throughout the whole process, being a synchronic and interactive exchange of information and opinions.

The hallmark of the European Union policy in the field of food safety is represented precisely by risk management. The European Union takes upon itself the responsibility for their inspection and/or prevention, based on the precautionary principle, for which proportioned measures—although temporary—should be taken whenever there is the possibility of harmful effects on health, even if a situation continues to be scientifically uncertain.

3 The Subject "Agriculture" in the Italian Constitution

Unlike the old Art. 117 Constitution, that included it among the subjects referred as competitors, the new Art. 117 of the Constitution does not talk about agriculture.

A plausible explanation for the silence of the constitutional reform of 2001 could be that agriculture has made a leap from a competitor matter under the old Art. 117 of the Constitution to unnamed matter of the fourth paragraph of the new Art. 117 of the Constitution and then of residual powers of the regions.

On the other hand, on this matter can transversally affect the legislative power of the State, because agriculture: (a) is carried out by traders or entrepreneurs, who are in competition (and the protection of competition is recalled, under the letter e, the second paragraph of Art. 117); (b) it has the ability to produce both positive and negative externalities, being able also to "compromise" the environment (and protection of the environment is drawn under the letter s from the second paragraph of Art. 117); (c) has *naturaliter* the destination to produce food and therefore falls within the wider context of the alimentation (and the alimentation is recalled from the third paragraph of Art. 117); (d) must ensure the health of agricultural products as food (and health protection is recalled by the third paragraph of Art. 117); (e) requires public interventions of support recognising them as economic activity (and innovation support for productive sectors is recalled from the third paragraph of Art. 117); (f) presents connections with the territorial government (recalled from third paragraph of Art. 117).

Conversely, there are matters that belong either to shared competence (support for innovation, the territorial government, health and nutrition) or to the exclusive jurisdiction of the State (environmental protection and competition), which interfere deeply, with prescriptive intensity, with agriculture matters. Thus, the responsibility of the State—especially in purpose-matters of environmental protection and competition—has a very large capacity to extend, through detailing rules, to the agriculture discipline.

The principal strategic asset of the agricultural sector, that is the land, is the subject of article 44 of the Constitution, an apparently exceeded provision, but of which in fact are possible multiple readings that allow an unexpected update.

It is evident the purposeful imprint that the Constituent wanted to confer to such provision, which has two objectives to be pursued: the rational use of the land and the creation of equal social relations. On the other hand, the Article 44 reveals numerous connections with the other provisions of the Italian Constitution and in particular with the Articles 9 and 32, that according to part of the doctrine represent the constitutional fundamentals of the environment law.[2]

According to another orientation the Article 44 is connected to the Article 117 paragraph 2 letter s of the Constitution and—in some respects—to the letter *m*. On one hand, the purpose of the rational land use has assumed an ambientalistic significance in relation to which the intervention of the state legislator is foreseen

---

[2]Greco (2002), Galloni (1998) and Libertini (1991).

under Article 117 paragraph 2 letter *s*: thus, are outlined two bordering spheres of competence, the protection of the environment and of the ecosystem in the hands of Parliament and the agriculture in the hands of the Regions.

On the other hand, equity in social relations—considered as a concrete application of the substantive equality principle—allows the connection of the Article 44 with Article 117 paragraph 2 letter *m* of the Constitution, according to which the State is exclusive owner of the legislative competences in matters of essential levels of services concerning civil and social rights.[3]

On the subject see sentence no. 10 of 2010 of the Constitutional Court, which has preserved the contested state discipline considering it necessary to ensure that the constitutional right to social assistance and to the satisfaction of the food needs remain guaranteed even in a difficult economic situation, under uniform conditions throughout the national territory.[4]

The interpretation of Article 44, which seems to have more innovative capacities, however, is the one that puts it on the service of future generations. Also, this evolutive reading highlights the ecological value of the provision that no longer seems bounded to the land only, but it seems to be able to include in its horizon of protection more complex profiles, related to the concept of environment.

The criterion of rationality demands a cautious use of natural resources to preserve them: Article 44 of the Constitution seems to recall here a concept that was developed many years later in an international context, that is the sustainable development.[5]

4 The Interference Between the Subjects "Alimentation" and "Agriculture"

It is evident that between the matter alimentation and the matter agriculture there is an interference, since the first is intended to give *naturaliter* food products. And in the current language, but also in numerous regulations, it is common to use the formula "agri-food".

A confirmation of this interference is located in the Regulation No. 178/2002, which requires to "consider all aspects of the food production chain as a single

[3]Angelini (2006).

[4]With such pronounce, moving from the normative that has introduced the so-called "social card" (Art. 81 of law decree n. 112 of 2008), the Constitutional Court has recognised the food right both as "fundamental right" and as "social right" based on the Articles 2, 3, second paragraph, and 38 of the Constitution. See Anzon Demmig (2010), Belletti (2010), Longo (2010), Padula (2010) and Ruggeri (2010).

[5]Rodotà (1975).

process, starting with primary production up to the sale of food to the consumer".[6] This legal definition of EU on alimentation—and as it is a regulation it also applies to the Italian law—is objectively all-encompassing, in the sense that encloses the entire chain of production, processing and marketing of agricultural products.

According to Art. 3 paragraph 17 of Regulation No. 178/2002, "for primary production shall mean all stages of production, breeding and growing of primary products including harvesting, milking and farmed animal production prior to slaughtering and includes hunting and fishing and gathering of wild products".

On the other hand, the interference between agriculture and health protection can be seen from the fact that agricultural products, as intended for humans and animals, must be healthy. Therefore, it must be recognised that this specific segment of the agriculture subject—the production and trade of safe and healthy food—does not fall entirely within the framework of this subject, but also falls within the other subject that to protect public health (joint jurisdiction under the third paragraph of Art. 117 of the Constitution).

To the Regions remain a full competence to instrumental aspects such as organisation and management of principals veterinary, pesticides and plant protection products, waste management agriculturally derived, storage and transportation of agricultural products, the healthiness of local preservation, processing and sale of food products.

By contrast, the determination of essential service levels to ensure respect of the right to health from the threats represented by unhealthy agricultural products can only be unified throughout the country and is therefore of State responsibility. If there are different regional standards in the production of plants and animals derivations in terms of the guarantee of healthy food it may cause distortion of competition, the protection of which belongs to the State exclusively.

It is interesting to observe how in the global disorder of the markets of the last years the land—always seen as an essential asset of the agricultural and even food sectors—have maintained its stable value, even with an upward trend.

Moreover, the availability of cultivable land has gained geo-economic importance, in the sense that there is generally a positive correlation between the quantity of cultivable land per capita of a Country and its importance in agricultural trade.

Many strive to have control of the land: not just sovereign wealth funds and State enterprises of the rising powers, but also large capitals strangers to the agricultural world. Investments in land of the pension funds represent a typical example of how the phenomenon is complex and involves multi-level global companies, with implications that can result in the convergence of destinies between a US retiree with those of small farmers in the Sub-Saharan Africa.[7]

The investment targets are mainly African Countries and of South-East Asia, Latin America and the former Soviet Republics. With the definition of land grabbing, the phenomenon has gained media attention in 2008, when a South Korean multinational corporation announced that it has reached an agreement with the

[6] 12° Considering Regulation n. 178/2002.

[7] De Castro (2015), p. 103.

Government of Madagascar for an exclusive use and at no cost to 1.3 million hectares (half the arable land of the country) to produce corn and palm oil for 99 years. The agreement never became operational, but the news spread in the middle of the food price crisis of 2008, has highlighted the huge "commercial pressures on land".[8]

Not all investments in developing countries are synonymous of grabbing, because the fact that an investment is classified as land grabbing depends on many variables. Large land acquisition approved without the consultation of the local populations is certainly a case of land grabbing (and in some target Countries this absence of procedure is the norm, as the deficit of participation in the democratic life in general is very high).

In some regions, subject of the land deals, are missing records of land ownership and often the very concept of property, which, however, becomes essential for protecting those who cultivate and breed animals for generations on land considered without owners and finds himself without support because its government has sold all or has rented that land for decades at a derisory price.

For this type of large projects, the preliminary estimations of the environmental and social impact—and therefore also of the transparency guarantees of financial governance—should provide the essential conditions to approve the investment.[9]

5 Quality as Part of Food Valorisation

The current global economic and financial circumstances are such as to catalyse the problems related to nutrition transforming them into a matter of safeguarding of the dignity of persons.[10]

Reading those Charters of supranational rights and those Constitutions of some States that reincluded the "right to food", it is not difficult in general to perceive the disconnection between these soft statements and the concrete condition of poverty in which however live a considerable number of inhabitants of the planet.[11]

And yet, by limiting the observation of high-income economies, the food has stopped being just food for nutrition, but it has become something that also relates to

[8]The definition is of the International Land Coalition (ILC), which brings together intergovernmental and non-governmental organisations that see in land access and in the control of land resources by local communities a primary way to defeat poverty and hunger.

[9]Acquisitions for agricultural purposes are only one of many interests that focus on land assets. The land is demanded to produce textile fibers and biofuels; to compensate the loss of fish stock of the aquaculture marine ecosystem; to protect ground water, sequester carbon and protect biodiversity; to provide building space. All this by limiting the examination of the request of the construction and neglecting that of underground resources, affected by mining activities. See Pimentel (2006).

[10]See Rodio (1990) on the concept of "non-affluent", p. 39.

[11]Casavola Caroli (2011) and Panzera (2010). For an overview of the Constitutions that incorporated the "right to food" please see Antonio Gusmai (2015).

the sphere of desire. In some contexts, drinking wine is not consuming one drink, but making an experience in history and culture. Eating can be simply an outline activity to other activities like watching a movie, or an expression of pure hedonism. The decision of which rice to buy can become an ethical issue and the selection of fruit purchase channels can turn into a form of social mobilisation and civil commitment.

In these realities the price of food commonly on sale often refers only to a minority stake to the value of the raw material; the rest relates to the economy of services such as packaging, pre-wash, pre-cooking etc. In other cases, the added services have even a "paramedical" function, as in the case of probiotics and other so-called "enriched" foods.

Thus, expressions such time saving and functional foods have entered the common lexicon of the modern consumer. On the other hand, the abundance of information related to the marketing of the products, including the exasperate recourse of the quality cues by companies, does not mitigate the growing distrust of the consumer towards a food producer perceived as increasingly distant and products felt like something less and less decipherable.

In this scenario are less economic and more increasing socio-cultural variables to influence the consumer in the expression of his freedom of choice. Among the motivations of consumption stand out those that go totally beyond the satisfying of functional needs and that induce to explore "different" dimensions, but always in the name of quality. "Quality" is the key word, the *passepartout* for the reinvention of food.

The Community guidelines, as well as the national food market disciplines, aim to ensure—in addition to safety—the quality of food, because the quality does not exist as an absolute; only the right or the convention can describe the characteristics that make a quality product, thus can establish production and marketing rules, the respect of which will allow the product to have ethical or environmental advantages that correspond to the concerns of civil society.

Being unable in a global market to delegate to the opinion of the single individual the decision whether a food is a quality product, there must be an authority that provides the relation of these "additional" advantages of objective basis allowing their verifiability and measurability and build standard regulations, with precise definitions instead of generic terms.

To identify the factors that can satisfy the subjective expectations of the consumer, the European law has been so far expressed by valuing the production area (DOP and IGP), traditional recipes (STG), the non-use of synthetic chemicals in the production and/or processing of agricultural products (the biological system) and by regulating specific forms of control and certification. Therefore, they are definitely of quality those food products with a designation of protected origin and protected geographical indication, as well as guaranteed traditional specialities and the organic farming products, all regulated in a punctual way.

This is the idea of preserving traits of traditional production models and certifying them, also to offer the perspective of a consumption identified in a place, in a landscape and in a tradition often centenary. There is a diverse chorus of local experiences in the name of entrepreneurship and social innovation in agriculture that standards need to be able to valorise.

Because food is an aliment, but in fact it is also a living history of the men not susceptible to be subrogated under global market forces and private-homogenising, destructive of human identities and local cultures that, historically, have also been able to provide the planet with irreplaceable cultural feed.

Just think that the same term "aliment" derives from Latin *alere*, which can be translated meaning "grow", "to develop", "breed" not only physically, but also spiritually and socially.[12]

On the other hand, promoting the sale of certain products, which are characterised for some reason because of their quality or their origin, it is a tool to facilitate greater movement of goods and competition between businesses. Competition needs a complete and accessible information, which justifies the establishment of quality brands from the public decision-maker, who through their intermediary provides a service of general interest aiming the proper functioning of the fundamental mechanisms of the free market (the legitimacy of public institutions to register collective trademarks is expressly sanctioned in Articles 5 and 66 of Regulation 207/2009/CE on the community trade mark and Article 19 of the Italian Code of industrial property. And since the Regions do not exercise direct entrepreneurial activities, it can be just collective quality and origin marks).

Also, according to the European Parliament resolution 2013/2098 (INI) of 14 January 2014, called "on the regional brand: towards best practice in rural economies", such brands of excellence can contribute to the competitiveness of the most vulnerable and remote areas through the coordination of quality goods and services anchored in the specificities of each territory and its heritage (environmental, cultural, historical etc.).

This certification system enables thousands of small and medium businesses to participate in the global economy, pushing them to make a decisive contribution to the maintenance of the vitality of entire local systems, thereby triggering a territorial valuation process within a perspective of identity, even compared to meta-economic parameters such as the environment, quality of life, economic and social cohesion and solidarity. In this perspective it could be envisaged to give a regulatory response—through the discipline of further distinctive signs as "fair trade", "animal welfare", "low carbon dioxide emanation", "sustainable fishing"—also to consumer requests seeking a particular value that will inspire an emotional or cultural correspondence to that specific food.

In this framework of rules, farmers who submit themselves to the prescribed mode of production and related control have an extra chance, in the sense that the quality becomes a mean to establish itself in the competitive race.

[12]Monteduro (2015).

6 The Constitutional Reform in Ongoing Process ("Alimentation" and "Valorisation of Specific Territories")

The upcoming reform of Part II of the Italian Constitution provides major changes at the institutional level, particularly through the redefinition of the structure and functions of the Senate, but also in relation with the legislative powers of the state and the regions, with the elimination of the competitor power in the third paragraph of Art. 117 of the Constitution in favour of a clearer distribution. Therefore, if it were confirmed (it is expected a referendum in the autumn of 2016), the draft of the constitutional reform could affect the regulatory framework analysed up to this point.

The idea that inspired the reform is that of making more fluid the relations between legislative powers and more flexible the allotment criteria of the legislative powers.

In the Government perspective, the withdrawal of the guarantees because of less rigid distribution is compensated by the role that the new face of the Senate ensures to the territorial autonomies; but, as in the legislative innovations—especially in constitutional matters—the phenomenon of heterogenesis of purposes is anything but sporadic, it cannot exclude the advent of a new season of uncertainty and contentiousness related with the allocation of powers between center and peripheries, with an increase of the constitutional court activity.

Thus, the declared aim of removing the overlays that have emerged following the constitutional reform of 2001, this proposal adopts a functional-teleological perspective, more oriented to the functions and the objectives of the action by the public authorities more than to the identification of rigid matters. In particular—for what is of interest here—it proposes the deletion of the concurrent legislative power and reformulate these matters. The suppression of competitor power is linked to the transit of the matters contained therein in the list of those that are subject to the exclusive authority of the state: for example, according to the draft law, it would become a matter of state-own exclusive food safety expertise.

Since the entry into force of such changes the legislative power of Regions,[13] may result greatly resized, but to compensate this consequence it is expected: on one hand, a list of matters that belong to the exclusive competence of the regional[14]; on the other hand, the possibility for the State to delegate the exercise of the legislative function in the areas subject to its exclusive authority.[15]

Moreover, the power of the Regions is related not only to matters not expressly reserved to the exclusive legislation of the State, but also to the functional areas "own" of the regional legislation (always with the aim of strengthening regional

[13] Pajno (2014) and Scaccia (2014).

[14] A similar proposal was contained in the text submitted to a referendum in 2005 (Art. 39, tenth paragraph, of the Project). See D'Atena (2005), p. 305.

[15] Art. 26 PG.

competence in front of the State pervasiveness).[16] In this way the constitutional legislator intended to identify "the main areas in which the regional legislator can establish regulatory systems aiming to the enhancement of the specificity of the territories, but coherent with the guidelines defined by the state legislature on the outcome of a legislative process in which the intervention of Senate of the Autonomy is called to play a fundamental role".[17]

This is not the venue for a review of the overall reform.[18] There is no doubt, however, that, in front of the affirmation contained in the accompanying report of the governmental Project were, "as close the power is to the citizens, the higher will be the quality of democratic life and the capacity of institutions to meet civil and social rights granted to them under the principle of vertical subsidiarity", raise some concerns the substance analysis of the reform that limits greatly the space reserved for regional legislation, without inserting appropriate counterweights in state decision-making processes.[19]

References

Angelini F (2006) Articolo 44. In: Bifulco R, Celotto A, Olivetti M (eds) Commentario alla Costituzione. Utet, Torino, pp 902–914
Anzon Demmig A (2010) Potestà legislativa regionale residuale e livelli essenziali delle prestazioni. Giur cost 1:155–165
Belletti M (2010) La questione della tutela del nucleo essenziale dei diritti nel nuovo assetto autonomistico. Perc cost 1:149–163
Bin R (2015) Oltre il velo d'ignoranza. Venti questioni su Regioni e riforme costituzionali. Le Reg 1:83–100
Casavola Caroli H (2011) Dignità della persona e diritto al minimo vitale nello Stato sociale tedesco. Gior dir amm 3:1–11
D'Atena A (2005) Le Regioni dopo il Big Bang: il viaggio continua. Giuffrè, Milano
De Castro P (2015) Cibo. La sfida globale. Donzelli, Roma
De Siervo U (2015) Risposte a "Venti domande sulle riforme costituzionali". Le Reg 1:139–152
Galloni G (1998) Le fonti costituzionali del diritto agro-ambientale. Dir giur agr amb 7:5
Greco N (2002) Connessioni ed eventuali conflitti funzionali tra bonifica, governo dell'acqua e difesa del suolo. St parl pol cost 35:9–28
Gusmai A (2015) Il diritto fondamentale al cibo adeguato tra illusioni e realtà. Dir fond 1:2–16
Libertini M (1991) La nuova disciplina del danno ambientale e i problemi generali del diritto dell'ambiente. In: Perlingeri P (ed) Il danno ambientale con riferimento alla responsabilità civile. ESI, Napoli, pp 21–44
Longo E (2010) I diritti sociali al tempo della crisi. La Consulta salva la social card e ne ricava un nuovo titolo di competenza statale. Giur cost 1:164–191
Monteduro M (2015) Diritto dell'ambiente e diversità alimentare. Riv quadr dir Amb 1:88–131
Padula C (2010) Considerazioni in tema di fonti statali secondarie atipiche. Dir pubb 1:365–408

[16] Art. 26 PG.

[17] APSD, 22.

[18] On this point see Bin (2015), De Siervo (2015) and Ruggeri (2015).

[19] APSD, 4.

Pajno S (2014) Considerazioni sulla riforma costituzionale in progress, tra Governo, Senato e Camera dei deputati. Fed 24:2–49

Panzera C (2010) I livelli essenziali delle prestazioni concernenti i diritti sociali. In: Campanelli G, Carducci M, Grasso N, Tondi della Mura V (eds) Diritto costituzionale e diritto amministrativo: un confronto giurisprudenziale. Giappichelli, Torino, pp 57–105

Pimentel D (2006) Soil erosion: a food and environmental threat. Environ Dev Sustain 7:119–137

Rodio R (1990) Difesa giudiziaria e ordinamento costituzionale. Cedam, Padova

Rodotà S (1975) Articolo 44. In: Branca G (ed) Commentario della Costituzione. Zanichelli, Bologna-Roma, pp 222–231

Ruggeri A (2010) Livelli essenziali delle prestazioni relative ai diritti e ridefinizione delle sfere di competenza di Stato e Regioni in situazioni di emergenza economica (a prima lettura di Corte cost. n. 10 del 2010). For cost 10:1–5

Ruggeri A (2015) Una riforma che non dà ristoro a Regioni assetate di autonomia. Le Reg 1:243–256

Scaccia G (2014) Prime note sull'assetto delle competenze legislative statali e regionali nella proposta di revisione costituzionale del Governo Renzi. Astrid 8:4–16

Tondi della Mura V (2007) Sussidiarietà ed enti locali: le ragioni di un percorso innovativo. Fed 20:1–13

Consumer Choice as a Pathway to Food Diversity: A Case Study of Açaí Berry Product Labelling

Hope Johnson, Christine Parker, and Rowena Maguire

1 Introduction

Forest lands and the rich social and ecological diversity contained within are being lost as demand for agriculture land expands globally.[1] During this process traditional cultivation practices are marginalised resulting in a loss of dietary diversity.[2] As Vira et al. observe 'Despite the huge potential of forest and tree foods to contribute to diets, knowledge on many forest foods, especially wild foods, is rapidly being lost because of social change and modernisation'.[3] Forest loss coupled with the associated declines in dietary diversity and traditional knowledge are a threat to the human right to food. This right requires diverse food production systems that are sustainable, support livelihoods (especially those who are most marginalised) and meet nutritional needs.[4] This chapter explores the exporting and labelling of a traditional food source the "açaí berry" and examines whether the production and sale of açaí has the potential to improve food diversity.

The marketing claims made for açaí berry products on their labels suggest they are a good news story for food diversity. Açaí berries are generally not produced on farms but rather harvested from public and privately-owned rainforests in the

[1] See, e.g., Thrupp (2000) and Gonzalez (2011).

[2] See, e.g., Waltner-Toews and Lang (2000).

[3] Vira et al. (2015), p. 24.

[4] De Schutter (2011).

H. Johnson (✉) · R. Maguire
Queensland University of Technology School of Law, International Law and Global and Governance Research Program, Brisbane, QLD, Australia
e-mail: H2.johnson@qut.edu.au; r.maguire@qut.edu.au

C. Parker
School of Law, University of Melbourne, Carlton, VIC, Australia

A. Isoni et al. (eds.), *Food Diversity Between Rights, Duties and Autonomies*, LITES – Legal Issues in Transdisciplinary Environmental Studies 2,
https://doi.org/10.1007/978-3-319-75196-2_19

Amazon. Thus, the purchase of açaí berry products promises the consumer that they can avoid complicity in the negative environmental impacts of industrialised food systems and instead help preserves the Amazon rainforest through their choice of purchase. Moreover, purchasing açaí products is depicted as a decision that makes a positive contribution to rural development, as Amazonian communities are commonly involved in the harvest of açaí. Brondízio, a leading anthropologist on rural populations in the Amazon, claimed that 'There may be no better example of an economic prospect for overcoming underdevelopment in rural Amazonia than the case of açaí palm fruit production system'.[5] This framing appeals to health-conscious middle and high-income consumers seeking to use market forces to create more sustainable and ethical food systems.[6]

This chapter critically examines these marketing claims and in doing so considers the extent to which autonomous consumers can promote food diversity through their purchasing choices. Food diversity is considered through the lens of biological diversity and dietary diversity. Specifically, food diversity is understood as encompassing: biodiversity at a macro level (that is, the biodiversity required food production), at a farm level (diversity of species on farms or in otherwise food producing areas) and at the individual and household level (variety of accessible foods and nutritional adequacy).[7]

This chapter finds that demand for açaí berries can contribute to food diversity; but the ability of consumers to ensure açaí production contributes to food diversity is limited by their geographic distance from production, misrepresentations in labelling and marketing and a lack of information regarding the standards behind the claims. In order to make this point, the chapter calls into question many of the claims made by açaí companies, finding that the berries are not necessarily contributing to the preservation of forest or rural development. Likewise, açaí berries are not being enjoyed in a manner consistent with traditional knowledge or diets and are not as nutritious as suggested by existing marketing.

The chapter starts by explaining the methodological approach adopted and the regulatory context. The chapter then examines five of the common claims made about the açaí berry: sustainably produced, part of a traditional diet, uniquely nutritious, fair trade and organic. The accuracy of these claims is used to consider the extent to which consumers can promote food diversity through their purchases, and the role of governance arrangements in constructing food diversity and food choices.

[5]Brondízio (2004), p. 339.

[6]Deaton (2004), p. 617; Guthman (2007).

[7]Berti and Jones (2013), pp. 186–187.

2 Background

2.1 Methodological Approach

Food labels and marketing claims involve discursive processes that construct and determine the type, accuracy and form of information conveyed to the consumer. Parker explains that the '...choices available to consumers have already been constructed and constrained by relationships between actors in the production, distribution, and exchange chain who bring products to retail'.[8] This chapter uses Parker's "backwards mapping" methodology to examine the claims made in the marketing and labelling of açaí berry products.[9]

"Backwards mapping" involves first identifying the food labeling and market claims to which consumers are exposed; and second working backwards through the supply and governance chains to investigate the production and supply practices behind the label and the public and private regulatory instruments that allow or create the claims on the label. This critical methodology reveals how the food label, broadly defined to include all the marketing and packaging on retail display, operates as a focal point for otherwise hidden supply and governance practices that limit and define the choices available to consumers at the point of purchase.

Following the backwards mapping methodology, thirty-eight (38) açaí berry products were identified in Australia during the year 2015. These 38 products were discovered by searching publicly available government and private databases followed by a Google search and retail site visits.[10,11] Twenty-nine

---

[8]Parker (2013), p. 54.

[9]See, e.g., Parker et al. (2013), Parker (2013) and Parker and Scrinis (2014).

[10]All trademarks registered in Australia with the terms 'açaí' or 'amazon' found on IP Australia were recorded and the Australian trademarked websites accessed. Details of açaí products advertised for sale on the trademark related websites were compiled. The 'açaí' term was applied in the search of the ASIC business and company names database. Entities with the word 'açaí' in the name not initially identified with a registered trademark were selected and searches of websites conducted to collate details on the açaí products for sale. The Australian Certified Organics (ACO) database was accessed using the keyword 'açaí' and search conducted. Company websites identified through ACO were searched and the details of further açaí products available for sale compiled. Google searches using the terms 'açaí' and 'buy' identified additional 'açaí' products available for sale in Australia. Products were also identified in searches of Australian sellers on ebay.com.au. Those products not previously identified were added to the Excel spreadsheet compilation.

[11]Images of packaging labels for each product available for review online were observed and information recorded. A total of five (5) Sydney health food stores and pharmacies were visited in a search of the pills, juices and powdered açaí products identified online to record any omitted information contained on labels at retail. Not all products were observed on the shelves of the visited retail stores. For the products located, photographs were taken and information presented on labels recorded with the compiled information. One cafe and one juice bar found to be selling the two frozen pulp products, respectively, was visited in Sydney and the product information detailed on the labels were recorded along with additional information provided in advertising pamphlets at point of sale. All information provided in the advertising (online content, label information and/or point-of-sale pamphlets) of the identified products was compiled into the Excel spreadsheet to facilitate the analysis.

(29) companies/businesses[12] were identified, responsible for selling the 38 açaí products. Freeze dried powder sold as a dietary supplement was the most common form in which açaí was sold, with pills or capsules and drinks and juices also common. We then coded the claims made on the food label and identified five common claims concerning the açaí berry:

1. Açaí berry production preserves the Amazon;
2. Açaí berries are part of a traditional diet and informed by traditional knowledge;
3. Açaí berries are uniquely nutritious;
4. Açaí berries are fair trade
5. Açaí berries are organic

The results of the first stage data collection are summarised in Table 1.

In the second stage we traced back the meaning of these claims in terms of production practices and the public and private standards that applied to them. This involves comparison of the claims on the labels with anthropological research about practice and history of açaí consumption, supply chain analysis and analysis of the public and private standards that govern the practices used and claims made. We do this for each of the five different types of claims.

2.2 *Laws That Apply to Açaí Berry Claims*

Australia provides a useful case study for two main reasons. Firstly, Australia's regulation of food labelling and marketing claims largely aligns with the relevant international standards.[13] This means that the standards applied by Australia are similar to standards applied elsewhere, and so the information provided (or not provided) on the label are likely to be comparable to other jurisdictions. Secondly, Australia is an example of a Western country with unsustainable consumption patterns and where consumers generally agree that there shopping choices can make 'a positive difference to farmers and workers in poor countries'.[14] In Australia, like many other developed or emerging economies, the role of personal choice in promoting food diversity and sustainability is emphasised instead of

[12]Including one specialist café.

[13]At the international level, food labelling and other claims are regulated by the standards of the Codex Alimentarius ('Codex'). The Codex formulates non-binding standards that set a benchmark for domestic laws especially in those countries that are members to the World Trade Organization. Several Codex guidelines target misleading, deceptive or false claims including claims. For example, the Codex General Guidelines on Claims states that 'no food should be described or presented in a manner that is false, misleading or deceptive or is likely to create an erroneous impression'. According to these guidelines, claims that cannot be substantiated should be prohibited and claims regarding how the food is grown (e.g. natural or organic) are determined by national laws but may be prohibited. See, Codex Alimentarius (1991), principle 1.2.

[14]Agriculture and Agrifood Canada (2012), p. 6.

Table 1 Açaí products and their marketing claims

Product	Claim 1	Claim 2	Claim 3	Claim 4	Claim 5
Amazon Power™ organic freeze dry açaí powder (Amazon Power Pty Ltd NSW)	✓	✓	✓		✓
Amazon Power™ Açaí Pulp (Amazon Power Pty Ltd NSW)	✓	✓	✓	✓	
Amazon Power™ organic açaí capsules (Amazon Power Pty Ltd)	✓	✓	✓	✓	✓
Amazonia frozen smoothie packs (Amazonia Pty Ltd)			✓	✓	✓
Organic Açaí Freeze Dried Packed (Amazonia Pty Ltd)					
Organic açaí capsules (Amazonia Pty Ltd)			✓		✓
Açaí Berry Organic Powder (Austral Herbs Pty Ltd)			✓		✓
Açaí + Berry Powder (Bioglan)			✓		✓
Emma Vita Açaí Berry Coffee (Emma Vita Pty Ltd)		✓			✓
Açaí Powder Raw Organic (Loving Earth Pty Ltd)	✓	✓	✓	✓	✓
AcaVie Certified Quality Premium Açaí™ (MonaVie)		✓	✓		
MonaVie Active® gel (MonaVie)		✓	✓		
MonaVie Active® juice (MonaVie)		✓	✓		
MonaVie Pulse® (MonaVie)		✓	✓		
MonaVie® health juice blends (MonaVie)		✓	✓		
Certified Organic Açaí Powder (Morlife Pty Ltd)		✓			✓
Organic Açaí Powder (Nature's Distributors Pty Ltd)		✓	✓	✓	✓
Açaí Power™ capsules 500 mg (Nature's Goodness Australia Pty Ltd)		✓	✓		
Açaí Power™ Juice Blend Concentrate (Nature's Goodness Australia Pty Ltd)		✓	✓		
Açaí Power Vacuum Dried Açaí Powder (Nature's Goodness Australia Pty Ltd)		✓	✓		
Açaí Powder (Nutradry Pty Ltd)		✓			✓
Super Açaí + Berries (Parmacare Laboratories Pty Ltd)	✓		✓		✓
Berry Power Açaí Powder (Power Super Foods)	✓		✓		
RioLife 100% organic freeze-dried açaí powder (Riolife)	✓	✓	✓	✓	
Açaí Berry (Organic) (Bulk Powders)	✓			✓	✓
Açaí Magic Berries™ (Sumabe Holdings Pty Limited)	✓		✓		✓
Açaí berry fruit juice drink made from concentrate (The Berry Company Pty Ltd)		✓	✓		

structural change to food systems, including regulatory shifts.[15] These reasons, coupled with the fact that Australian consumers rely on labelling to inform their choices, make Australia a suitable case study for examining consumer choice as an avenue to promote food diversity.[16]

The main Australian regulatory regimes affecting food labels and marketing claims are the Food Standards Australia and New Zealand regime (enforced by state food and health regulatory agencies) and the Australian Consumer Law prohibitions on misleading and deceptive conduct (enforced by the Australian Competition and Consumer Commission ('ACCC'), along with state Fair Trading Regulators. Food Standards Australia and New Zealand sets food labelling standards including general and specific labelling and information requirements that apply to food products.[17] The ACCC and state fair trading regulators investigate and prosecute false or misleading claims on goods and it regulates the use of certification trade marks (e.g. fair trade labels).[18] It requires that claims related to the social, nutritional or environmental benefits of a product be capable of being substantiated.[19] Broadly, false and misleading representations about a good (including food) are prohibited, particularly false or misleading representations about the quality or a particular history of goods is prohibited.[20]

3 Açaí Berry Case Study

For centuries, Indigenous Amazonians and then later traditional tribes have used the açaí palm (*Euterepeoleracea*) for construction and consumption purposes.[21] Açaí is native to the Amazon floodplains and more broadly Central and South America.[22] The production and consumption of the açaí has exponentially increased since the 1970s because of demand from urban Brazilian communities and Western consumers.[23] Because of this demand, açaí is generally harvested by rural communities and then delivered to processors. The majority of açaí processors are corporations that industrially process the açaí.[24]

[15] See, e.g., Hobson (2004), p. 124 where the author explains the emphasis on consumer choice as an "economy-first" approach that takes 'a win-win approach and non-threatening stance toward global markets, focusing on voluntary rather than prescriptive changes to practice'; For evidence of this approach in Australia, see e.g., Parker (2013, 2015) and Johnson (2015).

[16] For evidence of the reliance placed on food labelling by Australian consumers to make choices, see, Mhurchu and Gorton (2007).

[17] Food Standards Australia and New Zealand (2013).

[18] Australian Competition & Consumer Commission (2015); *Competition and Consumer Act* 2010 (Cth) part II.

[19] *Competition and Consumer Act 2010* (Cth) sch 2, s 18.

[20] *Competition and Consumer Act 2010* (Cth) sch 2, s 219.

[21] Brondízio (2004).

[22] Colapinto (2011).

[23] Hiraoka (1994), p. 332.

[24] Filho and Sousa (2013).

3.1 Claim 1: Açaí Berry Production Preserves the Amazon

Eleven of the 38 products examined emphasised that the way açaí is grown and harvested preserves the Amazon rainforest. For instance, one claim was that: 'This Açai is wild harvested from deep in the Amazon forests. This means the fruit grows in its native bio-diverse ecosystem maintaining its natural nutrient profile. Wild harvesting ensures that the delicate environment of the Amazon is protected' (Amazon Powder, Loving Earth Pty Ltd).

Generally, the harvesting process was described as 'wild-harvested', 'wild-gathered', 'hand-picked' and 'manually harvested'. These phrases are used to convey the message that açaí berries are grown in the wild as opposed to on farms; and that açaí berries are harvested by hands and not machinery. Some of these products referred to the role of indigenous Amazonians in the harvesting of acai. For instance, one product states '...we only use berries that have been manually harvested, ensuring the protection of the palm trees that berries grow on. The manual harvesting of the berries also provides hundreds of jobs for the Indigenous tribes around Brazil' (Amazon Power Pure Açaí Pulp, Amazon Power Pty Ltd).

Research into açaí production and harvesting shows a rather different picture. One of the greatest misconceptions about açaí harvesting concerns the social category of the people harvesting the fruit. Açaí is mainly harvested by Amazonian communities who tend to live subsistent lifestyles.[25] Generally termed Caboclos (which we will refer to as Amazonian peasants),[26] these communities have mixed ancestry (Latin American, European and African descent) and a diverse range of cultural practices and beliefs.[27] As such these groups do not necessarily identify as, or are identified within Brazil as being, indigenous or first nation people of the Amazon. Rather, this group of people are a social category distinguished by their geographical location and socio-economic activities. In broader Latin American society, Amazonian peasants are often depicted as lazy because of their subsistent lifestyles and class position.[28]

The food label also infers that the açaí berry grows naturally in the wild suggesting that the product is sustainably grown and harvested, and consistent with promoting biological diversity. Prejudicial views about the 'lazy' Amazonian peasants have perpetrated the myth that açaí berries occur naturally in quantities required to meet global demand without human intervention. This belief suggests the production of açaí berry does not involve a significant degree of human skill or labour. Research by anthropologists contradicts this belief showing that floodplain communities have a significant role in the management of açaí plants including through thinning, weeding, pruning, inter-cropping techniques and the development of seedlings.[29] Arguably then, the land and crop management by Amazonian

[25]Bussman and Zambrana (2012), p. 258; Pace (1998), p. 145.

[26]Caboclos is generally a derogatory term. Alternatives include "traditional Amazonian population", "Amazonian peasants" or "river people". See, Pace (1997).

[27]Fraser (2010), p. 652.

[28]See Harris (2005), p. 468.

[29]Brondízio and Siqueira (1997), p. 258.

peasants may fit some definitions of "agroforestry", and is certainly an example of small-scale production systems that tend to use less intensive harvesting methods.[30]

Through the application of local knowledge, Amazonian peasants have intensified the production of açaí without leading to deforestation or monoculture farming. Brondízio and Siqueria explain:

> Agroforestry systems that mimic native forests are "invisible" in the analysis of most researchers who employ conventional measures of [agricultural] intensification. The result is agroforestry's characterization as extensive, partially extractivist and non-dependent on labor and energy input other than for "gathering".[31]

Conversely, conventional ideas of what farming involves, combined with the marginalisation of Amazonian peasants, have fed into the claims that açaí is passively "gathered". One of the products claimed: 'Wild harvesting ensures the delicate environment of the Amazon is protected and the ancient traditions of the Indigenous people from this region are respected and preserved' (Açaí Powder, Loving Earth Pty Ltd). Referring to açaí management as "wild harvesting" overlooks the fact that traditional farming practices by Amazonian peasant groups have played a significant role in intensifying açaí production.

There is a growing body of work investigating the impacts of açaí production on the biological diversity of the Amazon rainforest. Unfortunately, production of açaí has intensified in the floodplains leading to large areas of diverse forest being converted into açaí agroforests and açai being grown on monoculture plantations.[32] A recent study by Freitas et al. examined the impact of the demand for açai on the biological diversity of Amazonian floodplain forests. They found a loss of 50% of tree species diversity and a 63% reduction in pioneer species (hardy varieties that are the basis for forest ecosystems).[33] These findings align with empirical studies that illustrate a negative correlation between increased international agricultural trade and declines in on-farm and off-farm biodiversity and dietary diversity.[34] More broadly, trends in the intensification of açaí production for export reflect the conflict between economic principles underpinning free trade laws that promote specialisation in agricultural exports and the need for food diversity.[35]

In line with the uptake of more specialised farming models, there is a documented push by Brazilian politicians towards attracting private investment into açai plantations and increasing the use of synthetic fertilisers in açaí management.[36] The scaling up of açaí production to meet increasing global demand means that industrial processes are being used to farm açaí. Neoliberal development paradigms focus on

[30]Clark (2014).

[31]Brondízio and Siqueira (1997), p. 245.

[32]Weinstein and Moegenburge (2004); Lewis (2008); Freitas et al. (2015); Moegenburg and Levey (2002); Smith (2015), p. 257.

[33]Freitas et al. (2015).

[34]Gale (2000), Boyce (1999), Johnstone (1995) and Norgaard (1987).

[35]Gale (2000), p. 286; Gonzalez (2011).

[36]Lewis (2008), pp. 297–298.

productivity over sustainability and biological diversity. While açaí may have been produced in harmony with nature in the past, current practices cannot be considered sustainable and the labelling of 'wild harvested' is misleading. Consumers of açaí are therefore limited in their ability to be agents of change, as production methods of açaí do not challenge the environmental harms associated with trade liberalisation or intensive agricultural production.[37]

3.2 Claim 2: Part of Traditional Diets and Informed by Traditional Knowledge

Açaí is also heavily promoted as being a "traditional food" suggesting that consumption of the fruit contributes to dietary diversity. One label asserts 'For countless centuries, the people of the Amazon have revered this unique fruit for its nutritional content and prized it as a source of health and vitality' (AcaVie Certified Quality Premium Açaí, Mona Vie). This claim is accurate as Indigenous Amazonians consumed (but did not rely on) açaí fruit before colonisation and later açaí became a staple for Amazonian peasants as Indigenous communities generally prefer other food staples.[38] However, the ways in which açaí is processed and designed to be consumed in Western and other international markets changes it from the traditional staple.

Customarily, açaí is consumed after being soaked in water, pulped, strained and then drunk, added to grains or served with fish or meats.[39] In contrast, recommended serving approach to Australian consumers is mixing açaí with other fruits. Traditional beliefs in Brazil associate the inter-mixing of açaí with other fruits as leading to ill-health, calling into question the marketing claims regarding the açaí product being part of traditional diets.[40]

The disconnection between traditional diets and knowledge and the claims made by açaí marketing is evident in other ways. For instance, the identified products often described açaí as a source of "vitality", a "natural energy boost", an "energizing superfood", a "sustained energy boost", and enhancing "energy production, endurance and strength". In contrast, according to traditional beliefs açaí has a reputation for causing fatigue and specifically for being '*a heavy food that weighs you down and makes you lethargic*'.[41] So, while consumption of açaí contributes to dietary diversity by increasing the variety of natural products consumed, it is misleading to suggest that the current method for enjoying açaí is traditional or that the fruit is a traditional energy booster.

[37]Except perhaps if the açaí is organic and fair-trade certified, but as will be discussed these schemes have their limitations in terms of ensuring equity and sustainable production methods.

[38]Bussman and Zambrana (2012), p. 260.

[39]Bussman and Zambrana (2012), p. 257; Fajans (2012), p. 120.

[40]Fajans (2012), p. 127.

[41]Fajans (2012), p. 130.

3.3 Claim 3: Especially and Uniquely Nutritious

Açaí is marketed as a commodity that assists in promoting dietary diversity by helping to address types of malnutrition associated with Westernised diets. An açaí berry fruit juice drink sold in a major supermarket chain in Australia refers to 'The açaí berry from the Brazilian rainforest' as 'the tree of life'. Another line of açaí berry products points out that western consumers have only recently been granted access to this traditional knowledge, with the help of branded exporters and retailers of processed açaí products:

> A staple of Amazon natives for hundreds of years Açaí berries were largely unheard of outside of Brazil until very recently. As its fruit loses its high nutritional content only 24 hours after harvest, this potent super food was only consumed within the close confines of the Amazon until the early 1990s. Now snap frozen upon harvest, the rest of the world can enjoy the taste and health benefits Açaí berries have to offer (Açaí Power™ Vacuum dried açaí powder, Nature's Goodness Australia Pty Ltd).

There is a sense on some of the labels that it is the mystical origins in the Amazon jungle itself that contributes to the superior scientific properties of the Açaí berry. One product refers to "Over 50 essential nutrients many unique to the Amazon soil" (Amazonia, frozen smoothie packs). Another label suggests that buying an expensive[42] açaí supplement powder can help the consumer "unlock the energy of the Amazon& better health every day" because it is "particularly rich in antioxidants, anthocyanins and polyphenols" *(BioglanAçaí + Berry Powder).*

An in depth New Yorker article about the açaí berry phenomenon quotes anthropologist Jane Fajans' observation that, according to its marketing, açaí berry could become,

> . . .a kind of balm for millennial anxieties. . . a miracle cure for, among other things, obesity, attention-deficit disorder, autism, arthritis, Alzheimer's disease, and erectile dysfunction. . . There are all these claims. . . that it takes away the toxicity of living in the First World and transports you back to the healthy, natural world of the rainforest.[43]

However, as Parker and Curll show,[44] no unique evidence supports the health and anti-ageing claims made for açaí berry products over many other nutrient dense foods. The labelling of these products conflates the well-accepted health benefits of antioxidants and other nutrients found in a variety of 'normal' fresh, unbranded fruits and vegetables with claims exaggerating the unproven benefits of particular

[42]The recommended retail price for 50 g of açaí berry powder purchased online is in the range of $15 to $20, where a kilo of powder retails at around $250.

[43]Colapinto (2011).

[44]This chapter is one outcome of a larger project entitled 'A sociological analysis of the anti-ageing treatment market: the dynamics and expectations' (2014–2016) funded by the Australian Research Council. As one part of this project, research has been carried out by Christine Parker and Janine Curll regarding the regulation of health and anti-aging claims on açaí products in Australia. Christine Parker, Janine Curll and Hope Johnson have also gathered data regarding eco-claims on açaí products.

phytochemicals apparently found in higher concentrations in açaí berries, and backed up with the "exotic backstory" of ancient Amazonian origins.[45]

3.4 Claim 4 Açaí Berries Are Fair Trade

Fair trade is a social movement that operates through third-party certification schemes. It aims to create incentives for companies to improve their social and environmental impacts and to inform consumers about trade-related human and environmental issues.[46] Raynolds explains that fairtrade is a '...labelling project where consumers are given information about the social and environmental conditions under which commodities are produced and then asked to pay to support more sustainable production and trade'.[47] Accordingly, the accuracy of fairtrade claims and the perceived legitimacy of the third-party certification system are critical to fair trade schemes fulfilling their objectives.

None of the 38 studied products had fair trade certification. Conversely, none of the identified açaí products had been subject to private, third-party auditing in which their compliance with several social, economic and environmental considerations is evaluated, monitored and communicated to consumers through labelling. To receive certification, a company would need to source açaí from certified co-cooperatives or certified companies.[48]

It is unclear why açaí cooperatives have not obtained fair trade certification. However, common barriers to fair trade certification identified in the literature include land tenure insecurity and a lack of government support for small-scale producers to enable them to formulate cooperatives.[49] Another possibility is that the harvesting of açaí does not meet fair trade standards related to workplace health and safety. This is because açaí harvesting involves climbing a tall palm while carrying a machete to cut down palm fronds that grow the berries. Once on the ground, people hand-strip the berries from their stalks. Raffles describes it as 'rough, dangerous work, hard on hands and feet, made worse by the relentless insects'.[50]

---

[45]Christine Parker, Janine Curll, Alan Petersen, Casimir MacGrgor, "Unlocking the energy of the Amazon? The regulation of health claims on açaí berry products", unpublished paper on file with authors.

[46]Generally, fair trade certification involves a valid non-governmental organisation (NGO) receiving an application from a company that is seeking the certification of a particular product. The NGO then audits and evaluates the particular processes behind a product against set criteria that concerns social and environmental factors. If certification is approved, the company can display the fair trade logo on its product and will be subject to two more audits within 3 years to monitor compliance. See, International Organization for Standardization. ISO/IEC 17065:2012(en)v. See, FLO-CERT (2015a).

[47]Raynolds (2000), p. 298.

[48]FLO-CERT (2015b, c).

[49]See, e.g., Tellman et al. (2011).

[50]Raffles (2002), p. 202.

Although official certification was not obtained, several identified products claimed to follow "fair trade practices", which implies compliance with a set of external standards. Accordingly, references to "fair trade" may be considered misleading if they infer compliance with fair trade certification standards. For instance, one fair trade claim was that '*The Açaí berries used in this product have been harvested by local families, which also means rural community and grower co-operative are empowered through fair trade pricing.*' Another claim was that '*Every step of our supply chain is closely monitored to ensure sustainable and Fairtrade practices are followed.*'[51] Nevertheless, research by Tiesl found that consumers fail to appreciate the differences between certification labels and sustainability claims, so the inference of being certified may not have been made to a reasonable consumer.[52]

3.5 Claim 5 Açaí Berries Are Organic

Nine of the 38 identified products were certified organic according to a third-party Australian organisation and seven of the products made organic claims but did not have certification. Like other countries, imported products in Australia can be labeled 'organic' if the Australian Standard for Organic and Biodynamic has been complied with.[53] The standard involves self-regulation in the sense that companies can certify themselves as meeting the management standards and do not have to be certified via external independent third-party assessment. This process of self-regulation does not ensure the integrity of the 'organic' label. This then leaves enforcement with the ACCC and consumers to raise complaints or concerns about the organic claim. In the case of açaí, the knowledge required to determine if an "organic" claim is misleading is difficult if not impossible to obtain given the geographic distance and complicated supply chain.

Despite this loophole, 14 of the products had received third party organic certification. Third party certification involves non-state actors setting management standards that meet and go beyond the Australian Standard for Organic and

[51]As outlined in Thampaplillai et al. (2015), pp. 475–479, Australian courts use a four-part methodology for determining whether a claim is misleading or deceptive. The first step is to identify the target audience, which can be a very broad group. The second step is to consider the various demographics within the group (e.g. from highly educated to poorly educated). The third step is to objectively assess the meaning of the words used in the claim (or images) and determine whether the claims are likely to mislead or deceive. The fourth step is to consider the wider context and whether that removes or reinforces the misleading nature of a claim. For more information, see *Taco Co of Australia Inc v Taco Bell Pty Ltd* (1982) 42 ALR 177. This methodology is often used by Australian courts in relation to misleading and/or deceptive conduct. See, e.g., *Australian Competition and Consumer Commission v Coles Supermarkets Australia* [2014] FCA 634; *Domain Names Australia Pty Ltd v au Domain Administration Ltd* [2004] FCAFC 247.

[52]Teisl (2003).

[53]Standards Australia (2015).

Biodynamic Products. There are many organic certifying bodies, however the non-for-profit entity "Australian Certified Organic" ('ACO') was the most common certification standard used for the 38 Açaí products examined. The ACO is the most established organic certification system in Australia.[54] ACO's Standard 7.5 relates to "wild harvesting", which is used by açaí processors to obtain certification.

To be certified organic under the ACO standards, açaí berries must have been harvested within a management system that is '*...verified, on an ongoing basis, as not degenerating to the natural ecosystem*'.[55] "The operator" undertakes this verification, which in this context is the açaí processing company. To not degenerate the natural ecosystems, açaí-harvesting methods must not involve clear felling or stripping of fruits in a way that adversely affects ecosystem functions. Additionally, the area in which the wild harvesting of açaí occurs must be capable of being clearly defined, identifiable on a map and be a "satisfactory distance" from conventional farms.[56] The processing company of the açaí is required to instruct "collectors and any local agents" about the defined area for harvest and the standards required for certification. These collectors must sign agreements evidencing that they have followed the instructions.

In reality, it would be difficult for a processing company to know with certainty that the instructions were followed. Likewise, it is difficult for a third-party certifier to confirm that the standards have been met given the standards largely rely on the evidence from açaí processors with vested interests in being approved. While consumers are not made aware of this,[57] there is limited objective or scientific evidence that the açaí product meets wild-harvesting organic standards. In the context of açaí then, organic certification is not confirmation that the product is sustainably harvested or confirmation that it has not been exposed to agricultural chemicals. Instead, organic certification in this context means that the processor has assured a third-party that it has done everything in their control to ensure the harvesting of açaí meets the organic standards; the processor may well have satisfied the requirements for certification, but they cannot provide assurance that the product meets the standards because they do not have control over those supplying the fruit. This is a somewhat difference message then the assurance consumers receive from an organic certification label, which is that the product is free from contamination and environmentally friendly.[58]

---

[54]Australian Certified Organic (2015).

[55]Australian Certified Organic Standard (2013), p. 73.

[56]Australian Certified Organic Standard (2013), p. 73.

[57]It seems unlikely that consumers will check the standards that applied in order for the product to be certified as Australians are increasingly trusting organic certification marks. A 2014 report revealed that 70% of all shoppers held that the certification logo increased their level of trust. See, Australian Organic Ltd (2014), p. 45.

[58]For a review of how Australian consumers view "organic" labelling, see Chang et al. (2003), p. 9. Generally, Australian consumers view food labelled as "organic" as being "green" in the sense of being environmentally friendly and "clean" in the sense of not being contaminated.

4 Conclusion

Demand for açaí is an opportunity for sustainable rural development in Brazil and the promotion of food diversity, in the sense of biological and dietary diversity worldwide. However, inaccurate or misleading information on foodstuffs undermines the contribution açaí consumption could make to food diversity and stronger regulation on food labelling and advertising is required to prevent such practices. However, the case study of açaí shows the limitations on consumers where they seek to promote food diversity. These limitations reveal that the consumer autonomy and personal responsibility refrains should not distract from addressing broader structural causes of declines in biological and dietary diversity.

References

Agriculture and Agrifood Canada (2012) "Socially Conscious Consumer Trends Fair Trade," Market Analysis Report. http://www5.agr.gc.ca/resources/prod/Internet-Internet/MISB-DGSIM/ATS-SEA/PDF/6153-eng.pdf. Accessed 21 Dec 2015

Australian Certified Organic (2015) About. http://aco.net.au/about/. Accessed 8 Dec 2015

Australian Certified Organic Standard (2013) Wild harvest. http://austorganic.com/wp-content/uploads/2013/11/ACOS-2013-final.pdf. Accessed 5 Dec 2015

Australian Competition & Consumer Commission (2015) False or misleading claims. https://www.accc.gov.au/consumers/misleading-claims-advertising/false-or-misleading-claims. Accessed 8 Dec 2015

Australian Organic Ltd (2014) "Australian Organic Market Report 2014". http://austorganic.com/wp-content/uploads/2015/05/AO_Report_2014_web.pdf. Accessed 21 Dec 2015

Berti PR, Jones AD (2013) Biodiversity's contribution to dietary diversity. In: Fanzo C, Hunter T, Boerelli F, Mattei F (eds) Diversifying food and diets using agricultural biodiversity to improve nutrition and health. Routledge, Abingdon, pp 186–206

Boyce J (1999) The globalization of market failure? International Trade and Sustainable Agriculture. Political Economy Research Institute. http://www.peri.umass.edu/236/hash/43dbd9bc4f82b433b3881dc845aa0318/publication/23/. Accessed 3 Dec 2015

Brondízio E (2004) From staple to fashion food: shifting cycles and shifting opportunities in the development of the acai palm fruit economy in the Amazon estuary. In: Zarin DJ, Alavalapati JRR, Putz FE, Schmink M (eds) Working forests in the neotropics. Columbia University Press, New York, pp 339–365

Brondízio E, Siqueira A (1997) From extractivists to forest farmers: changing concepts of agricultural intensification and peasantry in the Amazon estuary. Res Econ Anthropol 18:233–258

Bussman R, Zambrana NYP (2012) Facing global markets- usage changes in Western Amazonian plants: the example of Euterpe precatoria Mart. and E. oleracea Mart. Acta Soc Bot Pol 81 (4):257–261

Chang H-S et al (2003) University of New England "An Overview of the Organic Food Products Market in Australia" Working Paper Series, Agricultural Resource Economics. http://ageconsearch.umn.edu/bitstream/12928/1/wp030010.pdf. Accessed 21 Dec 2015

Clark EL (2014) Amazonia: the historical ecology of a domesticated landscape. In: Hecht S, Morrison KD, Padoch C (eds) The social lives of forests: past, present, and future of woodland resurgence. University of Chicago Press, Chicago, pp 157–183

Codex Alimentarius (1991) "General Guidelines on Claims, CAC/GL 1-1979". http://www.codexalimentarius.org/standards/list-of standards/en/?provide=standards&orderField=fullReference&sort=asc&num1=CAC/GL. Accessed 4 Dec 2015
Colapinto J (2011) Strange fruit: the rise and fall of acai. The New Yorker. http://www.newyorker.com/magazine/2011/05/30/strange-fruit-john-colapinto. Accessed 21 Dec 2015
De Schutter O (2011) The right to an adequate diet: the agriculture-food-health-nexus (A/HRC/19/59) presented at the 19th session of the United Nations Human Rights Council
Deaton BJ (2004) A theoretical framework for examining the role of third-party certifiers. Food Control 15:615–619
Fajans J (2012) Brazilian food: race, class and identity in regional cuisines. Berg, London
Filho MCF, Sousa JW (2013) Structuring of the acai pulp exporting agribusiness' organizational field. Rev Bus Manag 15(47):169–185
FLO-CERT (2015a) How it works. http://www.flocert.net/fairtrade-services/fairtrade-certification/how-it-works/. Accessed 1 Dec 2015
FLO-CERT (2015b) Compliance criteria list: trade certification, NSF Checklist TC 8.0 EN-GB. http://www.flocert.net/wp-content/uploads/2014/02/TC-PublicComplianceCriteria-ED-8.0-en.pdf. Accessed 5 Dec 2015
FLO-CERT (2015c) Public compliance criteria list- small producers' organizations, Checklist HL 7.2 EN-GB. http://www.flocert.net/wp-content/uploads/2014/02/PC-PublicComplianceCriteriaHL-ED-7.2-en.pdf. Accessed 5 Dec 2015
Food Standards Australia and New Zealand (2013) Food labelling. http://www.foodstandards.gov.au/industry/labelling/Pages/default.aspx. Accessed 8 Dec 2015
Fraser JA (2010) Caboclo horticulture and Amazonian dark earths along the middle Madeira river, Brazil. Hum Ecol 38(5):651–662
Freitas M et al (2015) Floristic impoverishment of Amazonian floodplain forests managed for acai fruit production. Forest Ecol Manag 351:20–27
Gale FP (2000) Economic specialization versus ecological diversification: the trade policy implications of taking the ecosystem approach seriously. Ecol Econ 34(3):285–292
Gonzalez C (2011) An environmental justice critique of comparative advantage: indigenous peoples, trade policy, and the Mexican neoliberal economic reforms. J Int Law 32(3):723–803
Guthman J (2007) The polyanyian way? Voluntary food labels as neoliberal governance. Antipode 39:457–478
Harris M (2005) Nature makes them lazy: contested perceptions of place and knowledge in the lower Amazon floodplain of Brazil. Conservat Soc 3(2):461–478
Hiraoka M (1994) Land use changes in the Amazon estuary. Global Environ Change 5(4):323–336
Hobson K (2004) Sustainable consumption in the United Kingdom: the "Responsible" consumer and government at "Arm's Length". J Environ Dev 13(2):121–139
Johnson H (2015) Eating for health and the environment. Aust Regul Resp Diet Change 15 (2):122–139
Johnstone N (1995) Trade liberalization, economic specialization and the environment. Ecol Econ 14(3):165–173
Lewis J (2008) The power of knowledge: information transfer and acai intensification in the Peri-Urban interface of Belem, Brazil. Agrofor Syst 74(3):293–302
Mhurchu CN, Gorton D (2007) Nutrition labels and claims in New Zealand and Australia: a review of use and understanding. Aust New Zeal J Publ Health 21(2):105–112
Moegenburg S, Levey D (2002) Prospects for conserving biodiversity in Amazonian extractive reserves. Ecol Lett 5(3):310–324
Norgaard RB (1987) Economics as mechanics and the demise of biological diversity. Ecol Model 38(2):107–121
Pace R (1997) The Amazon Caboclo: what's in a name? Lu-Braz Rev 34(2):81–89
Pace R (1998) The struggle for Amazon town: Gurupá revisited. Lynne Rienner Publishers, London
Parker C (2013) Voting with your fork? Industrial free-range eggs and the regulatory construction of consumer choice. ANALS, AAPS 649(1):52–73

Parker C (2015) Strawberry fields forever: can consumers see pesticides and sustainability as an issue? Sustain Sci 10:285–303
Parker C, Scrinis G (2014) Out of the cage and into the barn: supermarket power food system governance and the regulation of free range eggs. Griff Law Rev 23(2):318–347
Parker C et al (2013) The happy hen on your supermarket shelf: what choice does industrial strength free-range represent for consumers? J Bioeth Inq 10(2):165–186
Raffles H (2002) In Amazonia: a natural history. Princeton University Press, Princeton
Raynolds LT (2000) Re-embedding global agriculture: the international organic and fair trade movements. Agr Hum Val 17(3):297–309
Smith N (2015) Palms and people in the Amazon. Springer International Publishing, Switzerland
Standards Australia (2015) Our organisation. http://www.standards.org.au/OurOrganisation/AboutUs/Pages/History.aspx. Accessed 4 Dec 2015
Teisl M (2003) What we may have is a failure to communicate: labelling and environmentally certified forest products. Forest Sci 49(5):668–680
Tellman B et al (2011) Not fair enough: historic and institutional barriers to fair trade coffee in El Salvador. J Lat Am Geogr 10(2):107–127
Thampaplillai D et al (2015) Australian commercial law. Cambridge University Press, Australia
Thrupp LA (2000) Linking agricultural biodiversity and food security: the valuable role of sustainable agriculture. Int Aff 76(2):265–281
Vira B et al (2015) Introduction: forests, trees and landscapes for food security and nutrition. In: Vira B, Wildburger C, Mansourian S (eds) Forests and food: addressing hunger and nutrition across sustainable landscapes. Cambridge, UK, pp 9–28
Waltner-Toews D, Lang T (2000) A new conceptual base for food and agricultural policy: the emerging model of links between agriculture, food, health, environment and society. Global Change Hum Health 1(2):116–130
Weinstein S, Moegenburge S (2004) Acai palm management in the Amazon estuary course for conservation or passage to plantations? Conservat Soc 2(2):315–346

Critical Consumption and Ethical Purchasing Groups (GAS)

Roberto Franco Greco and Virginia Tascagni

1 Critical Consumption: Legal Framework of the Phenomenon

Traditionally, well-being and consumption have been intertwined in an interdependent relationship both in economic and symbolic terms. The economic growth following the Second World War, which accentuated the dissociation between production and consumption, in a certain sense has delineated or at least contributed to the increased social importance of the role of the consumer. With particular reference to agricultural products, this follows the idea of abundance as a condition for self-realisation, leading to that which is defined as the "*solitude of the global citizen*".[1] Once a certain level of prosperity has been reached, however, this tendency toward alienation through consumption leads consumers back to the necessity of re-identifying themselves in their activities and thus in their choices.[2] The concept of critical consumption therefore develops in this context, reversing organisational models and consumption choices: the variables that influence the consumers' selection may not in fact be merely utilitarian, but may refer to more noble values. The perspective of the citizen changes, moving out of a typically individualistic dimension, characterised by the tendency to selfishly satisfy one's own needs, or at most those of the family, from the perspective of economic and social growth, to support a vision that promotes more solidarity. Precisely in this

This chapter represents the results of a joint research effort; with reference to each author's contribution, Sects. 1 and 2 have been drawn up by Virginia Tascagni and Sects. 3, 4, and 5 by Roberto Franco Greco. The conclusions contained in Sect. 6 were jointly written by both authors.

[1]Bauman (2000).

[2]Minestroni (2006).

R. F. Greco (✉) · V. Tascagni
University of Salento, Lecce, Italy

A. Isoni et al. (eds.), *Food Diversity Between Rights, Duties and Autonomies*,
LITES – Legal Issues in Transdisciplinary Environmental Studies 2,
https://doi.org/10.1007/978-3-319-75196-2_20

sense, consumption becomes informed: it is no longer based solely on the assessment of the relationship between quality and price of a product, but it also considers the methods of production and distribution. It is no coincidence that this tendency is particularly accentuated in the agricultural sector, that is, in reference to those products that are indispensable and most directly reflect the level of quality of life, of oneself, as well as others. In fact, new criteria are becoming relevant regarding what inspires consumers' choices. In addition to the general necessity to limit consumption to what is really useful, they pay particular attention to the production techniques with low environmental impacts and energy consumption, which leads to a preference for organic production, and to the ability to immediately know the background of the products, for example, by means of direct sales. It has considerably facilitated the exercise of the consumer's right to transparency, especially about the quality and geographical provenance of agricultural products.[3] That is the consequence of the diffusion of a new culture, primarily fueled by a broader and sincere interpretation of solidarity, which no longer results in an ethical-moral value but rather poses a mandatory legal duty, as expressly defined in Article 2 of the Italian Constitution.[4] The consumer no longer looks to the immediate satisfaction of his individual needs, but rather becomes aware that the dynamics of consumption may also depend upon respect for human rights, including the right to food which is recognised as a fundamental right of everyone, or even the "*most fundamental*" of all.[5] The right to food, responding to the primary needs of the individual, is essential to the exercise of the right to health, expressly protected both at the individual and collective level in Article 32 of the Constitution.[6] "*The full development of the human person*" under Article 3 of the Italian Constitution, para. 2, cannot fail to consider the right to adequate quantitative, as well as qualitative nourishment.[7] In fact, "*food insecurity*" is the immediate consequence of insufficient access to food or its unequal distribution.[8] The issue can be resolved, however, by ensuring the development of production activities at the local level and securing *food safety*. For this, the right to health "*must also be considered essential and immanent within the entirety of the consumption problem*".[9] Not surprisingly, Article 2 of the Consumer Code (d.lgs. 206/2005) reserves the protection of health as first among the fundamental rights of the consumer and then positions the others in relation to how instrumental they are to the protection of health. Particularly, the right to information operates as an instrument to rebalance the relationship between producers and

[3]Scagliarini (2015).

[4]See *infra*, para. 4.

[5]Resolution of Parliamentary Assembly of the Council of Europe no. 1957/2013: "*Food is our most basic need and right. If we cannot secure access to sufficient, safe and nutritious food for present and future generations, our health, development and fundamental rights will be compromised*".

[6]Ferrara (2010).

[7]See *infra*, para. 4.

[8]Biggeri and Volpi (2006).

[9]Vignudelli (1984).

consumers, obliging adequate information about the goods produced to be provided. At the same time, this is also a mode of consumer protection in a different form than exclusively contractual, aiming to put them in the position to also know the scientific and technological advances in the food sector that could more or less influence their choice. That is why the Consumer Code expressly provides that the following information must also be indicated on the packaging of the product: "*the possible presence of materials or substances that may cause harm to persons, property or the environment*", as well as "*instructions, possible precautions and the intended use of the product, where beneficial to the use and safety of the product*" (Art. 6). In this way, the consumer can make conscious choices oriented towards products that on one hand are able to satisfy their primary needs, and on the other have a minor negative impact on their health and the environment.

2 The Evolution of Consumption and Production Towards Models of Environmental Sustainability Through the Protection of Land and the Protection of Biodiversity

Conscious consumption, however, does not automatically translate into critical consumption. If the consumer orients his choices to prefer one or more producers, he becomes aware, but is still unable to significantly influence the market: the consumer choice, while inspired by a "logical ethic", remains individual. Consequently, the consumer would remain passive in the production schemes. But not only that; the voluntary nature of adhesion to certain values could mislead one to consider the ethical behaviour of consumers only as a virtuous lifestyle, but legally irrelevant, especially considering the lack of legislation specifically dedicated to it. Because consumption is quite critical, it must present certain characteristics to offer a truly different model, a valid alternative to the traditional scheme. Considering the above, therefore, the right of free association should also be read in this sense as voluntarily and democratically guaranteed by the Consumer Code, not limiting consumer protection to the contractual sphere. The producer can effectively be subjected to the strength of the "*consum-actors*" in the choice of organisational models for their business operation, of raw materials utilised and production methods only if the mass stops being the sum of diverse individual choices and starts to emerge as a party that adheres to a different system of values, conscious of the shared objectives they intend to achieve.[10] Conscious consumption thus becomes critical when "*social formations*" (e.g., Const. Art. 2 and 118) are finally valued, in which the citizen-consumer fits.[11] It is in this transition that one may grasp the significance of the difference between the consumer's right to information and the right to education,[12]

[10]Minestroni (2006).

[11]See *infra*, para. 4.

[12]Marzanati and Mattioni (2007).

essential to the empowerment of a consumer revolt for care of the public interest.[13] In fact, Article 4 of the Italian Consumer Code attributes a public function to educational activity, namely "*contributing to an awareness of their rights and interests, the development of associations, participation in administrative procedures, and representation in exponential organizations*". The full "*autonomy of the consumer*"[14] could therefore really be pursued, translating into freedom of choice that, objectively, realises the peoples' right to self-determination regarding issues related to food and nutrition.[15]

As we can see, the "weak" position of the consumer can therefore in certain respects represents an important starting point to build and strengthen a system of protection that drives further, towards the need to ensure the principle of equality, in the form of equal treatment, between the two traditionally antithetical categories of producers and consumers. The field of action widens significantly: the right to health of the consumer is closely connected to protection of the environment; therefore, the consumer's right to information and education relates to the freedom of association, so that production choices can depend on the consumers' choices. The recognition of consumers as weak subjects with *inalienable* rights (Art. 143, Consumer Code) necessarily corresponds to a set of duties and their fulfillment is achieved through exercise of those rights.

If the Consumer Code definitely provides the framework that enables the definition and understanding of consumer awareness, one may catch the essence of critical consumption only by keeping in mind the changes in production models in the agricultural sector pushed at the European level. Particularly, if the fundamental objective of the legislators of the consumer code was a reordering of the rules and regulations to protect the weaker party,[16] the rural development policies of the last 20 years have actually had a far greater impact on the consumer. The reference is to the promotion of the "culture of localism". This is manifested in the agricultural sector through the European Community's public policies favouring the development of small local businesses, which includes a group of consumers located in the same area.[17] These policies consider agricultural activity as deserving of regulation,

[13]Pizzolato (2010).

[14]East (2003).

[15]Colombo and Onorati (2009).

[16]Tanza (2009).

[17]Besides the importance given in the second pillar of the Common Agricultural Policy to the enhancement of the resources and therefore rural areas, consider the aforementioned Resolution n. 1957/2013 of the Council of Europe, inasmuch as it attaches great importance to responsible consumption, food security and access to food. In particular, "*[. . .] as concerns affordability of food: strengthen solidarity mechanisms to combat poverty, which obstructs access to food by the population concerned; increase development aid devoted to agriculture and improved food preservation, and honour aid commitments entered into; [. . .] foster food security in fragile countries, particularly in sub-Saharan Africa and southern Asia, by improving small farms' resilience and rural livelihoods, assisting good regional governance in agriculture and food policies, and enhancing aid to vulnerable populations; support the minimum human rights principles applicable to large-scale land acquisitions or leases identified by the United Nations Human Rights Council*

not only as a productive activity, but also by virtue of its close link between development and protection of the territory.[18] Acknowledging the existence of such a link between the modes of production in the agricultural sector and the protection of the public interests that exist within a territory, it is easily understandable how consumer choice in this particular sector also presents the same capacity to project itself beyond the individual dimension. From the above it is evident, considering that critical consumption assumes the enhancement of food diversity, as an index of territorial identity[19] and declination of biodiversity, its conservation is amongst the fundamental objectives of international, European and national environmental law.[20] Food diversity is therefore an objective of the protection of access to food, so it constitutes the object of a fully exercisable right, necessitating the existence of multiple species and plant varieties.[21] From this perspective, we rank critical consumption as the most suitable instrument to protection agricultural and food biodiversity, also through the possibility to exploit "*market mechanisms*".[22] When speaking of agricultural products, for consumer choice, not only would the raw materials be identified, such as the sources of those goods, but also the different climatic conditions, as well as the territorial characteristics that attribute typicality to the products. For this reason, increasing the diversity means a parallel increase in the consumer's possibility to choose, rather than being theoretically free but strongly limited by concrete needs that make one indifferent when faced with absolutely interchangeable products.[23] The inseparable relationship between resources of the land and the local history, materially translated into the agricultural products, encourages the consumer to "demand" cultural diversity and therefore long-term food so he can truly be sovereign, in terms of a sense of belonging to the territory.[24] The rejection of standardisation of eating habits is more understandable if one considers the cultural aspect as a fundamental component of food diversity. This is why, is it also easier to grasp the legal dimension of critical consumption, in terms of the protection of cultural heritage of the area. Food diversity, assuming the conservation of different plant varieties, is a necessary condition because there is criticality in consumption. To exercise effective pressure on the market, consumers must be able to conduct an assessment between truly diverse products, which enables them to

---

[...]; promote food security by establishing a new universal framework for post-2015, integrating the Millennium Development Goals and the Sustainable Development Goals [...]".

[18]Benozzo and Bruno (2009).

[19]See the contribution by Monteduro in this volume.

[20]Benozzo and Bruno (2009).

[21]FAO (2004). The preference for monocultural agriculture, uncontrollably imposed at least since the beginning of the nineties, has significantly compromised the biodiversity, determining a rise in the homogenisation of food and consequently of eating habits. The UN Food and Agriculture Organisation (FAO) estimates that around three-quarters of the plant varieties have been lost because of globalisation.

[22]FAO (2015).

[23]East (2003).

[24]Arcidiacono (2013).

exercise a right of choice, first of all individual and only collective in the presence of a food plurality that increases their motivation for aggregation. Not surprisingly, Law 194/2015 laying down rules for the protection and enhancement of agricultural and food biodiversity provides the possibility of creating "*food communities*",[25] agreements that aim, *inter alia*, toward strengthening a background of traditional knowledge directly linked to the crops and the food.[26] Therefore, from a legal point of view, critical consumption rightly fits in the "*national system of protection and enhancement of agricultural and food biodiversity, aimed at protecting genetic resources for local food and agriculture from the risk of extinction and genetic erosion*" referred to in Art. 1 of L. 194/2015. This takes on a greater concrete relevance especially in reference to the promotion of divulging "*places where products are marketed that are connected to the same resources, including direct sale points*" and the express mention of ethical purchasing groups as necessary protagonists for the enhancement of food biodiversity.[27] Because of a recent cultural tendency to reclaim past values, albeit in the presence of different needs and resources, critical consumption means that the consumer's choice and also that of the producer are not the only ones possible.

3 Legal Recognition of Ethical Purchasing Groups (GAS) in Law n. 244/2007. Critical Profiles and Legal Perspectives of Protection

The phenomenon of critical consumption finds its roots in the experience of ethical purchasing groups (in Italian "*Gruppi di Acquisto Solidale*", hereinafter referred to as GAS), a specific legal declination. Already present for some time as free associations of consumers, GAS have been legislatively recognised and regulated by Law n. 244/2007 (Finance Law of 2008), Art. 1, paragraphs 266, 267 and 268.

Paragraph 266 defines them as "*non-profit associations set up to carry out collective purchasing and distribution of goods with ethical purpose, social solidarity and environmental sustainability, in direct execution of the institutional purposes and excluding administrative and sales activities*".[28]

---

[25]The same law defines them, in Art. 13, c. 2, as "*local areas resulting from agreements between local farmers, farmers and livestock breeders, purchasing groups, schools and universities, research centers, associations for the protection of agricultural and food biodiversity, school cafeterias, hospitals, caterers, commercial businesses, small and medium-sized artisan businesses for agricultural and food processing, as well as public bodies*".

[26]See the contribution of Monteduro in this volume.

[27]Compare Art. 12 and 13, Law 194/2015. See *infra*, para. 5.

[28]GAS is generally composed of families who come together to buy food and goods of common use, directly from the producer, to then distribute them. Direct contact with the producer is the central feature of such purchasing practices. They thereby cut out the intermediaries from the distribution to create a short supply chain. This provides clear opportunities for savings in the

Paragraph 267 deals with the taxation of GAS. In this sense, the activities performed by GAS are "*(. . .) limited to those facing the members, not considered commercial for the purposes of the tax scheme provided for in the Presidential Decree of 26 October 1972, n. 633, subject to the provisions of Art. 4, sixth paragraph, and for the purposes of the tax regime of the single text of the Republic Presidential Decree of 22 December 1986, n. 917*". Paragraph 268 deals with the financial coverage of the burden derived from application of the preceding paragraph.

The combined reading of these provisions allows one to understand the legal form and purpose of GAS, as well as the object of their activities and the tax regime that applies. From the introductory sentence of paragraph 266 it is noted that, as associations, they may assume the form of recognised associations or foundations and non-recognised associations or committees. Thus, the operation of GAS is regulated, according to the form chosen in each specific case, by provisions that extend from Art. 14 to Art. 42 of the Italian Civil Code.[29] Therefore, Law n. 244/2007 shall not be applicable to those consumers who operate without conforming their operation to one of the forms of association prescribed by the legislation.[30]

The absence of profit comports with the characterisation of GAS as non-commercial and allows one to frame the phenomenon in the largest legal category of *non-profit* (i.e., third sector). Consequently, they are exempt from value-added tax (VAT) and are subjected to legislation on income tax under Articles 143–150 of the Republic Presidential Decree 917/1986 (Italian General Tax Law, abbreviated T.U.I.R. in Italian).[31] The VAT exemption was enacted by the legislature "*without prejudice to the provisions of Art. 4, paragraph 7*" of Republic Presidential Decree n. 633/1972. This means that it depends on respect of the latter.

The civil law provisions under Articles 14–42 cited above do not define non-commercial entity, but it is clear that it concerns private legal persons under the same regulations. This definition is rather contained in the tax legislation, that

purchase of products. This has definitely represented one of the most decisive factors in the growth and rapid spread of the phenomenon, but the central element is provided for in the solidaristic approach to consumption. Even before it was legally recognized according to the formula of *ethical purpose, social solidarity and environmental sustainability*, the BASIC DOCUMENT OF GAS (1999) identified it as a fundamental paradigm of their activities. In this, it states that GAS are "one of the many situations in which you are making an alternative conception of the economy, now present in various parts of Italy in more or less structured forms. *We call them "Ethical" Purchasing Groups to distinguish them from the simple purchasing groups, which do not need to present ethical connotations, but just be an instrument of savings. The ethical aspect, or solidarity, of such groups seems the most important side, which characterises them as having experience in the field of critical consumption (. . .)*". Pirani et al. (2009).

[29]Title II. of legal persons.

[30]Visconti (2008).

[31]Particularly, Art. 148, paragraph 1, with reference to associative entities, provides that "*activity carried out for members or participants is not considered commercial, in accordance with the institutional purposes of associations, consortia and other non-commercial entities of an associative type (. . .)*".

under Art. 73, paragraph 1 subsection c) of the same Decree n. 917/1986 defines non-commercial entities as "*public and private entities other than businesses, residing in the State, which do not have the exercise of commercial activities as their sole or principal object*".[32]

The provisions in question also allow for distinguishing GAS from general purchasing groups (*gruppi di acquisto in generale* in Italian, hereinafter GDA), consumer formations spontaneously created that have been present for a long time in commercial practice. Devoid of real and proper official recognition by the legislature, they are voluntary associations of commercial operators, generally retailers, that come mainly to increase their bargaining power and competitiveness in the market. Despite joining, they retain their legal autonomy and delegate the supply operation to a single operator. The quantity of goods handled depends on the share of the supply economy that the operators access. Participation in a GDA allows members to compete with larger enterprises and to gain a competitive advantage over those of the same size.[33] Therefore, the distinction is represented by the "solidarity" attribute of GAS, in which the competitive advantages in terms of price are of secondary value compared to those from the ethical aims, solidarity and sustainability of the collective purchasing activities.

In addition, the exclusive reference to the purchase of goods but not services in paragraph 266 allows one to distinguish GAS from service purchasing groups, a growing phenomenon particularly in the energy sector. Like the GDA, this category was also created to increase the bargaining power of the members. Thus, with the aim of achieving a competitive advantage in terms of purchase price, for example, of electrical energy.[34] However, multiple experiences of services being purchased that include benefits for environmental protection purposes must be reported. One example is purchasing groups for certified green energy (in Italian, *Gruppi di*

[32]The point should be made clear that the tax category of non-commercial entity does not coincide perfectly with that of civil law subjects that pursue aims "*non-economically valuable*". The reference is precisely to associations, foundations and committees. The tax profile is relevant to the object of the activity, which cannot be exclusively or principally commercial (Art. 73, T.U.I.R.). Instead, under civil law, the non-commercial status of the entity depends on the purpose pursued. Therefore it is possible that a non-profit entity would come under the application of the tax provisions for commercial entities where, although the aims pursued are not economically valuable, the object of the activity concerned is principally or exclusively commercial. Ultimately, the absence of profit is necessary but is not, in and of itself, sufficient to render the entity non-commercial from a tax point of view. According to the same Art. 73, paragraph 4 "*the exclusive or principal object of the resident entity is determined according to the law, the articles of incorporation or charter, if they exist as public articles or a private agreement authenticated or registered. The principal object means the essential activity to directly accomplish the primary aims laid down by law, the articles of incorporation or the charter*". Colombo and Poletto (2012).

[33]Lorenzoni (1968) and Gennari (2009).

[34]Created following the liberalisation of the energy market, the energy service purchasing groups spread rapidly in Italy and Europe. Liberalisation has encouraged the development of economies of scale that allow access to much better price conditions the greater the demand for energy.

acquisto per l'energia verde certificata [R.E.C.S.]),[35] in which the energy supply comes *in total*, or nearly so, from renewable sources.[36] Although in these cases it seems that a compatibility of purpose may be configured between service purchasing groups and GAS, the possibility of equal legal compatibility between the forms is excluded because the object of the collective purchase differs.

From inspection of paragraphs 266, 267 and 268 of Law n. 244/2007, one may infer that the legislature, in recognising GAS for the first time, has dedicated more attention to ethical, solidarity and environmental tax considerations. Yet the aspect that most characterises these experiences, and which is more interesting to explore here, may be found in the purposes toward which collective purchasing activities must be oriented. As set forth, the same purposes are likely to be delineated as mere legal suggestions; the effective pursuit of which is required according to the discernment of the groups' members. In fact, the same system of tax incentives referred to in paragraph 267 does not depend on an ethical, solidaristic or sustainable approach to operation, but rather to the absence of profit and, therefore, the non-commercial dimension of the associations. This probably risks a mixture between GAS and purchasing groups that, while not effectively being in solidarity, simply do not present any profits. The "tax consultant approach" of the 2007 legislature is perhaps the legacy of the fragmentation of Italian legislation on *non-profits*. Still without a definitive statute for non-commercial entities, it draws in large part from the regulation of tax law. Not by chance, as has been seen, the definition of non-commercial entity is in this matrix. The constitutionally oriented interpretation of the phenomenon (which will be discussed in the next section), and in light of some recent environmental regulations, seems to move from these critical questions and contribute to their solution, defining in the GAS, in the future, a clear *best practice* for legal protection of the environment.

[35]Renewable Energy Certificate System.

[36]The certification is done through the purchase of Guarantee of Origin (G.O.) certificates, introduced at the European level by Directive 2011/77 CE. In Italy, this was implemented by D.M. of 6 July 2012 with the aim "to support the production of electrical energy from renewable sources by establishing incentives and modes of simple and stable access, that promotes effectiveness, efficiency and sustainability of the incentives in adequate relation to the pursuit of its objectives *(...)*" (Art. 1). As well as by natural persons, the service purchasing groups are often triggered by small and medium-sized businesses as part of the *Green Marketing* strategies promoted by different operators. *Green Marketing* includes the set of business activities directed to provide services, products, technical solutions and tariffs compatible with the environment. On this theme, Corradi (2011).

4 Ethical Purchasing Groups in Light of the Principles of the Italian Constitution. Solidaristic Approach to Operation Within the Inviolable Rights-Mandatory Duties Dialectic

When reflecting on the relationship between the fundamental principles of the Italian legal system and GAS, there are multiple assessments that allow for a constitutionally oriented interpretation of the phenomenon. The first and most direct reference is Art. 2 of the Constitution, which defines the principle of solidarity as a fundamental value of the Republic and as a basis within its social life. The attribute of solidarity in purchasing groups directly represents a declination. This is expressly confirmed in paragraph 266 of Law n. 244/2007, which aims the operation of GAS towards the pursuit of *social solidarity*.[37]

Particularly, Art. 2 establishes the inextricable link between *inviolable rights*,[38] recognised and guaranteed to humans[39] "*both as an individual and in social*

[37]Social solidarity is only one aspect of the solidarity attribute. It enables you to grasp, in the most intuitive way because it literally draws, its direct link with the constitutional principle of solidarity. As previously revealed, the solidaristic characteristic of GAS also consists of ethical and environmental sustainability aims.

[38]The Constitutional Court has on many occasions held that the inviolable rights constitute the irretractable heritage and, together with the fundamental principles, represent those values that underpin the entire system. In the past, *ex plurimis*, it has been observed in: judgment n. 30 from 2015, which states that they "*constitute the identification and inalienable elements of the constitutional order*"; decision n. 238 of 2014 that sets the "*boundaries of respect for the identification elements of the constitutional order, namely the fundamental principles and the inviolable rights of man*". The central role the Italian Constitution gives to the inviolable rights involves the impossibility to completely sacrifice them should they be subjected to balancing with other constitutionally protected interests. In these cases, it is necessary to ensure a minimum level of protection guaranteed to prevent their impairment. On this issue, the Constitutional Court, with ruling n. 252 of 2001, made clear the need to "*guarantee an irreducible core of laws (...) as inviolable to human dignity, which requires prevention of establishment of situations without protection*". Along the same lines, also see, *ex plurimis*, judgments n. 509 of 2000, n. 309 of 1999 and n. 267 of 1998.

[39]Note that the Italian Constitution recognises the inviolable rights of man and not to the citizen. This led the Constitutional Court to conclude that, with respect to the principle of equality (Art. 3), they must also be guaranteed to foreigners (already with ruling n. 120 of 1967). Therefore, it is the duty of the Republic to remove the economic and social obstacles that limit freedom and equality between men (Art. 3), both citizens and foreigners. For more background, see judgments n. 22 and 230 from 2015 of the Constitutional Court. On reflection, in reference to GAS, it is possible to identify the person as the recipient of the solidarity activity they carry out. This indication is not obvious if one thinks, for example, about the current handling of the migratory crisis in the European political sphere, in which the welfare state often proves inadequate to ensure decent living conditions for foreigners. In this sense, public-private cooperation and the pursuit of ethical, social and environmental objectives, attributes that already refer to a necessarily global dimension, assume, today perhaps more than ever, a particular political and social value.

formations where they express their personalities", and *mandatory duties*[40] "*of political, economic and social solidarity*".[41] That link shows that the Italian Constitution gives the individual the responsibility to promote the social dimension through the legal perspective of the duty of solidarity. The highly prescriptive value of Art. 2 implies the obligation, and not the choice, of every *citizen* to abide by social, political and economic solidarity. The GAS aims of ethics, social solidarity and environmental sustainability make these a means, available to the public, for the fulfillment of these mandatory duties. Moreover, moving from the three-dimensional value of constitutional solidarity (economic, political and social), GAS present a clear mode of participation to bring about substantial equality between citizens, as they allow for the "*full development of the human person and the effective participation of all workers in the country's political, economic and social organisation*" (Art. 3, para. 2).

In the middle of the analysis of GAS from the perspective of constitutional values, to better understand the legal significance of the inviolable rights/mandatory duties dialectic, it seems appropriate to point out, with only a slight digression, that it does not focus exclusively on private activities, but, with the same effect, is also embedded in administrative actions of public entities. On this point, Law 241/1990 binds administrative activity to the pursuit of public interests and provides that the public-interest qualification constitutes the indispensable precondition of administrative action. This is clear from the first article, which under paragraph 1 provides that "*administrative activity pursues the purposes determined by law (...)*". These correspond to the public interest in necessary protections which possession, and administration, of mandatory legal duties requires. Therefore, the administrative measures are configured as an "*act of exercising an instrumental power toward fulfillment of the duty*" and the epilogue to the sequence *public interest- duty- power-proceedings*.[42]

Returning to GAS, the reference to workers in Art. 3 of the Constitution is symptomatic of the direct relationship between the solidarity principle and the right to work outlined in Art. 4. Not surprisingly, some of the literature on *non-profits* has identified in the performance of individuals within organisations a way to make this right effective through fulfillment of the duty "to perform, according to their possibilities and choice, an activity or function that contributes to the material

[40]The theme of the role played by the system of constitutional duties located in Lombardi (1967) and Carbone (1968) is important, albeit dated, doctrine. Now it is subject of revived attention. *Ex plurimis*, see more recently Balduzzi et al. (2006); Violante (2014); Grandi (2014); and, in particular, Rimoli (2015).

[41]The relationship between inviolable rights and the solidarity principle, according to Rodotà (2006): "*(...) Article 2 of the Italian Constitution talks about a dialogue between subsidiaries, a social individual in which the alternative and the separation between rights and duties are overcome by (and embedded in) the link between inviolable rights and the solidarity principle: the Republic "recognises and guarantees rights" and, together, "it demands the fulfilment of mandatory duties of solidarity (...)*".

[42]Monteduro (2014).

or spiritual progress of the society" (Art. 4, para. 2). According to this interpretation, GAS (as a species of non-profits) would also instrumentally link to the implementation of the first article of the Italian Constitution, which provides that "*Italy is a democratic Republic, founded on work*".[43]

In addition, consider that GAS, as associative entities, is an expression of the citizens' right of freedom of association under Art. 18 of the Italian Constitution. The association is characterised by stability of participation and pursuit, by the members, of common objectives. In the present case, it is the pursuit of noted ethical, solidarity, and environmentally sustainable purposes. Particularly, the freedom of association was contemplated with reference to trade unions and political parties. But it is clear that the fundamental principles of the Italian Constitution, which humans see as their *ubi consistam*, require that the exercise of freedom not be regulated in the exclusive scope of these two forms. Rather, Art. 18 foreshadows further constitutional recognition of the social dimension of humans, understood in the sense of collective projection of individual freedom *in social formations where he expresses his personality*.

The interpretation of key constitutional provisions regarding the legal experience of GAS finds another point of reference in Art. 118, para. 4 of the Italian Constitution. It provides that local authorities "*promote citizens' autonomous initiatives, single and collective, to carry out activities of general interest, on the basis of the principle of subsidiarity*". The subsidiarity that stands out in this case is called horizontal, based on sincere cooperation between public bodies, citizens and intermediate bodies. The provision gives central importance to the initiative of private individuals in public action. They are to carry out activities related to administration, whose ownership is still kept in the public hands. The public does not directly provide the performance, but regulates access to them with a view toward cooperation with private entities aims to guarantee the optimal response to the needs of the citizens.[44] With reference to GAS, the aims they pursue are obviously included in the activities of general interest to which the provision in question refers. Certainly amongst other things politically occasioned by the crisis of the *welfare state*, it cannot be denied how recognition of horizontal subsidiarity has fueled the legal prism of solidarity, steering it further toward social empowerment of the individual.

The described framework outlines the existence of a kind of "dutiful solidaristic society", which has the individual as its protagonist, who yesterday inspired the activity of fellow constituents but today finds in GAS a legal form of modern and complete manifestation. The solidaristic approach to activities of collective purchase makes them a model of social training in which humans express their personalities, fulfilling the mandatory duties of solidarity that they possess. The Italian Constitution, in defining the system of principles as the basis for regulating human conduct *uti singuli* and *uti socius*, has in a certain way eliminated individualism and recognised the social value of every single man. In this sense, it seems that the

[43]On the general relation between the right to work and non-profits, see Alpa and Conte (2013).

[44]Troisi (2008) and Savarese (2014).

pursuit of ethical, social and environmental aims by GAS not only and easily configures the fulfillment of mandatory duties of solidarity, but it is also and above all the social projection of the individual as a natural manifestation of his being.

This reconstruction shows, for this author, an indisputable theoretical appeal, but it demonstrates weaknesses when one wonders about its degree of effectiveness.[45] In fact, when analysed from a substantive point of view, one notes the possibility, and not the certainty, of the effects of these products. The analysis showed the existence of mandatory duties of *social, political and economic solidarity* for everyone. For these GAS represents a form of performance. However, there is no person in charge of the trade union in control of the concrete fulfillment of these duties and it is not possible to identify the responsibility of the *citizen* possibly defaulting.

Unlike the case of defaulting public administrations, consider the recognised mechanisms of protection for private citizens. Conversely, the choice whether to collaborate is left to the exercise of individual freedom. Moreover, it seems inappropriate to legally force the members to necessarily collaborate between themselves. It follows that the mandatory nature of the duty does not result in its certain fulfillment. In essence, it is not only the choice of GAS as a form of fulfillment to be possible, but it is the performance in itself.

This chapter does not pretend to solve these (perhaps irresolvable) conflicts, but rather to raise awareness about the importance of a solidaristie approach to the legal relationships and daily life amongst the members. This importance seems that it may show more clearly through analysis of GAS from the point of view of the legal protection of the environment, the reference to mandatory duties of environmental solidarity and the consideration of humans, even before the imputation center of subjective legal situations, as factors of the "environment system" (to be discussed below).

4.1 The Paradigm of Environmental Solidarity as an Evolving Destination for the Concept of Solidarity Incorporated by the Italian Constitution

The recognition of the dimension of environmental law "centered on duties" allows one to identify in environmental solidarity the evolving destination of the Constitution, that, as we have seen, declines it in the three way: political, economic and social. To understand, we must initially focus on the legal concept of the environment and consider the effects of the application of the principle of sustainable development in the approach toward its protection.

[45]Not by coincidence, Rodotà (2014), in recognising the impossibility of a democracy devoid of solidarity, speaks of it as a "*necessary utopia*".

The legal definition of the environment is contained in Art. 5, para. 1, subsection. c) of the Legislative Decree 152/2006 (i.e. Environmental Code). It stipulates that the environment should be understood as "*the system of relations between anthropogenic, natural, chemical-physical, climate, landscape, architectural, cultural, agricultural and economic factors*". Even just focusing on textual data, the provision permits the conclusion that *the environment in the legal sense*: (i) does not only correspond to the *environment in an ecological sense* (natural, chemical-physical, climate factors) but encompasses it, also including *the human environment* (anthropogenic, landscape, architectural, cultural, agricultural and economic factors); (ii) it is a legal category with a complex structure that is not formed by the simple sum of factors that make it up, but rather from their relationships because they form a "*system*". Also, implicitly, the provision in question permits one to conceptualise the environment as a kind of "meta-system" given the interrelations between ecological systems and socio/cultural systems.[46] In it, man is a factor (the anthropogenic) that does not exist in itself, but only in terms of its relations with other factors. This legislative representation allows one to note how the objects of the three variations of constitutional solidarity exist within the legal concept of the environment and are an integral part of that "*system of relations between factors*" of which it is composed. In this perspective, environmental law is the legal branch that studies the complexity of the system-environment and that it sets "*(…) as its primary objective the promotion of levels of quality of human life*" (Art. 2, para. 1, of Legislative Decree 152/2006).[47]

Clarifying the object of legal protection of the environment, the principle of sustainable development allows you to identify in the legal canons of duty and responsibility the aspects that permit a legally conscious approach to it. In this way, it will be possible to go back to the reconstruction of the paradigm of environmental solidarity, identifying the entity from the synthesised concept in the sense indicated above.

The principle of sustainable development is the logical-legal cornerstone in the heart of environmental matters. The most famous statement is that of the *World Commission on Environment and Development* in the document *Our Common Future* of 1987, which defined as sustainable the "*development that guarantees the satisfaction of the needs of present generations without compromising the ability of future generations to satisfy theirs*". After this formulation, the principle has progressively become a priority consideration in the legal community at different levels. It was introduced into Italian law by Legislative Decree n. 4/2008, which dedicated the entire Art. 3-*quarter* of the Environmental Code. Here, this provision will not be the subject of further discussion. Respectively, it is enough to consider the third paragraph, which, for the best possible application of the principle, carries the express reference to the need to put the solidarity in the dynamics of production

[46]See the contribution of Monteduro in this volume.

[47]The provision states the purely athropocentric character of environmental protection. On this theme, see Lugaresi (2015).

and consumption. Rather, it seems useful to indicate the impact that this recognition has had on the classical conception of the environment. Identifying the latter as the subject of an individual right, it has been revolutionised by the "pact between generations" and converted into the legal dimension of duties.

For individual rights, it means the legal situation that allows the person who holds it to fully protect the good or interest to which it relates.[48] The independence of nature from man, the recurrent helplessness of man against it (in the case of natural disasters, for example) and the inability to protect non-human elements of nature through recourse to the category of individual rights evidently make inadequate the protection of environmental interests through its use. The protection of the environment better presents the legal paradigms of duties and responsibilities.[49]

This approach, when read in light of the Italian Constitution, can detect the existence of a functional link between the aim of environmental protection and the means of "*(...) fulfillment of mandatory duties of political, economic and social solidarity*" (Art. 2, Constitution) with respect to present and future generations. As a result, there is the possibility to link the tri-dimensionality of constitutional solidarity to the synthesising concept of environmental solidarity. Moreover, the combined reading of Art. 2 with Art. 3 of the Constitution makes the fulfillment of mandatory duties of environmental solidarity instrumental to the full development of the human person. In fact, the latter cannot be achieved if above all the survival of the species is not guaranteed: the effective aim of environmental law.[50] Ultimately, environmental solidarity is a coherent evolution of the legal potential already inherent in the traditional concept of political/economic/social solidarity in the wording of Art. 2 of the Constitution. The mandatory duties connected with it allow the environment to be described as a "right of duties": the right of environmental solidarity duties incidental to "*(...) the system of relations between the anthropogenic, natural, chemical-physical, climate, landscape, architectural, cultural, agricultural and economic factors*".

---

[48]Torrente and Schlesinger (2011), Chieppa and Giovagnoli (2011), Gazzoni (2013) and Trimarchi (2014).

[49]Fracchia (2015). Even the constitutional jurisprudence has progressively abandoned the approach toward the environment as the object of a subjective legal situation and has progressively elevated it to an interest of constitutional status. More particularly, the courts have stated that it is an intangible asset composed of elements *ex se* subject to discipline and specific protection without compromising its magnitude of unitary good. Morzenti Pellegrini (2008).

[50]Fracchia (2015).

5 Ethical Purchasing Groups and Legal Protection of the Environment, Particularly the "Principle of Sustainable Solidarity"

Today, the experience of GAS seems to move towards a "second childhood". In fact, 2015 was the year it became possible to declare their official entry into the discipline of legal protection of the environment.[51] The merit of this operation is Law n. 194 of 2015, with which Italy has established "*provisions for the protection and enhancement of agricultural and food biodiversity*". Thus, after 8 years of Law 244/2007, which primarily addresses the fiscal discipline of the group, the legislature intervened by moving the center of attention towards their environmental value.

Biodiversity is an environmental quality that allows one to understand the operational mechanisms of the ecosystems. It is now known that its conservation is central to the discussion of environmental protection. To understand the importance, just think of the negative impact on the resilience of the environmental system from its loss.[52] The term biodiversity is often only used to indicate diversity of species. Instead, it is correct to refer to any type of biological diversity. It precisely identifies the complexity of interdependencies between species diversity, genetic diversity and ecosystem diversity.[53]

Therefore, Law n. 194 of 2015 relates to two specific forms of biodiversity. More specifically, it "*(...) lays down the principles for the establishment of a national system of protection and enhancement of agricultural and food biodiversity*". This system's objective is "*(...) protection of genetic resources for local food and*

[51]Law n. 208 (Stability Law 2016) was also considered in 2015, and its Art. 1, para. 391 provides that "*(...) the National Family Card also functions to create one or more family purchasing groups or national ethical purchasing groups*". According to the same paragraph, Family Card means the Card "*targeting families consisting of Italian citizens or foreign citizens in Italian territory, with at least three dependent minor children (...) The card provides access to discounts on the purchase of goods or services or tariff reductions granted by public or private entities wishing to contribute to the initiative*". The fact that it can be dedicated to the creation of national GAS manifests the acquired awareness of the legislature with respect to the groups' operation and the intention to promote the establishment of new ones.

[52]In the ecological field, the resilience of environmental systems is the ability of the system to tolerate a disturbance and counteract the negative impacts caused by it. The resilience allows the system not to collapse because of disturbance events, to adapt to new levels of balance and maintain intact its original environmental features because of those changes. The higher the level of a system's resilience, the lower its vulnerability. Biodiversity, preserving the resilience of the environmental system, has the ability to mitigate the effects of climate change on the human species. See Bignami (2010); Benson and Garmestani (2011); Ruhl (2011); Voigt (2013); Arnold and Gunderson (2013); Allen and Garmestani (2014); Monteduro et al. (2015); Monteduro (2015). For further bibliographic references, see the contribution of Monteduro in this volume.

[53]Under Art. 2 of the Convention on Biological Diversity (1992) "*the term "biological diversity" means the variability amongst living organisms from all sources, including* inter alia *terrestrial, marine and other aquatic ecosystems, and the ecological complexity of which they are part; including the diversity within species, and between species of the ecosystems*".

agricultural interests from the risk of extinction and genetic erosion" (Art. 1, para. 1). This purpose is being pursued "*(...) also through the protection of rural areas, helping to limit depopulation and to preserve the territory from genetic contamination and loss of genetic heritage*" (Art. 1, para. 2). Particularly, GAS are addressed in paragraph 2 of Art. 13, establishing the "*Community of food and agricultural and food biodiversity*". With this formula, it covers "*(...) the local areas resulting from agreements*[54] *between local farmers, livestock farmers and breeders, ethical purchasing groups*", etc.[55] They are established "*in order to raise public awareness, to support agricultural and food production (...), and to promote behaviours that protect agricultural and food biodiversity*" (Art. 13, para. 1).

In the authors' opinion, the recognition of GAS as having an active role in the protection and preservation of biodiversity has made their environmental function explicit, as well as the linear relationship of instruments toward the pursuit of the aims of the Environmental Code. With this in mind, GAS: contribute to the promotion "*(...) of the levels of quality of human life, to be achieved through preservation and improvement of environmental conditions and the rational utilisation of natural resources*" (Art. 2, para. 1 of Legislative Decree 152/2006); given the "solidarity" attribute that distinguishes them, they encourage the inclusion of the principle of solidarity "*(...) in the context of the dynamics of production and consumption (...) to preserve and improve the quality of the environment also in the future*" (Art. 3 *quarter*, para. 3 of Legislative Decree 152/2006); participating in the preservation of biodiversity, they contribute to "*(...) preserve the proper functioning and evolution of natural ecosystems from negative changes that can result from human activities*" (Art. 3 *quarter*, para. 4 of Legislative Decree 152/2006). Therefore, consumption is configured, where critically oriented to the activity, for example, of GAS, as a vehicle of environmental protection. It is the same intimacy of relationship between nutrition and the right to environment to support it: it deals with nutrition to the extent that this affects the "*system of relations*" represented by the legal notion of the environment. In fact, food is a legal good in many different

---

[54]These agreements, in accordance with Art. 3, may concern: "*a) the study, recovery, and transfer of knowledge about genetic resources of local food and agricultural interest; b) the realisation of forms of short supply chains, direct sale, exchange and purchase of agricultural and food products within local circuits; c) the study and dissemination of organic agricultural practices and other cropping systems with low environmental impact and aimed at saving water, lowering carbon emissions, increasing soil fertility and lowering the use of containers for distribution and sale of products; d) the study, recovery and transfer of traditional knowledge related to agricultural crops, natural selection of seeds to cope with climate change and proper nutrition; e) the creation of educational, social, urban and collective gardens, which are instruments to promote local varieties, environmental education and agricultural practices, social gatherings, redevelopment of abandoned or degraded areas and unutilised farmland*".

[55]The same law defines them, in Art. 13, c. 2, as "*local areas resulting from agreements between local farmers, farmers and livestock breeders, purchasing groups, schools and universities, research centers, associations for the protection of agricultural and food biodiversity, school cafeterias, hospitals, caterers, commercial businesses, small and medium-sized artisan businesses for agricultural and food processing, as well as public bodies*".

senses, which manifests its "*binding*" importance by correlating life to the (a) individual, (b) social and (c) ecological dimensions: (i) because it determines the existence of the individual and the ability to grow, develop, and stay healthy; (ii) because it is the history, culture and tradition; (iii) considering, for example, the anthropogenic pressures involved with the production of ecosystem goods and services. Once again, the dialectic between rights and duties emerges, in this case between rights/duties related to life.[56]

The same dialectic, enriched by the results of the analysis, challenges the aims of GAS to mean "*ethical, of social solidarity and environmental solidarity*", allowing for renewed interpretations. From this perspective, it seems interesting to note that that paragraph 266 of Law 244/2007, in identifying the purposes above, provides a direct link between social solidarity and environmental sustainability. If in addition you consider the date of legal recognition of the environmental value of GAS, under this provision, it seems reasonable to assume there is an ongoing evolution of constitutional solidarity (political, economic and social), which, as we have seen, is found in the current evolution of the environmental solidarity concept, legally balancing the sustainability. From the association between solidarity and sustainability one could discern the birth of a new principle, which could be called the "principle of sustainable solidarity". This principle fully sums up the dialectic between inviolable rights and mandatory duties, signaling, perhaps permanently, an evolution in the direction of legal protection of the environment. In this sense, under the aegis of Art. 2 of the Italian Constitution this inextricable link between *inviolable rights*, guaranteed and recognised for present and future generations, and *mandatory duties of environmental solidarity* would be enshrined.

6 Conclusions

When discussing critical consumption, we often must deal with objections of those who, while recognising it as a virtuous practice, argue that consumers alone cannot significantly influence the patterns of production. In the past, such evaluations could be supported with some historical example.[57] However, in today's legal landscape,

[56] See also the contribution of Monteduro in this volume.

[57] Amongst the experiences that demonstrate the efficacy of critical consumption, two examples are presented. The first is emblematic of Altamura (Bari). In this central city of Puglia, a McDonald's fast food restaurant, after just a year of activity, was forced to close for lack of customers in 2002. Amongst the main causes of the failure was the opening of a traditional *focacceria* close to the fast food business. Consumer preferences have been oriented towards the typical "*focaccia of Bari*", rather than with respect the products offered at the biggest fast food chain in the world. The story was documented as a Docu-Fiction by Nico Cirasola in 2009, titled *Focaccia Blues*. The second example relates to a particular variation of critical consumption: the boycott. Specifically, it refers to the one addressed to Nestlè's products, which demanded from the government of Ethiopia (heavily burdened by famine) substantial compensation for the nationalisation of one of its businesses in 1975 following a *coup*. The boycott was so pervasive as to force Nestlè to accept symbolic

they are clearly refuted by the increasing interest demonstrated by the legislature to the phenomenon. The regulation of GAS constitutes a concrete demonstration. In fact, if before the environmental, justice, human rights aspects and ethical content of production simply represented parameters on which particularly sensitive consumers based their purchasing choices, today they are found in a legal reference to the "solidarity" aspect of GAS and result in the fulfillment of mandatory duties of environmental solidarity. In interpreting the experiences of GAS from the point of view of legal protection of the environment, questions regarding the critically oriented market are not a simple sum of individual questions, but something more. They are qualitatively characterised because they link in an instrumental way to the rise "*(...) of the levels of human life, to be achieved through the protection and improvement of environmental conditions and the prudent and rational utilisation of natural resources*": activities that present generations must address for themselves, as well as future generations.

Always with a view toward legal protection of the environment, if you think about the recognised role of GAS under the law in terms of protection and enhancement of agricultural and food biodiversity, it seems more appropriate to consider that these groups, with their collective purchasing activities, do not simply benefit from a different market demand, but are themselves co-producers of food diversity by imposing the differentiation of supply.

In this manner, the formulation of the "principle of sustainable solidarity" does not appear as a mere interpretive exercise, but forms part of the dialectic over inviolable rights and mandatory duties in a closer legal sense. According to this approach, consider, lastly, that the fulfillment of the mandatory duties of environmental solidarity connected to it could correspond, in perspective, to the affirmation of a new inviolable right within the open catalogue of Article 2 of the Constitution: the right to food sovereignty. The latter, as state of the art, is devoid of a concrete and defined recognition. It gravitates between declarations of principle and calls on the importance of guaranteeing it to the people, perhaps also because the phenomenon of corporatisation of the law[58] remains as such and does not translate into its effective realisation.

The many definitions of food sovereignty,[59] amongst them nearly homogeneously, consider that which is contained in the "Manifesto on the future of food"[60] drawn up by the International Commission for the Future of Food and Agriculture. It defines it as a "*fundamental principle at national, regional and community level. The bodies and local, national and regional communities at each level have the fundamental right and duty to protect, sustain and support all the necessary conditions to encourage food production which is abundant, healthy,*

compensation conditioned on the obligation to return it to the Ethiopian coffers, binding the destination to the realisation of public utility works. On the second example, see Antonelli (2004).

[58]Mattei (2011).

[59]Colombo (2002), Desmarais (2009), Martufi and Vasapollo (2009) and Mafrici (2016).

[60]Manifesto on the future of food (2003).

accessible to all and that conserves the land, water and ecological integrity of the places where it is produced, respecting and supporting producers' means of subsistence". The same definition also suggests the need to avoid interference with the exercise of the rights connected to it by the people. In fact, it states that "(...) *no international body or multinational corporation has the right to alter this priority*".[61]

Such a definition logically recalls the themes of short supply chains, the "custody" of biodiversity through agriculture, the criticality of collective purchases of agricultural products in the form of GAS and the cultural self-determination and cultivation of the people, which seem, today, to be expressions of an alternative approach to the production/consumption dialectic, when, in truth, they should be the norm; they should be declined as natural manifestations of the conception of agricultural goods as products of the land and not of the factory. Think of the perception of "organic" and the labeling paradox of what is natural, healthy and clean, an exception for a system that observes the prevalence of intensive production of clear unsustainability.[62] And so it is, in the scenario of agricultural globalisation, "*the principle of sustainable solidarity*" may constitute the vehicle of recognition for the right to food sovereignty, and both support the need to restore the fractured balance between product and land and between individual/community and production in the "environmental system"; helping to convert back into rule what is today an exception.

References

Allen CR, Garmestani AS (2014) Social-ecological resilience and law. Columbia University Press, New York

Alpa G, Conte G (2013) Enti senza scopo di lucro nel diritto civile e profili di responsabilità civile. http://www.altalex.com/. Accessed 26 Nov 2013

Arcidiacono D (2013) Consumatori attivi. Scelte di acquisto e partecipazione per una nuova etica economica. FrancoAngeli, Milano

Arnold CA, Gunderson L (2013) Adaptive law and resilience. Environ Law Report 43:10426–10443

Balduzzi R, Cavino M, Luther J (2006) I doveri costituzionali: la prospettiva del giudice delle leggi. Atti del convegno di Acqui-Terme Alessandria, 9–10 giugno 2006. Giappichelli, Torino

Bauman Z (2000) La solitudine del cittadino globale. Feltrinelli, Milano

Benozzo M, Bruno F (2009) La valutazione di incidenza. La tutela della biodiversità tra diritto comunitario, nazionale e regionale. Giuffrè, Milano

Benson MH, Garmestani AS (2011) Embracing panarchy, building resilience and integrating adaptive management through a rebirth of the national environmental policy act. J Environ Manage 92:1420–1427

Biggeri M, Volpi F (2006) Teoria e politica dell'aiuto allo sviluppo. FrancoAngeli, Milano

[61]Petrini (2009).

[62]See also Petrini (2009).

Bignami DF (2010) Protezione civile e riduzione del rischio disastri. Metodi e strumenti di governo della sicurezza territoriale e ambientale. Maggioli Editore, Santarcangelo di Romagna
Carbone C (1968) I doveri pubblici individuali nella costituzione. Giuffrè, Milano
Chieppa R, Giovagnoli R (2011) Manuale di diritto amministrativo. Giuffrè, Milano
Colombo L (2002) Fame. Produzione di cibo e sovranità alimentare. Jaca Book, Milano
Colombo L, Onorati A (2009) Diritti al cibo! Agricoltura sapiens e governance alimentare. Jaca Book, Milano
Colombo GM, Poletto M (2012) Enti non profit in pratica. Fisco, contabilità e bilancio. Wolters Kluwer, Milano
Commissione internazionale per il futuro dell'alimentazione e dell'agricoltura (2003) Manifesto sul futuro del cibo
Corradi V (2011) Le incerte vie della sostenibilità. Aziende di servizi pubblici e cittadini per l'ambiente. FrancoAngeli, Milano
Council of Europe - Resolution 1957 (2013) - Food security, a permanent challenge for us all
Desmarais AA (2009) La via Campesina. La globalizzazione e il potere dei contadini. Jaca Book, Milano
DOCUMENTO BASE DEI GAS. I gruppi di acquisto solidale. Un modo diverso di fare la spesa (1999)
East R (2003) Comportamento del consumatore. Apogeo, Milano
FAO (2004) Biodiversità, la nostra alimentazione ne dipende. Roma. ftp://ftp.fao.org/docrep/fao/006/y5418i/y5418i00.pdf
FAO (2015) Coping with climate change. The roles of genetic resources for food and agriculture. Roma. http://www.fao.org/3/a-i3866e.pdf
Ferrara R (2010) Salute e sanità. Giuffrè, Milano
Fracchia F (2015) Il principio dello sviluppo sostenibile. In: Rossi G (ed) Diritto dell'ambiente. Giappichelli, Torino, pp 175–182
Gazzoni F (2013) Manuale di diritto privato. Edizioni Scientifiche, Napoli
Gennari F (2009) Il controllo di gestione nella governance delle imprese commerciali. FrancoAngeli, Milano
Grandi F (2014) Doveri costituzionali e obiezione di coscienza. Editoriale Scientifica, Napoli
Lombardi G (1967) Contributo allo studio dei doveri costituzionali. Giuffrè, Milano
Lorenzoni G (1968) I gruppi di acquisto fra dettaglianti. Patron, Bologna
Lugaresi N (2015) Diritto dell'ambiente. CEDAM, Padova
Mafrici A (2016) Globalizzazione agricola e libertà di mercato. Gangemi Editore, Roma
Maietta OW (2004) Il consumatore etico e il marketing agroalimentare. In: Antonelli G (ed) Marketing agroalimentare. specificità e temi di analisi. FrancoAngeli, Milano, pp 187–211
Martufi R, Vasapollo L (2009) Futuro indigeno. La sfida delle Americhe. Educazione all'economia dei popoli. Jaca Book, Milano
Marzanati A, Mattioni A (2007) La fraternità come principio del diritto pubblico. Città Nuova, Roma
Mattei U (2011) Beni comuni. Un Manifesto. Laterza, Bari
Minestroni L (2006) Comprendere il consumo: società e cultura dai classici al postmoderno. FrancoAngeli, Milano
Monteduro M (2014) Il provvedimento amministrativo: natura e caratteri. In: Carlotti G, Clini A (eds) Diritto amministrativo. Maggioli Editore, Santarcangelo di Romagna, pp 105–128
Monteduro M (2015) Diritto dell'ambiente e diversità alimentare. Riv. quad. dir. amb. 1:88–131
Monteduro M, Buongiorno P, Di Benedetto S, Isoni A (2015) Law and agroecology. A transdisciplinary dialogue. Springer, Berlino
Morzenti Pellegrini R (2008) in Codice dell'ambiente. Commento al D.lgs. 3 aprile 2006, n 152, aggiornato alla L. 6 giugno 2008, n 101. Giuffrè, Milano
Petrini C (2009) Terra madre. Come non farci mangiare dal cibo. Slow Food Editore, Cuneo
Pirani A, Gaviglio A, Pedol ML (2009) Le filiere biologiche in Lombardia. II- Analisi economica delle fasi di trasformazione distribuzione e consumo. FrancoAngeli, Milano

Pizzolato F (2010) Scelte private e pubblici interessi: la regolazione dei consumi. In: Cocco G (ed) Diritti dell'individuo e diritti dei consumatori. Giuffrè, Milano, pp 109–144
Rimoli F (2015) Appunti per uno studio sulla dimensione funzionale dei doveri pubblici. http://www.federalismi.it. Accessed 1 Jul 2015
Rodotà S (2006) La vita e le regole: tra diritto e non diritto. Feltrinelli, Milano
Rodotà S (2014) Solidarietà: un'utopia necessaria. Laterza, Bari
Ruhl JB (2011) General design principles for resilience and adaptive capacity in legal system – with application to climate change adaptation. N C Law Rev 89:1374–1401
Savarese P (2014) La sussidiarietà e il bene comune. Edizioni Nuova Cultura, Roma
Scagliarini S (2015) Diritti dei consumatori. In: Morelli A, Trucco L (eds) Diritti e Territorio. Il valore delle autonomie nell'ordinamento repubblicano. Giappichelli, Torino, pp 253–263
Tanza A (2009) Contratti e consumatori. In: Viola L (ed) Il contratto: validità, inadempimento, risarcimento, 1. Cedam, Padova, pp 1497–1575
Torrente A, Schlesinger P (2011) Manuale di diritto privato. Giuffrè, Milano
Trimarchi P (2014) Istituzioni di diritto privato. Giuffrè, Milano
Troisi M (2008) La continuità dei principi della l. n. 328 del 2000 nella nuova stagione legislativa in materia di servizi sociali. Il caso della Regione Puglia. http://www.federalismi.it. Accessed 5 Mar 2008
Vignudelli A (1984) Il rapporto di consumo. Profili di rilievo costituzionale. Maggioli, Rimini
Violante L (2014) Il dovere di avere doveri. Einaudi, Torino
Visconti G (2008) I "gruppi di acquisti solidali" disciplinati dalla finanziaria per il 2008: uno strumento per la tutela del potere d'acquisto dei consumatori. In: Diritto.it
Voigt C (2013) Rule of law for nature. New dimensions and ideas in environmental law. Cambridge University Press, Cambridge

From Districts to Agricultural Enterprise Networks

Nicola Lucifero

1 Introduction

The subject of my report offers the chance to look at *food diversity*—a term that, concisely, expresses the concept of the diversity of agricultural production based on the local nature and quality of products characterised by their territory of origin—from a new viewpoint, namely that of relationships between companies, their position on the territory and the legislative instruments by which the value of local produce is enhanced.

The perspective is that of the relationships between enterprises, both farms alone and between farms and industries, through forms of cooperation and integration that can, on a stable basis, find the space to allow for increased competitive capacity on the market and for the rationalisation of the production and distribution process. This is a phenomenon driven by social-economic factors rather than legal factors, which lead entrepreneurs to resort to models of collaboration to increase competition on the level of the global economy, to rationalise production costs, and to increase investments in technological innovation by exploiting the complementary nature of know-how and of research. For these needs, which answer the traditional problem of the size of farms and the weakness of the agricultural sector because of the impossibility of controlling market choices, relevance has recently been acquired, also and above all in virtue of the restrictions of European law, by the obligations imposed on enterprises to comply with technical standards in the sphere of production, standards that must be met otherwise the entire production chain and, individually, each entity

N. Lucifero (✉)
Department of Agricultural, Food and Forestry Systems, University of Florence, Florence, Italy
e-mail: nicola.lucifero@unifi.it

A. Isoni et al. (eds.), *Food Diversity Between Rights, Duties and Autonomies*,
LITES – Legal Issues in Transdisciplinary Environmental Studies 2,
https://doi.org/10.1007/978-3-319-75196-2_21

of the chain, may bear responsibility to the end consumer for such lack of compliance.[1]

These factors, together with the complexity of today's markets, have opened up a new phase featuring the relationship between companies in which entrepreneurs who, considered individually, are actual or potential competitors, although maintaining their independence and their strategic functions, operate through stable, long-term agreements. The logic of the cooperation,[2] which is at the base of the relationships between companies, moves beyond the tendency to form associations that, in previous decades, was typical of the agricultural sector and which arose to face the natural precariousness of the position of the single entrepreneur by expanding corporate dimensions by means of practising agriculture in a collective manner. Cooperation began because the need was felt to belong to an organised group, for the more efficient use of a company's resources and for improved productivity by the integration of the production process, and thus to be more competitive on the market, responding to consumer's expectations and desires.[3]

In this situation, the phenomena of production districts and of networks of companies take on importance; they are the result of regulation developed through legislative initiatives that overlapped each other in a disordered manner at times. For some time now, production districts have represented agglomerates of several enterprises, usually small or medium sized, located in the same circumscribed area and specialised in one or more production steps; networks, however, represent forms of coordination based on stipulated agreements thanks to which companies, usually small, aim to develop greater competitiveness on the markets without having to merge or unite under the control of a single subject.

The purpose of this work is to outline the legislative developments that have been features of the law on districts, up to the more recent organisations contemplated by the network agreement, through the repeated legislatives innovations that have followed one on another, generally in a disorganised manner and resulting from grants and supporting measures, pointing out the particular nature of the implementation of the law on networks in agriculture and the needs of the sector in the promotion of the quality of local produce. From a methodological viewpoint, the legal provisions that have been introduced and the changes at the diverse levels and in the diverse areas of regulation will be reconstructed, also considering how they often overlap, and considered as a whole to thus define the development of the system of the districts,[4] to then arrive at company networks as an instrument for

[1]The reference is to the law on foodstuffs and, particularly, the requirements and obligations imposed on the operators of the food sector by Regulation 178/2002.

[2]As observed in legal theory, cooperation and competition are opposite concepts. However, in complex situations like today's markets, cooperation between independent companies, if viewed from a systematic viewpoint, can be an important tool for increasing the overall competitiveness of the markets. Thus Libertini (2014) p. 406.

[3]Jannarelli (2011a), p. 161 and foll.

[4]According to authoritative legal theory, the complex series of legislative measures that characterised the experience of the districts can be divided into three main areas: (a) the *direct*

companies to relate to each other and through which to pursue the purpose of reinforcing the organisational measures and the integration of the chain, the exchange and spread of best technologies and supporting services, and forms of collaboration between companies.[5]

2 Rural Districts and Quality Agri-Food Districts

In agriculture, districts bring to mind a complex regulatory situation that has evolved over the last 20 years with highly diversified legislative measures and which is still today devoid of any standardised regulatory framework. The interest in districts in the agricultural sector having developed based on the experience of the industrial districts, and on the level of analyses carried out by economists, has gone hand in hand with the political evolution of rural development as a method to support local development in rural areas.

In general terms, the Anglo-Saxon expression "cluster", often used to indicate such districts, brings to mind the interaction between the businesses that produce merchandise or which provide services within the same economic and social fabric, and recognises the locality as the source of enterprise rules, thus making it the context in which the agricultural enterprise is placed and develops, and in which "the old model constructed on the agricultural enterprise considered as a *hortus conclusus* of a larger, but distinct, rural space could appear inadequate".[6]

This model identifies as its own distinctive features those of the local area and of the common "know-how" traditionally belonging to the local community, and together with these, it gives significant importance to the particular organisational methods of production among several enterprises placed on the same level.[7] In fact, the concept of cluster, from the viewpoint identified by economic analysis,[8] leads to a single overall view of the production process through the localisation of the production, and to the particular organisational methods of the same, and it takes as the objective not the single enterprises but *local production systems* that, being in the same geographic area and because of the relationships and integration between

---

and declared legal bases, focusing on certain main regulatory nuclei, (b) the *indirect legal bases,* even if declared, present in the legal provisions regarding local agreements, which have differentiated sources, contents and effectiveness; (c) the *legal bases that are neither direct nor even declared*, which, although without using the word "district", have an impact on the districts and on rural development. Thus, Albisinni (2011), pp. 409–410; for a systematic reconstruction of the law see, by the same author (2003), p. 684; (2010), p. 1.

[5]On the same subject, for a general overview, Iamiceli (2009), Macario and Scognamiglio (2009), p. 915; Genovese (2013), p. 55.

[6]Germanò (2001), pp. 69–70.

[7]Grandori (1999), pp. 75–91; Grandori and Soda (1995), pp. 183–214.

[8]The first studies were those of Becattini (1979), p. 7; Becattini (1987, 2000). On the same subject, Romano (1999), p. 2; Pacciani (1998), pp. 1–14.

enterprises, become an internal element of the production structure and, together, an essential instrument capable of competing on the market. The cluster model does not only reflect the economic viewpoint, i.e. the efficiency created by the integration and the relationships established by a company in the production context, but it highlights its social and historic profile as a social-territorial entity featuring joint activities in a circumscribed geographic area, distinguished by its own nature and its own history, composed of a community of people and of various enterprises.

From the legal viewpoint, agricultural clusters represent a phenomenon that is hard to pin down, and the requirements of the cluster, i.e. those of the productive concentration of company systems and of the territory—as the expression of the historic and local identity and of the local traditions linked to production—already present in the economists' calculations, do not emerge, in a concrete and practical manner, from the provisions of reference. In this context, the reference to the *territory* becomes important—no longer merely understood as the perimeter of an area that has certain natural and environmental features, but rather in relation to the agricultural activity where the production vocation becomes the rule for the use of the territory—for some time now the focus of the attention of European and national legislation, regarding the rules of use and of the destination of the territory, as well as those rules for the appreciation of its products on the market.

As already mentioned, legislation of the agricultural clusters follows that which developed for the industrial clusters and it is consolidated through negotiated planning. Particularly, Article 36 of Law No. 317 of 5.10.1991[9] gives the definition of "*distretti*", i.e. districts or clusters, as *local production systems featuring a high concentration of industrial enterprises, as well as the productive specialisation of the enterprise systems,* and articulated legislation is contemplated within which the Regions are assigned the task of identifying the local production systems and the Industrial Development Consortia, created as public economic entities, are assigned the task of providing real services to the business enterprises. Conversely, the districts were created not only by identifying other subjects in addition to the local bodies, but also by exploiting contractual agreements between public subjects and companies for the implementation of public intervention in the sphere of the economy.[10]

[9]Article 36, paragraph 2: *Local production systems featuring a high concentration of industrial enterprises and the productive specialisation of the enterprise systems are defined as industrial districts.* It may be noted that the provision quoted recalls the definition of *local production systems,* indicated in paragraph 1 of the same provisions, as amended by Law No. 140 of 11 May 1999, as *similar productive contexts, with a high concentration of enterprises, mainly of small and medium size, and with a particular internal organisation.*

[10]To this regard legal theory has pointed out that the legislator repeatedly uses the expression *negotiated regulation,* behind which various legal bodies are concealed, even if defined as agreements or contracts or understandings between public bodies and private subjects, to underline the public component of the planning activities aimed at the development of a certain geographic zone, where the use of mutually accepted instruments of intervention should be traced back to the diverse formula of programmed negotiation precisely to underline the private-contractual logic and

In this system, within the scope of the guideline legislation of 2001,[11] provision is made for the *agricultural, agri-food and fishery districts,* not mentioned in the original text of the law on industrial districts, Law No. 317/1991.[12] In fact, the assignment to the delegated legislator of the task to provide for the identification of the conditions for the creation of quality agri-food, rural and fishery districts[13] answers the long felt need to regulate, also in the primary sector, the issue of districts and the introduction of new rules for production activities in a manner that would integrate them at the local level. More in general, in the guideline legislation, agriculture and territory share an important point of cohesion in the search for rules of law that can give value to the territorial dimension considering the natural and human aspects that characterise agri-food productions, according to prospects of sustainability aimed at combining the quality of the environment and the quality of the production processes, progressively extended to project these elements, inasmuch as expressions of the local area and of the local interconnected enterprise systems that operate in that geographic area, onto the features of the products and services obtained.

With Article 13 of Legislative Decree No. 228 of 18 May 2001, the delegated legislator defines the "*distretti rurali e agroalimentari di qualità*", i.e. the quality rural and the agri-food districts, including them within the scope of the *local production systems* and distinguishing rural districts as those *having a single historical and territorial identity deriving from the integration of agricultural activities and other local activities, as well as the production of goods or services of a specific nature, consistent with the natural and territorial vocations,* while the quality agri-food districts are defined as those with *a significant economic presence and productive interrelation and interdependence of the agricultural and agri-food enterprises,* also at inter-regional level, *and with one or more articles of produce certified and protected pursuant to Community or national laws in force, or with traditional or local produce.*

If prima facie the said provision brings to mind precisely the features of the district model, such as the historic and territorial identity and the interrelationships between enterprises, as well as the specifically local nature of the products, in fact it contains strong contradictions and some uncertainties in respect of its application. In this sense, it is sufficient to point out that while districts are included in the scope of *local production systems* as contemplated for the industrial system,[14] rural districts, for their very nature, do not necessarily present a high concentrations of enterprises,

technique that govern, in a substantially preponderant manner, the framework and the effects of the programming tools. Thus Adornato (1999), p. 7; (2000), p. 1125.

[11]See Legislative Decrees nos. 226, 227 and 228 of 18 May 2001.

[12]Successively amended by Law No. 140 of 11 May 1999, which transferred therein the models of *territorial agreements and local development* that had developed, adapting the law to the innovations introduced in State organisation and introducing the concept of *local production systems.*

[13]See Article 7, paragraph 3, letter (a) of Delegated Law No. 57 of 3 March 2001.

[14]See Article 36, paragraph 1 and 2, of Law 317/1991, as amended by Law No. 140/1999.

and their distinctive features, on the contrary, are found in the territory and in those values that are not only economic but also historical, cultural and social that are also expressed in productive activities *deriving from the integration between agricultural activities and other local activities.* Furthermore, Article 13 gives no indication whatsoever for the external recognition of the activities and produce of the district, such as a mark of the identity of the district (such as the *label* of the *Pays* in France), which would obviously not be equivalent to the distinctive PDO and PGI marks since these latter marks override district rules and they refer to a specific product, even if coming from a same territory. Conversely, what should be expressed is the fact that the product of the district, for the special relationship between the farms and the territory, is the result of human and natural factors and, therefore, it represents the quality of the territory that wishes to be recognised on the market by means of specific models of regulatory provisions.

In short, the national legal framework relative to districts for agriculture has remained at the level of a theoretical hypothesis without any systematic consolidated framework, the contents having been referred to action by the Regions.[15] In fact, the conditions for the recognition of the districts have been indicated—for rural districts, the *same historical and territorial identity, the production of goods or services of a specific nature, consistent with the natural and local traditions and vocations,* while quality agri-food districts are distinguished by traditional or specifically local high quality produce—without introducing elements for a specific discipline relative to the application of the district model to agricultural areas, and without specifying the elements of the scheme for the regulation, not only in economic terms, of the local communities, underlying the definitions thus introduced.

It must be said that the Regions, also on the strength of the far-reaching extent of their powers attributed by the reform of Title V of the Constitution pursuant to Constitutional Law 3/2001, have introduced various measures to regulate the districts in their own territories, thus giving rise to a composite and diversified situation. But above all, with the rural development policy, and particularly with Regulation 1698/2005, the districts have become a relevant component in the overall policy for governing the primary sector.[16] This regulation focused on the many subjects of the programming, identified inasmuch as being complementary to or belonging to the sector, and it invited the Member States to involve, in their national strategic plans and in their regional rural development programmes, not only the local public bodies and the other competent public authorities, but also the economic and social parties and any other body representing civil society, requiring the participation of the

[15]See Article 13, paragraph 3 of Legislative Decree No. 228 of 18 May 2001 according to which *the Regions must provide for identifying rural districts and food and agricultural districts.*

[16]This was quite different from Regulation (EC) 1257/1999, previously in force on rural development, which first indicated sustainable measures and only in the final articles gave a few provisions on programming profiles, on this point referring to Regulation (EC) 1260/1999 of 21 June 1999 containing general provisions on structural funds, later abrogated by Regulation (EC) 1083/2006 of 11 July 2006, in turn abrogated by Regulation (EU) No. 1303/2013 of the European Parliament and Council of 17 December 2013.

subjects belonging to the sector in the development and the supervision of the national strategic plans, as well as in the preparation, implementation, supervision and assessment of the rural development programmes (Article 6).

The currently effective Regulation (EU) 1305/2013 of 17 December 2013 on rural development, unlike the preceding programming period, indicates *cooperation* in agriculture as the tool to stimulate innovation and to spread innovative processes, which is a necessary condition for their knowledge.[17] There are many forms of cooperation, depending on the best methods of cooperation that can respond to the diversified economic and territorial situations of the Member States, adapted to achieve the basic strategic objectives (environmental, competitive, territorial rebalancing) and divided according to the possible subjects involved. Article 35 contemplates various forms of cooperation, namely: (1) cooperation between the operators of the agricultural and forestry sector and of the food production chain in the EU, and the other subjects that contribute to the achievement of the targets and of the priorities of the CAP (the Common Agricultural Policy); (2) the creation of centres and of networks; (3) the establishment and management of operating groups of the EIP (the European Innovation Partnership) for productivity and sustainability in agriculture.

Without doubt, the approach of the new rural development plan represents, to a certain extent, a break with the previous plan, not only for the objectives but above all for the instruments by which to pursue the set targets; even if the reference remains the local system and the development of the territory, the use of contractual tools to pursue cooperation between the operators, and of partnerships for involving both public and private subjects, acquire importance. The reference to the "network", mentioned many times by EU legislation, highlights a tool other than the district, consisting of relationships between companies disciplined by contract; the cooperation achieved through enterprise networks does not depend on the territory, nor it is aimed and savings achieved by aggregation but rather at savings achieved by the fact that the enterprises are complementary to each other.

3 From Production Districts to Enterprise Networks

More recently important innovations regarding the discipline of agreements between enterprises have been introduced within the sphere of the finance acts of 2006[18] and of 2007,[19] amended by urgent provisions (2008[20] and 2009[21]), and of the finance act

[17]See *Recital* no. 29.

[18]See Article 1, paragraph 366–368, Law No. 266 of 23 December 2005.

[19]See Article 1, paragraph 889 and following, Law No. 296 of 27 December 2006.

[20]See Article 6-*bis* of Decree Law No. 112 of 25 June 2008.

[21]See Article 3 of Decree Law No. 5 of 10 February 2009, converted, with amendments, by Law No. 33 of 9 April 2009.

of 2013,[22] above all for the recognition of tax, administrative and financial concessions and regarding research and development. Particularly, Law 266/2005 has introduced the definition of "*production districts*", as *free aggregations of enterprises organised on the territorial level and on the functional level*,[23] with the aim of fostering the development of the *areas* and of the sectors of reference and of improving the organisational and production efficiency, according to principles of vertical and horizontal subsidiarity, also by identifying collaboration models with the business category associations,[24] and the law mentioned above addressed to the same special tax, administrative and financial measures. Without entering into the merits of the many and specific concessions, also applicable to rural and agri-food districts,[25] it is worth noting that the legal measures introduced place both territorial districts and functional districts, i.e. not linked to any specific territorial community, on the same plane and that they identify the disciplinary fulcrum with the production base, accentuating their aspects of organisation and of determination, and at the same time highlighting the subjective elements with respect to a previous approach that appeared to emphasise an organic viewpoint.

In recent years, new tools have become available in support of the innovation of production structures in the form of enterprise networks and network agreements, with contents and formulas modified over just a few months by a flurry of successive legal provisions, and most recently by Article 3, paragraphs 4-*ter* and following, of Decree Law No. 5 of 10 February 2009, converted by Law No. 33 of 9 April 2009, entitled "*Distretti produttivi e reti di impresa*" [Production Districts and Enterprise Networks].[26] The said provision defined "network agreement" as an agreement between several entrepreneurs who pursue the purpose of increasing, individually and collectively, their capacity for innovation and their competitiveness on the market, and for this end they undertake, under a common network programme, to collaborate in predetermined forms and spheres regarding the practice of their own businesses or to exchange industrial, commercial, technical or technological information or services or to practice jointly one or more activities that are included within their own core businesses.

---

[22]See Article 1, paragraph 583, letter (c), of Law No. 147 of 27 December 2013.

[23]See Article 1, paragraph 366, of Law No. 266 of 23 December 2005.

[24]See Article 1, paragraph 366 and following, Law No. 266/2005, *cit.* The Constitutional Court, with decision no. 165 of 11 May 2007, has declared the illegitimacy of the last part of the paragraph that does not contemplate the definition of the features and methods of identifying production districts by decree of the Ministry of the Economy and Finance, in concert with the Ministry of Productive Activities, the Ministry of Agricultural and Forestry Policies, the Ministry of Education, Universities and Research and the Ministry for Innovation and Technologies, after agreement with the Standing Conference for Relations between the State, the Regions and the Autonomous Provinces of Trento and of Bolzano, and after hearing the opinion of the Regions concerned.

[25]See Article 1, paragraph 369, of Law No. 266 of 23 December 2005.

[26]Later amended by Decree Law No. 78 of 31 May 2010, converted by Law No. 122 of 30 July 2010; by Decree Law No. 83 of 22 June 2012, converted into Law No. 134 of 7 August and, lastly, by Article 36, paragraph 4, letter (a), of Decree Law No. 179 of 18 October 2012, converted, with amendments, by Law No. 221 of 17 December 2012.

The legislative innovations of 2009 abrogated Article 6-*bis* of Law No. 112/2008 and they have thus restricted the extension of the enterprise networks in terms of the provisions previously contemplated for production districts, limiting said extension to those inherent to the administrative simplifications, the financial concessions and the measures on research and development; above all, they have differentiated the phenomenon from that of the districts in a more marked manner, linking the application of the provisions on networks to the existence of a contractual agreement of which the limits are outlined. In this way, the legislator has marked the legal confine between the concept of district and that of network; indeed, the comparison between Law No. 266/2005, paragraph 336, and Law No. 33/2009, as most recently amended, shows that the first provision relegates to a successive regulation the task of identifying the features of the districts inasmuch as it refers to a purely physical fact (i.e. *the free aggregation between enterprises*), and the correspondence of the said physical fact to the notion of district is entrusted to a recognising provision of the public administration; the second, however, defines the network contract and indicates certain contents of the same to grant benefits and simplifications to enterprises that are, in fact, parties to such a contract.[27] Conversely, the notion of district is an empirical concept based on economic and social data such as the high concentration of enterprises in the same place and their production specialisation, which is the basis for defining relational "clusters" that can involve private and public subjects; the notion of network is, in turn, based on highly elastic elements of fact and it indicates an organisational model of enterprises, not necessarily located in the same place, which coordinate with each other pursuant to legal agreements either ensuing from the fact that they all hold certain common organisational elements or from contractual bonds.

It is worth noting, albeit very briefly, that the text of Law No. 33/2009, as successively amended, indicates the features of the network contract as an agreement between enterprises, by which the "network" between the enterprises is determined,[28] that produces obligatory effects (not of conveyance), and which, on the causal level, refers to the joint practice of economic activities aimed at increasing the reciprocal capacity of innovation and competitiveness on the market; it also has a

[27]In fact, doctrine, rationalising the empirical data resulting from research in the field, has already identified a difference between the two concepts, defining the network as the system of contractual and/or proprietary relationships, featuring a strong level of productive interdependence, stable relationships, generally based on trust, and with the specification that "*the networks of small and medium enterprises are not industrial districts, for two differences: (i) they can develop in non-local contexts and (ii) when they are formed in a district system, they represent a sub-system of relationships between enterprises, which is not necessarily the main sub-system*". Thus Cafaggi (2004), p. 34. On the same subject, see Mosco (2010), p. 839; Villa (2009), p. 103; Mosco (2010), p. 944.

[28]The enterprise networks referred to by this article are subject to the provisions of Article 1, paragraph 368, letters (b), (c) and (d), of Law No. 266 of 23 December 2005, and successive amendments, and subject to authorisation issued by a decree of the Ministry of the Economy and Finance in concert with the Ministry of Economic Development, to be adopted within 6 months from the relative request (Article 3, paragraph 4-*quinquies*, of Decree Law No. 5/2009).

pre-determined content and it is a multi-lateral contract with a common purpose and, lastly, it is subject to a cost of establishment inasmuch as it must be drawn up by public deed or authenticated private deed (or by digitally signed deed) and it is subject to publication by the registration of the contract in the Companies Register of the place where the network has its head office; the contract implies the creation of a body that is attributed with the management of the network programme and, regarding the means destined for the execution of the programme, the companies can normally constitute a joint fund or they can operate through the constitution of assets destined to the single business deals pursuant to Article 2447-*bis* of the Civil Code.

Pursuant to paragraphs 4-*ter*, 4-*quater* and 4-*quinquies* of Article 3 of Decree Law No. 5/2009,[29] the network organisational models are outlined separately according to whether the network is a complex network with external entities with legal personality (so-called "subject-network"), or a network without legal personality (so-called "contract-network"), or lastly a simple network. The acquisition of legal personality,[30] i.e. an organisation that can be an independent centre holding rights and obligations, depends not only on meeting certain requirements,[31] but also on a precise expression of such an intention by the contracting parties; the network thus becomes a third subject in respect of the participating companies, with significant consequences regarding various aspects, especially in terms of taxation,[32] concessions and bankruptcy. A network without legal personality, on the contrary, is not an *ens tertium* in respect of the parties, so that the deeds carried out produce effects legally and subjectively directly on the participants. The choice of one or the other model will depend on the specific interests that the parties wish to pursue, being thus able to join to each other in a more or less binding contractual relationship.[33] In addition, the first-mentioned type will have a common fund of assets and a

---

[29]See Article 45 of Decree Law No. 83 of 22.6.2012, converted, with amendments, into Law No. 134 of 7 August 2012; Article 36 of Decree Law No. 179 of 18 October 2012, converted, with amendments, into Law No. 221 of 17 December 2012.

[30]On the notion of legal personality, BIANCA (2002), p. 311.

[31]There is debate in doctrine on whether, in addition to common funds, a common governing body is also necessary to acquire legal personality; however, Article 3, paragraph 4-*ter*, letter e), ruling that "*the common body acts in representation of the network when it acquires legal personality*", gives rise, according to certain doctrine, to recognising the implicit existence of a common body for the performance of an activity with third parties; on this point: Cafaggi et al. (2012), p. 492 and foll. *Contra* Caterino (2014), p. 173; Granelli (2013), p. 833, do not maintain it to be necessary to also institute such a common body to become a legal subject.

[32]With circular No. 20/E of 18.06.2013, the Revenue Agency recognised that the subject-network was a taxpayer; the subject-network has its own VAT number, it is obliged to keep accounting records in the form required according to the activity performed and, lastly, it is subject to the consequent declaration obligations. In addition, the circular specifies that adhesion to a contract-network, on the contrary, has no influence on the companies of the network regarding taxes; the contract-network, albeit having a tax code, does not become an independent taxpayer.

[33]For the distinction between "*strong networks* with legal personality and *weak networks* without legal personality", see Cuffaro (2013), p. 25 and foll.

body destined, among other things, to perform a commercial activity with third parties, and it will have a head office and an official name and it will be registered in the ordinary section of the Companies Register of the district in which it has established its head office. By this registration, the network acquires legal personality. The second-mentioned type does not require registration as an independent subject in the ordinary section of the Companies Register, but only in the same section in which each adhering company is registered. Similar publicity is required for the third kind, which, however, does not have a common provision of assets and/or a common body.[34]

Flexibility, adaptability and full decisional independence are the features of these networks, which are not located in the same place and often involve several sectors, with the capacity of developing innovation and competitiveness according to market logics, and which, although mainly made up of small and medium enterprises, can also involve companies of all sizes. A network serves for pursuing the purposes of aggregation and collaboration between companies with the aim of rationalising and exploiting the complementary nature of the product manufacturing and distribution processes on the market. The purpose of the network lies precisely in the creation of forms of collaboration between enterprises concerning complementary activities that are carried out in a single phase or which comprise several phases of the production chain, and they are created when market relationships are found to be inadequate for organising the complementary aspects, and when a vertically integrated company would be excessively costly and insufficiently flexible.

4 The Network Contract in Agriculture

Although the above considerations outline the content of the law on the network contract in general, albeit very briefly, it is worth noting that the legislator, by several legal measures regarding the network contract between agricultural enterprises, or between agricultural and industrial enterprises, has introduced specific provisions resulting from differentiated treatment reserved to the agricultural sector that, if considered within the system of relations between agri-food companies, indicates special legislation for the network contract in agriculture compared to the general network agreement[35] thus confirming the special nature of the agricultural sector.[36]

[34]On this subject, and especially on the effectiveness of the registration of the enterprises, see Caprara (2015), p. 113.

[35]This conclusion is also reached by Russo (2015), p. 191 and foll.; on this subject, for an initial examination of the law, see Costantino (2013), p. 668; Caputo (2014), p. 46 and foll.

[36]In general, on the "exception" of agriculture, see lastly Jannarelli (2013), p. 405.

The specific provisions that discipline and foster the adoption of network contracts in agriculture[37] responds to the logic of promoting and sustaining the reorganisation and modernisation processes of the agricultural sector and they are reflected in formal concessions or in provisions regarding the sphere of labour or of taxation.

To this regard, with reference to publicity obligations, the provision of Article 36, paragraph 5, of Decree Law No. 179 of 18 October 2012, as amended by Law No. 221 of 17 December 2012, takes on importance, pursuant to which the network contract of the agricultural sector can be signed by the parties with the assistance of one or more category organisations of the agricultural sector that are most representative at national level and which have participated in drawing up the final version of the agreement. Conversely, the obligations of the afore-mentioned Article 4-*ter*, of Article 3, of Decree Law No. 5/2009 no longer hold firm, since a public deed is not necessary and only the supervision of the aforesaid category associations is sufficient. However, the implementation of the provisions in question is limited, since the said provision does not specify when a network contract refers or does not refer to the agricultural sector.[38] Furthermore, pursuant to Article 36, paragraph 2-*bis,* of Decree Law No. 179 of 2012, the network contract in agriculture may contemplate the creation of a financial provision for mutual assistance between the contracting parties that adhere to the *ISMEA [Agricultural and Food Market Services Institute] National Mutual Fund for the Stabilisation of the Income of Agricultural Enterprises*. The Fund is constituted by the voluntary contributions of farmers and it can also benefit from public contributions as far as compatible with European laws on State aid. Pursuant to paragraph 2-*ter*, the possibility of constituting a mutual fund between the contracting parties of an enterprise network is also contemplated.

Article 45, paragraph 3, of Decree Law No. 83 of 22 June 2012, then converted into Law No. 134 of 7 August 2012, also contemplates the inapplicability to the network contract of the provisions of Law No. 203 of 3 May 1982. This is an autonomous provision not included in Article 3 of Decree Law No. 5 of 2009, which

---

[37]The innovations of the "2016 Agricultural Act", now in the process of being approved by the Senate, include two provisions concerning the enterprise network in agriculture: Parliament has expressed its favour, with the introduction of Article 12 on the joint hiring of employees by agricultural enterprises linked by a network contract, reducing the requested percentage of agricultural enterprises within such networks that decreases from 50% to 40% (for such employment to be possible); Article 11-*bis*, pursuant to which, for network contracts in the agricultural, forestry and food and agriculture sectors, the obligation of drawing up a balance sheet, within 2 months from the closure of the financial year, concerns only the enterprise networks that have acquired legal personality.

[38]This point is answered, albeit partially, by the note of the Ministry of Economic Development of 4 June 2014 regarding the filing at the Companies Register of a contract between agricultural enterprises and a firm that provides tax advisory service where it is specified that the network belongs to the agricultural sector, inasmuch as the firm's non-agricultural activity was, in any case, "instrumental and ancillary" to agriculture.

has relevance for the possible problems of coordination with the discipline of Law No. 203/1982.[39]

Additional tax concessions are provided in Decree Law No. 91 of 24 June 2014, amended in Law No. 116 of 11 August 2014, where, among other things, it is contemplated that for the small-medium agricultural enterprises the agricultural production deriving from the joint practice of the activities, according to the common network programme, "*can be divided among the contracting parties in nature with the attribution to each, as if originally owned or produced by each, of the quota of product agreed in the network contract*". The division and the attribution of the products among the network members as if "originally owned" by the same involves important tax advantages relative to the application of land yields instead of enterprise incomes and a wider possibility of also considering as connected activities certain activities practised by the same that, in the absence of the attribution of original entitlement, would not be included among those contemplated by the third paragraph of Article 2135 of the Civil Code. Concerning tax benefits, it is worth mentioning, particularly, Article 3, of Decree Law No. 91 of 2014 (*Measures to support Made-in-Italy products*) as converted into Law No. 116 of 2014,[40] which recognises a tax credit for encouraging the constitution of new enterprise networks or new activities within the pre-existing networks in the case of enterprises that produce agricultural products. Under Article 6-*bis* of the aforesaid provision, emending Article 1 of Law No. 311 of 2004, it is ruled that the resources referred to under paragraph 354 of the 2005 Financial Act, may also be used for subsidised loans for research and technological innovation carried out by agricultural, forestry and agri-food companies that are parties to a network contract; the successive paragraph 2 rules that "*except for the limits imposed by European law, agricultural, forestry and agri-food enterprises organised pursuant to a network contract [...] acquire priority in access to the loans contemplated by the measures of the regional and national rural development programmes relative to the 2014-2020 planning*".

The legislator's favour regarding concessions within the sphere of labour law for enterprise networks has resulted in general provisions contemplating the institution

[39]The reference is to L. Russo, ibidem, 194.

[40]Article 3 of Decree Law No. 91 of 24 June 2014: "*In order to foster the creation of new enterprise networks or the performance of new activities by the already existing enterprise networks, the enterprises which produce agricultural products, fish and aquaculture products referred to in Annex 1 of the treaty on the functioning of the European Union, as well as the small and medium enterprises, as defined by Regulation (EC) no. 800/2008 of the Commission, of 6 August 2008, which produce food and agricultural products, fish and aquaculture products not included in the afore-mentioned Annex I, even if constituted in cooperative societies or united in consortia, are allowed, within the expenditure limit pursuant to paragraph 5, letter b), a tax credit of 40% of the expenditure for new investments sustained for the development of new products, practices, processes and technologies, and for cooperation in chains, however and in any case of not more than Euro 400,000, in the tax period in progress at 31 December 2014 and the two successive tax periods*".

of the transfer system and of joint-employment by companies of a network.[41] Pursuant to Article 30, paragraph 4-*ter*, of Legislative Decree No. 276 of 10 September 2003, subsequent to the amendments introduced by Decree Law No. 76 of 28 June 2013 (then converted by Law No. 99 of 9 August 2013), the joint employment of the "employees hired" is allowed for all the companies that have signed an enterprise network agreement that is valid (therefore conforming to the discipline of Article 3, Decree Law No. 5 of 2009 as successively amended and supplemented) provided it is disciplined in the contract itself.

In addition, in the case of a network of agricultural enterprises, subsequent to the amendments of Article 31 of Legislative Decree No. 276 of 2003,[42] the possibility is contemplated for "*agricultural enterprises, including those constituted in the form of cooperative societies, belonging to the same group referred to under paragraph 1, or with the same owner or controlled by subjects linked to each other by family relationships up to the third degree, to proceed jointly to hire employees to work for the relative enterprises"; the joint hiring of employees pursuant to the preceding paragraph 3-bis is also possible for enterprises linked by a network contract, at least when 50 percent of them are agricultural enterprises*". Lastly, pursuant to paragraph 3-*quinquies*, "*the employers answer jointly for the contractual, social security and legal obligations which ensue from the work relationship established according to the methods disciplined by paragraphs 3-bis and 3-ter*". For joint hiring, in the case of networks of which at least 50% of the enterprises are agricultural enterprises, the above-mentioned discipline, unlike other joint employment, makes a general reference to the network contract and, furthermore, there is no need for a specific indication in the network programme.

It must lastly be observed that apart from the above-mentioned provisions, clearly aimed at encouraging and favouring the primary sector, network contracts are legitimate within the limits allowed by EU law for the consequences of such agreements from the viewpoint of the law on competition.[43] The network links

[41]The evolution of the labour market in recent years has led the legislator to introduce provisions that contemplate a disassociation between the employer and the subject, which takes avail of the employee's performance. On this point, of the various institutes, see Perulli (2014), p. 463. Joint employment is an institute similar to transfer, but different from the latter (in which the transferring employer always remains the employee's employer) in as much as the employee is employed by two or more employers, according to the practical discipline of the relationship as indicated by the parties in the network contract. However, there is no specific legal discipline for joint employment, leaving the parties to the contract free to regulate it as they think fit. According to the Ministry of Labour's Circular No. 35 of 29 August 2013, "with regard to possible criminal, civil and administrative liability – and therefore regarding sanctions imposed for unlawful behaviour – it will be necessary to refer to the network contract, therefore without 'automatic' joint liability on the part of all the parties to the contract".

[42]Amendments contemplated by Decree Law No. 76/2013, converted by Law No. 99/2013.

[43]In general terms, these agreements can be legitimate within the limits allowed by EU law, which precisely for the sake of technological development, admits them provided the consequent advantages are to a suitable extent passed on to the consumers (pursuant to Article 101, paragraph 3, of the TFEU); but this consideration merits a specific comment since it is necessary to distinguish, in the

within which the agricultural enterprise can be positioned, through a process of aggregation between enterprises, are potentially many and they contribute to resizing both the main activities and the connected activities that move in a duplicate and indistinct direction: horizontally towards the production of goods and the exploitation of the rural dimension within the complex territorial situation; and vertically for the projection of the company onto the market. It is in this perspective that the main usefulness of the network can be found, different and far from the pre-existing types of structures, such as the professional organisations, both regarding the structure—the professional organisations must adopt a corporate form while this is not required of the networks—and regarding the objectives and activities that, in the case of professional organisations, are different (Article 2 of Legislative Decree 102/2005) from those of the enterprise networks even if they are of a horizontal nature.[44]

These brief considerations outline a potential use of network contracts in agriculture that, without doubt, can turn out to be useful in the production of the product because of the complementary nature of the enterprises involved, in terms of economic efficiency, but even more for the rationalisation of the process of placing the products on the market, with the hoped-for aim of leading to an increase in the value achieved from the sale because of the decrease in the operators along the production chain. This is a very important matter in the primary sector because of the features that have long been typical of the contractual relationships in the chain of the food and agriculture sector, because of the delay on the part of the national legislator to structure a modern food and agricultural system. There are widely differing cases with a great number of models that can be adopted in practice, considering the specific aims that are to be pursued, by sharing and reducing production costs, by rationalising the management of the resources, and by optimising the production means.

The most recent statistics show growth in the interest on the part of the agricultural world for the network form of collaboration.[45] And this is the direction in which the network contract is registering success, as a tool that can offer a valid answer to the need to react to the difficulties connected to the economic crisis and to the complexity of the global markets. The cases of network contracts in the agri-food sector, registered to date, are marked by the complexity of the subjects involved, by the diversification of the types of products and by the diverse geographic areas involved. There are networks formed solely of agricultural entrepreneurs, both individuals and associations, with variable numbers of participants that operate

primary sector, whether the agreements operate "horizontally" or "vertically": while the agreements between farmers, in view of the exception of agriculture pursuant to Article 42 of the TFEU, should not represent a problem concerning the rules of competition, other types of contracting parties that resort to network contracts could create such a problem and, therefore, they must observe the limits of EU law.

[44]On this subject Jannarelli (2011b), pp. 462–473.

[45]Out of a total of more than 2500 enterprise network contracts registered in the whole of Italy, more than 400 regard agriculture, Lombardy being the Region where more contracts are stipulated. In this respect refer to *www.contrattidirete.registroimprese.it*, data updated at February 2016.

within the same production chain, and there are more differentiated networks of enterprises that operate in differing sectors (vine cultivation and wine production,[46] cultivation of olives,[47] animal husbandry, fruit and vegetable cultivation, cultivation of cereal crops, as well as in the sectors of dairy produce, forestry, fishing, bee keeping, pasta making[48] and rice farming), without any link with the territory of origin. Other examples can be found in the sector of farm house tourism in which the enterprise network tool absolves a strategic role. The "*Rete Agriturismo e Fattorie Didattiche della Sardegna*" [the Network of Farm House Tourism and Didactic Farms of Sardinia], to which several enterprises belong, has set itself the goal of increasing the capacity of enterprises to enter the national and international market through an offer of farm house tourism and didactic farm tourism through close collaboration and information exchange. In the sector of organic farming, as contemplated by Regulation (EC) 834/2007, there is an enterprise network named "Mondobio" formed of four companies specialised in the sector that, from the supply of raw materials to the transformation and successive marketing of the finished product, certified as strictly organic, aims to expand the market of reference and to find new outlets for the production thanks precisely to the capacity of coordination and of integration between the various segments of the agri-food chain.

It is worth mentioning, lastly, an example of how an enterprise network can be used through a vertical integration of agricultural enterprises and companies that transform the agricultural produce, of chemical and oil-chemical industries, of companies that produce biofuels and of companies specialised in the management of biogas plants. The purpose of "Rete CR 2050" in this case is to research new techniques and to develop new technologies using solely biomass materials, no food-feed and production waste products with the aim of increasing production yield, creating added value throughout the entire chain, increasing the agricultural income, and reducing environmental impact to a minimum; the network has set itself the target of promoting renewable energy sources by identifying technological and production solutions capable of optimising the biogas yield of residues that can no longer be processed and of purifying bio-methane. Interest in the Rete CR 2050

---

[46]In the vine and wine sector, there is a network named "Net Wine Italy" with the aim of improving the production process to favour entry to the global market and to increase competitiveness. This network includes several wine cellars, many of which are producers of the *Ex Vinis* trademark, as well as other companies specialised in packaging, plant, marketing, etc. IS a network open to the adhesion of various subjects and it counts on the skills of the various participants in their respective fields of practice, to increase the competitiveness and the innovative capacity of each of them.

[47]In the olive-oil production sector, there are small networks, like the "*Olio dei Due Mari*" network that counts only four enterprises but which pursues the aim of offering a product of a high level and quality standard in respect of rigid parameters, and larger networks like the "*Frantoi di Maremma*", which includes farms, as well as non-industrial oil mills.

[48]An interesting example is in the pasta sector, namely the "*Rete della Pasta dei Coltivatori Toscani*", to guarantee the quality of Tuscan pasta and to support the competitiveness of the local companies of the food and agricultural chain through joint investments in innovation and forms of stable collaboration in the processes of the production and transformation of durum wheat (*www.lapastadeicoltivatoritoscani.it*).

project at European level,[49] the collaboration between local institutions and category associations and the many subjects involved, show that the network contract can be modelled according to the specific network programme to be pursued.

5 Conclusions

The considerations briefly illustrated above, which start from the law on districts to reach the more recent law on enterprise networks, lead to two main conclusions. The above-described phenomenon of the aggregation of companies in the agricultural sector, more than in others, has a significant weight because of the large number of enterprises needed to achieve cost containment and revenue growth. The territorial nature of the production processes, in consideration of the delocalisation of the production processes and of the phenomenon of the network contracts, could reveal a discontinuity of the network, compared to the district model, the network having lost the local dimension, which is a feature of the district. In fact, the evolution of legislation has modified the tool by which to pursue relations between enterprises also on a local basis in which the prevailing aspect is the contractual element linked to initiatives that place a number of enterprises in direct contact with each other, as the simplest, most flexible, temporary tool that is also the most easily adaptable to the fast changes to which the enterprises that compete on international markets are exposed.

In any case, placing the issue in question in context, and specifically in the wider context dictated by the reform of the Common Agricultural Policy, therefore in the context of the progressive convergence of objectives and forms of intervention between the two keystones of the CAP, the territorial management model that can mobilise resources and favour an expansion of the economic opportunities in a manner compatible with the community's quality of life by the decentralisation of the actions and the adoption of decisions specifically chosen in view of the regional scale, which model answers the strategic configuration of rural development, always holds firm. From this perspective, it is important for the subject of the cooperation to remove marginal forms of companies in respect of the market, and to implement production development and technologies, by means of agreements along the chain also with the involvement of non-local enterprises, in the awareness that the territory and its protection is also pursued by giving value to local productions and by defending the geographic features. In fact, by marketing local products, the territory

[49]One of the main actions that the European Commission intends to carry out to implement the objectives of the bioeconomy strategy, based on the 7th Framework Project of the 2020 Horizon Programme, is to encourage the institution of *"networks that offer the necessary logistics for integrated and diversified bio-refineries and pilot and demonstration plants throughout Europe, including the logistics and the procurement chain necessary for the use in "cascade" mode of the biomass and of the waste flows"* [*Innovation for sustainable growth: a bioeconomy for Europe,* Press Release of the Commission of 13 February 2012, COM(2012) 60].

does not disappear but lives on, and with it the sustainable use of the resources and the production activities performed. From this perspective, the *contract* and the *market* maintain importance in the more recent legislative innovations.

Examination of the evolutionary process that moves from the district to the enterprise network shows that production rooted in the territory and, in general, the relationship between agriculture and territory, becomes no longer the rule pursued by the district, but it remains as an element of cultural identity, an expression of local wisdom and history, and of traditions and of natural and local vocations, expressed through the activities carried out in the area and through its products, and also through the cooperation of the chain, both horizontally and vertically, evoked by the European legislator within the sphere of Regulation (EU) 1305/2013. Conversely, network contracts can represent a form of aggregation between enterprises by which to meet the needs of the localisation of production processes considered particularly worthy of protection by the law (i.e. short chains[50]), but also to connect enterprises to exploit more profitably the placing of the products on the market and their sale, in which the territory and the values expressed by the same, not only economic but also historic, cultural and social values, are expressed through the product, and such values, through the product, are recognised as belonging to the territory thanks to their geographic signs, i.e. protected designation of origin (PDO) and protected geographical indication (PGI), pursuant to Regulation (EU) 1151/2012, linked to the product on the market, and the provisions relative to the food information to consumers under Regulation (EU) 1169/2011.

References

Adornato A (1999) Evoluzione dell'intervento pubblico e contrattazione programmata in agricoltura. Giuffré, Milano

Adornato A (2000) La contrattazione programmata in agricoltura. In: Agricoltura e diritto. Scritti in onore di Emilio Romagnoli. Giuffré, Milano

Albisinni F (2003) Distretti in agricoltura. Utet, Turin, p 684

Albisinni F (2010) Distretti e sviluppo rurale: elementi per una lettura delle regole di diritto. Agrireg 20:21–25

Albisinni F (2011) Distretti e contratti di programma in agricoltura. In: Trattato di Diritto Agrario, directed by Costato L, Germanò A, Rook Basile E, Il Diritto agrario: circolazione e tutela dei diritti, vol I. Turin, pp 409–410

Becattini G (1979) Dal "settore" industriale al "distretto" industriale: alla ricerca dell'unità di indagine dell'economia industriale. Riv ec pol ind 5:1–7

Becattini G. (1987) Mercato e forze locali: il distretto industriale. Il Mulino, Bologna, 1987

Becattini G (2000) Dal distretto industriale allo sviluppo locale. Bollati Boringhieri, Torino

[50]The "short chain" is a procurement chain formed by a limited number of operators who undertake to promote the cooperation, the local economic development and the close socio-territorial relationships between producers, transformers and consumers (Article 2, paragraph 1, letter *m*, of Regulation 1305/2013) and which represents a tool to foster the appreciation of the features of the territory and which is also important in respect of the environment.

Bianca CM (2002) Trattato di Diritto civile, La norma giuridica. I soggetti, vol I. Giuffré, Milano
Cafaggi F (2004) Reti di imprese tra regolazione e norme sociali. Il Mulino, Bologna
Cafaggi F, Iamiceli P, Mosco GD (2012) Gli ultimi interventi legislativi sulle reti. In: Il contratto di rete per la crescita delle imprese. Giuffré, Milano
Caprara A (2015) Il contratto di rete e gli adempimenti pubblicitari: le pubblicità del contratto. Giur comm 1:113–131
Caputo N (2014) Contratti di rete in agricoltura. Fisco 46:4521
Caterino D (2014) Appunti critici in tema di governance nei contratti di rete. Riv dir soc 2:164–173
Costantino L (2013) Il contratto di rete tra imprese nel settore agricolo. Riv dir agr 1:668–684
Cuffaro V (2013) I contratti di rete. Riv Dir alim 1:25–46
Genovese A (2013) Riflessioni sul contratto di rete. Profili privatistici e fiscali. Cacucci, Bari
Germanò A (2001) Manuale di diritto agrario. Giappichelli, Torino
Grandori A (1999) Il coordinamento organizzativo tra imprese. Svil org 171:75–91
Grandori A, Soda G (1995) Inter-firm, networks: antecedents, mechanism and forms. Org Stu 2:183–214
Granelli C (2013) Il contratto di rete: prime applicazioni pratiche. Contr 8–9:833–864
Iamiceli P (2009) Le reti di imprese e i contratti di rete. Giappichelli Torino
Jannarelli A (2011a) Profili giuridici del sistema agro-alimentare tra ascesa e crisi della globalizzazione. Cacucci, Bari
Jannarelli A (2011b) I contratti nel sistema agroalimentare. Trattato di Diritto Agrario, directed by Costato L, Germanò A, Rook Basile E. Il Diritto agroalimentare, vol II. Utet, Torino, pp 462–473
Jannarelli A (2013) I prodotti agricoli tra alimenti e merci: alle radici moderne dell'eccezionalità agricola. Riv dir agr 1:405
Libertini M (2014) Contratto di rete e concorrenza. Giust Civ 2:406
Macario F, Scognamiglio C (2009) Reti di imprese e contratti di rete: spunti per un dibattito. Contr 18:915
Mosco G (2010) Frammenti ricostruttivi sul contratto di rete. Giur comm 1:839
Pacciani A (1998) Il progetto "Maremma distretto rurale d'Europa". Atti dell'Accademia dei Georgofili, Firenze, pp 1–14
Perulli A (2014) Contratto di rete, distacco, codatorialità, assunzioni in agricoltura. In: Fiorillo L, Perulli A (eds) La riforma del mercato del lavoro. Giappichelli, Turin, p 463
Romano D (1999) I sistemi locali di sviluppo rurale. CNEL, Rapporto sull'agricoltura. L'agricoltura tra locale e globale. Distretti e filiere. CNEL, Rome
Russo L (2015) Il contratto di rete in agricoltura. Riv dir civ 1:184–191
Villa G (2009) Il coordinamento interimprenditoriale nella prospettiva del contratto plurilaterale, in P. Iamiceli Le reti di imprese e i contratti di rete. Giappichelli, Turin

Company Management Oriented Towards Sustainable Development: An Indirect Form of Protection of Food Diversity?

Serenella Luchena

1 Preface

Oscar Wilde said: "I can't stand people who do not take food seriously" and in fact, food is a very serious subject, especially considering the close link between food and life.

According to one doctrinal approach food is a bond that connects the various dimensions of life: life in an individual sense (an individual eats to exist, grow, achieve or maintain his health); life in a social sense (traditions, customs, techniques and dietary habits are all cultural products and the complex economic systems of food chains are built around food); and life in the ecological sense (the food pressure of humans as a whole determines the exploitation of goods and ecosystem services, but at the same time, the food phenomenon is one of the greatest examples of coevolution between society and nature).[1] Now, if food is life and if an essential characteristic of life in all of its dimensions is diversity; "the importance of food diversity (as a life tool) is evident".

However, the right to food is inserted into a European and international legal framework mainly aimed at maximizing commerce and free trade; a legislative framework that sees food as a "business". The rules and practices that govern the food system seem to be, in fact, dictated by an almost exclusive view of food as a "commodity", as a business source. It has been noted[2] that strong economic interests have often led to the loss of the characteristics of the irreplaceability of territories and cultures in which the food has its roots reducing it to a fungible commodity that is

[1]Monteduro (2015).
[2]Monteduro (2015).

S. Luchena (✉)
Department of Economic Sciences, University of Salento, Lecce, Italy
e-mail: serenella.luchena@unisalento.it

A. Isoni et al. (eds.), *Food Diversity Between Rights, Duties and Autonomies*, LITES – Legal Issues in Transdisciplinary Environmental Studies 2,
https://doi.org/10.1007/978-3-319-75196-2_22

replicable on an industrial scale with the subsequent convergence towards uniform diets of consumers who, very often, are inclined to purchase products from distant countries, not because they are products different than those obtained locally, but probably because they have a lower price. Here by price we mean in the economic sense, because that paid "in terms of health" (of the consumers) could be (and perhaps, is) very high.[3] Evidence of this can be seen in the big scandals involving the large multinational companies operating in the food industry. We can consider, only for illustrative purposes, Nestlé and artificial milk for newborns[4] and the use of genetically modified organisms in products for infants[5]; or the Coca-Cola Company responsible for serious contamination and unauthorized use of water[6]; or the Monsanto Company, leader in the agricultural and agro biotechnology sector,[7] known for being the largest producer of genetically modified and traditional seeds and involved in numerous legal proceedings in which they were convicted[8] for

[3]Tommasi (2017).

[4]Nestlé has been accused of an aggressive and irresponsible commercial policy in developing countries where powdered milk for newborns is concerned. The milk is distributed in hospitals for free. UNICEF has declared that the substitution of mother's milk for powdered milk in third world countries has caused the death of one and a half million babies each year because of problems with the sterilization of water and bottles. Furthermore, despite the adoption of the International Code of Marketing of Breast milk substitutes on behalf of the World Health Organization to which the multinational has adhered, verification has shown repeated infractions of the same. Added to this is the scandal that directly involved Italy in 2009, the year that the Italian industry was criminally convicted and ordered to pay damages, along with Tetra Pak International, for the contamination of Nidina milk with a chemical component used in the ink for packaging of Tetra Pak. In essence, the component had contaminated the milk contained in the packaging.

[5]In 2004, a test carried out by Greenpeace showed the presence of genetically modified organisms in a package of Nesquik. In 2010, in Brazil, after an investigation carried out by Senacon, a consumer protection agency, the Brazilian Minister of Justice fined Nestlé, Pepsi and the Mexican company Grupo Bimbo for having hidden the presence of genetically modified organisms in their products.

[6]The Coca Cola Company was ordered to pay India 350 euro compensation after having caused serious environmental and health damages near its bottling plant in Plachimada in the State of Kerala (Southern India. A framework of multi-sectorial damage was identified. Particularly, the multinational "caused environmental degradation through the overexploitation of the aquifer and its relative impoverishing and by the irresponsible practice of dispersing waste. The latter was presented to the farmers as fertilizer thereby contaminating the land and water with toxic metals such as: cadmium, lead and chrome.

[7]Monsanto was founded in 1901 in St. Louis, US, by John Queen and soon became the top US producer of saccharine. During its history, the company went from producing ingredients for the pharmaceutical and food industries to a company involved in the production of chemical products and phosphates, until its entrance, in the 1960s, into the agricultural sector; and finally, at the end of the 1970s, the biotechnological field in agriculture. In 1987, Monsanto decided to separate the chemical business from the others. In 2000, the current Monsanto Company was born with agriculture as its core business.

[8]The most well-known incidents regard the following: Agent Orange produced by Monsanto e Dow Chemical for the US army involved in Vietnam. The substance was sprayed by the US aviation in Southern Vietnam from 1961 to 1971. The official objective of the operation, called Operation Ranch Hand, was to destroy the forests where the Vietcong were hiding, thus depriving them of

serious damages to the environment and health, etc.; or multinational fast food McDonald's whose hamburgers, nuggets and fries have been involved in a series of scandals.[9]

It seems, however, that in recent years socio-environmental questions have continued to take on more importance. We are seeing, in fact, an ever-increasing demand for quality, wholesomeness, and authenticity of food products. Climatic and energy *shocks* and social and environmental issues related to the topic of sustainable development have all contributed to the acceleration of this process.

The figure of the uninformed, indifferent consumer is moving towards that of the more attentive "consumer"[10] oriented towards more informed purchases. So, they are people who "believe they should be given an answer to their concerns, as they no longer wish to be or feel they are a part of unjust situations and desire to play a decisive role in today's world" through "responsible consumption" with informed campaigns to better future consumption. Current consumers are no longer simply "targets" of production, but instead, are "real players in the food policy of the entire supply chain".[11]

At the same time this change has led us towards a different concept of business, that is, a business whose success lies not only in its ability to produce revenue, but also in its ability to produce healthy genuine foodstuff while contributing to the protection of natural resources and a balanced territorial development, creating jobs and giving greater importance to work quality.[12] A company who, if operating in the agri-food sector, must pay particular attention to the territoriality and quality of products, favoring ties with local reality, since it is the consumer himself who is demanding the identification of the territoriality of local food products.

---

cover and reducing their possibility to procure food. The result was 400 thousand dead, 500 thousand children born with serious malformations and 4 million people struck by tumors in the following years. Another well-known scandal was the use of polychlorinated biphenyls (PCB) in materials for repairing buildings, electronic devises, and paint. Thousands of buildings were contaminated. Particularly well-known is the episode of the "bovine growth hormone". This genetically modified hormone was created to stimulate milk production in bovines, but the effects on livestock health were devastating. And that is not all, there were consequences on consumer health too; a link between the hormone and breast cancer, colon cancer, and prostate cancer.

[9]In 2012 the Hamburger scandal in the United States when it was discovered that they used hamburger meat with an added substance called pink slime: a mush obtained from remains of the slaughter and deboning of livestock that was then washed with ammonia and put in the ground meat in various percentages before being frozen. In 2014 the scandal of expired rotten meat imported from China and used in the fast food chains: McDonald's, KFC and Pizza Hut.

[10]Musumeci (2005).

[11]Durazzo (2010).

[12]This new concept of business seems to be shared by some multinationals operating in the food industry, just think of the most important event with the theme of diet and food, I am referring to Expo 2015, whose motto was "Feeding the planet, Energy for life" and whose main themes were, among others, bettering the quality and safety of foods, the protection of biodiversity, respect for the environment as an ecosystem for agriculture and the enhancing the value of cultural and ethnic elements as "culinary tradition"; yes, indeed, it had among its sponsors McDonald's and Coca-Cola.

This has spurred what is often defined as an integral approach (multi-stakeholders) on the part of enterprises; that is, an approach that considers not only economic aspects, but also environmental and social ones.

Thus, the scenario before us today is a global market mainly aimed at maximizing commerce and free trade in which the actors move according to their own logic and interests. On one hand, are businesses that, despite being mostly moved by "egotistical" interests connected to producing revenue, cannot fail to consider the interests and needs of the other players they deal with. On the other, are all the other parties who interact with the company motivated by interests tied to health and life in general and who think standardized food is incompatible with what is a considered a fundamental aspect of life, that is..."diversity".[13]

All of which points to the need to find a point of equilibrium that would allow for a balancing of interests in which the topic of food diversity, although not fully protected by positive law,[14] would have a certain importance in the operation of businesses, while keeping in mind that the running of a business requires the use not only of traditional production factors (capital and labor), but also natural capital: air, water, land and biodiversity.[15]

Often, we hear talk of corporate social responsibility and sustainable development. One example of this is the 'Carta di Milano'. The 'Carta' was introduced on

[13]Monteduro (2015), according to whom "the individual life of an organism is made possible by the differentiation of cells, tissues, organs and apparatus. The diversity of living species, that is biodiversity, is a precondition to the maturity, complexity, wealth and resilience of ecosystems. For socio-systems, that precondition is essentially cultural diversity".

[14]Monteduro (2015).

[15]In 2004 the FAO organized "World Food Day" in which it underlined that "A rich variety of cultivated plants and domesticated animals are the foundation for agricultural biodiversity. Yet people depend on just 14 mammal and bird species for 90% of their food supply from animals. And just four species—wheat, maize, rice and potato—provide half of our energy from plants. Apart from the absolute number of species, it is also essential to conserve genetic diversity within each species. Modern agriculture has encouraged many farmers to adopt uniform high-yielding types of plant or animal. But when food producers abandon diversity, varieties and breeds may die out—along with specialized traits. This rapidly diminishing gene pool worries experts. Having a broad range of unique characteristics allows plants and animals to be bred to meet changing conditions, while giving scientists the raw materials they need to develop more productive and resilient crop varieties and breeds. Rather than a single crop variety that guarantees a high yield, farmers in developing countries are more likely to need an assortment of crops that grow well in harsh climates, or animals with resistance to disease. For the poorest farmers, the diversity of life may be their best protection against starvation. Consumers also benefit from diversity through a wide choice of plants and animals. This contributes to a nutritious diet, particularly important for rural communities with limited access to markets." The FAO estimates that "three-quarters of the genetic diversity of agricultural crops have been lost over the last century. And of 6300 animal breeds, 1350 are endangered or already extinct. Global efforts to conserve plants and animals in gene banks, botanical gardens and zoos are vital. But an equally important task is to maintain biodiversity on farms and in nature, where it can evolve and adapt to changing conditions or competition with other species. As custodians of the world's biodiversity, farmers can develop and maintain local plants and trees and reproduce indigenous animals, ensuring their survival. The document can be found at: http://www.fao.org/docrep/006/y5418i/y5418i00.HTM.

April 28, 2015 at the World Expo 2015 and is addressed to citizens, associations and companies. Its purpose is to become "a collective manifesto towards global awareness on the topic of the role of food", and is meant to be signed by everyone who intends to share the commitments therein. The 'Carta' gives companies the burden of "contributing to the objectives of sustainable development both through the innovation of processes, products and services and through adoption and fulfillment of codes of social responsibility".

But what does it mean to manage a business following principles of social responsibility to ensure sustainable development? What are the rules of *governance* company managers need to follow? And finally, does company management geared towards sustainable development represent a tool for the protection of food diversity?

2 Sustainable Development and Corporate Social Responsibility

Looking at the prospect outlined above, the first question that comes to mind is what is sustainable development and when is a company sustainable?

According to the definition suggested by the Brundtland Commission in 1987 development is sustainable when it "meets the needs of the present without compromising the ability of future generations to meet their own needs".[16] This affirmation evokes the idea of "the satisfaction of essential needs/human rights of people on an intra-generational and intergenerational two-fold scale".[17] In it we can find the expansion of the list of human rights recognized at a national and international levels, the limits of nature and, above all, the necessity to adapt economic development to these limits.

Seen from this perspective, it becomes necessary to shift "the attention from the quantity to quality of economic growth which, in order to be sustainable, must respect the constraints based on the capacity of regeneration and absorption of the natural environment". Therefore, sustainability is founded on the need to "guarantee the conditions that allow and have allowed life to develop over time, monitoring all of the negative effects of human activities on nature (air, water, land, biodiversity)

[16]The concept of sustainability, in this sense, is connected to the compatibility of the development of economic activities and environmental protection. The ability to satisfy basic needs implies, therefore, the creation of economic development that has the environment, as its main objective, but at the same time also sees richer countries adopt production processes and lifestyles compatible with the ability of the biosphere to absorb the effects of human activities and developing countries grow in demographic and economic terms at a pace that is compatible with the ecosystem. The concept of sustainable development was drawn up by the Brundtland Commission from two fundamental elements: the environment as a key dimension of economic development and intergenerational responsibility in the use of natural resources.

[17]Vercelli (2005a); Mascia (2007).

threatened by a use that is greater than its ability to renovate/regenerate. The problem is not one of maintaining a static equilibrium, which in itself does not exist in nature, but one of safeguarding and not compromising the dynamic processes of the self-organization of a bio-ecological system.[18]

Sustainable development, therefore, is configured as a multi-dimensional and cross-sectorial approach that involves fundamental transformations on an ethical level and institutional level, from local to international, through an integration of economic policy with environmental, social and cultural policies.[19]

It is obvious that a macroeconomic development will not be sustainable if it is not based on the existence of sustainable businesses.[20] So it is affirmed that, on an ideological level, for "a firm to gain citizenship in the social system" it must commit to the care, protection and prevention of the society in which it operates.[21] Since the 1980s we have witnessed an assertive affirmation of the theory of social responsibility that tends to give firms the task of directly caring for interests other than the shareholders. These elements can be identified in: (1) voluntary commitment beyond the given legal and regulatory framework[22]; (2) the selection of productive choices based on the criteria of social and environmental sustainability; (3) advanced labor policies and customer satisfaction.[23]

Therefore, a concept of corporate social responsibility arises and is tied to the notion of sustainable development, whose guiding principles are enshrined in framework documents issued by organizations or public and private institutions of national status, aimed at promoting the respect for the general rules and regulations of social responsibility among businesses operating in the transnational environment. Consider, among others, the principles contained in the Universal Declaration of Human Rights of the United Nations in the Rio Declaration on Environment and

---

[18]Mascia (2007).

[19]Libertini (2006).

[20]Vercelli (2005b).

[21]In reality, entrepreneurship if, on one hand, brings utility and welfare to society as the driving force of economic development; on the other hand, it could cause damage to the basic assets of individuals and communities (health, environment, human dignity, biodiversity) and moreover, economic power could lead it to engage in opportunistic behavior. Businesses, given the increasing globalization, are often encouraged to profit from the possibility of establishing their operations in countries with backward regulatory systems. These countries are often lacking in the most basic rules for the protection of fundamental human and social rights and consequently, can offer traders a particular freedom of action and cost savings. The topic of business ethics is quite vast and we can find ample economic, business-economic and sociological literature on the subject. See: Sen (1998), Nobel prize winner in Economic Science. See also Sacconi (1991), D'Orazio (2002), Corno (2002), Rossi (2003) and Buonocore (2004).

[22]All those activities imposed by laws that protect interests outside the company or the interests of those who enter into contractual relations with it are excluded. See Calandra Buonaura (2011).

[23]Libertini (2009).

Development of 1992,[24] the UN Global Compact of 1999,[25] and the OECD guidelines for multinational companies of 2011.[26] Also in the European Union we can find motions for corporate social responsibility in the form of the issuance of non-binding acts aimed at supporting the dissemination of a culture of corporate social responsibility and acts related to the profile of the disclosure of non-financial information that resulted in Directive 2013/34/EU on "annual financial statements, consolidated financial statements and related reports of certain types of undertakings, amending Directive 2006/43/EC of the European Parliament and of the Council and repealing Council Directives 78/660/EEC and 83/349/EEC".[27]

But what are the interests included in the concept of social responsibility?

Social Responsibility is defined by the European Commission as integrating "social and environmental concerns in their business operations and in their interaction with their stakeholders on a voluntary basis." And further states that "being

[24]The United Nations Conference on the Environment and Development held in Rio de Janeiro in 1992, strengthened the principle of sustainable development through the formalization of the acts adopted after the Summit: The Rio Declaration on Environment and Development, Agenda 21 and Non-Legally Binding Authoritative Statement of Principles for a Global Consensus on the Management, Conservation and Sustainable Development of All Types of Forests. The notion of sustainable development was also embraced in the environmental treaties to be signed in Rio: specifically, the Convention on Climate Change that came into force in 1994, and the Convention on Biological Diversity that went into effect in 1993. Particularly, Article 2 of the Convention on Biological Diversity contains the concept of 'sustainability', defining the 'sustainable' use of biological resources in a manner and at a pace that does not imply a long-term reduction and that maintains the capacity to meet the needs of both present and future generations. The Acts of Rio and subsequent world conferences promoted by the United Nations, especially the Johannesburg Conference in 2002, confirm a principle of sustainable development based on three interdependent factors: environmental protection, economic growth and social development. Since the UNCED, sustainable development has been consolidated as a principle of international law and has contributed to the evolution of international environmental law through the conclusion of global environmental treaties and several regional agreements. In the European Union, sustainable development is fundamental to the actions and policies of the Union in environmental matters.

[25]Global Compact is the UN initiative on corporate social responsibility launched by the UN Secretary General Kofi Annan at the World Economic Forum in Davos—Switzerland-1999.The Global Compact is the UN initiative on corporate social responsibility launched by the UN Secretary General, Kofi Annan, at the World Economic Forum in Davos—Switzerland- in 1999. The aim is to invite companies, agencies of the UN, trade unions and society to make a "Global Pact" to promote corporate social responsibility through voluntary compliance and the promotion of ten fundamental principles relating to human rights, labor, environment and the fight against corruption. It is not a legislative instrument, but is a reference framework that is based on the voluntary choice of companies to commit themselves to responsible corporate citizenship; trusting in public accountability, transparency and a new awareness of the individual interests of companies and other entities engaged in social work.

[26]For a vast discussion of the principles of OCSE 2011 see: Gulotta (2012).

[27]Green Papers—Promoting a European framework for corporate social responsibility COM (2001) 366; Communication from the Commission to the European Parliament, the Council and the European Economic and Social Committee of 22 March com 2006 136; Communication from the Commission to the European Parliament, the Council and the European Economic and Social Committee and the Committee of the Regions COM/2011/0681.

socially responsible means not only fulfilling the applicable legal obligations, but going beyond and investing more in human capital, the environment and stakeholder relations".[28] In this aspect, the ethical conduct that the company undertakes does not merely consist in the compliance with abstract moral criteria, but instead is based on the practical consideration of human and social interests and needs that may conflict with economic activities. Thus, the significant interests, for these purposes, are those of the firm's employees, its suppliers, its consumers and the community in which it operates who can be summed up in the term *stakeholders*, the interests different from those brought by *shareholders,* that is, by those who own company stock.[29]

In fact, the raison d'être of the problems related to corporate social responsibility lies in the fact that the interests of shareholders and managers do not necessarily coincide with the interests of other stakeholders.[30] That is, the former pursue the interest of maximizing profit and this often conflicts with the interests of consumers, employees, suppliers etc. How to resolve such a conflict?

3 Brief Reflections of Corporate Responsibility and Italian Corporate Law

From what we have just seen emerges a certain vagueness in the concept of corporate social responsibility that influences the identification of what are or what should be the imposed conduct of the principles contained in it. Such vagueness is accentuated by the fact that their definition depends, as authoritatively claimed, on "the level of sharing of ethical corporate values which develop in a given community".[31] Now, without neglecting the recommendations that come from the documents of the transnational entities previously mentioned, we will try to verify, from a standpoint of existing law, the true feasibility of such responsibility in our legal system.

---

[28]Calandra Buonaura (2011), according to whom "from the vast literature and, above all, from the documents drawn up by international bodies that have dealt with corporate social responsibility, we get the impression of such a generality and vagueness in the definition of not only the scope and content, but also in the inspiring principles of that responsibility as to create a noticeable bewilderment in the jurist who wants to find indications on which to define rules of governance to which administrators must adhere".

[29]Costi (2006), according to whom in the range of stakeholders it is necessary to distinguish between "stakeholders that have contractual relations with the business (consumers, suppliers and employees) and stakeholders who are not tied to it through a specific relationship, such as the community in which the business operates. In the first case the stakeholders give their contribution to the company in the various roles of employees, consumers and suppliers; in the latter stakeholders' interest can be positive (community wellbeing), but often has a negative sign, that is, the interest of the community is to come out unscathed (so-called negative externalities) which might come from the business operations (think of environmental damage)".

[30]Costi (2006).

[31]Calandra Buonaura (2011).

The issue of corporate social responsibility, when viewed in the context of corporate organization, cannot disregard a fundamental element that is represented by social interest, which would seem to be the sole parameter for identifying the interest administrators must pursue and could be identified as an interest of the partners with a money-making aim[32] or a cooperative aim[33] for which the company was formed.[34]

Now, in the absence of a norm that allows administrators to pursue the interests of stakeholders other than the shareholders themselves can they put those interests on the same level? Probably not. In light of the existing provisions, they could make decisions designed to meet the interests of external stakeholders only if such decisions, at least in the long run, were beneficial to the shareholders,[35] or when the company has voluntarily acquired this approach as part of its internal organization through the inclusion in statutes in a clause that establishes, for example, that "the pursuit of the objective profit must respect the interests of consumers, suppliers and the community"; that is, when the company has adopted codes of conduct with the prior consent of its shareholders.[36]

As can be seen from one of my previous works the institutionalization of corporate social responsibility, in the terms seen above, would seem to place the interests of the various *stakeholders* on the same level (including those of the *shareholders*) and it would also seem that, in this way, administrators are required to pursue, in a balanced way, all of the interests involved in business operations.[37] It is as if there were a redefining of the role of administrators, no longer seen as fiduciaries of the shareholders, but instead, as "mediators"[38] among the different categories of subjects who contribute in various ways to the operating of the business.[39]

---

[32]Art. 2247 cod. civ.

[33]Art. 2511 cod. civ.

[34]Calandra Buonaura (2011).

[35]The debate among jurists related to the nature of social interests ended with the substantial victory of the contractual positions exalting the maximization of "shareholder value" as the main objective of a business in a context characterized by an increasing autonomy of managers and "by an increasing pressure of financial markets and institutional investors on the performance of the same managers". See Libertini (2009). In this way the company is conceived as an "autonomous center of financial programming of the adding value to invested capital, with great autonomy in the choice of productive tools, found on global markets with the possibility of decentralization". See still: Costi (2006); Calandra Buonaura (2011).

[36]One could ask if a majority is enough or is it necessary for it to be a unanimous decision.

[37]Luchena (2011).

[38]The term cannot be taken in the strict sense of the word because a mediating body should be either absolutely disinterested or adequately representative of the different parties to be trustworthy. See Denozza (2005).

[39]Romagnoli (2002), according to whom anyone involved in the economic analysis of organized business tends to consider simultaneously the interests of members (*shareholders*) and the third parties involved in the initiative of the issuer (*stakeholders*) in a perspective of continuity that seems to dilute the differences between those who are part of a contract and those who are not.

Amidst all this, however, we must not forget that shareholders, in any case, retain the power, approximately mediated, to appoint and dismiss administrators; therefore, the latter, while exercising discretionary power, could not legitimately fail to pursue the interests of their principals as a priority. It follows that, in terms of legal liability, managerial decisions oriented exclusively or predominately towards the maximization of shareholders' interests, even at the expense of the interests of the other *stakeholders,* could not be criticized if they are in compliance with the law.[40]

Basically, administrators will tend to consider, in their managerial activities, interests different than those of the shareholders only if the former are instrumental in the pursuit of adding increased value for the shareholders that is viable in time. They might feel obligated to satisfy those interests when there is a concrete benefit for the company and its members; a benefit that might not necessarily be immediate and which could be economic or reputational and, in the latter case, it is presumed that the benefit may be transformed into an economic one.[41]

Now, after reaching these conclusions and keeping in mind that international law is traceable, almost entirely, to the vast category of soft law, it is legitimate to wonder with what tools a business can concretely reach its goals of social responsibility.

4 Tools for Social Responsibility: Ethical Codes

The tools with which you can find a concrete fulfilling of corporate social responsibility can be traced back to ethical business codes adopted voluntarily and to social reporting (which, under latest legal norms, for some businesses is mandatory). The adoption of a code of ethics determines the beginning of a dialogue with the *stakeholders* through a process of *stakeholder engagement* at the end of which a company makes a social report where the business highlights the social consequences of its actions.[42]

The code of ethics can be called a "Constitution", that is, a fundamental charter of rights and moral obligations of each person who participates in a business organization. It aims at dignifying profit "value" through "values".[43]

---

[40]Denozza (2005), according to whom "the normative framework puts administrators in a position in which the complete satisfaction of the interests of partners protects them both from losing their job and from legal liability; while the satisfaction of the other stakeholders' interests, if it conflicts with the satisfaction of that of the partners, exposes them, at best case, to risks and, in the worst case, to the certainty of negative consequences. In this situation you could not even ask a saint to behave as a mere mediator".

[41]Calandra Buonaura (2011).

[42]Horn (1980); Hamilton (1984); Muchlinski (1995).

[43]See Luchena (2011).

The prevision of an apparatus of behavioral rules was encouraged by various legal provisions at a national, a European and international level.[44] Among others, the d. lgs. 8 June 2001 n. 231 (regarding the administrative liability of legal entities, companies and associations without legal status) grants organizations, including businesses, exemption from liability for offenses committed in their interest or to their advantage by subjects placed in senior positions or under direction or supervision, if they have adopted "organizational models" to prevent illegality, models that can be adopted based on codes of conduct drawn up by associations representing organizations, then notified to the Ministry of Justice that, in conjunction with the competent ministries, has 30 days to formulate any observations concerning the suitability of the models to prevent crimes.[45] Even if, in this case, the object is to avoid "illegal behavior and not ensure socially responsible behavior". It is evident that at an organizational level a connection is feasible.[46]

The structure of codes of conduct varies according to business choices. It might be made up of: general ethical principles that reassume and guide the company *mission*, defining the reference values (for example: honesty, transparency, social responsibility, importance of the individual—diversity and inclusion, labor protection, safeguarding of the environment, respect for laws, codes and regulations); or of specific principles of behavior that analytically define the guidelines and norms to which the recipients of the Code must conform and also rules for managing relations with the various *stakeholders*; or still, of indications for the procedures of implementation and control with specifics on the methods of monitoring and reporting of violations of the code of ethics and indications on related disciplinary actions.

The code, therefore, is a sort of container for the rules: rules regarding social responsibility commitments entrusted to those with some discretional power and rules entrusted to all those who work in and for the organization.[47] It is interesting to observe that the same can take on legal significance if they are the subjects of appropriate advertising tools.[48]

5 Effectiveness of Ethical Commitments

The profile of the effectiveness of ethical commitments stirs particular interest. In fact, there is no lack of people who express their concern that the multiplying of initiatives in social responsibility is only an opportunistic choice that is a façade for businesses who need to salvage their image, perhaps because of a past scandal. The

[44]See: Luchena (2011), note n. 7.

[45]Luchena (2011).

[46]Costi (2006).

[47]Rossi (2008).

[48]Irti (2003). See still: Tommasi (2012).

effectiveness of the regulations contained in ethical codes depends both on the balance of interests through the provision of adequate organizational models whose function is to limit the discretion of administrators, and on the adoption of systems of incentives, monitoring, sanctions whose function is to make a concrete impact on corporate behavior through ethical regulations.[49]

The commitment of administrators to balance interests must come from the rules of the ethical code. This happens now when the social or environmental objectives pursued require that administrators adhere to sufficiently specific rules or criteria (for example, when an administrator is asked to respect the applicable rules of a Western country or of the European Community, even if the activity of the company takes place in a country without such laws) and when an ethical commitment translates into a commitment of means. Therefore, administrators are called on "to prepare all necessary measures to ensure that each phase of social life, whether it is pertinent to the decisional moment or the executive one, is sufficiently structured so that the result produced corresponds to a parameter the administration considers to be correct".[50]

Thus, it is necessary to specify and contextualize the commitments taken on in general terms in the ethical code through the provision of coherent organizational models, which are structured around contingency plans entrusted to specific offices and procedures (for example, the activation and financing of research aimed at maximizing the reduction of environmental impact where the company has made a commitment to protect the environment even beyond what is prescribed by law). This approach focuses on the organizational factors of a business as a tool for prevention and containment of the risk of negative externalities and bases their adequacy on the evaluation of the diligence required of administrators in their management activity.[51]

In essence, the organizational model contained in a code of ethics must: (1) indicate the techniques and methods that the company shall adopt to pursue its objectives; (2) represent the venue in which the company shall establish its content and the limits of the ethical or social commitment it has voluntarily taken on, dictating the parameters within which administrators shall use their discretion when balancing different interests.

Organizational rules might also set a limit to the costs a business intends to take on to carry out its voluntary commitments.[52] It may happen that the task of balancing falls into the hands of the administrators in cases in which it escapes the organizational dimension and so, the preventive organization; what is important is that it is

[49]Luchena (2011), note n. 25.

[50]Toffoletto (2008).With the reform of corporate law in 2003 the obligation of an "adequate organizational structure" became a main feature on which to base the modern concept of corporate social responsibility for organizational risk both from a standpoint of public law (criminal liability) and from a standpoint of civil law (damage payment). See: Pecoraro (2005).

[51]Pecoraro (2005).

[52]Rossi (2008).

possible to find a criterion that allows for the identification of priorities and resolution of potential conflict in the code itself.

Clearly, the articulation of the adopted organizational model will depend on the commodity sector and on the size of each individual firm.[53]

Of vital importance for the effectiveness of the plan, is the provision of disciplinary sanctions that an internal organ responsible for ensuring compliance with the code should put into place.[54] A self-regulatory structure, even if it is very precise and detailed, loses credibility if it does not have, as a counterpart, a proportionate disciplinary apparatus.[55] Otherwise, it is nothing but a system that, even if perfect, is totally unarmed, destined to remain on paper and devoid of effectiveness. The risk, therefore, is the construction of code-alibis, cleverly created to reassure the associates about the morality and purification of the corporate *humus,* but which in reality lack a genuine goal of self-regulation.

---

[53]Pecoraro (2005), according to whom "a universally valid *compliance program* does not exist".

[54]Costi (2006).

[55]In America a study by the Department of Justice in 1969 had highlighted that, in a year, more than 60% of the companies examined had been convicted. Later, research done in 1980 by *Fortune* magazine showed that over a 10-year period, 11% of the "most respected" companies had been caught red-handed committing what the magazine defined as "flashy offenses" (rigged contracts, large scale scams, corruption of senior officials). In the 1980s all sorts of crimes occurred (fraud committed by the most well-known brokerage companies, savings institutions, health firms and insurance companies). The situation continued into the 1990s, a time when fines were imposed for antitrust violations and rigged contracts. Edward Kennedy said "corporate crime robs the nation. They rob our children of clean air and water, they rob our elders in terms of mental health, enticing them with ill-advised investments; they rob people whose lives depend on medication that is effective changing it with dangerous and inefficient drugs. This is how corporate crime undermines public confidence in our free market system". Moral of the story: despite the extraordinary apparatus realized to protect transparency in 1933; despite the extraordinary work done by the SEC in the 1970s and 1980s, it became clear that the system of preventative controls was partially in vain. And so, in 1991 the *U.S. Sentencing Commission,* established in 1984, launched a reform inspired by the *carrot-stick* philosophy, establishing high fines (stick) and simultaneous reduction of the same, up to 80% and more (carrot) for firms that established self-regulatory programs (so-called *compliance programs*), thus giving "an injection of ethics to business". However, businesses had preceded the reform by launching the idea that any "good" company should have a code of ethics, but this expedient soon revealed its limits. Since the ethical codes did not foresee any sanctions, they became a mere alibi created to reassure partners of the morality and purification of company culture, but without any authentic self-regulatory measures. Basically, a fraud that led some commentators to conclude: "business ethics has become the business of ethics." The situation changed completely with the reform of 1991. Different than the codes of ethics, violation of the *compliance programs* is hit hard as hard as possible by the law, including the 'death' of the company. The reference to individual ethics is the premise on which *compliance programs* are built.

6 Social Accounting

While the adoption of codes of conduct aims at defining, *ex ante,* the rules to which the company must adhere in its relations with stakeholders, social reporting identifies, *ex post,* the actual behavior.[56]

For some time, the world of corporate reporting has been undergoing a profound change.[57] They have been discussing the autonomy a social accounting report (so-called "non-financial") must take on in relation to the traditional financial statements. If for some it should have ample autonomy, for others it could have a place in the financial statement or the documents that accompany it.[58] The latter seems to be the path followed by the EU with Directive 2014/95/EU of the European Parliament and the Council of 22 October 2014, previously mentioned, which had modified directive 203/34/EU regarding communication of a non-financial nature by some businesses and some larger sized groups.[59]

According to what was established in the Directive "Large undertakings which are public-interest entities exceeding on their balance sheet at the closing dates the criterion of the average number of 500 employees during the financial year shall include in the management report a non-financial statement containing information to the extent necessary for an understanding of the undertaking's development,

[56]Costi (2006).

[57]For a vast panorama on the progressive expansion of corporate reporting see Incollingo (2014).

[58]See: Castellani (2015).

[59]From an examination of the recitals it emerges that the directive represents the final moment of a path started a while ago in the European Union that concerns the ethical, social and environmental topics, a path aimed at bettering the quality, transparency and comparability of non-financial information. In April 2011, the Commission, in its communication entitled: "Single Market Act. Twelve levers to boost growth and strengthen confidence. Working together to create new growth", underlined the need to bring transparency in social and environmental information furnished by firms in all sectors to a comparable level in all Member States. "The need to improve undertakings' disclosure of social and environmental information, by presenting a legislative proposal in this field, was reiterated in the Commission communication entitled 'A renewed EU strategy 2011–2014 for Corporate Social Responsibility', adopted on 25 October 2011. In its resolutions of 6 February 2013 on, respectively, 'Corporate Social Responsibility: accountable, transparent and responsible business behavior and sustainable growth' and 'Corporate Social Responsibility: promoting society's interests and a route to sustainable and inclusive recovery', the European Parliament acknowledged the importance of businesses divulging information on sustainability such as social and environmental factors, with a view to identifying sustainability risks and increasing investor and consumer trust. Indeed, disclosure of non- financial information is vital for managing change towards a sustainable global economy by combining long-term profitability with social justice and environmental protection. [. . .] Thus, the European Parliament called on the Commission to bring forward a legislative proposal on the disclosure of non-financial information by undertakings allowing for high flexibility of action, in order to take account of the multidimensional nature of corporate social responsibility (CSR) and the diversity of the CSR policies implemented by businesses matched by a sufficient level of comparability to meet the needs of investors and other stakeholders as well as the need to provide consumers with easy access to information on the impact of businesses on society." The Directive must be transformed into national law by Member States by 6 December 2016 and its implementation must take effect starting from 1 January 2017 (or during the year 2017).

performance, position and impact of its activity, relating to, as a minimum, environmental, social and employee matters, respect for human rights, anti-corruption and bribery matters, including:

(a) a brief description of the undertaking's model;
(b) a description of the policies pursued by the undertaking in relation to those matters, including the due diligence processes implemented;
(c) the outcome of those policies;
(d) the main risks related to those matters linked to the undertaking's operations including, where relevant and proportionate, its business relationships, products or services which are likely to cause relevant impacts in those areas, and how the undertaking manages those risks;
(e) non-financial key performance indicators relevant to the particular business".

The provision for the insertion in the report of non-financial information regarding the social and environmental profile has some consequences on some rather important aspects we consider here. Primarily, it implies that administrators take on responsibility for the accuracy of the information contained in such documents. Furthermore, given the required notification of such documents (together with the financial statement) to the supervisory board and the statutory auditors or the auditing firm, 30 days prior to the date for their approval at the shareholders meeting (according to art. 2429 cod. civ.), results in the burden of supervision and verification by the latter organs of the work of the administrators in terms of social responsibility with the consequences that ensue regarding the responsibility of such bodies. Moreover, the reporting of non-financial information should be deposited and presented at the shareholders meeting with the publicity that such a presentation gives and the consequent relevant legal expectations that may result in favor of the various interested stakeholders.[60]

It is interesting to note that, during the transposition, the Member States, in calling for the disclosure of non-financial information, ensure that "companies can base themselves on national, union or international standards".[61] Therefore, in Europe, the legal recognition of the principles of integrated reporting seems imminent, thus inserting the social and environmental dimensions into the area of economic-financial information, proposing three different bottom line of reporting (profit accounts, people accounts, planet accounts). And so, reporting will not only concern economic results, but also those of social and environmental responsibility.[62]

[60]Costi (2006).

[61]Internationally, the most common methodology of reporting is GRI (Global Reporting Initiative). It is an independent organization whose purpose is to design, promote and disseminate guidelines for voluntary reporting of economic, environmental and social reporting to make the reporting as credible as financial reporting in terms of severity, verifiability and comparability.

[62]Yongvanich and Guthrie (2006), Perrini and Tencati (2006) and Hopwood et al. (2010).

7 Closing Remarks

It is now a widespread conviction that a good reputation and social legitimacy have an important role in the development of, especially, large corporations because of the increasing sensitivity of consumers and investors. But if this is true, it follows that codes of conduct and social reporting tools are of great importance in legal terms.[63]

Does company management oriented towards sustainable development indirectly protect food diversity?

Based on the previous analysis, it probably does. If reputation is a fundamental entrepreneurial element, if a good business reputation on the market comes from assuming social responsibility aimed at sustainable development, if talking about sustainable development means allowing life to develop over time while keeping an eye on the negative effects of human activities on nature (air, water, land, bio-diversity), then it follows that such an approach to business management represents an indirect form of protection of food diversity.

In other words, company management aimed at sustainable development indirectly protects food diversity through the adoption on behalf of the company of codes of ethics and through reporting tools whose function is, in fact, to actualize and externalize that protection.

References

Buonocore V (2004) Etica degli affari e impresa etica. Giur Comm II:181–200

Calandra Buonaura V (2011) Responsabilità sociale dell'impresa e doveri degli amministratori. Giur Comm I:526–548

Castellani G (2015) Responsabilità sociale d'impresa. Ragioni, azioni e reporting. Maggioli, Santarcangelo di Romagna

Corno F (2002) L'etica nel governo dell'impresa. Convergenza tra pensiero laico e dottrina sociale. Guerini & Associati, Milano

Costi R (2006) La responsabilità sociale dell'impresa e il diritto azionario italiano. Quaderno di giur. comm. In: La responsabilità dell'impresa. Giuffrè, Milano, pp 83–116

D'Orazio E (2002) L'etica degli affari in Italia: dalla riflessione teorica agli sviluppi recenti nell'istituzionalizzazione dell'etica nelle imprese. Politeia LXVI:112–147

Denozza F (2005) L'interesse della società e la responsabilità sociale dell'impresa. Bancaria, 18–35

Durazzo GM (2010) Governance dell'azienda alimentare ed etica della legislazione. CONFAPI, Brescia

Gulotta C (2012) Etica e globalizzazione: riflessioni sui principi OCSE 2011 per le imprese multinazionali e sui principi ONU in tema di affari e diritti umani. Dir del Comm Inter, 539–563

Hamilton G (1984) Les entreprises multinationals: effets et limites des codes de conduit internationaux. PUF/IRM, Paris

Hopwood AG, Unerman J, Fries J (2010) Accounting for sustainability. Practical insights. Earthscan, London

Horn N (1980) Legal problems of codes of conduct for multinational enterprises. Kluwer, Deventer

[63]Costi (2006).

Incollingo A (2014) Le prime esperienze di bilancio integrato. Analisi e riflessioni. Giappichelli, Torino

Irti N (2003) Due temi di governo societario (responsabilità amministrativa – codici di autodisciplina). Giur Comm I:693–699

Libertini M (2006) La responsabilità d'impresa e l'ambiente. Quaderni di giurisprudenza commerciale. Atti Convegno per i 30 anni di "Giurisprudenza commerciale", Milano, pp 199–228

Libertini M (2009) Impresa e finalità sociali. Riflessioni sulla teoria della responsabilità sociale dell'impresa. Riv Soc I:1–33

Luchena S (2011) Codice etico e modello organizzativo-sanzionatorio nel d. lgs. N. 231/01: legittimità ed efficacia. Giur Comm I:245–258

Mascia M (2007) Crescita, sviluppo, ambiente: il ruolo dell'impresa per uno sviluppo sostenibile. Avalaible at: http://www.diocesipadova.it/s2ewdiocesipadova/allegati/1407/ mmascia_sviluppos_08_09.pdf

Monteduro M (2015) Diritto dell'ambiente e diversità alimentare. Riv Quad Dir Amb I:88–131

Muchlinski PT (1995) Multinational enterprises and the law. Oxford University Press, Oxford

Musumeci U (2005) L'impatto sociale delle multinazionali. In: Sacconi L (ed) Guida critica alla responsabilità sociale e al governo d'impresa: problemi teorie e applicazioni della CSR. Bancaria Editrice, Roma, pp 559–570

Pecoraro C (2005) Gestione del rischio da ignoto tecnologico: prime riflessioni sull'adeguatezza dell'organizzazione. Scritti in onore di Vincenzo Buonocore II:1425–1455

Perrini F, Tencati A (2006) Sustainability and stakeholder management: the need for new corporate performance evaluation ad reporting systems. Bus Strategy Environ 5:296–308

Romagnoli G (2002) Corporate governance, shareholders e stakeholders; interessi e valori reputazionali. Giur Comm I:350–370

Rossi G (2003) Il conflitto epidemico. Adelphi, Milano

Rossi S (2008) Luci ed ombre dei codici etici d'impresa. Riv Dir Soc I:23–41

Sacconi L (1991) Etica degli affari. Individui, imprese e mercati nella prospettiva dell'etica razionale. Il Saggiatore, Milano

Sen A (1998) On ethics and economics. Blackwell, Oxford

Toffoletto A (2008) Amministrazione e controlli. In: AA.VV. Diritto delle società [Manuale breve], Milano

Tommasi S (2012) Pratiche commerciali scorrette e disciplina dell'attività negoziale, Cacucci Editore

Tommasi S (2017) Food diversity and consumer protection. Eur Food Feed Law Rev 12(3):217–223

Vercelli A (2005a) Sviluppo sostenibile. Guida critica alla responsabilità sociale e al governo d'impresa. Bancaria Editrice, p 51

Vercelli A (2005b) L'impresa sostenibile. Guida critica alla responsabilità sociale e al governo d'impresa. Bancaria Editrice, p 361

Yongvanich K, Guthrie J (2006) An extended performance reporting framework for social and environmental accounting. Bus Strategy Environ 15:309–321

Part III
Food Diversity as a Legal Value: Principles, Aspects and Problems

The Multifaceted Nature of "Food Diversity" as a Life-Related Legal Value

Massimo Monteduro

1 The State of Play

International and Italian legal literature has dedicated considerable worthy research to food law[1] in relation to questions such as food security,[2] quality[3] and typicality (along with the related topics of indication and guarantee of origin),[4] the right to food, as well as food sovereignty.[5]

[1]Parisio (2005), MacMaoláin (2015), Masini (2015), Masini (2014), Alemanno and Gabbi (2014), Germanò et al. (2014), Lambek et al. (2014), Costato et al. (2013), Costato and Albisinni (2012), Ferrari and Izzo (2012), Jannarelli (2011a) and Costato et al. (2011). For a recent public law perspective see the systematic proposal by Perfetti (2014), p. 3 et seqq.

[2]Chen (2014), Rook Basile and Carmignani (2013), Collart Dutilleul (2013), Bevilacqua (2012), Adornato (2012), p. 405 et seqq; Rook Basile and Germanò (2011), D'Addezio (2010), p. 379 et seqq; Jannarelli (2010), p. 565 et seqq; Hospes and Hadiprayitno (2010) and Sollini (2006).

[3]Germanò and Strambi (2014), Cordini (2013), Petrelli (2012), Ricci (2012), Germanò (2009) and Germanò (2008).

[4]Capelli (2014), p. 487 et seqq; Capelli (2001), p. 177 et seqq; Angelicchio (2014), p. 345 et seqq; Trapè (2012), Lucifero (2011), p. 321 et seqq; Canfora (2011), p. 73 et seqq; Canfora (2010), p. 91 et seqq; Rubino (2010), p. 255 et seqq; Rubino (2007), p. 121 et seqq; Arfini et al. (2010), Cristiani and Massart (2008), Lucchi (2008), p. 49 et seqq; Strambi (2007), p. 746 et seqq; Visser (2007), p. 179 et seqq; Pisanello (2006), p. 556 et seqq; Strambi and Alabrese (2005).

[5]Rinella and Okoronko (2015), p. 89 et seqq; Andree et al. (2014), Lupone et al. (2013), FAO (2012), McDermott (2012), p. 543 et seqq; De Schutter (2012), p. 1 et seqq; De Schutter and Cordes (2011); Paoloni (2011), p. 159 et seqq; Adinolfi (2010), p. 125 et seqq Certomà (2010), p. 1 et seqq; Kent (2008) and Haugen (2007). Of great importance for their depth are the reflections of Jannarelli (2011b), p. 33 et seqq.

M. Monteduro (✉)
Department of Legal Studies, University of Salento, Lecce, Italy
e-mail: massimo.monteduro@unisalento.it

A. Isoni et al. (eds.), *Food Diversity Between Rights, Duties and Autonomies*, LITES – Legal Issues in Transdisciplinary Environmental Studies 2,
https://doi.org/10.1007/978-3-319-75196-2_23

The same cannot be said, however, in the matter of *food diversity* as a significant value in law, with the exception of a few recent research initiatives underway.[6]

One need only think that the "Carta di Milano", presented at the 2015 Expo in Milan "*Feeding the Planet, Energy for Life*", contains neither explicit nor general calls for food diversity as a legal value.[7]

Nor does the recent, and in many ways, very important law approved in Italy at the end of 2015, which contains the "Measures for the protection and valorisation of the biodiversity of agricultural and food interest", that is, the L. 1 December 2015, n. 194 (published in the Gazzetta Ufficiale n. 288 on 11 December 2015) deal with food diversity in a systematic and comprehensive manner. Notwithstanding its expansive title, this law is, in fact, focused on the protection of the genetic biodiversity of the plant, animal, and microbial resources that furnish the primary materials for agro-food production. Meaning, the Law does not directly protect diversity amongst foods, but rather the "biodiversity of agricultural and food interest" or the genetic diversity of living organisms that constitute the *input* from which different foods are obtained (see *infra*, par. 3). This is shown very clearly in the passages in which the Law declares as its object the "protection of the local genetic resources of agricultural and food interest from the risk of extinction and genetic erosion" (art. 1, comma 1), and clarifies that "for the purposes of this law, "genetic resources of agricultural and food interest" means genetic material of vegetable, animal and microbial origin, with a real value or potential for food and agriculture" (art. 2, comma 1).[8] The genetic diversity of the living sources

[6]Among these, note that on April 8, 2014, Université Laval (Québec, Canada) launched a "Law Research Chair for Food Diversity and Security", currently held by Professor Geneviève Parent. See https://chaire-diversite-alimentaire.ulaval.ca/ and https://chaire-diversite-alimentaire.ulaval.ca/wp-content/uploads/2016/08/fiche-chaire.pdf (accessed on August 2, 2017). See also the valuable contribution by Petrillo (2012), p. 211 et seqq, especially p. 222 et seqq.

[7]The Carta di Milano, presented by the Minister for Agricultural, Food and Forestry Policies on 28 April 2015, contains within its many sections a series of references, for example, to "biodiversity", to the "the diversification of agricultural production and livestock keeping so as to safeguard biodiversity and animal welfare", to the promotion of activities for the "knowledge and the exchange of different food cultures, starting with typical, local and organic products" with respect to "dietary, physical, and environmental education programmes as instruments of health and prevention" to be developed "in schools and school meal services". An express, specific and wide-ranging reference to the legal value of food diversity is therefore missing. However, the Carta di Milano recognizes that "food plays an important role in defining each person's identity and is a cultural component that describes and gives value to a territory and its inhabitants" and that it is necessary to consider "food as a cultural patrimony, and as such [...] highlighting the value of its origin and originality with transparent regulatory processes". The full text of the Carta is available at http://carta.milano.it/wp-content/uploads/2015/04/Italian_version_Milan_Charter.pdf (accessed on August 2, 2017).

[8]The Law 1 December 2015, n. 194 establishes (art. 1) the "principles for the establishment of a national system for the protection and enhancement of the biodiversity of agricultural and food interest, aimed at the protection of the local genetic resources of agricultural and food interests from the risk of extinction and genetic erosion", founded on four pillars: the national Registry for the biodiversity of agricultural and food interest (art. 3); the national Network for the biodiversity of agricultural and food interest (art. 4); the national Portal for the biodiversity of agricultural and food interest (art. 5); the permanent Committee for the biodiversity of agricultural and food interest (art.

of food is a value that is certainly implicated in food diversity, but, as this paper seeks to show, the latter is a value of synthesis that, despite encompassing animal, vegetable and microbial biodiversity, is not exhausted only within it (see. *infra*, par. 3).

What seems useful therefore is an endeavour to investigate the problem of food diversity from a broader perspective so as to delineate some frames of reference.

2 A Reference Paradigm: Social-Ecological Systems (SES) Model

The well-known conceptual model of *social-ecological systems* (SES)[9] can constitute an excellent vantage point from which to examine, in legal terms, the question of *food diversity*.[10]

---

8). Moreover, the Law envisions: the recognition of the figures of "custodian farmers" and of "custodian breeders" (engaged in the conservation of local genetic resources that are subject to the risk of extinction or of genetic erosion) (art. 2); the adoption of a "Plan" and of "National Guidelines" for the conservation of the biodiversity of agricultural and food interest; in reality however referring only to (the Plan) the biodiversity of "agricultural interest" and (the National Guidelines) to the "conservation *in situ*, on farm and *ex situ* of the vegetable, animal and microbial diversity of agrarian interest" (art. 7); the institution of a Foundation for the protection of the biodiversity of agricultural and food interest, destined solely however "to support the actions of farmers and breeders in the implementation of this law, as well as to support the public authorities concerned, exclusively for the purposes of growth, in the production and conservation of seed varieties at risk of genetic erosion or extinction" (art. 10); the promotion of "biodiversity of agricultural and food interest routes" (art. 12); the institution of the National Day for the biodiversity of agricultural and food interest, fixed for 20 May each year (art. 14); support and participation in research concerning the biodiversity of agricultural and food interest (art. 16). Among all the provisions of the law, very few touch on food diversity beyond the simple perspective of agricultural biodiversity. Among these, it is necessary to draw attention to: the promotion of the "Food community and of the biodiversity of agricultural and food interest" (art. 13), which correspond to "local areas arising from agreements between local farmers, custodian farmers and breeders, ethical purchasing groups, academic institutions and universities, research centres, associations for the protection of the quality of the biodiversity of agricultural and food interests, school meal services, hospitals, bars and restaurants, retailers, small scale and medium-sized artisan enterprises in the agricultural and food sectors, as well as public bodies", and can also have as its object "the establishment in particular of short supply chains, direct sales, and the exchange and sale of agricultural and food products within the local circuit", as well as "the study, recovery and transmission of traditional knowledge concerning … a good diet"; school initiatives, through projects promoted by the Regions to sensitize young people to and foster awareness of local agro-food products (art. 15) (Author's own translation).

[9]Hinkel et al. (2014), p. 51 et seqq; Levin et al. (2013), p. 111 et seqq; Epstein et al. (2013), p. 432 et seqq; Boyd and Folke (2012), Glaser et al. (2012), Crane (2010), p. 19 et seqq; Armitage et al. (2009), p. 95 et seqq; Norberg et al. (2008), p. 46 et seqq; Ostrom and Janssen (2005), p. 239 et seqq; Berkes et al. (2003), Holling (2001), p. 390 et seqq.

[10]Ericksen (2008), p. 234 et seqq, p. 237, who observes that "Berkes and Folke (1998), Folke et al. (2003) and Holling (2001) describe coupled social–ecological systems as co-evolved, with mutually dependent and interacting social and ecological components and highly uncertain and

Combining the perspectives of the ecological and the social sciences,[11] ecological systems and social systems are configured as sub-systems whose mechanisms of inextricable interaction and reciprocal interdependence are revealed only when integrated and studied as "parts" of the "whole", that is, the meta-systems that include them, identified precisely as socio-ecological systems. In the SES, human societies exist in their own right "within" ecosystems and "coexist" with the limits of expansion that these last impose on human behaviors, receiving, conversely, the benefits of *ecosystem services*,[12] which in large part are socially attributed to the category of the *commons*.[13] Socio-ecological systems are complex, dynamic, non-linear systems with multiple equilibria, are subject to chaotic rather than regular and gradual alterations, have non-uniform attributes, but are differentiated and distributed according to diverse and variable spatial scales depending on different temporal horizons, as well as varying levels of complexity. For this reason, an abstract "ideal equilibrium" or a primary "optimal state" of socio-ecological systems does not exist.[14]

For some years the SES model has also begun to be recognized in the field of the legal sciences, both at the international level[15] and in Italian doctrine.[16]

This model likely partly inspired the innovative legal definition of "environment" that was introduced for the first time into the Italian legal system in art. 5, para. 1, lett. c), of d.lgs. 152/2006 (so-called Environmental Code: definition in force under

unpredictable outcomes. This conceptualization of human–environment interactions is useful for food systems, although the links between the social and environmental components may be indirect in many cases."

[11]Besides the contributions cited in note 9, see also Ostrom (2009), p. 419 et seqq; Young et al. (2006), p. 304 et seqq; Ostrom (2005). Some of the tributes honouring the memory of Elinor Ostrom are, concerning this matter, particularly significant: cfr. Anderies and Janssen (2012).

[12]Cf. for a broad exploration of the subject, Potschin et al. (2016); see also Ruhl (2015), p. 306 et seqq.

[13]Lant et al. (2008), p. 969 et seqq.

[14]Matthews and Huggett (2013), p. 326 et seqq; Krasny et al. (2011), Folke et al. (2011), p. 719 et seqq; Jax (2010), particularly p. 170; Gotts (2007), p. 24 et seqq; Folke (2006), p. 253 et seqq; Rohde (2005), Walker et al. (2004), p. 5 et seqq; Gunderson and Holling (2002). For the inclusion of the point of view of the law in the transdisciplinary debate between the ecological sciences and social sciences, see Monteduro et al. (2015).

[15]Cfr. the acts of the international symposium "*Law for Social-Ecological Resilience Conference*" held on 17–19 November 2010 in Stockholm, published in 2013 in the journal *Ecology and Society* (*Law and Social-Ecological Resilience, Part I: Contributions from Resilience 2011*, Issue 2; *Law and Social-Ecological Resilience, Part II, Contributions from Law for Social-Ecological Resilience Symposium Stockholm, Sweden, 2010*, Issue 3). Besides the pioneering study by Tarlock (1994), p. 1121 et seqq, as well as in Percival and Alevizatos (1997), p. 25 et seqq, also see Ebbesson (2010), p. 414 et seqq; Benson and Garmestani (2011), p. 1420 et seqq; Ruhl (2011), p. 1373 et seqq; Ruhl (2012), p. 31 et seqq; Voigt (2013), Monteiro De Lima (2013), p. 695 et seqq; Westra et al. (2013), Arnold and Gunderson (2013), p. 10426 et seqq; Barnes (2013), p. 6 et seqq; Cumming (2013), p. 15 et seqq; Armitage (2013), p. 11 et seqq; Garmestani et al. (2013), p. 37 et seqq; Ebbesson and Hey (2013), p. 25 et seqq; Garmestani and Allen (2014), Green et al. (2014), p. 3534 et seqq; Robinson (2014), p. 19 et seqq; Monteduro et al. (2015).

[16]For the debate in Italy, see Cafagno (2007) and Monteduro (2014). For the perspective of *sustainability science*, see Montini and Volpe (2011), p. 157 et seqq.

Italian law for eleven years, from 29 April 2006 until 21 July 2017, the date in which this was repealed by art. 2 of d.lgs. 104/2017). This regulatory provision (inserted into an Environmental Code article entitled, significantly, “Definitions”) had the following formula: “environment, understood as a *system of relations* among anthropic, naturalistic, chemico-physical, climatic, landscape, architectonic, cultural, agricultural, and economic factors.”[17]

This legal definition gave a series of significant textual indications:

- the *environment in legal terms* (“a system of relations among factors”) does not completely coincide with the *environment in ecological terms* (“naturalistic, chemico-physical, climatic factors”) inasmuch as it is broader, encompassing the latter without being exhausted in it;
- the environment in legal terms (“a system of relations among factors”) also encompasses in itself, beyond the environment in the ecological sense, the *environment in the social sense*, that is, the human environment (“anthropic . . . landscape, architectonic, cultural, agricultural, and economic factors”);
- the environment in the legal sense is a “system” composed not simply of factors (that is, it is not a quantitative and static algebraic *sum* of properties, nor a type of *universitas rerum*), but above all of “relations among factors”: that is, it is a dynamic and qualitative *synthesis* of the interactions among properties integrated into a “whole” that is much more than the sum of its “parts”,[18] as a higher level aggregation of interests that expresses emergent properties, that is, new and original properties that are impossible to observe or predict at the level of the system’s individual components.[19]

---

[17]Author’s own translation.

[18]The Italian Constitutional Court, in its sentence of 14 November 2007, n. 378, affirmed: “It is necessary, in other words, to view the environment as a "system", considered, that is, in its dynamic aspects, which in reality it is, and not only from a static and abstract viewpoint [. . .] the environment as an organic entity, prescribed, that is, by standards of protection that have as their aim its entirety as well as individual components considered as parts of the whole [. . .] the uniform and comprehensive discipline of environmental heritage pertains to the public interest in primary (sentence n. 151 of 1986) and absolute (sentence n. 210 of 1987) constitutional terms, and must ensure (as EU law requires) a high a level of protection, unwaivable by other sectoral laws. It should be emphasized, however, that alongside unified environmental heritage law other legal interests can coexist, their object being components or aspects of environmental heritage but concerning diverse interests that are under legal protection. In this respect, the environment is "transverse matter", in the sense that diverse interests make demands on the very same object: that of environmental protection and those pertaining to its uses”. The same constitutional court, in its sentence of 13 April 2011, n. 128 and of 12 March 2015, n. 32, affirmed that the “environmental protection” concerns “the interactions and balance between the different parts of the "biosphere" as a "system"” [...] in its dynamic aspect” (Author’s own translation).

[19]Cf. Swannack and Grant (2008), p. 3477 et seqq, p. 3478, where the different levels of life aggregation are organized according to a hierarchy of increasing complexity are called to mind: (1) the cell; (2) the tissue; (3) the organ; (4) the system; (5) the individual organism; (6) the population—living systems composed of individual organisms from the same species –; (7) the community or biocoenosis—living systems composed of organisms from different populations –; (8) the ecosystem—living systems composed of diverse communities that interact with a biotope,

That positive definition of "environment" seemed to assume precisely the conceptual lines of the socio-ecological systems model: the environment construed as a legal concept, with the notion furnished by Italian legislature being "taken seriously", is therefore not exhausted in ecosystems nor in social systems considered in isolation and separately, but serves as a meta-system (socio-ecological) of systems (ecological and social) that interact by organizing themselves into a whole that is greater than the sum of its parts.[20]

In light of what has been stated thus far, environmental law is called upon to establish principles and binding regulations so that the interrelations between *nature* and *society*[21] take place, ensuring that, in both space and time, the diverse

---

that is, with abiotic components represented by a set of physical and chemical factors such as climate, light exposure, the presence of water, soil type and substrate; (9) the landscape, which in ecology is simply a living system composed of ecosystems dynamically interrelated amongst themselves; (10) the biome or eco-region, a system of landscapes; and finally, (11) the ecosphere, that is, the planet composed by different biomes. Each of these levels of organization of biological life can be considered a "*living system*", that is, a complex system (open, dissipative, autopoietic, capable of self-organization and self-renewal) that has an *autonomous existence and biological relevance distinct from the individual existence of the components that compose the system.* The conceptualization that organizes life by increasing levels of complexity is founded on the very principle of "emergent properties": each time biological life passes from one aggregation level to another that is higher on the hierarchical pyramid, the higher level system exhibits new properties, or original characteristics that cannot be foreseen by observing individual components on the preceding level. The new properties of the higher level living system are not the result of the quantitative sum of the properties of the individual components but of their qualitative synthesis, inasmuch as the emergent properties arise *ex novo* from the collaboration and close interaction of all the co-existing lower level living systems. Cfr. Jordán and Jørgensen (2012), Russell et al. (2011), p. 2 et seqq; Thomas and Cebrian (2010), p. 380 et seqq; Campbell et al. (2008), p. 2 et seqq; Naveh and Carmel (2003), p. 35 et seqq, p. 37; Lévêque (2003), particularly p. 134. For a transdisciplinary biosociological approach, see Pumain (2006). An attempt at a general theory of "*living systems*" that transcends the scope of the biological sciences, proposing a heuristic framework that is also valid for the social sciences was advanced in the past by the powerful work of Miller (1978).

[20]See Monteduro (2016), p. 269 et seqq. For a conceptual perspective favorable to the configuration of the environment in the legal sense as a complex "system" of relations, in Italian legal doctrine, see in particular Cafagno (2007) and Farì (2013), particularly p. 3 et seqq e 24 et seqq.

[21]Note that, in the ecological sciences, there are those who have recently proposed an approach that transcends the SES model: namely "anthroecology theory", on which see Ellis (2015), p. 287 et seqq. Even in the abstract of this essay, the author affirms that "though still at an early stage of development, anthroecology theory aligns with and integrates established theoretical frameworks including social–ecological systems, social metabolism, countryside biogeography, novel ecosystems, and anthromes. The "fluxes of nature" are fast becoming "cultures of nature." To investigate, understand, and address the ultimate causes of anthropogenic ecological change, not just the consequences, human sociocultural processes must become as much a part of ecological theory and practice as biological and geophysical processes are now". According to the author (pp. 316–317) "the call to integrate humans into ecology is older than the discipline itself and has been loud and clear for generations (e.g., Darwin 1859; Tansley 1935; Hawley 1944; Odum 1953; Pickett and McDonnell 1993; Redman et al. 2004; Collins et al. 2011). Major progress has been made by theory on social—ecological systems (Redman et al. 2004; Folke et al. 2005; Folke 2006; Hornborg and Crumley 2007; Alessa and Chapin 2008; Carpenter et al. 2009; Chapin et al. 2011; Collins et al. 2011; Levin et al. 2013), social metabolism (Baccini and Brunner 2012; Fischer-Kowalski et al. 2014; Malhi 2014), coupled human and natural systems (Liu et al. 2007), human ecology (Boyden 2004; Dyball and Newell 2014), urban ecology (McIntyre et al.

interconnected factors (ecological, cultural, economic) continue to be able to exist together (coexistence) and to be able to evolve together (coevolution) as has occurred in previous eras. The purpose of environmental law is therefore that of

---

2000; Pickett et al. 2001; Grimm et al. 2008), agroecology (Tomich et al. 2011; Gliessman 2015), countryside biogeography (Daily et al. 2001; Mendenhall et al. 2014), novel ecosystems (Hobbs et al. 2006), anthromes (Alessa and Chapin 2008; Ellis and Ramankutty 2008), and Anthropocene island biogeography (Helmus et al. 2014). Anthroecology theory builds on these with the aim of going further. To advance the science of ecology in an increasingly anthropogenic biosphere, it is useful to begin with "The First Law of the Anthropocene": the ecological patterns, processes, and dynamics of the present day, deep past, and foreseeable future are shaped by human societies. Anthroecology theory takes this law as given, and integrates human societies into ecology globally across geologic time through an evolutionary framework of human sociocultural niche construction directed at explaining ecological pattern, process, and change within and across an increasingly anthropogenic terrestrial biosphere. In this way, human societies are integrated into ecological science in much the same way as climate systems, which powerfully shape the patterns, processes, and dynamics of ecology, while also being influenced, but less so, by these interactions". Further, the author affirms (p. 320) that "for ecologists, the meaning should be very clear. The forces of humanity are now akin to those of climate geophysics or biology and therefore as fundamental to understanding the processes that shape life on Earth as the sciences of climate, soils, or biology. To engage in scientific study of ecological pattern, process, and change as it exists today and for the foreseeable future demands a firm grasp of the human sciences and their deep integration into ecological theory and practice. It is no longer adequate merely to study the consequences of human transformation of ecological pattern and process: Ecology must become a science of their ultimate causes. In the teaching of ecology, humans are generally presented as operating entirely within a biological world, sustained by natural ecosystems, in statements like "humanity is a biological species in a biological world" (Wilson 2012) and "that man is, in fact, only a member of a biotic team is shown by an ecological interpretation of history" (Leopold 1949). Human alteration of ecology tends to be depicted as a recent crisis brought on by modern industrial societies and their rapid population growth, disturbing fragile natural ecosystems, and threatening both humanity and nonhuman nature, with such framing usually accompanied by a call to return to or maintain some prior balance of nature (Rockstrom et al. 2009; Simberloff 2014). As with the seemingly perpetual need for ecology to reject the balance of nature concept, these romantic notions should have no place in ecological science (Cronon 1983; Briggs et al. 2006; Pickett 2013) [...] Humans are a sociocultural species living in a sociocultural world on a used planet. It is time to go beyond balances of nature and even fluxes of nature to embrace the "cultures of nature" in ecology. The paradigm must shift. Cultures create and sustain natures. Individual humans act intentionally, but they do so within their social contexts and depend on cultural values, perceptions, and actions (Dyball and Newell 2014; Ives and Kendal 2014; Mace 2014; Medin and Bang 2014). The question is not how to stop "others" from destroying nature, or finding a way to "get back to nature," but how to engage societies toward shaping nature more beneficially for both humans and nonhumans (Mace 2014; Palomo et al. 2014). Socio-cultural niche construction in an increasingly anthropogenic biosphere is neither new nor disastrous, but the perpetual activity of human societies engaged in the intentional cooperative engineering of ecosystems since prehistory (Smith and Wishnie 2000; Ellis et al. 2013b). To incorporate human sociocultural niche construction at the core of ecological pedagogy, the framing of humans as destroyers of nature must transition to narratives of societies as nature sustainers (Chapin et al. 2011). In moving toward this goal, the work of archaeologists, natural historians, agroecologists, urban ecologists, conservationists, engineers, and designers all have much to offer and much to gain. The teaching of ecology has always appealed to a sense of wonder about the natural world. As educators we must build a new sense of wonder and discovery about the "tangled bank" of human sociocultural systems and the diversity of ecosystems they create and sustain together with the traditional ecologies of cultures, cultural landscapes, and cities. By embracing sociocultural evolution and teaching it, current and future generations of ecologists and the public will be better equipped to guide societies toward better outcomes for both people and nonhuman nature."

safeguarding, at all territorial levels, the durability of the conditions that are indispensable for the survival of socio-ecological systems.

3 Social-Ecological Systems in the Anthropocene Era: The Transgression of Planetary Boundaries

The role of *environmental law* becomes crucial when considering the more recent discoveries of the *environmental sciences*, which highlight the ever more dramatic acceleration of the ecological crisis.[22] In the first place, the theory of the "Anthropocene" confirms the dawn of a new geological era for humankind, subsequent to the Holocene, when for the first time in history, human beings would become the main and decisive factor in the transformation of terrestrial environmental conditions, surpassing the combined impact of all external forces, be they natural, astronomical or geophysical, that had determined the level of change in the Earth-system during our planet's 4.5 billion years in existence.[23] In the second place, a now famous study published in 2009 stresses that human societies have already transgressed some of the "Planetary Boundaries", thus leaving the "Safe Operating Space for Humanity" and entering the "Danger Zone", where the risks of ecosphere destabilization from uncontrollable, non-linear type processes are extremely high, with outcomes that could prove to be catastrophic for the human species.[24] In the study, nine large-scale Earth-system processes that regulate the fundamental ecological systems that sustain life on earth were identified; quantitative limits (to be precise: the planetary boundaries) for seven of these were identified,[25] and three of

---

[22]Westra et al. (2013), Bennett (2016), p. 55 et seqq. The Encyclical *Laudato Si'* of Pope Francis dated 24 May 2015 constitutes a trenchant and passionate testimonial, in its cry of alarm launched at humanity (§ 54 and § 161): cfr. for comment Tracey (2016), p. 207 et seqq.

[23]For a recent framework, Gaffney and Steffen (2017), p. 53 et seqq; McNeill and Engelke (2016), Biermann et al. (2016), p. 341 et seqq; Lewis and Maslin (2015), p. 171 et seqq; see seminal contributions by Crutzen (2001), p. 23 and Crutzen and Stoermer (2000), pp. 17–18.

[24]The founding study is the very well-known one by Rockström et al. (2009a), p. 472 et seqq; Rockström et al. (2009b), Article 32; cf. also Folke et al. (2011), Wijkman and Rockström (2012), Galaz et al. (2012), p. 1 et seqq; Biermann (2012), p. 4 et seqq; Barnosky et al. (2012), p. 52 et seqq; Hughes et al. (2013), p. 389 et seqq. See recently Häyhä et al. (2016), p. 60 et seqq; Newbold et al. (2016), p. 288 et seqq (according to these last authors, in the 58% of the earth's surface where 71.4% of the human population live, biodiversity loss has already exceeded the planetary limit, having reached levels capable of undermining the capability of ecosystems to sustain the human communities).

[25]These seven were: *[1] climate change* (CO_2 concentration in the atmosphere <350 ppm and/or a maximum change of +1 Wm^{-2} in radiative forcing); *[2] ocean acidification* (mean surface seawater saturation state with respect to aragonite ≥80% of pre-industrial levels); *[3] stratospheric ozone depletion* (<5% reduction in O_3 concentration from pre-industrial level of 290 Dobson Units); *[4] interference with the biogeochemical nitrogen (N) cycle* (limit industrial and agricultural fixation of N_2 to 35 Tg N yr^{-1}) *and the phosphorus (P) cycle* (annual P inflow to oceans not to exceed 10 times the natural background weathering of P); *[5] rate of biodiversity loss* (annual rate

those limits were shown to have already been crossed, namely those concerning climate change, rate of biodiversity loss, and changes to the global nitrogen cycle.[26] A more recent update to this study, published by the same team of scientists in 2015,[27] has refined the theoretical model and shown that as many as four planetary boundaries have already been breached. Other than the three already cited, that of land-system change has also been crossed, primarily through deforestation[28] as trees

---

of <10 extinctions per million species); *[6] global freshwater use* (<4000 km^3 yr^{-1} of consumptive use of runoff resources); *[7] land-system change* (<15% of the ice-free land surface under cropland). The two additional planetary boundaries for which the authors had not able to determine a boundary level are: *[8] atmospheric aerosol loading*, and *[9] chemical pollution*.

[26]Rockström et al. (2009b), pp 14–15. With particular reference to biodiversity, the authors' warning even in 2009 was already troubling: "this relatively safe boundary of biodiversity loss is clearly being exceeded by at least one to two orders of magnitude, indicating an urgent need to radically reduce biodiversity loss rates" (p. 15); "the world cannot sustain the current rate of loss of species without resulting in functional collapses" (p. 20).

[27]Steffen et al. (2015). The nine large-scale Earth-system processes remained the same in 2009, but the 2015 study has produced notable updates. The planetary boundary that was called, in 2009, "loss of biodiversity" was renamed "change in biosphere integrity" in 2015: the concept of biosphere integrity includes the conservation of biodiversity but is not limited to it, inasmuch as it means to consider not only the structural impact on the biosphere (that is, on the biosphere as genetic reserve of diversity), but also the functional impact (that is, on the biosphere as a regulator of the cycle of nutrients and the flow of materials, waste and energy through the Earth system). The planetary boundary called, in 2009, "interference with the global Phosphorus and Nitrogen cycles" was renamed "biogeochemical flows" in 2015. The planetary boundary named, in 2009, "chemical pollution" became much broader in 2015, assuming the new name of "introduction of novel entities": it deals not only with toxic synthetic substances but more generally with all the substances that nature has not evolved to assimilate, for example radioactive materials and nanomaterials. The planetary boundary "land-system change" was assessed in 2015 in terms of changes to surface areas covered by forest, rather than agricultural terrain, owing to the strong impact of forests on another planetary boundary, that of "climate change". In general, the 2015 study strongly emphasized the interdependence and the interrelationships between the nine macro-processes, which cannot possibly be considered in isolation, but which create feedback effects of various types. Another matter of note from the 2015 study is the fact that some macro-processes (biosphere integrity, biogeochemical flows, land-system change, and freshwater use) were assessed also at the regional level, with the thresholds then combined globally. Finally, perhaps the most important change in the 2015 study with respect to that of 2009 is the fact that the authors identified two of the nine macro-processes as "core boundaries": these are the planetary boundaries of climate change and biosphere integrity. These two planetary boundaries are considered of more importance and specific weight than the other seven, which are influenced by a two-level hierarchy scheme.

[28]Cf. FAO, *The Global Forest Resources Assessment 2015* (a synthesis is available at http://www.fao.org/news/story/en/item/326911/icode/ and at http://www.slideshare.net/FAOoftheUN/how-have-forests-changed-results-from-fra-2015, accessed on January 16, 2016); see further MacDicken (2015). From the FAO's *The Global Forest Resources Assessment 2015* it can be seen that from 1990 to the present 129 million hectares of forest have been lost, an area equivalent in dimension to almost the entirety of South Africa. While forests constituted 31.6% of world territory in 1990, or around 4128 million hectares, in 2015 this figure decreased to 30.6%, meaning, to around 3999 million hectares. Forests on the global scale therefore continue to diminish while population increases and forests are converted into agricultural land and destined for other uses. Becasue of the rise in world population the average area of forest per capita has in fact decreased from 0.8 ha per capita in 1990 to 0.6 ha per capita in 2015. Furthermore, the percentage of planted forest rose

have been heavily logged in past decades[29] to allocate the land for other, *in primis* agricultural, uses. Furthermore, other planetary limits (the acidification of the oceans, for example) although not yet breached, are not far from the threshold.

---

compared to that of natural forest (the latter, in line with the FAO's approach, includes both areas of primary forest, where there is minimal human interference, and areas secondary forest, which, despite human interference have regenerated naturally); planted forests currently account for 7% of the world's overall forest area, having increased by over 110 million hectares since 1990. Nevertheless, in the face of such negative data there are some underlying positive trends: the net annual rate of loss in forest areas has in fact slowed down, going from 0.18% in the early 1990s to 0.08% in the 2010–2015 period. During this period, Africa and South America registered the highest net annual loss of forest, with 2.8 million hectares and 2 million hectares respectively, even though the rate of loss had diminished compared to the preceding 5 years. From 1990 to the present the majority of deforestation has taken place in the tropics. To the contrary, the net area of forest has increased in temperate regions, while there have been no significant changes in the boreal and subtropical regions. The forest areas earmarked principally for the conservation of biodiversity currently represents 13% of world forests, meaning, 524 million hectares, with the largest areas in Brazil and in the United States. The FAO finally estimates that between 2001 and 2015 the total carbon emissions from forests were reduced by more than 25%, due precisely to the slowing down of the rate of global deforestation.

[29]Steffen et al. (2015). Concerning the planetary boundary for "land-system change," the 2015 study proceeded according to the following methodology. At the global level, this has evaluated the area of forested land that is maintained on the ice-free land surface, expressed as a percentage of the potential area of forested land in the Holocene (that is, the area of forest assuming no human land-cover change). At the level of the "biome," the study considered the area of forested land that is maintained in each of the three major forest biomes—tropical, temperate, boreal—expressed as a percentage of the potential forest area in each of these three biomes. At the level of the biome, the boundary calculation was based on (1) the relative potential of land cover change within each biome to influence the climate system remotely, especially at the global level; and (2) the potential for a threshold within each of the forest biomes in which land-cover change beyond a certain area activates self-reinforcing feedbacks that lead to land-cover change across a much larger area. That being said,, globally, the planetary boundary proposed by the authors is the following: 75% of potential forest cover should be maintained (or approximately 47.9 million km^2 of the ice-free land surface of Earth, based on areal estimates). This boundary has been constructed as a weighted aggregate of the following three individual biome boundaries. Tropical Forest: 85% of potential forest cover should be maintained (approximately 19.3 million km^2). Temperate Forest: 50% of potential forest cover should be maintained (approximately 9.5 million km^2). Boreal Forest: 85% of potential forest cover should be maintained (approximately 19.1 million km^2). The database used to define the potential area of the forest biomes and that used to estimate the area of forest remaining do not use identical definitions of various forest types or what constitutes a forest compared to a woodland. In using the ESA GlobCover 2009 database [Arino et al. (2012)] to estimate current forest cover, the authors used the 100-40% cover category of remaining forest to define where forest was present in a given area. This category would also include some degraded or partially cleared forests as "remaining forest", as well as some plantation forests such as palm oil. This category thus probably overestimates the actual amount of original forest cover remaining. This would lead to somewhat high percentages of remaining forest and thus to an optimistic estimate of the actual position of the control variable with respect to the boundary. Aggregating the forest remaining compared to the potential forest for all of the biomes gives a global value of 62% forest remaining. The global boundary is 75% with a zone of uncertainty between 75% and 54%. Thus, the current value transgresses the boundary but lies within the zone of uncertainty.

The study's conclusions were shared and also received in the legal field, particularly by the 7th Environmental Action Programme (EAP) of the European Union, approved by Decision No 1386/2013/EU of the European Parliament and of the Council of 20 November 2013. The Programme has been significantly named "Living well, within the limits of our planet", and is expressly founded precisely on the recent ecological theory of "Planetary Boundaries".[30]

The Programme[31] stresses, among other things, that almost two-thirds of world ecosystems are in decline and that the constant degradation and erosion of natural capital could jeopardize two centuries of improvements in our standard of living.

4 Life as Fundamental Legal Value: The Binding Duty to Protect Life at the Ecological Scale as a Requirement of Life on the Social and Individual Scales

Hence the confirmation of the crucial and strategic role of environmental law in defending the fundamental value of *life* in its various levels of relevance for the law.

First of all, life at the individual scale, referring (in ecological terms) to the organism, that is, (in legal terms) to the subject. This is the first and fundamental dimension of life as the object of an inviolable individual right (to live) and of a collective duty (not to make life vulnerable and to protect it).

Progressively moving forward, life at the social scale, referring (in ecological terms) to populations and communities, that is, (in legal terms) to the social groups where the individual's personality develops, to the human communities united by the binding duties of political, economic and social solidarity. Social life entails the involvement of individual life not only in political organization, but also in networks of social relationships that are cultural (socio-cultural systems) and economic (socio-economic relations). The former provide "nourishment" to the community and its members principally on the intangible level, the latter on the material plane.

On an even more complex level, ecological life, that is, ecosystems and the biosphere. In ecological systems, human species coexist with non-human species (plants, animals, microorganisms) and with biotopes (non-living spaces—rock, water, air—with certain chemical, physical, climatic, light and temperature characteristics).

---

[30]Point 8, note 5, of the 7th Environmental Action Programme (EAP): "Thresholds associated with nine 'planetary boundaries' have been identified which, once crossed, could lead to irreversible changes with potentially disastrous consequences for humans, including: climate change, biodiversity loss, global freshwater use, ocean acidification, the nitrogen and phosphorus cycles and land-use change (Ecology and Society, Vol. 14, No 2, 2009)".

[31]Monteduro (2014), particularly p. 8 et seqq e 31 et seqq.

As is well known, ecological science assumes the existence of *different levels of aggregation of life, organized according to a hierarchy*[32] *of increasing degrees of complexity*: it is fundamental to understand that each level of organization is considered a "living system",[33] that is, a complex system (open, dissipative, autopoietic, capable of self-organization and self-renewal[34]) that has an autonomous existence and biological relevance[35] distinct from the individual existence of the components that compose the system. The conceptualization that organizes life by increasing levels of complexity is moreover based on the principle of "emergent properties"[36]: each time biological life moves from an aggregation level to one higher on the hierarchical pyramid, the higher level system exhibits new properties, or original characteristics that cannot be predicted by observing individual components on the preceding level. The new properties of the higher level living system are not the result of the quantitative sum of the properties of the individual components but of their qualitative synthesis, inasmuch as the emergent properties arise *ex novo* from the collaboration and close interaction of all the co-existing lower level living systems. The human being as a living individual organism is not positioned at the top of the biological hierarchy but rather halfway down the hierarchical scale of living systems, preceded by the lower levels of (i) the cell, (ii) the tissue, (iii) the organ and (iv) the system. Living systems of a higher level on the hierarchical scale than the individual organism then follow, consisting of (vi) the population—living systems

[32]Jordán and Jørgensen (2012); for a transdisciplinary biosociological approach, cfr. Pumain (2006). Recently, see Eldredge et al. (2016).

[33]Scheiner and Willing (2011), p. 333 et seqq, particularly p. 334 et seqq: "Our hierarchical perspective makes clear that ecological theory is directed at understanding biological entities at or above the level of individuals [...] one way of organizing living systems is as a hierarchy that extends from molecules to biomes. At each level, biological entities (e.g. cells, individuals, communities) interact with matter and energy to form living systems." In ecology, an opposition has historically existed between holistic guidelines (also in explicitly systematic variants, such Clements' paradigm of the "superorganism" and Havelock's "Gaia Hypothesis", or in less radical versions, such as the Odum brothers'—by now classic—energy) and reductionist guidelines (such as Gleason's individualistic concept), giving rise to a debate that that has still not been fully resolved; for a reconstruction see Bergandi (2011), p. 31 et seqq and Trepl and Voigt (2011), p. 45 et seqq, as well as Lefkaditou (2012), p. 46 et seqq. An attempt at a general theory of "*living systems*" that transcends the scope of the biological sciences, proposing a heuristic framework that is also valid for the social sciences was advanced in the past by the powerful work of Miller (1978).

[34]Kay (2000), p. 135 et seqq.

[35]Existence which, although qualitatively different from the individual organism's life, is nevertheless configured as a form with an autonomous biological significance as a higher level organization of life that transcends the lower levels and results in "*complex entities*": this expression is used by Lévêque (2003), p. 134; with specific reference to communities or biocoenosis, see Morin (2011), p. 165; for ecosystems, see Thomas and Cebrian (2010), p. 380; in relation to landscapes, cfr. Naveh and Carmel (2003), p. 35 et seqq, p. 37, according to whom (our italics) "because of their emergent organizational systems properties, landscapes are more than the sum of their measurable components. They become *an entirely new entity* as an ordered whole or "Gestalt" system, in which, *like in organisms* (or a melody) all their parts are related to each other by the general *state of the whole*". Cfr. also Campbell et al. (2008), p. 2 et seqq.

[36]Solomon et al. (2013), p. 5 et seqq; Russell et al. (2011), p. 2 et seqq.

composed of individual organisms from the same species, (vii) the community or biocoenosis—living systems composed of organisms from different populations, (viii) the ecosystem—living systems composed of diverse communities that interact with a biotope, that is, with abiotic components or a set of physical and chemical factors such as the climate, light exposure, the presence of water, soil type or substrate; (ix) the landscape—in ecology a living system composed of dynamically interrelated ecosystems; (x) the biome or eco-region, a system of landscapes; and finally, (xi) the eco-sphere, that is, the planet composed of diverse biomes.[37] Biology is traditionally concerned with living systems from the cell to the organism; ecology as a science is the specific study of living systems at higher levels than the individual organism, from populations to the biosphere, structuring itself under different disciplinary sub-areas (the ecology of populations, the ecology of communities, the ecology of ecosystems and the ecology of the landscape).

The human being, both as individual subject and as a member of the social groups wherein his or her personality is developed—that is, in ecological terms, as both organism and population—can nevertheless survive only on the condition that levels of life aggregation that are higher on the hierarchical scale survive, from communities (including non-human populations) and ecosystems: a potential collapse of these biological systems, in fact, would inexorably lead to the extinction of the lower levels that comprise them, human beings included. It follows that ecological systems are of an "inalienable existence" for humankind, under natural law, because human beings cannot destroy or harm them without simultaneously annihilating themselves or degrading their quality of life. It further follows that ecological systems are a necessary belonging for humankind, still under natural laws, through the relationship between "part" to "whole" that runs between the organism and the levels higher to it in the organization of life.[38]

5 Life, Bio-Cultural Diversity and Food Diversity

Diversity is an essential characteristic of life in all its dimensions (individual, social, ecological), and is the key that makes life possible, giving it constant nourishment and permitting movement, evolution, and the flow of energy and matter.

The individual life of the organism is made possible through the differentiation of cells, tissues, organs, and the apparatus.

The diversity of living species, that is, *biodiversity*, is a precondition for the maturity, the complexity, and the richness and resilience of ecosystems.

For sociosystems, this precondition is essentially *cultural diversity* (of knowledge, traditions, values, methods, opinions, tastes, systems of organization, production, and of relationships).

---

[37] Swannack and Grant (2008), p. 3477 et seqq, especially p. 3478.

[38] Cfr. on this point Monteduro and Tommasi (2015), p. 161 et seqq.

Environmental law, therefore, to protect life within the complex relationships that structure socio-ecological systems, must necessarily safeguard the value of *bio-cultural diversity*, which may not be reduced to biological diversity alone, nor solely to cultural diversity, but is a synthesis of both.[39]

The considerations developed thus far confirm the fact that food is not an unknown matter in environmental law, provided that one is willing to view food not as a "factor" to be considered individually, but rather at the limits in which it influences the "system of relations" of the environment in the legal sense.

Food, in fact, is noteworthy in its dimension as a "binder" that connects *life at the individual scale* (the individual eats to exist, grow, achieve or maintain its state of health), *life at the social scale* (traditions, customs, techniques and food styles are cultural products and complex economic systems from agro-food supply chains are constructed around food), and *life at the ecological scale* (the pressure for food from the human community determines the exploitation of the ecosystem's goods and services, but, at the same time, the phenomenon of food is historically one of the greatest paradigms of co-evolution between society and nature).

What becomes apparent is the connection between *rights/obligations relating to food* and *rights/obligations relating to life* as a key to understanding the fundamental role played by food within socio-ecological systems.

The reciprocal involvement between life and diversity, and between food and life, reveals the protection of food diversity as a value that is capable of assuming legal significance at the systemic level.

This is a value that is gradually emerging, but which is simultaneously exposed to serious dangers.

[39]Cf Pepe (2007), p. 33 et seqq; Maffi and Woodley (2010), pp. 5–6; Cocks (2010), pp. 67–77; Serrelli (2010), pp. 143–148; Petrillo (2012), especially p. 222 et seqq; Agnoletti and Emanueli (2016), Sobo (2016). As Dunn (2008), p. 315 underlines: "As biological diversity is eroded, key elements of cultural traditions, practices and language are lost. Conversely, as cultures and languages are lost, we lose irreplaceable information about the natural world, as well as notions and philosophies of place, time and humanity. Biological conservation and cultural conservation must therefore be considered simultaneously within a 'biocultural diversity' framework". Petrillo (2012), p. 228, adds that "the challenge facing legislators around the world is the introduction of mechanisms aimed at the protection, safeguard and enhancement of the entirety of the biological and cultural diversity present within a community. An only partial approach to the phenomenon of biocultural diversity may likewise result in serious damage to biodiversity [...] in view of the close relationship between the biological and cultural constituents of a given community, it is now unthinkable to want to safeguard biological diversity at the expense of the traditions, cultures, rites, social practices that constitute its vitality [...] to preserve biological diversity without assessing the impact of such actions on the cultural diversity of the territory in which it operates would mean, in other words, to destroy biodiversity even further" (Author's own translation).

The FAO has repeatedly warned the global community about the existence of deep processes that, for decades, have been threatening food diversity by eroding its natural, social and cultural capital at an increasingly more troubling rate.[40]

It is worth noting that the progressive depletion of food diversity, caused by the *trend* towards the disappearance of traditional agro-food products and the loss of culinary traditions and secular gastronomic knowledge, does not only concern the

[40]More than ten years ago, on the occasion of World Food Day on 16 October 2004 and followed by the International Treaty on Plant Genetic Resources for Food and Agriculture coming into effect in the same year, the FAO underlined in a document [FAO (2004)] that modern agriculture has encouraged many farmers to use uniform species of plants or animals that produce high yields, but when food producers abandon diversity, variety and species may disappear along with their respective genetic characteristics. According to the FAO, this rapid decrease in genetic diversity is extremely troubling if one considers that over the last century three quarters of the genetic diversity in agricultural crops have disappeared, and that, of 6300 animal varieties, 1350 are either at risk or are already extinct. The 2004 document concludes by observing that no sooner do regions abandon traditional local food production than food variety is reduced. According to the document, the human species utilizes only 14 mammal and bird species for 90% of its food requirements from animals, and only 4 species—grain, corn, rice and potato—supply the individual with half of its energy from plants; for this reason, besides the number of species, it is essential to also conserve genetic diversity within each species. Subsequently, the FAO, after a period of 10 years [FAO (2015), p. 1 et seqq (introduction by Collette L et al) and 11 et seqq (chapter by Jarvis A et al)], recorded a further deterioration of the situation. According to the FAO, there would arise the necessity of increasing by at least 60%, over the coming decades, the present global production of food to be able to ensure the sustenance of the additional three billion inhabitants on the planet in 2050; on the other hand, over the next 50 years between 16 and 22% of current wild plant species will simultaneously be at risk of extinction. Finally, in the face of the devastating effects of climate change on the sustainability of food systems, it would now be a compelling necessity to safeguard and promote the diversity of the traditional and typical crops in each territory to foster resilience as a key factor in combatting climate change. The FAO again, in a previous study in 2009 on the food systems of indigenous populations [Kuhnlein et al. (2009)], had admitted that technological developments in agriculture over the six decades in existence of the FAO have led to a vast disconnection between populations and their food. Globalization and homogenization have replaced local food cultures; high-yield cultivation and monoculture farming have taken the place of biodiversity; industrial and high input agricultural methods have degraded ecosystems and damaged agroecological zones; the modern food industry has led to chronic diseases linked to diet and other forms of malnutrition. The same study, which analyzes 12 food systems from indigenous populations, has drawn attention to (p. 20) the fact that traditional and local food cultures "contain treasures of knowledge from long-evolved cultures and patterns of living in local ecosystems" and that "the dimensions of nature and culture that define a food system" must be protected against the risk of destruction. The FAO returned to the issue with another document in 2013 [Kuhnlein et al. (2013)], bringing to light how traditional food systems are rapidly disappearing and calling on the international community to protect them: "global human food systems have been created and supported by a combination of the earth's multitudes of life forms and ecosystems and by human ingenuity, developed and shared over many thousands of years. Today, however, both the cultural diversity and the global biodiversity that gave rise to human food systems are threatened in many places" (p. 25).

food systems of indigenous populations: the problem also arises with respect to the food traditions of European civilization,[41] Italy included.[42]

6 Food Diversity as Organic Synthesis of Different Fundamental Legal Values

Regarding legal analysis, this paper aims to reconstruct food diversity as an organic synthesis of various fundamental legal values.

A foremost protected core value that "constitutes" food diversity is the *biological diversity* of food sources, meaning, of agricultural crops, seeds, plant varieties,

---

[41]Cfr the document entitled "A recipe for protecting Europe's culinary heritage" (latest update dated 21 October 2013), available at http://cordis.europa.eu/news/rcn/36175_en.html (accessed on August 2, 2017) and relative to the project funded by the European Union *EuroFIR* ("European food information resource network"), with a funding of 12 million euros, finished in 2010. Therein can be read "EuroFIR included traditional and ethnic foods, and in particular on the preparation of dishes closely linked to a country or region. People have often passed on recipes and methods from generation to generation by word of mouth, effectively turning these foods into a facet of cultural identity. Throughout Europe, however, many traditional foods and ways of cooking are at risk of disappearing, due partly to changing lifestyles and habits. It is, therefore, important that traditional foods are documented, because they represent a direct link to our cultural heritage".

[42]Concerning Italy, one need only mention that the Ministry for Agricultural, Food and Forestry Policies has adopted a specific decree (the D.M. of 9 April 2008, entitled "Identification of Italian agro-food products as an expression of Italian cultural heritage") with which: it recalled the preceding ministerial decree 18 July 2000, concerning the "Elenco nazionale di Prodotti Agroalimentari Tradizionali" (PAT) and the relative decrees relating to the annual revisions (with which, in realization of art. 3, comma 3 of the ministerial decree 8 September 1999, n. 350, have been provided for in the publication of the aforesaid national list of traditional agro-food products); it stressed that Italy being "a custodian of great agricultural and agro-food traditions", makes "an act of *defence* of the Mediterranean diet … as the basis for a healthy diet" necessary; it therefore established that traditional Italian agro-food products contained in the list by ministerial decree 18 July 2000, as well as successive integrations "constitute the expression, not only of inventiveness, ingenuity and the socioeconomic evolutionary process of the collective Italian territory, but also of the traditions and cultures of the regions, of the provinces, and generally of the communities that are spread throughout the Italian territory and as such these must be subject to the protection and safeguard of Italian institutions", as "an expression of Italian cultural heritage" (Author's own translation). On this point, and also for all the appropriate bibliographical references, besides the other authors cited *supra* in footnote 4, see the analysis of Strambi (2010), p. 1 et seqq. The "Quattordicesima revisione dell'elenco nazionale dei prodotti agroalimentari tradizionali" (PAT) was approved by Directorial Decree MIPAAF of 5 June 2014. However, these measures for safeguard were not by any means sufficient, such that in April 2015, on the eve of the Expo di Milano, the Confederazione Italiana Agricoltori (CIA) launched through the press the warning that one in four typical and traditional food products (in total, around a thousand foods from Italian culinary traditions) is presently at risk of disappearance: the news in relation to the CIA communication is available at http://www.adnkronos.com/sostenibilita/tendenze/2015/04/13/sapori-antichi-nicchia-prodotto-tipico-rischio-estinzione_qfyiDsDqEhLMA8O7BlhpiK.html, consulted on 15 January 2016.

animals, and microbial communities. Whether foods are identified directly with the fruits of nature or with the fresh animal or plant produce supplied by agriculture, farming or fishing, or whether they derive from the transformative processes of the latter, in any case it is evident that food diversity (*ex post*) is co-determined by the biological diversity (*ex ante*) of the natural "raw materials" from which foods are sourced. From this perspective, what is in play, therefore, is the protection of not only agrarian biodiversity[43] but also of farming, fishing, etc, as *species* of the broadest *genus* of biodiversity, with its legal status protected to the utmost by environmental law at the international, European and national levels.[44]

A second core value that constitutes food diversity in law is *territorial and landscape diversity*,[45] that is, the differences between human communities established in food's different territories of origin, of which the latter are heirs and custodians. The point is crucial, because it shows how "food heritage"[46] is linked not only to a specific "terrain" or portion of material space, with its particular physical-chemical, climatic and agronomic characteristics, but also and above all to a "territory",[47] understood as the place of establishment and life of a given community over a long period of time.[48] The relationship between the territorial diversity of the regions and the differentiation of the agro-food products that they produce is

[43]Besides the recent Italian law of 1 December 2015, n. 194, containing "Measures for the protection and valorisation of the biodiversity of agricultural and food interest", which is broadly referred to in the Introduction and in the preceding note 8, one need only recall, in the extensive bibliography: Head (2016), Halewood (2016), Kochupillai (2016), p. 19 et seqq; Timmermann and Robaey (2016), p. 285 et seqq; EFSA (2014), Vezzani (2013), p. 433 et seqq; Ministry for Agricultural, Food and Forestry Policies (2013a, b, c); Genghini et al. (2013), Santilli (2012), Russo (2012), p. 187 et seqq; Sirsi (2011), p. 493 et seqq; Paoloni (2005). Of great interest, in the extra-legal field, is the volume edited by Gepts et al. (2012). See also Monteduro (2015), p. 57 et seqq; Di Benedetto (2015), p. 99 et seqq; Pierri (2015), p. 451 et seqq.

[44]For a broad synthesis, see Prip (2017), Bowman (2016), p. 3 et seqq; Morgera (2014), p. 983 et seqq; De Vido (2014), p. 803 et seqq; Marfoli (2012), p. 155 et seqq. Cfr. also Gillespie (2011), Ministero dell'ambiente e della tutela del territorio e del mare (2011), Benozzo and Bruno (2009) and Benozzo (2008).

[45]Hamdouch and Zuindeau (2010), p. 243 et seqq.

[46]Contreras Hernández and Ribas Serra (2014), p. 84 et seqq; Bessière and Tibère (2011).

[47]See *supra*, footnote 4.

[48]As Strambi (2010), p. 13, observes "the concept of tradition is intuitively linked to that of origin, in a certain territory, in a clearly defined geographical area, where, to be exact, the community in which one came into being exists" (Author's own translation).

manifest in the fact that the territories claim the right to self-represent their "food identities"[49] through their different typical products, by defending the roots of their own models of food culture and by valuing food diversity itself as an instrument of territorial development (one considers the subject of food and wine tourism, for example). There appears, therefore, a dimension of territory/identity based on the collective self-determination of food models: this hinges, in law, on principles of primary importance that are recognized by various constitutions—by the Italian Constitution for instance—such as the autonomy principle[50] and the differentiation principle.[51] This is also linked to the "right to regional cultural identity"[52] and to the "right to the development of local food heritage".[53] Furthermore, since "*landscape*" is "an integral part of the territory whose characteristics are derived from nature, the history of humanity or from their reciprocal interrelationships",[54] what becomes evident is the close ties between the local food and wine traditions and the agrarian landscape, as the second, at the same time, visible result of the shaping action of the first in the course of the centuries and the assumption that these can be perpetuated in the future. Hence, the constitutional root[55] of this "identity" and collective dimension of the diversity of food territories. The importance of territorial differences between rural communities is also recognized at the EU level by the CAP, particularly in its second pillar: it suffices to cite Regulation (EU) No. 1303/2013, according to which the CAP must operate "taking into account the key territorial challenges of the various types of territories" (Whereas n. 16 and art. 10 and 18) and develop approaches that are "tailor-made to fit the specific territorial needs" (Whereas n. 18); or Regulation (EU) n. 1305/2013, which means "achieving a balanced territorial development of rural economies and communities" (art. 4). Of great importance is also the European Parliament Resolution of 14 January 2014 on "Regional Branding: Towards Best Practice in Rural Economies" (2013/2098(INI)), which highlights the importance of a "recognition of the identity of each territory, as defined and informed by its heritage," in reference to "sets of high-quality goods (food and non-food) and services which are inextricably interlinked and which embody the specific characteristics of each territory and in particular its heritage (historical, cultural, geographical, etc.)," and from this perspective means to foster a "territorial value creation process seeking to encompass products and services within a perspective of identity and social responsibility and to complement, by forming a

[49]Ramli et al. (2016), p. 207 et seqq; Brulotte and Di Giovine (2016), Tortorella and Traclò (2008), Tregear (2007).

[50]See e.g. Italian Constitution, Art 5 and Art 114.

[51]See e.g. Italian Constitution, Art 118.

[52]See Ruggiu (2014), p. 486 et seqq, particularly p. 490.

[53]Conte (2014), p. 453 et seqq, particularly p. 461.

[54]Legislative Decree No. 42 of January 22, 2004 (Italian "Code on Cultural Heritage and Landscape"), Art 131, para. 1 (Author's own translation).

[55]Cf Italian Constitution, Art 9, which protects the landscape as a fundamental constitutional value.

unified whole and generating synergies with, the existing origin-related agri-foodstuff quality labelling schemes".

A third core value is that of *cultural diversity*,[56] protected not only at the constitutional level,[57] but even more so by EU (art. 3 TEU; art. 167 TFEU; Preamble and art. 2 of the Charter of Fundamental Rights of the European Union) and international law (one need only mention the UNESCO Conventions for the Protection of the World Cultural and Natural Heritage of 1972, for the Safeguarding of the Intangible Cultural Heritage of 2003 and for the Protection and Promotion of the Diversity of Cultural Expressions of 2005). What is in play, particularly, is the diversity of food, culinary and gastronomic arts and traditions, that is, the complex of techniques, practices, knowledge (often only orally transmitted from generation to generation) that transforms raw materials to food on the table, which is defined as an intangible cultural heritage. This heritage must be preserved in time so that its cultural testimony reaches future generations, regardless of the past and present roots of this legacy in specific territories.[58]

Fourthly, a further and no less important legal value in food diversity is the *protection of human health through the guarantee of a nutritionally varied diet,* which is of definite constitutional importance.[59] This is a fundamental aspect: the disappearance of some typical foods or food traditions, together with the growing homogenization and standardization of food products caused by the mass industrialization of the processes of food production and by the reduction of the range of food varieties[60]

[56]Brulotte and Di Giovine (2016), Anderson et al. (2016), p. 2 et seqq; Montanari (2014), Kuhnlein et al. (2009), Sabbattini (2006), p. 647 et seqq; Strambi and Alabrese (2005), Montelione (2000), p. 463 et seqq.

[57]Cf. Italian Constitution, Art 9. Referring to Italian regulations, besides being covered by Art. 9 Const, these techniques of knowledge can come under, in part, the safeguard of provisions for the protection of craftsmanship (protected by the Italian Constitution at Art 45 Const.).

[58]As is known, in 2010 the Mediterranean Diet was included by UNESCO in the List of Intangible Cultural Heritage of Humanity: on the point, see Colella (2013), p. 583 et seqq. More recently, the Italian Commission for UNESCO has nominated "The Art of the Neapolitan Pizzaioli" for the same list (the news is available on https://www.politicheagricole.it/flex/cm/pages/ServeBLOB.php/L/IT/IDPagina/8511, accessed on August 2, 2017).

[59]In Italy, for example, health is protected by the Italian Constitution as a fundamental value in Art 32.

[60]El Bilali et al. (2017), p. 1 et seqq, who state: "Monotonous diets based on starchy staples lack essential micronutrients and contribute to the burden of malnutrition and micronutrient deficiencies. Food-based strategies have been recommended as the first priority to meet micronutrient needs. An essential element of food-based approaches involves dietary diversification - consumption of a wide variety of foods across nutritionally distinct food groups - as a way to meet recommended intakes of nutrients. The simplification of human diets associated with the increased accessibility of inexpensive agricultural commodities and the erosion of agro-biodiversity is partially responsible for nutrient deficiencies and excess energy consumption in many parts of the world [...] there is a link between biodiversity and food and nutrition security, which includes a focus on dietary diversity [...] unfortunately, globalization of agricultural markets and changes in lifestyles are having a profound impact on the conservation and use of these resources leading to their irreplaceable loss. Maintaining knowledge of indigenous communities is relevant for the conservation and

simultaneously increases the risks of allergies and intolerance and facilitates the onset of some health issues.[61]

Fifthly, food diversity is a precondition for *the freedom to decide on and choose one's own food style.* This value is linked to the sphere of fundamental liberties[62]: acts of food choice are in fact an exercise in personal liberty and a demonstration of the thought of the individual, who has the right to cultivate his or her own tastes and preferences (expressions of self-identity and self-determination)[63] provided that they do not conflict with other primary public rights and interests (such as the protection

---

sustainable use of biological diversity. Indigenous knowledge on how to recognize, cultivate and use local crops is being lost at unprecedented rate. The genetic diversity of food crops and animal breeds is diminishing rapidly. Since the 1980s the need to protect and preserve biological diversity has been a global priority. At the same time, however, it appears clear that the biological diversity of ecosystems cannot be protected without concurrently preserving cultural diversity. By its definition, biocultural diversity means being aware of the close correlation between the loss of cultural and linguistic diversity and loss of biological and genetic diversity, and vice-versa [...] The loss of agricultural diversity occurring around the Mediterranean basin is having negative repercussions on the food security and livelihood of populations living in the region. An exacerbation of the genetic erosion of agro-biodiversity due to globalization trends and climate change is reducing the sustainability of local production systems and along with it the ability to safeguard the Mediterranean diet through the continued use of indigenous species and varieties. At the same time, silent cultural erosion is also affecting the diversity of food cultures. Such a phenomenon is undermining also the identity of millions of Mediterranean people whose traditions are so intimately linked to food cultures [...] This paper suggests that biodiversity is necessary for dietary diversity and nutrition security".

[61]See Faber and Wenhold (2016), p. 22 et seqq; Hasan-Ghomi et al. (2015), p. 11 et seqq; Vadiveloo et al. (2015), p. 555 et seqq; Narmaki et al. (2015), p. 722 et seqq, according to whom increasing dietary diversity might be associated with a decrease in oxidative stress and incidence of chronic diseases; Skypala and Vlieg-Boerstra (2014), p. 442 et seqq; Roduit et al. (2014), p. 1056 et seqq, who conclude that "an increased diversity of food within the first year of life might have a protective effect on asthma, food allergy, and food sensitization and is associated with increased expression of a marker for regulatory T cells"; Allen et al. (2014), p. 498 et seqq; Fanzo et al. (2013), Kaiser (2011), p. 62 et seqq; Drescher (2007), Wahlqvist (2004), pp. CRH16–CRH18; Wahlqvist and Specht (1998), p. 314 et seqq, who write that "the maintenance of biodiversity is important to human health for several reasons: (i) a varied food supply is essential to maintain the health of the omnivorous human species; (ii) a range of diverse food sources is necessary to safeguard against climatic and pestilent disasters which may affect one or more of the food sources; (iii) a diversity of plants and animals may provide a rich source of medicinal material, essential for the extraction of undiscovered therapeutic compounds; (iv) intact ecosystems of indigenous plants and animals appear to act as a buffer to the spread of invasive plants and animals, and of pathogens and toxins, thus contributing to the health of populations nearby; and (v) the 'spiritual' values of exploring the diversity of plants, animals and ecosystems in an area appear to have a beneficial effect on mental health, strengthening the feeling of 'belonging to the landscape'".

[62]Cf. e.g. Semands (2014), p. 149 et seqq, who states (p. 189): "The right of a person to choose what food she consumes should be considered fundamental;" (2013), p. 173 et seqq. But see the criticism of Wiseman (2015), p. 1299 et seqq. See also Berg (2013).

[63]Chieffi (2015), who affirms the existence of an authentic "right to food self-determination," construed as the "right of every individual to freely choose for themselves the diet for their own sustenance" (p. 5 and p. 7) (Author's own translation).

of life or health).[64] These are often associated with the exercise of other liberties, such as, for example, religious freedom[65] (where the professed religion prevents the consumption of some foods and privileges others, or imposes the use of certain techniques of food production and consumption). However, such freedoms would be empty or merely superficial if the individual were not faced with a real choice, that is, a variety of options from which to decide; hence the legal importance of food diversity, which must be guaranteed as a condition in the exercise of the individual's freedom to choose from different food traditions, as well as different foods.[66]

Finally, the diversity of food styles is also linked to the constitutional prerogatives of the family,[67] specifically to the parents' educative task. In fact, the task of

[64]See e.g. Martin (2008), p. 21 et seqq; in rejecting paternalistic approaches, Neto (2016), p. 337 et seqq; from a paternalistic point of view, see Gostin (2010), p. 33 et seqq, in response to Resnik (2010), p. 27 et seqq; in the perspective of a "secondary paternalism", see Bonotti (2014), p. 31 et seqq.

[65]Ceserani (2016), p. 369 et seqq; Fuccillo (2016) and Fuccillo et al. (2016), p. 1 et seqq; Fuccillo (2015) and Milani (2015), p. 349 et seqq.

[66]Cf. e.g. In the same line Rencher (2012) who states (p. 426): "food choice can be a means of self-expression and self-identity [...] through choice of food, people express their views of the world. Health, religion, culture, self-identity, and even politics are expressed by and play important roles in a person's food choice, and may support some level of constitutional protection to the right to choice of food". The author points to (p. 433) "the important role that food can take within a religious context. In light of the fact that the Bill of Rights gives particular deference to freedom of religion, food choice may deserve protection as a right because of its religious importance. On the other hand, it is well established that "freedom of religion" does not mean that a person may act or refuse to act for religious reasons and be absolutely shielded from government intervention. To the contrary: while the Constitution protects the freedom to believe, it does not necessarily protect the freedom to act in accordance with those beliefs. Specifically, a person's choice is protected from government actions that target the person's religious beliefs, but the same choice may not be protected if the government action only incidentally interferes with a religious practice". According to the author (p. 430), "the right may also have some connection to the fundamental right to personal liberty"; further (p. 436), "because food is an important part of cultural expression, there may yet be reason to give food choice heightened protection"; finally (p. 437), "food is both a means of self-expression and, at the same time, a means of speech. Because self-identity influences food choice, it is of little surprise that people choose food based on the personal values with which they identify". In the conclusion (p. 442), the author asks "Is there a right to food choice, and if there is, what level of constitutional protection does the right deserve?", and answers this question in the sense that "these are difficult questions, and are currently unanswered. Yet, when considering the right to choice of food, the law should look to additional considerations beyond food safety [...] Meanwhile, food choice is important for a number of reasons beyond safety, including its impact on health, its importance to religion, its value in cultural identity, and its importance as selfexpression and a form of speech. Furthermore, these reasons compare favorably to established fundamental rights and protected choices. Given these considerations, it may well be that food choice does indeed deserve some level of heightened protection."

[67]Cf Italian Constitution, Art 30.

educating children[68] concerning food would not be fulfilled[69] if parents did not have a choice between different foods and different styles of food.

In conclusion, food diversity can be said to be a synthesis of multiple diversities. It is a value system where numerous legal values of primary importance under constitutional protection converge and are contained: environment/biodiversity, territorial autonomy and differentiation, landscape, cultural heritage, human health, personal and religious freedom, and the educational choices of the family.

The strengths of these basic value aggregate organically, conferring to food diversity a role of absolute primary importance in law.

7 Conclusion. Food Diversity in the Face of Economic Pressures

Thus far, no provisions have been prescribed by positive law for the full and general protection of the value-system of food diversity, either at the international and European levels or in Italy: legislature therefore does not yet seem to have recognized food diversity as a single, autonomous value that transcends the individual dimensions connected to it.

Complying with changes in economic conditions, supranational law has to date aimed at simply making consumer access to food easier, imposing competition law as a dominant paradigm and being intolerant of significant market distortions be they direct or indirect. Yet from the outset this approach risks "amputating" food from factors that are manifestly "anti-economic" in production and trade (the patient work of the hands, the irreplaceable role of the person and their "touch", long periods of time, ancient arts that are transmitted orally, etc), and in such a manner emptying it of history, culture, identity.

Genuine competition between "food commodities" would theoretically only be possible on a level playing field; however, becasue of structural weight posed by the personal, time and economic costs necessary to their production, and because of the absence of economies of scale, foods that perpetuate and convey a historical memory of place would be destined for gradual marginalization when faced with global food industry giants in the epoch of globalization.

If the right to food continued to be inflected exclusively according to the consumer's subjective right, food would remain imprisoned in the single economic dimension of *thing* that can be the subject of rights *in rem* and, therefore, of contracts

[68]Cf de Wit et al. (2015), p. 22 et seqq; Russell et al. (2014), p. 1018 et seqq; Kümpel Nørgaard and Brunsø (2011), p. 141 et seqq; Wansink (2010), p. 313 et seqq.

[69]Obviously, the role of the family, and of parents particularly, in the food education of children cannot lead to a complete denial of the freedom of children to choose their own food style or even worse, to the exercise of choices that are harmful to the health of children. Cf Ouellette (2010), p. 955 et seqq.

of exchange. Food that is reduced to *res*, which, to circulate without barriers and obstacles in the global market, would be constrained to slowly renouncing the non-fungible and unique characteristics of the very territories and cultures from which it draws its roots, and to adapt itself, instead, to the *status* of fungible and identically reproducible goods on an industrial scale, to a serialized, standardized, normalized object, worked according to market necessities and of worth essentially because of its price, aesthetic presentation, labelling, and advertising through *marketing* channels.

Instead, as this paper proposes, recognizing cultural diversity as a complex, systemic and life-related legal value, that is, as a "condensation" of the normative energy originating from the combination of multiple constitutionally fundamental legal values that are bound up with the supreme value of life in its differing scales, which in their turn are inextricably interrelated, permits the attribution to food diversity of a much stronger (respecting the situation to date) "resistance" in law against purely economic interests.

From this perspective, should conflict arise between food diversity and economic interests associated with the protection of competition, free enterprise and the market, it will be necessary to pursue, as a first option, the strategy of balance between the different interests, according to reasonableness and proportionality. However, in the event that in the effort towards a balanced outcome, solutions capable of partially satisfying all the values in play do not arise, so that one must necessarily yield to the other, the conceptual reconstruction proposed here leads to prevalence being given to the safeguard of food diversity from threats—deriving from socio-economic dynamics—that may cause its depletion or disappearance, thus leading to a decline in interests opposed to the maintenance of food diversity.

This would certainly be a radical paradigm shift, but one that seems, by now, in the era of the Anthropocene, no longer avoidable even by the law.

If the right to food were to become completely (or even just predominantly) *thingified* and standardized for anonymous circulation in the market, it would therefore be, in the judgement of the writer of this contribution, irreconcilably in conflict with food diversity as a preconditional and overriding "value of values" which must be legally preserved in its qualitative integrity and passed on to future generations.

References

Adinolfi G (2010) Alimentazione e commercio internazionale nel rapporto del 2009 del relatore speciale delle Nazioni Unite sul diritto al cibo. Dir umani e dir int:125 et seqq

Adornato F (2012) La sicurezza alimentare tra primato della tecnica e crisi del diritto. Riv dir agr:405 et seqq

Agnoletti M, Emanueli F (eds) (2016) Biocultural diversity in Europe. Springer International Publishing, Switzerland

Alemanno A, Gabbi S (eds) (2014) Foundations of EU food law and policy. Ten years of the European Food Safety authority. Ashgate, London

Allen T et al (2014) Agricultural biodiversity, social-ecological systems and sustainable diets. Proc Nutr Soc 73(4):498 et seqq

Anderies JM, Janssen MA (2012, 16 Oct) Elinor Ostrom (1933–2012): pioneer in the interdisciplinary science of coupled social-ecological systems. PLOS Biol. https://doi.org/10.1371/journal.pbio.1001405

Anderson L et al (2016) Europe on a plate: food, identity and cultural diversity in contemporary Europe. Aust N Z J Eur Stud 8(1):2 et seqq

Andree P et al (eds) (2014) Globalization and food sovereignty: global and local change in the new politics of food. Toronto University Press, Toronto

Angelicchio G (2014) Le indicazioni geografiche. In: Scuffi M, Franzosi M (eds) Diritto industriale italiano. Tomo 1. Diritto sostanziale. Cedam, Padova, p 345 et seqq

Arfini F, Belletti G, Marescotti A (2010) Prodotti tipici e denominazioni geografiche. Strumenti di tutela e valorizzazione. Tellus, Roma

Arino O et al (2012) Global Land Cover Map for 2009 (GlobCover 2009). European Space Agency & Université Catholique de Louvain. https://doi.org/10.1594/PANGAEA.787668

Armitage D (2013) Resilience and administrative law. Ecol Soc 18(2):11 et seqq

Armitage DR et al (2009) Adaptive co-management for social–ecological complexity. Front Ecol Environ 7(2):95 et seqq

Arnold C, Gunderson L (2013) Adaptive law and resilience. Environ Law Report News Anal 43:10426 et seqq

Barnes RA (2013) The capacity of property rights to accommodate social-ecological resilience. Ecol Soc 18(1):6 et seqq

Barnosky AD et al (2012) Approaching a state shift in Earth's biosphere. Nature 486(7402):52 et seqq

Bennett J (2016) Global ecological crisis: structural violence and the tyranny of small decisions. In: Brauch HS, Oswald Spring U, Bennett J, Serrano Oswald SE (eds) Addressing global environmental challenges from a peace ecology perspective. Springer, Berlin, p 55 et seqq

Benozzo M (2008) La tutela degli ecosistemi e della biodiversità nel codice ambientale: gli strumenti ripristinatori e risarcitori del pregiudizio ecologico. Aracne, Roma

Benozzo M, Bruno F (2009) La valutazione di incidenza: la tutela della biodiversità tra diritto comunitario, nazionale e regionale. Giuffrè, Milano

Benson MH, Garmestani AS (2011) Embracing Panarchy, building resilience and integrating adaptive management through a rebirth of the national environmental policy act. J Environ Manag 92:1420 et seqq

Berg DJ (2013) Food choice is a fundamental liberty right. J Food Law Policy 9:173 et seqq.

Bergandi D (2011) Multifaceted ecology between organicism, emergentism and reductionism. In: Schwarz A, Jax K (eds) Ecology revisited. Reflecting on concepts, advancing science. Springer, Dordrecht, p 31 et seqq

Berkes F, Colding J, Folke C (eds) (2003) Navigating social-ecological systems: building resilience for complexity and change. Cambridge University Press, Cambridge

Bessière J, Tibère L (2011) Editorial: Patrimoines alimentaires. Anthropology of Food 8. http://aof.revues.org/6782. Accessed 2 Aug 2017

Bevilacqua D (2012) La sicurezza alimentare negli ordinamenti giuridici ultrastatali. Giuffrè, Milano

Biermann F (2012) Planetary boundaries and earth system governance: exploring the links. Ecol Econ 81:4 et seqq

Biermann F et al (2016) Down to earth: contextualizing the Anthropocene. Glob Environ Change 39:341 et seqq

Bonotti M (2014) Alimentazione non salutare, autonomia individuale e principio del danno. Notizie di POLITEIA XXX(114):31 et seqq

Bowman M (2016) Law, legal scholarship and the conservation of biological diversity: 2020 vision and beyond. In: Bowman M, Davies P, Goodwin E (eds) Research handbook of biodiversity and law. Edward Elgar, Cheltenham UK, Northampton MA, p 3 et seqq

Boyd E, Folke C (eds) (2012) Adapting institutions: governance, complexity and social-ecological resilience. Cambridge University Press, Cambridge
Brulotte RL, Di Giovine MA (eds) (2016) Edible identities: food as cultural heritage. Routledge, London
Cafagno M (2007) Principi e strumenti di tutela dell'ambiente. Come sistema complesso, adattativo, comune. Giappichelli, Torino
Campbell NA, Reece JB, Simon EJ (2008) L'essenziale di biologia, 3rd edn. Pearson Italia Spa, Milano
Canfora I (2010) Marchi e denominazioni di origine protetta: Bavaria e Bayerisches bier. Dir giur agr alim amb:91 et seqq
Canfora I (2011) Le "specialità tradizionali garantite". In: Costato L, Germanò A, Rook Basile E (dir) Trattato di diritto agrario. Vol 3: Il diritto agroalimentare. Utet giuridica, Torino, p 73 et seqq
Capelli F (2001) La protezione giuridica dei prodotti agro-alimentari di qualità e tipici in Italia e nell'Unione Europea. Dir comunitario e scambi internaz:177 et seqq
Capelli F (2014) Valorizzazione dei prodotti agroalimentari italiani tipici e tradizionali. Dir comunitario e scambi internaz:487 et seqq
Certomà C (2010) Diritto al cibo, sicurezza alimentare, sovranità alimentare. Riv dir alim:1 et seqq
Ceserani A (2016) Cibo 'religioso' e diritto: a margine di quattro recenti pubblicazioni. Quaderni di diritto e politica ecclesiastica 2:369 et seqq
Chen Y (2014) Trade, food security, and human rights. The rules for international trade in agricultural products and the evolving world food crisis. Ashgate, Farnham, Surrey/Burlington VT
Chieffi L (2015) Scelte alimentari e diritti della persona. Tra autodeterminazione del consumatore e sicurezza sulla qualità del cibo. Diritto Pubblico Europeo Rassegna online, No 1. http://www.edizioniesi.it/dperonline/data/uploads/articoli/chieffi-scelte-alimentari.pdf. Accessed 2 Aug 2017
Cocks M (2010) What is biocultural diversity? A theoretical review. In: Bates DG, Tucker J (eds) Human ecology: contemporary research and practice. Springer, New York, pp 67–77
Colella L (2013) La "dieta mediterranea" come patrimonio dell'umanità: dalla tutela dell'Unesco alla legge regionale della Campania n. 6 del 2012. Dir giur agr alim amb:583 et seqq
Collart Dutilleul F (ed) (2013) Legal dictionary of food security in the world. Larcier, Bruxelles
Conte VL (2014) Autonomie territoriali e cultura. In: Morelli A, Trucco L (eds) Diritti e autonomie territoriali. Giappichelli, Torino, p 453 et seqq
Contreras Hernández J, Ribas Serra J (2014) Sobre la construcció social del patrimoni alimentari. Revista d'Etnologia de Catalunya 39:84 et seqq
Cordini G (ed) (2013) Domestic protection of food safety and quality rights. Giappichelli, Torino
Costato L, Albisinni F (eds) (2012) European food law. Cedam, Padova
Costato L, Germanò A, Rook Basile E (dir) (2011) Trattato di diritto agrario, Vol 3: Il diritto agroalimentare. Utet giuridica, Torino
Costato L, Borghi P, Rizzioli S (2013) Compendio di diritto alimentare, 6th edn. Cedam, Padova
Crane TA (2010) Of models and meanings: cultural resilience in social–ecological systems. Ecol Soc 15(4):19 et seqq
Cristiani E, Massart A (eds) (2008) Prodotti tipici e turismo sostenibile come strumento di sviluppo del territorio. ETS, Pisa
Crutzen PJ (2001) Geology of mankind – the Anthropocene. Nature 415(6867):23 et seqq
Crutzen PJ, Stoermer EF (2000) The Anthropocene. Glob Change Newsl 41(1):17–18
Cumming G (2013) Scale mismatches and reflexive law. Ecol Soc 18(1):15 et seqq
D'Addezio MR (2010) Sicurezza degli alimenti: obiettivi del mercato dell'Unione europea ed esigenze nazionali. Riv dir agr:379 et seqq
De Schutter O (2012) Agroecology, a tool for the realization of the right to food. In: Lichtfouse E (ed) Agroecology and strategies for climate change. Springer, Dordrecht, p 1 et seqq

De Schutter O, Cordes KY (eds) (2011) Accounting for Hunger. The right to food in the era of globalisation. Hart, Oxford

De Vido S (2014) Tutela della biodiversità e rispetto dei diritti umani. Le sentenze CGUE nei casi "Cascina Tre Pini" e deviazione del fiume "Acheloo". Riv giur amb:803 et seqq

de Wit N et al (2015) Food culture in the home environment: family meal practices and values can support healthy eating and self-regulation in young people in four European countries. Appl Psychol Health Well-Being 7(1):22 et seqq

Di Benedetto S (2015) Agriculture and the environment in international law: towards a new legal paradigm? In: Monteduro M, Buongiorno P, Di Benedetto, Isoni A (eds) Law and agroecology: a transdisciplinary dialogue. Springer, Heidelberg, p 99 et seqq

Drescher LS (2007) Healthy food diversity as a concept of dietary quality: measurement, determinants of consumer demand, and willingness to pay. Cuvillier Verlag, Göttingen

Dunn CP (2008) Biocultural diversity should be a priority for conservation. Nature 456(7220):315

Ebbesson J (2010) The rule of law in governance of complex socio-ecological changes. Glob Environ Change 20(3):414 et seqq

Ebbesson J, Hey E (2013) Introduction: where in law is social-ecological resilience? Ecol Soc 18 (3):25 et seqq

EFSA (2014) Biodiversity as protection goal in environmental risk assessment for EU agro-ecosystems (Summary report, Scientific Colloquium 27–28 November 2013, Parma, Italy). Publications office of the European Union, Luxembourg

El Bilali H et al (2017) Exploring relationships between biodiversity and dietary diversity in the mediterranean region: preliminary insights from a literature review. Am J Food Nutr 5(1):1 et seqq

Eldredge N, Pievani T, Serrelli E, Temkin I (eds) (2016) Evolutionary theory: a hierarchical perspective. University of Chicago Press, Chicago

Ellis EC (2015) Ecology in an anthropogenic biosphere. Ecol Monogr 85(3):287 et seqq

Epstein G et al (2013) Missing ecology: integrating ecological perspectives with the social-ecological system framework. Int J Commons 7(2):432 et seqq

Ericksen PJ (2008) Conceptualizing food systems for global environmental change research. Glob Environ Change 18:234 et seqq

Faber M, Wenhold F (2016) Food security, dietary diversity, and biodiversity. In: Temple NJ, Steyn N (eds) Community nutrition for developing countries. AU Press, Athabasca University, Canada, Unisa Press, University of South Africa, p 22 et seqq

Fanzo J et al (eds) (2013) Diversifying food and diets: using agricultural biodiversity to improve nutrition and health. Routledge, Earthscan, London

FAO (2004) Biodiversità: la nostra alimentazione ne dipende, Roma. http://www.fao.org/docrep/006/y5418i/y5418i00.HTM. Accessed 2 Aug 2017

FAO (2012) Guidance note: integrating the right to adequate food into food and nutrition security programmes. Food and Agriculture Organization of the United Nations, Rome

FAO (2015) Coping with climate change. The roles of genetic resources for food and agriculture, Rome. http://www.fao.org/3/a-i3866e.pdf. Accessed 2 Aug 2017

FAO, The Global Forest Resources Assessment 2015. http://www.fao.org/news/story/en/item/326911/icode/ and http://www.slideshare.net/FAOoftheUN/how-have-forests-changed-results-from-fra-2015. Accessed 16 Jan 2016

Farì A (2013) Beni e funzioni ambientali. Contributo allo studio della dimensione giuridica dell'ecosistema. Jovene, Napoli

Ferrari M, Izzo U (2012) Diritto alimentare comparato. Regole del cibo e ruolo della tecnologia. Il Mulino, Bologna

Folke C (2006) Resilience: the emergence of a perspective for social–ecological systems analyses. Glob Environ Change 16:253 et seqq

Folke C et al (2011) Reconnecting to the Biosphere. Ambio 40(7):719 et seqq

Fuccillo A (2015) Il cibo degli dei. Diritto, religioni, mercati alimentari. Giappichelli, Torino

Fuccillo A (2016) Diritto e religioni nelle scelte alimentari. Stato, Chiese e pluralismo confessionale, No. 18. http://riviste.unimi.it/index.php/statoechiese/article/view/7137/6944. Accessed 2 Aug 2017
Fuccillo A, Sorvillo F, Decimo L (2016) Law, religions and food choices. Calumet. Intercultural law and humanities review:1 et seqq. http://www.windogem.it/calumet/upload/pdf2/mat_49.pdf. Accessed 2 Aug 2017
Gaffney O, Steffen W (2017) The Anthropocene equation. Anthropocene Rev 4(1):53 et seqq
Galaz V et al (2012) Global environmental governance and planetary boundaries: an introduction. Ecol Econ 81:1 et seqq
Garmestani AS, Allen CR, Benson MH (2013) Can law foster social-ecological resilience? Ecol Soc 18(2):37 et seqq
Garmestani A, Allen CR (eds) (2014) Social-ecological resilience and law. Columbia University Press, New York
Genghini M, Innocenti S, Ferretti M (2013) Multifunzionalità agricola, biodiversità e fauna selvatica: indagine e proposte di miglioramento della normativa partendo dalla Regione Toscana. ISPRA, Roma
Gepts P et al (2012) Biodiversity in agriculture. Domestication, evolution and sustainability. University Press, Cambridge
Germanò A (2008) Il libro verde della Commissione europea del 15 ottobre 2008: alla ricerca di una definizione di alimenti di qualità. Riv dir agr:480 et seqq
Germanò A (2009) Le politiche europee della qualità alimentare. Riv dir alim:3 et seqq
Germanò A, Strambi G (eds) (2014) Il nuovo diritto agrario dell'Unione Europea: i regolamenti 1169/2011 e 1151/2012 sull'informazione e sui regimi di qualità degli alimenti, e i regolamenti del 17 dicembre 2013 sulla PAC (atti dei Seminari, Firenze, 12 settembre 2013, 28 maggio, 6 e 13 giugno 2014). Giuffré, Milano
Germanò A, Ragionieri MP, Rook Basile E (eds) (2014) Diritto agroalimentare. Le regole del mercato degli alimenti e dell'informazione alimentare. Giappichelli, Torino
Gillespie A (2011) Conservation, biodiversity and international law. Edward Elgar, London
Glaser M, Krause G, Ratter BMW, Welp M (eds) (2012) Human-nature interactions in the Anthropocene: potentials of social-ecological systems analysis. Routledge, New York
Gostin LO (2010) Trans fat bans and the human freedom: a refutation. Am J Bioeth 10(3):33 et seqq. http://remedy.tcs.tulane.edu/portal/files/open-access/fim-1/ethics/session-4/reading-3.pdf. Accessed 2 Aug 2017
Gotts NM (2007) Resilience, panarchy, and world-systems analysis. Ecol Soc 12(1):24 et seqq
Green OO, Garmestani AS, Hopton ME, Heberling MT (2014) A multi-scalar examination of law for sustainable ecosystems. Sustainability 6:3534 et seqq
Gunderson LH, Holling CS (eds) (2002) Panarchy: understanding transformations in human and natural systems. Island Press, Washington DC
Halewood M (ed) (2016) Farmers' crop varieties and farmers' rights: challenges in taxonomy and law. Routledge, London
Hamdouch A, Zuindeau B (2010) Introduction. Diversité territoriale et dynamiques socio-institutionnelles du développement durable: une mise en perspective. Géographie, économie 12(3):243 et seqq
Hasan-Ghomi M et al (2015) Food security is associated with dietary diversity: Tehran Lipid and glucose study. Nutr Food Sci Res 2(1):11 et seqq
Haugen HM (2007) The right to food and the TRIPS agreement. With a particular emphasis on developing countries' measures for food production and distribution. Martinus Nijhoff, Leiden
Häyhä T et al (2016) From planetary boundaries to national fair shares of the global safe operating space - how can the scales be bridged? Glob Environ Change 40:60 et seqq
Head JW (2016) International law and agroecological husbandry: building legal foundations for a new agriculture. Routledge, London
Hinkel J, Bots PWG, Schlüter M (2014) Enhancing the ostrom social-ecological system framework through formalization. Ecol Soc 19(3):51 et seqq

Holling CS (2001) Understanding the complexity of economic, ecological, and social systems. Ecosystems 4(5):390 et seqq
Hospes O, Hadiprayitno I (eds) (2010) Governing food security: law, politics and the right to food. Wageningen Academic Publishers, Wageningen
Hughes TP, Carpenter S, Rockström J, Scheffer M, Walker B (2013) Multiscale regime shifts and planetary boundaries. Trends Ecol Evol 28(7):389 et seqq
Jannarelli A (2010) La nuova Food insecurity: una prima lettura sistemica. Riv dir agr:565 et seqq
Jannarelli A (2011a) Profili giuridici del sistema agro-alimentare tra ascesa e crisi della globalizzazione. Cacucci, Bari
Jannarelli A (2011b) Cibo e democrazia: un nuovo orizzonte dei diritti sociali. In: Goldoni M, Sirsi E (eds) Il ruolo del diritto nella valorizzazione e nella promozione dei prodotti agro-alimentari (Atti del convegno di Pisa, 1–2 luglio 2011). Giuffrè, Milano, p 33 et seqq
Jax K (2010) Ecosystem functioning. Cambridge University Press, Cambridge
Jordán F, Jørgensen SE (eds) (2012) Models of the ecological hierarchy. From molecules to the ecosphere. Elsevier, Amsterdam
Kaiser ML (2011) Food security: an ecological–social analysis to promote social development. J Community Pract 19:62 et seqq
Kay JJ (2000) Ecosystems as self-organised holarchic open systems: narratives and the second law of thermodynamics. In: Jørgensen SE, Müller F (eds) Handbook of ecosystem theories and management. CRC Press, Boca Raton, p 135 et seqq
Kent G (ed) (2008) Global obligations for the right to food. Rowman & Littlefield, Lanham
Kochupillai M (2016) The international legal framework for the protection of plant varieties and agrobiodiversity. In: Kochupillai M (ed) Promoting sustainable innovations in plant varieties. Springer, Berlin, p 19 et seqq
Krasny ME, Lundholm C, Plummer R (eds) (2011) Resilience in social-ecological systems: the role of learning and education. Routledge, London
Kuhnlein HV, Erasmus B, Spigelski D (eds) (2009) Indigenous Peoples' food systems: the many dimensions of culture, diversity and environment for nutrition and health. Food and Agriculture Organization of the United Nations, Centre for Indigenous Peoples' Nutrition and Environment, Rome
Kuhnlein HV et al (eds) (2013) Indigenous Peoples' food systems and well-being. Interventions and policies for healthy communities. Food and Agriculture Organization of the United Nations, Centre for Indigenous Peoples' Nutrition and Environment, Rome
Kümpel Nørgaard M, Brunsø K (2011) Family conflicts and conflict resolution regarding food choices. J Consum Behav 10(3):141 et seqq
Lambek NCS et al (eds) (2014) Rethinking food systems: structural challenges, new strategies and the law. Springer, London
Lant CL, Ruhl JB, Kraft SE (2008) The tragedy of ecosystem services. BioScience 58(10):969 et seqq
Lefkaditou A (2012) Is ecology a holistic science, after all? In: Stamou GP (ed) Populations, biocommunities, ecosystems: a review of controversies in ecological thinking. Bentham Science Publisher, Sharjah, UAE, p 46 et seqq
Lévêque C (2003) Ecology: from ecosystem to biosphere. Science Publishers, Enfield, NH
Levin S et al (2013) Social-ecological systems as complex adaptive systems: modeling and policy implications. Environ Dev Econ 18(2):111 et seqq
Lewis SL, Maslin MA (2015) Defining the Anthropocene. Nature 519(7542):171 et seqq
Lucchi N (2008) Il dibattito transatlantico sulla tutela delle indicazioni geografiche e delle denominazioni di origine. Riv dir agr:49 et seqq
Lucifero N (2011) La comunicazione simbolica nel mercato alimentare: marchi e segni del territorio. In: Costato L, Germanò A, Rook Basile E (dir) Trattato di diritto agrario. Vol 3: Il diritto agroalimentare. Utet giuridica, Torino, p 321 et seqq
Lupone A, Ricci C, Santini A (eds) (2013) The right to safe food towards a global governance. Giappichelli, Torino

MacDicken K (ed) (2015) Changes in global forest resources from 1990 to 2015. For Ecol Manag 352:1–146
MacMaoláin C (2015) Food law. European, domestic and international frameworks. Hart, Oxford
Maffi L, Woodley E (2010) Biocultural diversity conservation: a global sourcebook. Earthscan, London
Marfoli L (2012) Biodiversità: un percorso internazionale ventennale. Rivista quadrimestrale di diritto dell'ambiente:155 et seqq
Martin R (2008) The role of law in the control of obesity in England: looking at the contribution of law to a healthy food culture. Aust N Z Health Policy 5:21 et seqq
Masini S (2014) Diritto alimentare: una mappa delle funzioni. Giuffrè, Milano
Masini S (2015) Corso di diritto alimentare, 3rd edn. Giuffré, Milano
Matthews JA, Huggett RJ (2013) Ecosystem concept. In: Matthews JA (ed) Encyclopedia of environmental change, Vol 1 [A-F]. Sage, London, p 326 et seqq
McDermott MJ (2012) Constitutionalizing an enforceable right to food: a new tool for combating hunger. Boston Coll Int Comp Law Rev:543 et seqq
McNeill JR, Engelke P (2016) The great acceleration: an environmental history of the Anthropocene since 1945. Harvard University Press, Cambridge
Milani D (2015) Le scelte alimentari nelle società multireligiose: una prospettiva di libertà. In: Chizzoniti AG (ed) Cibo, religione e diritto: nutrimento per il corpo e per l'anima. Libellula Edizioni, Tricase, p 349 et seqq
Miller JG (1978) Living systems. McGraw-Hill, New York. Italian edition: La teoria generale dei sistemi viventi. Franco Angeli, Milano
Ministero dell'ambiente e della tutela del territorio e del mare (2011) Breve guida alla Strategia nazionale per la biodiversità. Ministero dell'Ambiente, Roma
Ministero delle politiche agricole alimentari e forestali (2013a) Linee guida per la conservazione e la caratterizzazione della biodiversità vegetale di interesse per l'agricoltura. Piano nazionale sulla biodiversità di interesse agricolo. INEA, Roma
Ministero delle politiche agricole alimentari e forestali (2013b) Linee guida per la conservazione e la caratterizzazione della biodiversità animale di interesse per l'agricoltura. Piano nazionale sulla biodiversità di interesse agricolo. INEA, Roma
Ministero delle politiche agricole alimentari e forestali (2013c) Linee guida per la conservazione e la caratterizzazione della biodiversità microbica di interesse per l'agricoltura. Piano nazionale sulla biodiversità di interesse agricolo. INEA, Roma
Montanari M (2014) Il cibo come cultura, 7th edn. Laterza, Roma
Monteduro M (2014) Per una "nuova alleanza" tra diritto ed ecologia: attraverso e oltre le "aree naturali protette". GiustAmm.it
Monteduro M (2015) From agroecology and law to agroecological law? Exploring Integration Between Scientia Ruris and Scientia Iuris. In: Monteduro M, Buongiorno P, Di Benedetto, Isoni A (eds) Law and agroecology: a transdisciplinary dialogue. Springer, Heidelberg, p 57 et seqq
Monteduro M (2016) La notion juridique de l'"environnement" à la lumière de la définition donnée par le législateur italien. Revue Juridique de l'Environnement 40(2):269 et seqq
Monteduro M, Tommasi S (2015) Paradigmi giuridici di realizzazione del benessere umano in sistemi ecologici ad esistenza indisponibile e ad appartenenza necessaria. In: Benessere e regole dei rapporti civili. Lo sviluppo oltre la crisi (Atti del 9° Convegno Nazionale SISDiC, Napoli, 8-9-10 maggio 2014, in ricordo di Giovanni Gabrielli). Edizioni Scientifiche Italiane, Napoli, p 161 et seqq
Monteduro M, Buongiorno P, Di Benedetto S, Isoni A (eds) (2015) Law and agroecology: a transdisciplinary dialogue. Springer, Heidelberg
Monteiro De Lima LH (2013) The principle of resilience. Pace Environ Law Rev 30(2):695 et seqq
Montelione E (2000) La produzione agroalimentare di qualità come bene culturale. Riv dir agr:463 et seqq
Montini M, Volpe F (2011) La scienza della sostenibilità e la necessità di regolazione. Riv giur ambiente:157 et seqq

Morgera E (2014) Against all odds: the contribution of the convention on biological diversity to international human rights law. In: Alland D, Chetail V, de Frouville O, Viñuales JE (eds) Unity and diversity of international law. Essays in Honour of Professor Pierre-Marie Dupuy. Martinus Nijhoff, Leiden, p 983 et seqq

Morin J (2011) Community ecology, 2nd edn. Wiley-Blackwell, Oxford

Narmaki E et al (2015) Dietary diversity as a proxy measure of blood antioxidant status in women. Nutrition 31(5):722 et seqq

Naveh Z, Carmel Y (2003) Landscape complexity versus ecosystem complexity: implication for landscape planning and management. In: Farina A, Santolini R, Penna A (eds) La complessità in ecologia, Atti del XII Congresso Nazionale della S.It.E., Urbino 16–18 settembre 2002. Società Italiana di Ecologia:35 et seqq

Neto ML (2016) Nutrition as public policy: still the guarantee or already the restriction of one's rights? In: Olsson IAS, Araújo SM, Vieira MF (eds) Food futures: ethics, science and culture. Wageningen Academic Publishers, The Netherlands, p 337 et seqq

Newbold T et al (2016) Has land use pushed terrestrial biodiversity beyond the planetary boundary? A global assessment. Science 353(6296):288 et seqq

Norberg J, Wilson J, Walker B, Ostrom E (2008) Diversity and resilience of social-ecological systems. In: Norberg J, Cumming GS (eds) Complexity theory for a sustainable future. Columbia University Press, New York, p 46 et seqq

Ostrom E (2005) Understanding institutional diversity. Princeton University Press, Princeton

Ostrom E (2009) A general framework for analyzing sustainability of social-ecological systems. Science 325:419 et seqq

Ostrom E, Janssen MA (2005) Multi-level governance and resilience of social-ecological systems. In: Spoor M (ed) Globalisation, poverty and conflict. Springer, Dordrecht, p 239 et seqq

Ouellette AR (2010) Shaping parental authority over children's bodies. Indiana Law J 85:955 et seqq

Paoloni L (2005) Diritti degli agricoltori e tutela della biodiversità. Giappichelli, Torino

Paoloni L (2011) I nuovi percorsi della food security: dal "diritto al cibo adeguato" alla "sovranità alimentare". Dir giur agr alim amb:159 et seqq

Parisio V (ed) (2005) Food safety and quality law: a transnational perspective. Giappichelli, Torino

Pepe V (2007) La tutela dell'ambiente e la biodiversità culturale. Riv giur amb:33 et seqq

Percival RV, Alevizatos DC (eds) (1997) Law and the environment: a multidisciplinary reader. Temple University Press, Philadelphia, p 25 et seqq

Perfetti LR (2014) Principi della disciplina pubblicistica dell'alimentazione. Riv dir agr:3 et seqq

Petrelli L (2012) I regimi di qualità nel diritto alimentare dell'Unione Europea: prodotti DOP IGP STG biologici e delle regioni ultraperiferiche. Editoriale Scientifica, Napoli

Petrillo PL (2012) La dimensione culturale del patrimonio agro-alimentare italiano in ambito UNESCO. Strumenti e procedure. In: Scovazzi T, Ubertazzi B, Zagato L (eds) Il patrimonio culturale intangibile nelle sue diverse dimensioni. Giuffrè, Milano, p 211 et seqq

Pierri M (2015) Agrobiodiversity, intellectual property rights and right to food: the case of Andean countries. In: Monteduro M, Buongiorno P, Di Benedetto, Isoni A (eds) Law and agroecology: a transdisciplinary dialogue. Springer, Heidelberg, p 451 et seqq

Pisanello D (2006) La riforma del sistema comunitario di tutela delle denominazioni d'origine e delle indicazioni geografiche dei prodotti agricoli e alimentari. Contratto e impresa Europa:556 et seqq

Potschin M, Haines-Young R, Fish R, Turner RK (eds) (2016) Routledge handbook of ecosystem services. Routledge, London

Prip C (2017) The Convention on Biological Diversity as a Legal Framework For Safeguarding Ecosystem Services. Ecosystem Services. https://doi.org/10.1016/j.ecoser.2017.02.015 (available online 30 March 2017)

Pumain D (ed) (2006) Hierarchy in natural and social sciences. Springer, Dordrecht

Ramli AMD et al (2016) Determinants of food heritage towards food identity. Environ Behav Proc J 1(1):207 et seqq

Rencher KL (2012) Food choice and fundamental rights: a piece of cake or Pie in the Sky? Nevada Law J 12(2):418 et seqq. http://scholars.law.unlv.edu/nlj/vol12/iss2/9. Accessed 2 Aug 2017
Resnik D (2010) Trans fat bans and human freedom. Am J Bioeth 10(3):27 et seqq
Ricci C (ed) (2012) La tutela multilivello del diritto alla sicurezza e qualità degli alimenti. Giuffrè, Milano
Rinella A, Okoronko H (2015) Sovranità alimentare e diritto al cibo. Dir pubbl comp eur: 89 et seqq
Robinson NA (2014) The resilience principle. IUCN Acad Environ Law eJournal 5/2014:19 et seqq
Rockström J et al (2009a) A safe operating space for humanity. Nature 461(7263):472 et seqq
Rockström J et al (2009b) Planetary boundaries: exploring the safe operating space for humanity. Ecol Soc 14(2):Article 32
Roduit C et al (2014) Increased food diversity in the first year of life is inversely associated with allergic diseases. J Allergy Clin Immunol 133(4):1056 et seqq
Rohde K (ed) (2005) Nonequilibrium ecology. Cambridge University Press, Cambridge
Rook Basile E, Carmignani S (eds) (2013) Sicurezza energetica e sicurezza alimentare nel sistema UE: profili giuridici e profili economici (Atti del Convegno, Siena, 10–11 maggio 2013). Giuffrè, Milano
Rook Basile E, Germanò A (eds) (2011) Agricoltura e in-sicurezza alimentare, tra crisi della PAC e mercato globale (Atti del Convegno IDAIC, Siena, 21–22 ottobre 2010).Giuffrè, Milano
Rubino V (2007) Le Denominazioni Comunali d'Origine (De.C.O.) e la loro protezione nel quadro della disciplina comunitaria sulle denominazioni geografiche dei prodotti alimentari. Dir un eur:121 et seqq
Rubino V (2010) Indicazioni geografiche indirette e denominazioni di origine dei prodotti alimentari nella sentenza 'Bud II'. Dir comunitario e scambi internaz:255 et seqq
Ruggiu I (2014) Identità culturale. In: Morelli A, Trucco L (eds) Diritti e autonomie territoriali. Giappichelli, Torino, p 486 et seqq
Ruhl JB (2011) General design principles for resilience and adaptive capacity in legal systems — with applications to climate change adaptation. N C Law Rev 89(5):1373 et seqq
Ruhl JB (2012) Panarchy and the law. Ecol Soc 17(3):31 et seqq
Ruhl JB (2015) In defense of ecosystem services. Pace Environ Law Rev 32(1):306 et seqq. http://digitalcommons.pace.edu/pelr/vol32/iss1/6. Accessed 2 Aug 2017
Russell P, Hertz P, Mcmillan B (2011) Biology. The dynamic science, 2nd edn. Brooks/Cole, Cengage Learning, Belmont, p 2 et seqq
Russell CG, Worsley A, Liem DG (2014) Parents' food choice motives and their associations with children's food preferences. Public Health Nutr 18(6):1018 set seqq
Russo L (2012) Agricoltura e tutela della biodiversità. In: Germanò A, Viti D (eds) Agricoltura e "beni comuni" (Atti del Convegno IDAIC, Lucera-Foggia, 27–28 ottobre 2011). Giuffré, Milano, p 187 et seqq
Sabbattini M (2006) Tradizione alimentare e diversità culturale. Dir giur agr alim amb:647 et seqq
Santilli J (2012) Agrobiodiversity and the law. Regulating genetic resources, food security and cultural diversity. Earthscan, London
Scheiner SM, Willing MR (2011) The state of theory in ecology. In: Scheiner SM, Willing MR (eds) The theory of ecology. University of Chicago Press, Chicago, p 333 et seqq
Semands E (2014) Food choice: should the government be at the head of the table? Okla Law Rev 67:149 et seqq
Serrelli E (2010) Diversità biculturale. In: Eldredge N, Pievani T (eds) Ecosphera. Il Futuro della Terra. Atlante vol 1 (A-L). UTET-DeAgostini, Torino, pp 143–148
Sirsi E (2011) La tutela delle risorse genetiche in agricoltura. In: Costato L, Germanò A, Rook Basile E (dir) Trattato di diritto agrario. Vol 2: Il diritto agrombientale. Utet giuridica, Torino, p 493 et seqq
Skypala I, Vlieg-Boerstra B (2014) Food intolerance and allergy: increased incidence or contemporary inadequate diets? Curr Opin Clin Nutr Metab Care 17(5):442 et seqq
Sobo EJ (2016) Dynamics of human biocultural diversity: a unified approach. Routledge/Taylor & Francis, New York

Sollini M (2006) Il principio di precauzione nella disciplina comunitaria della sicurezza alimentare: profili critico-ricostruttivi. Giuffré, Milano
Solomon EP, Berg LR, Martin DW (2013) Elementi di biologia, 6th edn. EdiSES, Napoli
Steffen W et al (2015, 13 Feb) Sustainability. Planetary boundaries: guiding human development on a changing planet. Science 347(6223). https://doi.org/10.1126/science.1259855
Strambi G (2007) Agriturismo e valorizzazione dei prodotti agro-alimentari tipici. Dir giur agr alim amb: 746 et seqq
Strambi G (2010) I prodotti tradizionali e la politica di qualità dell'Unione europea. Riv dir alim:1 et seqq
Strambi G, Alabrese M (eds) (2005) I prodotti agro-alimentari tipici e tradizionali come beni culturali. Nuove forme di tutela e valorizzazione del territorio. Campano, Pisa
Swannack TM, Grant WE (2008) Systems ecology. In: Jørgensen SE, Fath BD (eds) Encyclopedia of ecology, vol 4 [P-S]. Elsevier, Oxford, p 3477 et seqq
Tarlock AD (1994) The nonequilibrium paradigm in ecology and the partial unraveling of environmental law. Loy Los Angeles Law Rev 27(3):1121 et seqq
Thomas SA, Cebrian J (2010) Ecosystem pattern and processes. In: Jørgensen SE (ed) Global ecology. A derivative of encyclopedia of ecology. Elsevier, Amsterdam, p 380 et seqq
Timmermann C, Robaey Z (2016) Agrobiodiversity under different property regimes. J Agric Environ Ethics 29(2):285 et seqq
Tortorella W, Traclò F (2008) Governo del territorio e identità alimentari: un patrimonio della tradizione, un valore del futuro. Cittalia Fondazione Anci ricerche, Roma
Tracey MJ (2016) Pope Francis on the ecological crisis: its causes, nature, and urgency. In: Manca L, Kauth JM (eds) Interdisciplinary essays on environment and culture: one planet, one humanity, and the media. Lexington Books, Lanham, p 207 et seqq
Trapè I (2012) I segni del territorio. Profili giuridici delle indicazioni di origine dei prodotti agroalimentari tra competitività, interessi dei consumatori e sviluppo rurale. Giuffré, Milano
Tregear A (2007) Proximity and typicity: a typology of local food identities in the marketplace. Anthropology of Food S2. http://aof.revues.org/438. Accessed 2 Aug 2017
Trepl L, Voigt A (2011) The classical holism-reductionism debate in ecology. In: Schwarz A, Jax K (eds) Ecology revisited. Reflecting on concepts, advancing science. Springer, Dordrecht, p 45 et seqq
Vadiveloo M et al (2015) Dietary variety is inversely associated with body adiposity among US adults using a novel food diversity index. J Nutr 145(3):555 et seqq
Vezzani S (2013) Le risorse fitogenetiche per l'alimentazione e l'agricoltura nel dibattito sui "global commons". Riv crit dir priv:433 et seqq
Visser S (2007) La protezione delle indicazioni geografiche e delle denominazioni d'origine dei prodotti agroalimentari a seguito del rapporto del Panel dell'Organizzazione mondiale del commercio. Agr ist merc:179 et seqq
Voigt C (ed) (2013) Rule of law for nature. New dimensions and ideas in environmental law. Cambridge University Press, Cambridge
Wahlqvist ML (2004) Requirements for healthy nutrition: integrating food sustainability, food variety, health. J Food Sci 69:CRH16–CRH18
Wahlqvist ML, Specht RL (1998) Food variety and biodiversity: Econutrition. Asia Pac J Clin Nutr 7(3–4):314 et seqq
Walker B, Holling CS, Carpenter SR, Kinzig A (2004) Resilience, adaptability and transformability in social–ecological systems. Ecol Soc 9(2):5 et seqq
Wansink B (2010) Nutritional gatekeeping: empowering parents to help guide food choices for their children. Child Obes 6(6):313 et seqq
Westra L, Taylor T, Michelot AS (eds) (2013) Confronting ecological and economic collapse. Ecological integrity for law, policy and human rights. Routledge, New York
Wijkman A, Rockström J (2012) Bankrupting nature: denying our planetary boundaries. Routledge, New York

Wiseman SR (2015) The dangerous right to food choice. Seattle Univ Law Rev 38:1299 et seqq. http://digitalcommons.law.seattleu.edu/cgi/viewcontent.cgi?article=2286&context=sulr. Accessed 2 Aug 2017

Young OR, Berkhout F, Gallopin GC, Janssen MA, Ostrom E, Van Der Leeuw S (2006) The globalization of socio-ecological systems: an Agenda for scientific research. Glob Environ Change 16:304 et seqq

Cultural Heritage, Food Diversity and International Law

Saverio Di Benedetto

1 Food Diversity and International Law. Two (or Three) Possible Approaches

Food diversity is a term that is not frequently addressed by international lawyers, unlike food security and food sovereignty. Other chapters of this volume provide for a careful explanation of the meaning and scope of a concept—food diversity—which in last years has been receiving a growing and important attention at the European and international level. Suffice it to say that this term involves not only the existence of different kinds of food and agricultural products, but also the many practices and traditions that are diffused throughout the world with respect to the production, preparation and diffusion of food and food products. Accordingly, the concept encompasses both traditional recipes or culinary practices and the production of raw food ingredients.

Food diversity therefore owns a double axiological importance. On the one hand, diversity of foods means diversity of animal and vegetable species, that are used by human beings to feed on. In this sense, it is close to the concept of biodiversity, thus relating to an ecological value. On the other hand, diversity of food means diversity of agricultural activities and culinary practices as experienced by human societies, communities and groups. According to this second acceptation, it is an expression of a social practice and knowledge, and is thus related to the value of culture.

International law does not deal with food diversity through a specific convention, treaty or protocol. Yet, some existing international instruments could be relevant to address, or even regulate, food diversity, according to either of the values it embeds.

S. Di Benedetto (✉)
Department of Legal Studies, University of Salento, Lecce, Italy
e-mail: saverio.dibenedetto@unisalento.it

A. Isoni et al. (eds.), *Food Diversity Between Rights, Duties and Autonomies*,
LITES – Legal Issues in Transdisciplinary Environmental Studies 2,
https://doi.org/10.1007/978-3-319-75196-2_24

And food diversity, from an inverted perspective, could influence the application of such instruments.

Firstly, food diversity, as based on, or composed of, a variety of animal and vegetable species, is part of the broader concept of biodiversity, which is promoted and increasingly protected under the framework of the Convention on Biological Diversity (CBD). This leads to consider from a legal viewpoint the emerging (and fundamental) category of agrobiodiversity (and more broadly agroecology), whereby the intrinsic value of biological resources is reshaped according to a human perspective, that of farmers and rural communities having modelled since centuries and millennia territory and landscape.[1] Studying food diversity from this legal perspective involves the fundamental issue of regulating the ownership of, and the access to, natural resources producing food, mainly as to their genetic codes.[2] Other Volume's chapters deal with the issue of regulating genetic natural resources and food diversity, as well as geographical indications and traditional knowledge.

The second perspective to address the concept of food diversity through the lens of international law is related to the fundamental cultural value of food activities and culinary practices for the peoples of the world. Some international instruments regulating different aspects of culture interact with food diversity, starting from those dealing with cultural diversity. Moreover, the UNESCO Conventions on cultural heritage—particularly that on intangible heritage—have both rules and institutional programs that highlight the importance of culinary practices and agricultural activities for different human populations, communities and groups.

This chapter precisely intends to deal with the concept of food diversity in international law from such perspective, by relying on the practice of the above instruments, to investigate whether such concept may assume an autonomous relevance in the contest of international cultural law.[3] In this sense, a specific line of analysis regards the possibility to define the identity of some groups and communities in function of the peculiarity of their activities and practices related to food.

Finally, it is worth noting that food diversity could be also dealt with according to a further perspective, that of human rights law. The heading 'food diversity and human rights' would easily lead to the right to adequate food, set forth by Art. 11 of the International Covenant on Economic, Social and Cultural Rights, raising the question whether such a right could be influenced by the concept of food diversity. This perspective is not followed in this chapter. In fact, there exists a clear nexus between food diversity and right to food, but this does not seem to structurally shape

[1]On this point cf. Monteduro et al. (2015).

[2]There are two main international instruments addressing this issue: the FAO International Treaty on Plant Genetic Resources for Food and Agriculture adopted in Rome on 3 November 2001; the CBD Protocol on Access to Genetic Resources adopted in Nagoya on 29 October 2010. Cf. on these issues Santilli (2012).

[3]The choice of such a cultural approach to food diversity also allows to come back to the first value of agroecology, which does conciliate ecological paradigm with human modification of nature and landscape, according to a transdisciplinary approach enhancing the role of social sciences, such as anthropology and sociology.

the content and scope of such a right. The human right to adequate food requires the existence of sufficient food—in a given society—to grant populations with a safety diet. The element of diversity is basically relevant as a *material* variety of the diet that is apt to grant an adequate and safe nutrition. Accordingly, such variety may be concretely limited, because mainly nutritional elements from a biological point of view are to be considered. Ultimately, the right to food could be even antagonist with respect to food diversity—intended as embracing the multiform kinds of food existing in myriads of human populations and groups—because the reasons to struggle against starvation could (abstractly) justify food policy promoting massive food production against traditional practices less productive.[4]

2 Cultural Heritage and Cultural Diversity in International Law

The cultural element characterising food diversity—as just noted—consists of the practices, traditions and knowledge that have led from generation to generation to the extraordinary variety of foods and recipes currently existing in human civilizations. Food diversity thus has a prevailing intangible character, within the broader cultural experiences of given human populations and groups, while also some tangible elements—such as particular rural landscapes—may be related to the cultural value of food practices. Therefore, the playground of the proposed analysis is basically related to the universal value of culture, as encompassing both tangible outputs and intangible experiences of human creativity.

The approach to culture of international law in the past century mainly regarded the preservation and enhancement of "the arts", primarily understood as a tangible output of human and people creativity, as the early implementation of the 1972 UNESCO Convention for the Protection of the World Cultural and Natural Heritage clearly attests.[5] In such a perspective, a broader meaning of 'culture' encompassing intellectual, moral and spiritual elements (indeed intertwined with—and

[4]In fact, the Committee on Economic, Social and Cultural Rights, when commenting Art. 11 of the ICESCR), addressed the possible conflict between state measures aimed at fulfilling the need to a safety diet and the cultural value of food, by underlining that "the core content of the right to adequate food implies: The availability of food in a quantity and quality sufficient to satisfy the dietary needs of individuals, free from adverse substances, and *acceptable within a given culture*;" and that cultural acceptability "implies the need also to take into account, as far as possible, *perceived non nutrient-based values* attached to food and food consumption" (*General Comment n. 12. The right to adequate food* (Article 11 ICESCR), UN doc. E/C.12/1999/5, of 12 May 1999, para. 8 and 11, emphases added). As rightly noted, the terms 'take into account' and 'as far as possible' reduce the relevance of the cultural element for the scope of the right to adequate food (Maffei 2012a, p. 85 f). On the right to food see in this volume PIERRI.

[5]On this fundamental Convention see Francioni and Lenzerini (2008).

underlying—tangible cultural heritage)[6] remained quite neglected, being a matter devoted to state sovereignty or, at the most, dealt with by specific international instruments regulating intellectual property rights. Even the Covenant on Economic, Social and Cultural Rights offered a reduced room for cultural rights, without clearly addressing culture in its overall meaning.[7]

At the outset of the current century a fundamental shift towards a broader meaning of culture in international law occurred, starting from the adoption of the Declaration of Cultural Diversity in 2002.[8] The Convention for the Safeguarding of Intangible Cultural Heritage (hereinafter CSICH) adopted in 2003 represents the real turning point for considering culture in its overall meaning at the international level.[9]

There are two basic points that arise from the definition of intangible cultural heritage set out by the CSICH: the broader meaning of culture that is implied in such definition, as encompassing intellectual, moral and spiritual elements; the direct relevance of a plurality (diversity) of human entities, different from states, namely groups and communities. This latter point is reinforced by the two main international instruments regarding cultural diversity, namely the above 2002 Declaration and the 2005 Convention on Cultural Diversity (CCD).[10] These two points are fundamental to address the issue of relevance of food diversity in international law from the angle of the protection of culture. On the other hand, according to a circular mechanism, food diversity may provide further insight on the latitude of the protection of intangible cultural heritage and on some particular human communities such as rural communities (see last section).

[6]Cf. Lizinsky (2013): "all manifestations of heritage have an intangible element" (p. 2), then putting into discussion the absoluteness of the distinction (dichotomy) between tangible and intangible culture, by referring to the unitary concept of 'heritage', as distinct from that of 'property' (p. 7 f).

[7]In fact, the important right to take part in cultural life (Art.15 of the ICESCR) is limited to the individual dimension of every human person ("right of everybody"), without giving relevance to intermediate dimension of human entities (groups, communities, peoples) that contribute in a decisive way to produce and reproduce culture. Moreover, the broader dimension of culture as encompassing also intellectual, moral and spiritual elements, thus involving the same 'lifestyle' of individuals and groups, remained rather neglected until the recent comment of the Committee on Art. 15 (*General Comment N. 21*, adopted by the Committee on Economic, Social and Cultural Rights on 21 December 2009, UN doc. E/C12/GC/21). On the "endorsement of culture in its anthropological meaning" made by the General Comment 21, cf. Pineschi (2012), pp. 33–36.

[8]The *UNESCO Declaration on Cultural Diversity* was adopted by the 31st Session of the General Conference of UNESCO on 2 November 2001.

[9]It is worth noting that in the same period the practice of the 1972 UNESCO Convention started to include in its safeguarding lists also some cultural properties sharing tangible and intangible elements, such as cultural landscapes (see Sect. 5 below).

[10]The *Convention on the Promotion and Protection of the Diversity of Cultural Expressions* was adopted by the 33st Session of the General Conference of UNESCO on 20 October 2005.

3 The Definition of Intangible Cultural Heritage and Food Diversity

According to Art. 2.1 CSICH, which gives a definition of intangible cultural heritage:

> [T]he "intangible cultural heritage" means the practices, representations, expressions, knowledge, skills – as well as the instruments, objects, artefacts and cultural spaces associated therewith – that communities, groups and, in some cases, individuals recognize as part of their cultural heritage. This intangible cultural heritage, transmitted from generation to generation, is constantly recreated by communities and groups in response to their environment, their interaction with nature and their history, and provides them with a sense of identity and continuity, thus promoting respect for cultural diversity and human creativity. [. . .].

There are two basic elements in this definition of intangible cultural heritage: an extrinsic element, that is the exteriorization of the human creativity (practices, representations, skills, etc.); a subjective element, that is the collective character of such practices and representations, as created and re-created by a human community or group.[11] Both these elements may be concretely characterized with respect to food and culinary practices. Particularly, a key point is that the subjective component intrinsically entails a plurality of collective subjects being the bearers—and makers—of intangible cultural heritage. Such necessary plurality (or diversity) reinforces the inclusion of food diversity within the rationale of the protection offered by the rules on intangible cultural heritage, as argued in this section.

3.1 The Extrinsic Component of Intangible Cultural Heritage and Food Practices

It may be easily noted that food practices and culinary habits may well be encompassed within the "practices, representations, expressions, knowledge, skills" characterizing the objective element of the definition of intangible cultural heritage. In fact, it is common sense that such culinary practices, habits or skills are not limited to a domestic, or familiar, dimension, but may be individuated and described in a given social environment.

In this sense, it is possible to detect three main different categories of human activities related to food, each of which could assume in the concrete case a cultural character, commonly characterized by a social exteriorization: the production of

[11] According to Scovazzi, in addition to an "objective component" and a "subjective or social component", there would exist a third component of intangible cultural heritage, a spatial one, which would be constituted by the "cultural spaces" where the cultural experiences are developed: Scovazzi (2012), p. 180.

food, related to agricultural practices and skills (farming; breeding; harvesting)[12]; the preparation of foods and dishes, according to shared traditions and recipes; the communitarian consumption of meals, following given modalities, procedures or rituals.

This relationship between culinary practices and the extrinsic element of intangible cultural heritage is further confirmed by Art. 2.2, setting forth a list of domains that exemplify the human "practices, representations, expressions, knowledge, skills" composing the intangible cultural heritage.[13] Now, food practices are typically cultural manifestations that may belong to two or more of these domains, such as "oral traditions and expressions", "social practices, rituals and festive events", "practices concerning nature". Also, an important study realized by a panel of experts under the auspices of UNESCO confirmed that food practices may well be part of the domains of Art. 2.2.[14]

3.2 *Intangible Cultural Heritage and the Plurality (or Diversity) of Human Communities*

The second (subjective) element of the definition is the existence of a human community or group, which creates (and recreates) the element qualified as intangible cultural heritage.

It is worth noting that the character of diversity is implicit in the plural qualification of the 'makers' of intangible cultural heritage. According to Art. 2.1 CSICH, "intangible cultural heritage [. . .] is constantly recreated by communities and groups". Thus, a plurality of basic human entities (communities and groups) create and recreate intangible cultural heritage.

Moreover, such heritage is primarily composed by *the* cultural heritage *of each* human community and group that continuously create and recreate it. This crucial

[12]In this category it should be also included the practices of selection of species and crossbreeding.

[13]Art. 2(2): The "intangible cultural heritage", as defined in paragraph 1 above, is manifested inter alia in the following domains: (a) oral traditions and expressions, including language as a vehicle of the intangible cultural heritage; (b) performing arts; (c) social practices, rituals and festive events; (d) knowledge and practices concerning nature and the universe; (e) traditional craftsmanship.

[14]*Compte-Rendu des journées de Vitré sur les pratiques alimentaires*, 3 April 2009 (available at http://www.iiac.cnrs.fr/article1007.html): "*les pratiques alimentaires ont une dimension transversale vis-à-vis des domaines explicités à l'article 2 alinéa 2* en tant qu'elles s'intègrent à des systèmes articulés de relations sociales et de significations collectivement partagées. Les pratiques alimentaires concernent donc aussi bien les traditions et expressions orales, les arts du spectacle, les pratiques sociales, rituels et événements festifs, les connaissances et pratiques concernant la nature, ainsi que les savoir-faire liés à l'artisanat traditionnel" (§ 2, emphasis added). It is worth noting that, according to the panel of experts, "les pratiques alimentaires ne devraient pas être considérées uniquement comme une réponse aux besoins biologiques, mais comme des expériences culturellement élaborées par les groupes humains tout au long de leur histoire" (§ 1). Cf. Maffei (2012b), p. 232.

point is made clear by another part of the definition provided by Art. 2.1. All the "practices, representations, expressions, knowledge, skills" may be considered as intangible cultural heritage only when such communities and groups (and in some cases individuals)[15] recognize them "as part of *their* cultural heritage" (emphasis added). Accordingly, such intangible expressions (practices, skills, etc.) are part of a unitary category protected by international law—intangible cultural heritage—only if they firstly are part of the heritage of one of such communities and groups.[16]

This underlying characterization of intangible cultural heritage as an expression of the cultural diversity of human peoples and communities is explicitly confirmed by the CSICH. In its Preamble it is written "*Considering* the importance of the intangible cultural heritage as a mainspring of cultural diversity and a guarantee of sustainable development [. . .]".[17] And according to Art. 2.1 "intangible cultural heritage [. . .] provides them [communities and groups] with a sense of identity and continuity, thus promoting respect for cultural diversity and human creativity".

In sum, intangible cultural heritage, as a unitary category under international law, is composed by the intangible cultural heritage of each human community and group, and accordingly the protection offered by international law is directed to each intangible cultural heritage, as the result of a general interest[18] of international community related to the preservation of the identity of such communities and groups (and of individuals belonging to them).[19] Therefore the category of intangible cultural heritage essentially contributes to the affirmation of cultural diversity at an international level.

[15]The role of individuals is on the whole marginal, because the creation of intangible cultural heritage is a prerogative of communities and groups. Cf. Scovazzi (2012), p. 192.

[16]It is worth noting that the CSICH, by referring to groups and communities, rather than peoples and states, does "transcend the traditional notion of cultural heritage as a component of the nation state and an element of national sovereignty" Francioni (2008), p. 15. This point is particular relevant also in the perspective of human rights, because it supports the idea that cultural rights may be hold by collective entities that are different from states (and their internal partitions), such as indigenous peoples and rural communities.

[17]This further reference to sustainable development is very important when dealing with the issue of food diversity from a cultural perspective, as it is argued below.

[18]"Cultural heritage is of general interest to humanity" (Art. 19 CSICH). According to Francioni (2008), p. 6, "the international community as a whole has an interest in the conservation and protection of cultural heritage as part of the safeguarding of human dignity".

[19]The issue of cultural identity, of both individuals and groups, is complex and highly debated in international law, and would require an autonomous analysis. Very shortly, if the existence of an autonomous right to cultural identity is very controversial (cf. Donders 2008), a different discourse is possible as to the concept of cultural identity of given groups, with respect to the protection offered to them by international law (see Jovane 2007). In this chapter we assume that a (cultural) identity of communities and groups does exist, by relying on the same Art. 2.1 CSICH, which speaks of sense of identity of such collective entities. Besides, for collective entities different from states and legal persons, the concept of identity seems to be almost necessary for the same existence of such entities. The problem is rather to define the criteria to determine the identity (and thus the existence) from a legal point of view. This latter point is touched in the final Section of this chapter.

Against this backdrop, food practices and traditions, to the extent that they play an identitarian role for relevant groups, communities and peoples, may be protected by virtue of their distinctive character as intangible cultural heritage. Moreover, since intangible cultural heritage is by itself an expression of the diversity of each cultural community and group, food diversity—in its cultural value—could play an important role within the framework build up by the CSICH and other similar instruments of international law. This 'deductive' assumption seems to be confirmed by the practice of the CSICH.

4 Intangible Cultural Heritage Involving Food Practices. A 'Descriptive' Relevance of Food Diversity Within the CSICH Experience

The protection afforded by the CSICH at an international level is based on a system of two lists (Safeguarding List and Representative List) including the concrete practices, representations, expressions, knowledge, skills that constitute the "elements"[20] of the intangible cultural heritage of humanity. Such lists are drawn up and yearly updated by the Intergovernmental Committee for the Safeguarding of Intangible Cultural Heritage (hereinafter, the Committee), upon the proposal of a specific element by the States parties interested.[21] In the practice of the Committee we may find many cases of intangible cultural heritage involving in a less or more relevant way food practices and traditions.[22]

In the most part of such cases, and always in the first 2 years of activities of the Committee (2008 and 2009), food practices are mentioned only as a part of a broader cultural manifestation. This is the case, for instance, of several festivals included in either of the lists: food, beverages, meals and others food traditions are only a factor of the ceremony included within the list.[23] The same reasoning may be reproduced for cases of carnival manifestations, whereby food practices are present together

[20] This is the official name of each expression or practice that is inscribed in either of the lists by the Committee.

[21] This is a structural limit of the lists system, because "as states are the only ones which can nominate manifestations of heritage for the list, it is very easy to see that states will prefer those manifestations of heritage that are not politically controversial" Lizinsky (2011), p. 82.

[22] On the decisions of the Committee on intangible cultural heritage involving food practices (until 2011) see Maffei (2012b), pp. 225–240.

[23] Cf. Maffei (2012b), p. 227 f. referring in this sense to festivals inscribed in the Representative List in 2008 and 2009 (with bibliographic references), namely *Gangneung Danoje Festival* of Republic of Korea (drinking a sacred liquor and eating traditional crackers); *Qiang New Year Festival* of China (drinking a particular wine); *Dragon Boat Festival* of China (drinking a ruby wine and eating particular rice dumplings). For a very recent case see *Argungu international fishing and cultural festival* of Nigeria, ITH/14/9.COM/ Decision of 2 December 2016 (preserving specific fishing techniques).

with other cultural manifestations.[24] Also in cases of ritual and religious practices some foods and drinks are included as typical elements of such practices.

For these and other similar cases it has been correctly argued that the safeguard of the ceremony, festival or ritual "preserves indirectly also the food but the latter is not the main target of the protection".[25] It could be added that also a direct protection of some foods and culinary (or agricultural) practices could be required precisely to protect the broader cultural element (for instance, a kind of fish or drink that is essential for a civil or religious ceremony). In any case, the possible role of food diversity in these cases seems to be essentially descriptive, as a form of recognition of several situations in which international law grants, directly or indirectly, a protection to food products and practices by virtue of their cultural importance and peculiarity.

Some elements included in either of the Lists have an explicit culinary dimension, referring to specific meals, recipes, diets. Some of them are very significant for the approach here developed, and are analysed in next section. However, in other cases, the reasons justifying the inclusion in the list, and thus the concrete protection afforded, seems not to involve further consequences with respect to the autonomous relevance of the category of food diversity.

A rather famous case is the Gastronomic Meal of France, which was included in the Representative List in 2010 as a social practice.[26]

> The gastronomic meal of the French is a customary social practice for celebrating important moments in the lives of individuals and groups, such as births, weddings, birthdays, anniversaries, achievements and reunions. It is a festive meal bringing people together for an occasion to enjoy the art of good eating and drinking. [. . .] Important elements include the careful selection of dishes from a constantly growing repertoire of recipes; the purchase of good, preferably local products whose flavours go well together; the pairing of food with wine; the setting of a beautiful table; and specific actions during consumption, such as smelling and tasting items at the table. The gastronomic meal should respect a fixed structure, commencing with an apéritif (drinks before the meal) and ending with liqueurs, containing in between at least four successive courses, namely a starter, fish and/or meat with vegetables, cheese and dessert. [. . .].

This element clearly constitutes a food practice. The rationale of the afforded protection lies in the particular social significance of the structure and organization of the meal. By recalling the three different categories of human activities related to food (see above Sect. 3.1), the cultural character is thus related to the communitarian consumption of food, with a prevailing note related to the aesthetic combination of the different factors, including the scenography. On the other hand, food and recipes

[24] See for instance *Annual carnival bell ringers' pageant from the Kastav area* of Croatia, ITH-09-4. COM-CONF.209.

[25] Maffei (2012b), p. 229.

[26] ITH-10-5.COM-CONF.202 Decision of 19 November 2010, p. 21.

seem to remain in the background, by mainly counting in reason of the proportion between the different dishes and drinks.[27]

It is worth noting that the communitarian element is only referred to occasional events (*births, weddings, birthdays, anniversaries*), without further characterizing a concrete social group or community. The Gastronomic Meal is ultimately part of the tradition of the entire France, even if it is concretely practiced only in parts of the territory. But these parts are not characterized in an autonomous way, as communities and groups.[28]

Another example of a food practice that has been inscribed in the Representative List is the Gingerbread craft from Northern Croatia.[29]

> [...] The process of making gingerbread requires skill and speed. The recipe is the same for all makers, utilizing flour, sugar, water and baking soda – plus the obligatory spices. The gingerbread is shaped into moulds, baked, dried and painted with edible colours. Each craftsperson decorates gingerbread in a specific way, often with pictures, small mirrors and verses or messages. The gingerbread heart is the most common motif, and is frequently prepared for marriages, decorated with the newlyweds' names and wedding date. [...] Gingerbread has become one of the most recognizable symbols of Croatian identity. Today, gingerbread makers are essential participants in local festivities, events and gatherings, providing the local people with a sense of identity and continuity.

In this case, the element qualified as intangible cultural heritage is a traditional skill regarding the preparation of a particular food. The scope of the protection thus involves both the recipes traditionally handed down and the decorations made on the gingerbread. The cultural element is based on the generational transmission of this peculiar activity and its social function within the local communities, related to fundamental moments of the social life, such as marriages.

The sense of identity embedded into the gingerbread craft, which is crucial for the qualification of a given social practice or skill as intangible cultural heritage, is not ultimately referred to a specific community or group. In fact, while the craftsmen activities provide local people with a sense of identity, such local people are not further characterized by the report of the Committee. In contrast, "gingerbread has become one of the most recognizable symbols of Croatian identity" (without even referring to the Northern part of the country). Thus, the identitarian factor regards the entire nation involved. On the other hand, no concrete reference is made by the

---

[27]In fact, there are only generic references to category of dishes (*starter, fish and/or meat with vegetables, cheese and dessert*), which are not per se susceptible of being characterized in terms of food diversity. Also, the reference to a *constantly growing repertoire of recipes* remains quite generic, absorbed in the prevailing social structure of the meal.

[28]This point is confirmed by the Committee, in the part of the Report regarding the specific criteria for including the proposed element in the List. "The gastronomic meal of the French plays an active social role within its community and is transmitted from generation to generation as part of its identity" (criterion R1). Its community implies only one entity, corresponding to the entire French nation, and the identitarian function is referred to the entire France (as demonstrated by the use of the singular 'its').

[29]ITH-10-5.COM-CONF.202, Decision of 19 November 2010, p. 18.

Committee to an identitarian function of such skill and practice for a sort of community of gingerbread makers.[30]

These two examples, showing a direct qualification of a food practice as intangible cultural heritage, ultimately reinforce the idea that the system set out by the CSICH may play an important function to protect culinary practices having a distinctive cultural character. In this sense, we could say that the existence of several important cases of protection of different traditional culinary and food practices may lead to an overall effect of safeguarding food diversity at the international level. This could perhaps generate a further, positive dynamic in terms of creating a goal to autonomously protect food diversity, for instance under the umbrella of the CCD.

However, in this way the concept of food diversity would simply serve a descriptive function in the context of CSICH, by at the most representing an important factor to systematize a cluster of cultural practices and representations, and perhaps to enhance international cooperation in this crucial field. Nevertheless, a further, more pregnant function for food diversity in this legal context may be drawn from other examples provided by the UNESCO practice on cultural heritage.

5 Agricultural and Culinary Practices Having a Peculiar Importance in the UNESCO Experience

This section highlights some crucial cases of food and culinary practices protected by the two UNESCO conventions, as both intangible and tangible cultural heritage, by showing their peculiarities with respect to the cases analysed in the previous section.

5.1 Significant Cases of Food Practices Qualified as Intangible Cultural Heritage

At the end of the past section we have noted that also the important cases of French Meal and Croatian Gingerbread were not able to effectively attribute a really normative function to food diversity. In fact, in such cases it lacked a specific community that was distinct from the generic population of the involved state and found its identity in function of the peculiarity of its food practices. Conversely, the following cultural elements included in the CSICH Lists are precisely referred to an autonomous community, thus making it possible a different line of reasoning.

[30]Cf. Maffei (2012b), p. 234, noting that the Committee has dropped the reference made in the Croatian nomination to the community of gingerbread craftspersons.

In 2009 the Committee inscribed in the Representative List the element of *Oku-noto no Aenokoto*, nominated by Japan.[31]

> Oku-noto no Aenokoto is an agricultural ritual transmitted from generation to generation by the rice farmers of the Noto Peninsula [. . .]. The twice-yearly ceremony is unique among the harvest rituals of Asia in that the master of the house invites the deity of the rice field into his home, behaving as though the invisible spirit were really present. In December, to express gratitude for the harvest, the farmer draws a bath and begins to prepare a meal, summoning the deity from the field with the sound of pounding rice cakes. Welcoming his guest in formal clothes with a lantern, the farmer allows it to rest in a guest room before assisting it with a bath and offering a meal of rice, beans and fish. [. . .] A similar ritual is performed before planting in February to ensure an abundant harvest. Performed with individual variations throughout the region, Oku-noto no Aenokoto reflects the everyday culture of the Japanese, who have cultivated rice since ancient times, and serves as a marker of identity for the area's farmers.

This cultural practice is primarily a religious ritual, thus differing from the previous examples of the French Meal and Croatian Gingerbread. In this sense, *Oku-noto no Aenokoto* could seem to be like the above examples of rituals involving food aspects: in fact, during the ritual the master of the house pounds rice cakes to summon the deity, then prepares and offers a meal according to a standard recipe.

But the crucial aspect is that the Committee characterizes such ritual as an agricultural or harvest one. The practice is linked to the activity of farming of the human group. It is a food practice, because it is functional to the overall agricultural activity, centred onto the production of rice from generation to generation. Moreover, and chiefly, such cultural practice serves the decisive function to identify the concrete agricultural community. By using the Committee's words, it "serves as a marker of identity for the area's farmers".

In 2010, two specific culinary practices were included within the Representative List, namely the *Traditional Mexican Cuisine* and the *Mediterranean Diet*. Such elements may well be inserted in the proposed line of reasoning, by differing from the above analysed culinary practices (French Meal and Croatian Gingerbread) for being ascribable to specific communities differing from the national communities of the involved states.

The *Traditional Mexican cuisine – ancestral, ongoing community culture, the Michoacán paradigm* is described by the Committee as.

> a comprehensive cultural model comprising farming, ritual practices, age-old skills, culinary techniques and ancestral community customs and manners. It is made possible by collective participation in the entire traditional food chain: from planting and harvesting to cooking and eating. The basis of the system is founded on corn, beans and chili; unique farming methods [. . .]; cooking processes [. . .]; and singular utensils [. . .]. Native ingredients such as varieties of tomatoes, squashes, avocados, cocoa and vanilla augment the basic staples. Mexican cuisine is elaborate and symbol-laden, with everyday tortillas and tamales, both made of corn, forming an integral part of Day of the Dead offerings. Collectives of female cooks and other practitioners devoted to raising crops and traditional cuisine are found in the State of Michoacán and across Mexico. Their knowledge and techniques express community

[31]ITH-09-4.COM-CONF.209, Decision of 2 October 2009.

identity, reinforce social bonds, and build stronger local, regional and national identities. Those efforts in Michoacán also underline the importance of traditional cuisine as a means of sustainable development.[32]

In this case, the cultural element inscribed in the Representative list is an entire traditional cuisine of a big State. The concept of cuisine is twofold here. On the one hand, the Committee refers to a vast range of native ingredients, characterizing the typical Mexican dishes and recipes. But 'cuisine' is first and foremost the way to produce food according to a traditional and holistic cultural model, related to the strict link with the land and its products according to the collective participation to farming activities. And it is worth noting that such cuisine, by virtue of its traditional character (thus having slowly adapted to the rural environment), plays an ecological function according to the principle of sustainable development.

Against this background, while this particular 'farming paradigm' is qualified as Mexican, it is not common to all Mexican population. It has a precise origin in a State, Michoacán, and is cross-cutting the entire Mexican society. It regards only those communities of farmers that keep following such traditional agricultural methods and use "native ingredients", often relying on the work of collectives of female cooks.

A similar reasoning is possible for the *Mediterranean Diet*,[33] whose nomination was originally proposed by Greece, Italy, Morocco and Spain.[34]

Even if the Mediterranean Diet is a culinary practice (or lifestyle) largely diffused in both urban and rural Mediterranean societies, the Committee has chosen a stricter acceptation, by which such diet "constitutes a set of skills, knowledge, practices and traditions". Such cultural manifestations regard the three main categories of human activities related to food, namely production, preparation and consumption. As for the Traditional Mexican Cuisine, also Mediterranean diet assumes a broader cultural meaning, since it is "the cornerstone of social customs and festive events" and

[32]ITH-10-5.COM-CONF.202, Decision of 19 November 2010, p. 35.

[33]"The Mediterranean diet constitutes a set of skills, knowledge, practices and traditions ranging from the landscape to the table, including the crops, harvesting, fishing, conservation, processing, preparation and, particularly, consumption of food. The Mediterranean diet is characterized by a nutritional model that has remained constant over time and space, consisting mainly of olive oil, cereals, fresh or dried fruit and vegetables, a moderate amount of fish, dairy and meat, and many condiments and spices, all accompanied by wine or infusions, always respecting beliefs of each community. However, the Mediterranean diet (from the Greek *diaita,* or way of life) encompasses more than just food. It promotes social interaction, since communal meals are the cornerstone of social customs and festive events. It has given rise to a considerable body of knowledge, songs, maxims, tales and legends. The system is rooted in respect for the territory and biodiversity, and ensures the conservation and development of traditional activities and crafts linked to fishing and farming in the Mediterranean communities that Soria in Spain, Koroni in Greece, Cilento in Italy and Chefchaouen in Morocco are examples. Women play a particularly vital role in the transmission of expertise, as well as knowledge of rituals, traditional gestures and celebrations, and the safeguarding of techniques." (ITH-10-5.COM-CONF.202, Decision of 19 November 2010, p. 45).

[34]In 2013 other states have been included among those conserving the Mediterranean Diet, namely Croatia, Cyprus and Portugal.

underlies main experiences of the collective imagination (songs, maxims, legends). Moreover, its traditional methods of production and consumption are "rooted in respect for the territory and biodiversity".

This strict and pregnant acceptation of Mediterranean Diet (indeed necessary to qualify it as intangible cultural heritage) entails that its bearers and makers are not the entire societies of the involved states, but more concrete communities, primarily devoted to the farming and fishing activities that are at the outset of the traditional food chain. This crucial point is clearly highlighted by the Committee, when it refers to concrete Mediterranean communities (one for each involved state) as the effective realities where cultural practices and representations, reassumed in the formula Mediterranean Diet, are created and recreated.

In all last three analysed cases the value of food diversity may be appreciated in its entirety. Diversity of vegetal and animal species, often endemic, selected and protected according to a traditional model respecting the environment, as underlined for both Mediterranean and Mexican cuisine. And diversity of food production models, which prove to be different from other farming and harvesting practices, thus contributing to create food diversity from a more specifically cultural point of view. But the most important aspect to construe a legal meaning around food diversity is represented by the relationship between peculiarity of the food chain and underlying human communities, according to an identitarian paradigm.

5.2 The Protection of Rural Landscapes Under the 1972 World Heritage Convention

The distinction between tangible and intangible cultural heritage has a practical function, and appears to be valid to grant an effective protection to a broad range of cultural expression.[35] On the other hand, such distinction unavoidably produces overlapping zones, as it is precisely the case of cultural landscapes. Several examples of rural practices, qualified by the Committee as intangible cultural heritage, could well be seen from the point of view of the landscapes modelled by generations of farmers. However, cultural landscapes are classified and protected under the UNESCO Convention Concerning the Protection of the World Cultural and Natural Heritage of 1972.[36] As it is known, this Convention is likewise based on a system of lists, where the elements ("properties") of cultural or natural heritage are inscribed. We analysed here three significant examples of rural landscapes recently qualified as world cultural heritage.

The *Landscape of Grand Pré*, on the Atlantic coast of Canada, has been inscribed in 2012 on the World Heritage List, as a cultural property.[37]

[35]Lizinsky (2013), p. 2 ff.

[36]On this point see Last (2008).

[37]Decision: 36 COM 8B.27 (http://whc.unesco.org/en/decisions/4798), adopted in 2012.

This is a "cultural landscape bearing testimony to a remarkable effort, over many centuries, using the polder technique to develop agricultural farmland". The cultural element is thus firstly based on the agricultural activity, skilfully developed in an area dominated by extreme tides, through hydraulic drainage system creating a system of polders hosting crop farming. Likewise, the World Heritage Committee underlines the "community-based management system" established by the original colony community, the Acadians, and continued by successive inhabitants. It is worth noting that, according to the committee, this cultural practice has enabled "continuous and sustainable agricultural development". Finally, the community-based management system is "still in use today".

We may easily find the main characters that we have highlighted with respect to the precedent cases analysed in this section. There is an agricultural activity, having a peculiar cultural character (here firstly related to the ability to develop a system of dikes and polders according to traditional techniques). And there exist a rural community, which finds its identity in having modelled the territory and realized its farming activity. In this case, the relevant food practice is that of producing vegetable food according to traditional and peculiar techniques, by respecting the environment.

The *Bassari Country: Bassari, Fula and Bedik Cultural Landscapes*, located in Senegal, has been inscribed in 2012 in the World Heritage List, as a cultural property.[38]

It is composed by three cultural landscapes, each of which inhabited by a different community, but all sharing a traditional agricultural activity that likewise owns a sacred character. Also, in this case the rural landscape is characterized by a peculiar agricultural activity "including crop rotation and manuring, communal sowing, weeding and harvesting and commuting practices imposed by traditional agricultural systems and by the relative scarcity of resources". At the same time, the three communities consider cultivated lands as sacred, thus realizing an "intertwined complex of practices, social rules, rites and beliefs", in a vulnerable environment. In sum, also in this case, a peculiar and traditional agricultural activity is carried out in a sustainable way and represents a cultural and identitarian element for the involved human communities.

Finally, the *Cultural Landscape of Honghe Hani Rice Terraces*, on the Chinese department of Yunnan (south-west), has been inscribed in 2013 on the World Heritage List, as a cultural property.[39]

The landscape is composed of "an extraordinarily complex system of irrigated rice terraces that flows around the contours of the mountains", which were carved out of tropical forests by Hani farmers during more than a millennium. From the rather long text elaborated by the World Heritage Committee to explain and justify the inscription of *Honghe Hani* on the List, the following passage is very significant for the issue here discussed.

---

[38]Decision: 36 COM 8B.16 (http://whc.unesco.org/en/decisions/4787), adopted in 2012.

[39]Decision: 37 COM 8B.24 (http://whc.unesco.org/en/decisions/5152), adopted in 2013.

> Red rice, the main crop of the terraces is farmed based on a complex, integrated farming and breeding system within which ducks fertilise the young rice plants, while chickens and pigs contribute fertiliser to more mature plants, water buffalo slough the fields for the next year's planting and snails growing in the water of the terraces consume various pests. The rice growing process is sustained by elaborate socio-economic-religious systems that strengthen peoples' relationship with the environment, through obligations to both their own lands and to the wider community, and affirm the sacredness of nature. This system of dual interdependence known as the 'Man-God Unity social system' and its physical manifestation in the shape of the terraces together form an exceptional still living cultural tradition.

The strict connection between this 'tangible cultural heritage' and the above elements of 'intangible cultural heritage' related to food is evident. The landscape represents, as a synecdoche, the broader "living cultural tradition" of the community. Such cultural tradition is firmly based on an agricultural activity, the production of red rice, according to typical and original farming models, which also integrate breeding activities. At the same time, the production of food is part of a broader socio-economic-religious system, which is ecologically integrated with the natural environment,[40] including tropical forests.[41] Finally, this complex farming culture is the identitarian element for the rural community that has created such a peculiar cultural landscape.

6 A Proper Legal Function for Food Diversity in International Cultural Law

At the end of (Sect. 4) we offered a first conclusion regarding the possible role of food diversity with respect to a first set of cultural practices, approximately directly involving food aspects, characterized as intangible cultural heritage. In such cases, food diversity would simply play a role of systematization among different, and often heterogeneous, forms of protection of peculiar cultural manifestations at the international level.

On the contrary, the six cases of international protection of cultural heritage, referred to in the past section, allow to develop a different discourse, by which food diversity would serve a distinctive (distinguishing) function to individuate relevant communities, particularly rural communities.

[40]In other part of the World Heritage Committee decision is explained that such a system is "based on symbiotic relationships between plants and animals that reinforce communal obligations and the sacredness of nature".

[41]"The landscape reflects an integrated four-fold system of forests, water supply, terraces and houses", thus showing an agro-ecological approach.

6.1 Cultural Heritage and Communities

As noted above, the definition of intangible cultural heritage, as set out by Art. 1 CSICH, relies on two main elements: an extrinsic element, consisting of the exteriorization of the human creativity through representations, practices, expressions, knowledge, skills; and subjective elements, by which a human community or group creates and recreates such expressions and representations, recognizing them as an identitarian part of its cultural heritage.

Communities and groups are thus essential to qualify a human representation or practice as intangible cultural heritage.[42] At the same time, their intangible cultural heritage serves a certain function of self-identification for such communities.[43] In this sort of circular mechanism (whose actual existence depends on the intensity of the identitarian, or distinctive, function of cultural heritage) food diversity (as part of cultural diversity) may play an important role, particularly for some kinds of communities.

6.2 A Distinctive Function for Food Diversity?

Food diversity may be defined as the existence of different foods and food practices, each of which has its own characteristics. A particular species of wheat is not equivalent to other species of wheat. Likewise, a given agricultural product, practice or tradition is different from others. This quite obvious consideration leads to a potential endless amount of different agricultural products, practices and experiences, each of which could be abstractly deemed to have a social value. To give a legal sense to the concept of food diversity is thus necessary to define some criteria to select among such amount, that is to qualify a food or food practice as effectively peculiar and thus worthy of legal protection.

As to the biological side of food diversity, the selection of relevant species and natural products should be made according to scientific parameters, based on the biological classification of species and on the ecological function of each of them.

[42]It has been correctly noted that this "social component of the intangible cultural heritage marks a clear-cut distinction" with the tangible cultural heritage, as defined by the 1972 UNESCO Convention, "where this component is lacking or much less important" Scovazzi (2012), p. 189 f. According to Scovazzi, the key point of difference is that the elements of intangible cultural heritage "do not need to present an 'outstanding universal value' as required for inscription on the list drawn up under the World Heritage Convention" (Ibid.). On the interpretation of the concept of outstanding universal value in the World Heritage Convention see Yusuf (2008). This distinction has become less sharp after the recent introduction, within the elements of tangible cultural heritage, of cultural landscapes, as noted above.

[43]The same reasoning could be reproduced for cultural landscapes—and other manifestations of (tangible) cultural heritage of humanity—when they are created by a given human community and serve an identitarian function for such community.

This approach is well susceptible to produce legal consequences, particularly under the CBD regime.

But what criteria could guide such a selective operation, if food diversity was considered according to its cultural dimension? The suggestion here proposed is mainly focused on the normative basis of Art. 2.1 CSICH. Therefore, only those food products and practices that have a distinctive character for a given population or community (that is providing it "with a sense of identity and continuity") should deserve protection under international (cultural) law.

At first, this point expresses in more theoretical terms what has been suggested at the end of (Sect. 4). Food diversity, in its cultural meaning, is a way to categorize (systematize) a part of cultural expressions and representations that are worthy of protection under the CSICH. Such systematization could also be useful for further policies or programs to enhance and protect food diversity, by creating a bridge with the value of cultural diversity as dealt with (also) by the CCD.

But in some cases, the involved human community or group may find its *main* or *sole* identitarian basis precisely in a traditional food practice. In this hypothesis food diversity could play a more pregnant, and autonomous, role. The cases respectively dealt with under (Sects. 4 and 5) well illustrate the distinction just suggested.

The Gastronomic Meal is not obviously the main identitarian character of French population. On the other hand, as demonstrated above, it is not able to identify an autonomous category of population living in France. It is simply a (limited) component of the national identity of Frenchs, among many others. Likewise, Gingerbread Craft is not the main identitarian element of Croats. Nor it can characterize gingerbread craftspersons as a community (unlike the Croat Government sustained).

On the contrary, the Traditional Mexican Cuisine can identify a particular group (or groups) of persons among Mexican people. Moreover, the broad acceptation of 'cuisine' assumed by the Committee—as encompassing an entire cultural model related to cultivation of land—leads to affirm that its qualifying, or distinctive, function is decisive to identify some communities or groups. The same reasoning may be replicated for the Japanese community of rice farmers of the Noto Peninsula.[44] By the same token, the cultural landscapes of *Grand Pré* in Canada or *Honghe Hani Rice Terraces* in China, expressing a civilization of an area based on a rural dimension and involving other main cultural aspects, might be considered as the decisive element to qualify the farmers of such zones as a community.

In all these examples the involved human community or group find its *main* or *sole* identitarian basis precisely in its particular food practices and experiences. The question is to avoid that each human entity may be qualified as a 'community' by reason of a common food practice or experience. For instance, it would seem

[44]Probably a similar reasoning is also possible for the Mediterranean Diet and its local communities evoked by the Committee. In fact, this is a borderline example, as the concrete identification of such communities left some doubts as to the criteria chosen (Maffei 2012b, p. 238 f). In any case, as seen above, the Committee recognizes this identitarian character of Mediterranean Diet (in its broad meaning) for some local communities conserving "traditional activities and crafts linked to fishing and farming".

difficult to affirm that the fast-food consumers may constitute a community according to the meaning of the CSICH, even if a part of them asserted to be a community by reason of the supposed identitarian function of going each Saturday to eat hamburgers.[45]

In such a context, food diversity could be the key concept to determine when a food practice or experience is so peculiar (so 'diverse') as to serve the crucial function of detecting and qualifying a set of human beings as a group or community (thus 'diverse' from anyone else).

We could thus say that food diversity should be intended, from a legal point of view, as the key concept to assess the degree of cultural difference of a given food and agricultural practice, in function of qualifying the 'makers' of such practice as a community.

For this task, it is necessary to find some criteria that could underlie such delicate assessment of the degree of cultural difference. Of course, this is a kind of legal activity that should be, as such, highly transdisciplinary, by firstly relying on anthropological studies. The effective difference of a given cultural practice, such as a food one, is an assessment for social sciences as a whole.

However, a first 'legal' criterion may be found in the same text of Art. 2.1, whereby it requires that intangible cultural heritage is "transmitted from generation to generation". The existence of a series of generations, maintaining or evolving the same identitarian practice, is a valid criterion to avoid a proliferation of professed communities based on a food habit. Moreover, it expresses the element of tradition, which is implied in the same concept of intangible heritage.

A second 'legal' criterion could be drawn from the peculiarity of food diversity, as characterized by a main biological and ecological meaning. The criterion could thus be that the food practice is respectful of the environment, or even symbiotic in a given ecosystem, as for the cultural landscape of *Honghe Hani Rice Terraces* in Southern China. Such a criterion may be drawn from substantially all the referred cases, as highlighted under (Sect. 5).[46] In this way, the concrete operating of food diversity in the context of international cultural law could lead to a mutually supportive relationship with other international instruments involving food diversity, as the Biodiversity Convention.

This proposed 'distinctive' (or identitarian) function of food diversity seems to be particularly useful to give some subjectivity (at least in the terms of the UNESCO Conventions)[47] to those rural communities that do not own other characteristics (such as a peculiar language or religion) allowing them to autonomously stand out in the panorama of international law.

---

[45]This issue is part of the broader, and complex, issue of defining the (cultural) identity through the law. On this point, cf. Horváth (2008), pp. 188–191.

[46]As to intangible cultural heritage, this is consistent with the 'requirement of sustainable development', provided for by Art. 2.1 CSICH.

[47]It is worth recalling that among the criteria to qualify a practice as intangible cultural heritage, the CSICH requires a prior informed consultation of involved populations.

References

Donders Y (2008) A right to cultural identity in UNESCO. In: Francioni F, Scheinin M (eds) Cultural human rights. Brill, pp 317–340

Francioni F (2008) Culture, heritage and human rights: an introduction. In: Francioni F, Scheinin M (eds) Cultural human rights. Brill, pp 1–15

Francioni F, Lenzerini F (eds) (2008) The 1972 World Heritage Convention: a commentary. Oxford University Press, Oxford

Horváth E (2008) Cultural identity and legal status: or, the return of the right to have (particular) rights. In: Francioni F, Scheinin M (eds) Cultural human rights. Brill, pp 169–191

Jovane M (2007) The universality of human rights and the international protection of cultural diversity: some theoretical and practical considerations. Int J Minor Group Rights 14:231–262

Last K (2008) Article I, cultural landscape. In: Francioni F, Lenzerini F (eds) The 1972 World Heritage Convention: a commentary. Oxford University Press, Oxford

Lizinsky L (2011) Selecting heritage: the interplay of art, politics and identity. Eur J Int Law 22 (1):81–100

Lizinsky L (2013) Intangible cultural heritage in international law. Oxford University Press, Oxford

Maffei MC (2012a) Food as a cultural choice; a human right to be protected? In: Borelli S, Lenzerini F (eds) Cultural heritage, cultural rights, cultural diversity. Martinus Nijhoff, Leiden-Boston, pp 83–106

Maffei MC (2012b) Culinary traditions as cultural intangible heritage and expressions of cultural diversity. In: Borelli S, Lenzerini F (eds) Cultural heritage, cultural rights, cultural diversity. Martinus Nijhoff, Leiden-Boston, pp 223–249

Monteduro M, Buongiorno P, Di Benedetto S, Isoni A (eds) (2015) Law and agroecology. A transdisciplinary dialogue. Springer

Pineschi L (2012) Cultural diversity as a human right? General Comment No. 21 of the committee on economic, social and cultural rights. In: Borelli S, Lenzerini F (eds) Cultural heritage, cultural rights, cultural diversity. Martinus-Nijhoff, Leiden-Boston, pp 29–53

Santilli J (2012) Agrobiodiversity and the law – regulating genetic resources, food security and cultural diversity. Earthscan, Abingdon

Scovazzi T (2012) The definition of intangible cultural heritage. In: Borelli S, Lenzerini F (eds) Cultural heritage, cultural rights, cultural diversity. Martinus Nijhoff, Leiden-Boston, pp 179–200

Yusuf A (2008) Article I, definition of cultural heritage. In: Francioni F, Lenzerini F (eds) The 1972 World Heritage Convention: a commentary. Oxford University Press, Oxford

Food Diversity Between Human and Cultural Rights, Food Sovereignty and Protection of Intangible Cultural Heritage

Maurizia Pierri

1 Introduction. The Complex Nature of the Right to Food and Its Relationship with Food Diversity

The relationship between the right to food, understood as alleged legally relevant and food diversity, observed from the cultural point of view and in its connections with the right to health and with the personal freedom of the individual, is not immediately perceptible and requires a reflection from a semantic point of view.

Food is a natural/material element/ but is also *res non naturales* because it is the outcome and the representation of "*cultural processes that involve the domestication, transformation, the reinterpretation of what is present in nature*".[1] Food is also "culture" because it is produced, prepared and consumed: people do not restrict themselves to use what is available but *create their* own food, selecting plant varieties that they intend cultivating and rotate, replace or favour the production of crops instead of predation. Firstly, they choose what to eat based on economical, nature-nutritional and symbolic criteria[2]: in fact the socio-cultural context about food implies that not everything that is edible and nutritious is also socially, culturally and psychologically acceptable. Then the base-products are processed through the more simple or sophisticated culinary techniques, even those closely related to cultural and/or religious traditions.[3]

All these considerations can be summarised in the concept of the omnivore's dilemma, that dates to the writings of Rousseau and Brillat-Savarin, but it was

[1]Di Giacinto (2012), p. 179.

[2]Montanari (2004), pp. VII–VIII; Almerico (2014), p. 4.

[3]Anderson (2005).

M. Pierri (✉)
Department of Legal Studies, University of Salento, Lecce, Italy
e-mail: maurizia.pierri@unisalento.it

A. Isoni et al. (eds.), *Food Diversity Between Rights, Duties and Autonomies*, LITES – Legal Issues in Transdisciplinary Environmental Studies 2,
https://doi.org/10.1007/978-3-319-75196-2_25

officially singled out and identified as such by Paul Rozin. In his article entitled "The selection of Foods by Rats, Humans, and Other Animals",[4] he compared the existential condition of omnivores, such as rats and man, with that of animals with selective eating habits. Not omnivores animals have no doubt about what to eat since their preferences are genetically-determined and the natural and instinctive mechanisms function perfectly because their digestive systems are able to extract everything the body needs from a small range of foods. Omnivores (such as man), on the other hand, must dedicate time and thought to try to understand which of the innumerable foods offered by nature can be eaten without risk. The "feelings" of neophobia (the fear of eating an unknown substance) and neophilia (the desire to try a new tast) are totally unknown in animals with a specific diet.

The fact of being omnivorous and, therefore, generalist, is both an advantage and a challenge because the "flexibility" has allowed human beings to colonise all the habitats of the globe and adapt to all the different types of food offered but on the other hand, omnivores must spend time and energy understanding what to eat based on "what is good", and "what is bad". Individuals base themselves on culture and traditions that preserve the accumulated knowledge and experience of innumerable "tasters" who have gone before them. Culture codifies the rules of a prudent diet with a complex series of taboos, rituals, recipes, rules and traditions so it permits human beings to solve the omnivore's dilemma. It was emphasised that "Eating is a daily reaffirmation of [one's] cultural identity", to address the influence of food habits on an individual's self-identity.[5]

Therefore, it is clear that *ab origine* the cultural/immaterial element in food is indissolubly tied in with the material one.

The interlacement between *matter* and *culture* complicates the relationship between food and person also in the perspective of a legal framework that can be configured in terms of legitimate claim. Following the classification more accredited by philosophers of law,[6] the right to food can be classified between those of the fourth or fifth generation: which generates additional complexity, because it could condemn it to replace others, instead of integrating or specifying them.[7] It is obvious that it is not a "new" right but rather a growing awareness about a claim linked to an inviolable and consolidated right, such as the right to life and to personal integrity, as will become clearer when the explicit or implicit constitutional recognition of the right to food is discussed. The difficulty of pigeon-holing it within a "generation" of rights adds the problem of its physiologically composite structure: on the one hand, a fundamental human right that is instrumental to the enjoyment of other fundamental rights; on the other, a cultural right (closely related to food choice—qualitative aspect); finally an economic right in the dual sense of performance (with reference to the right to receive the minimum nutrition at subsistence level- quantitative

[4]Rozin (1976).

[5]Kittler et al. (2012), p. 4.

[6]Bobbio (1990).

[7]Rodotà (2012), p. 199.

aspect) and economic freedom, linked to trade in food stuffs, intended as an exchange commodity.

Under the latter profile, some have seen yet again that the right to food should be the most fundamental and concrete of economic rights[8] since most of the food consumed has a market value, also the food would be worth double to producers in connection to use (self-supply) and the value of the exchange. According to Spitz, "this dual nature is reflected in pairs of opposing elements at different levels of reality and analysis: commodity and basic needs, forces of extraction and forces of retention, market and self-provisioning, extortion performed by some social groups (landlords *v* tenants ... *omissis*) and redistribution within certain social configurations ... *omissis*, exploitation and solidarity, economic realism and moral outrage, economic laws and human rights".[9]

There is no doubt that the complexity of the relationship between human beings and nutrition has gradually been recognised by law: in fact, a conception of the right to food because of the guarantee of its fundamental components (see infra chapter "The Right to Food and Food Diversity in the Italian Constitution"): sufficiency and adequacy is deeply-rooted at international level. The first has a quantitative nature, even if articulated, being correlated to time coordinates—how many times a day you must eat—and physiological—the amount depends on constitution, age, sex and the type of activity carried out by an individual. It evokes the act of feeding (versus eating), closely connected with the right to survival (more than to life n.d.r.).[10] The second allows into the discussion on the right to food both the socio-cultural elements mentioned above (socio-cultural adequacy) and the qualitative elements (adequacy of diet) that presuppose the recognition of food diversity as a necessary condition for the development of adequate food. In fact, the impoverishment of food diversity results in an impoverishment of the right to food under the qualitative profile because it reduces the possibility of choice and therefore the possibility of identifying a varied diet, but it even stifles the cultural dimension.

This basic consideration constitutes a fundamental connection between the issue of the right to food and the protection of the environment and, in particular, of biodiversity, which represents the necessary presupposition to food diversity, an environmental frame functional to the elaboration of a varied diet and corresponding to the complex needs of individuals. On the one hand, in fact, food diversity responds to the need of "adequacy" under the qualitative profile of food (leading therefore to the principle of food health), whereas on the other hand it refers to a cultural requirement in the broader sense (and therefore even a religious one) that induces the development of food choices of a certain type. Some authors[11] coined the term "food habits" (also known as food culture or foodways) to describe the way humans use food, including everything from how it is chosen, acquired, and

[8]Spitz (1985), p. 306.

[9]Spitz (1985), p. 307.

[10]Almerico (2014), p. 3.

[11]Kittler et al. (2012), p. 3.

distributed to who prepares, serves, and eats it. They stated that the significance of the food habits process is that it is unique to human beings. They pondered why people spend so much time, energy, money, and creativity on eating. In this sense the theme is intertwined with that of agro-biodiversity i.e. diversity of cultivated species in agriculture, to which the cultural factor contributes heavily and the need to protect the agro/cultural heritage of certain communities who traditionally cultivate and consume certain products deemed appropriate to their socio-cultural roots. But it is also the connecting link with the issue of protection of agrobiodiversity because the decrease of plant varieties interweaves perversely with spontaneous or forced abandonment of local agricultural techniques, often heritage of minority territorial communities, inevitably results in the impoverishment of food diversity, with an important impact on human health.[12]

In this essay the dangers deriving from the impoverishment of biodiversity and their effects on supplì can only be mentioned, this danger has been underlined several times and in numerous documents from FAO.[13] Modern agriculture has the tendency to standardise species of plants or animals with great efficiency, which results in a huge loss of genetic variety: in the last century three quarters of the genetic diversity of agricultural crops and out of 6300 animal varieties 1350 are in danger of extinction or are already extinct. The same Agency observes how the abandonment of the production of local traditional foods often translates into a reduction of the variety of food. Partially, but only partially, the trend can be justified by invoking the objective of "nourish the planet," which is one of the Millennium objectives,[14] because biodiversity in relation to food and agriculture is essential to eradicate hunger in a permanent manner, to improve the quality of life and to ensure sustainability in food while increasing the production of food.

2 What Protection for a Complex Right? The Debate at International Level Between RtAF (the Human Right to Adequate Food), Food Sovereignty and the Incorporeal (Intangible) Cultural Heritage of Humanity

The instruments for the protection of food diversity that single countries have adopted remain differentiated, probably because of the physiological complexity, and will remain so until the concept of a holistic right to food is definitively consolidated, including both quantity and quality and, finally, the cultural aspect. However, the choices at national level coincide with the international debate on

[12]Monteduro (2015), p. 111 ss.

[13]See, inter alia: FAO (2004) and FAO (2015), p. 1 ss.

[14]Resolution adopted by the General Assembly in September 2000, available at http://www.un.org/millennium/declaration/ares552e.pdf.

concepts such as the human right to adequate food, food sovereignty or once again with the protection of food as incorporeal heritage of humanity.[15]

2.1 The RtAF

To fully understand the meaning of the RtAF (the Human Right to Adequate Food) it is necessary to only briefly recall that the first international recognition of the right to food dates to the Universal Declaration of Human Rights that in Art. 25 declares that it is a fundamental human right, instrumental to a dignified life. It implies the possibility of every person to have access to healthy, nutritious, safe and adequate food both from a quantitative and qualitative point of view. In fact, the right to food is composed of two complementary elements: that of food availability and that of the accessibility of the same.[16] The availability, according to the definition that offers the Committee for economic, social and cultural rights, must be understood as the possibility to draw its food supply directly from the ground or from other natural resources, or to have available distribution, processing and market systems operating and capable of transporting food products from the place of production to the one in which they are needed in the light of the question. Regarding the profile of accessibility, the same committee has stated that it is the physical and economic possibility of every individual to procure food in sufficient quantity. The sufficiency is not simply and reductively the quantitative dimension (minimum ration of calories) but qualitative, with reference to the adequacy of food compared to their own needs and their own culture, to the need to ensure a combination of elements that guarantee physical and mental growth, development and sustenance, physical activity. Everything in accordance with the physiological needs of human beings at all stages of life, according to sex and occupation. In this perspective food safety becomes fundamental, not only in the sense of certainty of access to food, enduring in time but also of the necessity of every individual to have access to healthy and secure food. In the International Covenant on Economic, Social and Cultural Rights adopted by the UN General Assembly on 16 December 1996, which came into force 10 years later, the Member States made the commitment to take every measure aimed at improving the methods of production, conservation and distribution of foodstuffs by the full application of scientific and technical knowledge, the spread of ideas relating to the principles of nutrition and the development or the reform of the agrarian regimes to achieve growth and use natural resources more efficiently. Subsequently, during the World Food Summit of the United Nations, the Rome Declaration on food safety was signed and a plan of action set out that reiterated the need for a commitment of member countries to implement policies aimed at ensuring a nutritionally adequate supply of food. The Committee for Economic, social and

---

[15]Beuchelt and Virchow (2012).

[16]Mechlem (2004), p. 631 ss; Pierri (2015), p. 455 ss.

cultural rights in Observation générale n. 12 of May 1999 interpreted Article 11 of the International Covenant on Economic, social and cultural rights by attributing a cultural connotation to the expression "adequate supply": "availability of food in sufficient quantities and quality to meet physiological needs, free of harmful substances and acceptable within each culture" (§6).

The cultural element, referred to the right to food, is further strengthened in the debate within the United Nations Commission on Human Rights, in the years following the resolution 2000/10 of April 2000, with which the right to food obtains its definitive consecration as the "right to have regular, permanent and free access, both directly and through monetary purchase, to an adequate and sufficient supply, qualitatively and quantitatively, corresponding to the cultural traditions of the people of whom the consumer is part and is capable of ensuring a physical and mental life, both individual and collective, which is free from anguish, satisfactory and worthy of living".

On the other hand, the implementation of the right to food involves the implementation of policies to promote the ability of the state communities to "autonomously provide for their food needs", improving the methods of production, preservation and distribution but also by improving employment prospects through agricultural reforms for the benefit of those who do not possess land.

The connection between the international recognition of the right to food and the protection of agro-biodiversity is well shown in the Declaration of Cordoba, submitted on 10 December 2008, during the 60th anniversary of the Universal Declaration of Human Rights, which refers explicitly to the results of the evaluation on international science, knowledge and agricultural technology for development adopted by the World Bank and the FAO and signed by 60 countries in April 2008. To fight the food crisis in the world and for its sustainability, the need for a complementary approach and diversified agriculture is recognised in this document, and it stresses that agricultural models based on small plantations may represent the right alternative to a food security based on human rights. Among the factors the declaration indicates as more worthy of attention there is: the lack of protection of small community landowners and indigenous peoples against the agro-industrial plantations; the agricultural reforms are insufficient to guarantee the lands in favour of poor rural communities, of indigenous peoples, agricultural workers, especially women; the lack of support for small scale production of foodstuffs in relation to access and control of seeds, water, infrastructure, information, credit and marketing; the excessive emphasis on international trade in agricultural products, to the detriment of local markets for local crops, which respond to the local needs and eating habits; the lack of safeguards to prevent abuse and the negative consequences of the excess of intellectual property rights on seeds; the excessive emphasis on the forms of agricultural production based on a high level of external inputs to the detriment of local crops; the lack of recognition and valorisation of food cultures and of traditional cultivations, which determine the marginalisation and the underutilisation of the same; the lack of adequate protection against the loss of bio-diversity caused by the expansion of monoculture in the production of food; the lack of recognition of the need for interventions/individual solutions for the distinct agro-ecological conditions, the distinct local cultural traditions and different levels and types of national development.

These last considerations justify the tendency to recontextualise international policies about the right to food and thus the fight against hunger, which is one of the eight Millennium Goals: the real problem is not food scarcity but the un fair distribution of the same, besides its inadequate quality, which, associated with the phenomenon of malnutrition, cannot be contrasted simply by introducing massive cultivations to the detriment of indigenous agricultural systems.

2.2 Food Sovereignty

The concept of RtAF has recently been compared and contrasted to that of Food Sovereignty,[17] which is to be found in some constitutions or national legislation (Venezuela-2008, Senegal-2004, Mali-2006, Nicaragua-2009, Ecuador-2009, Nepal-2009, Bolivia-2009). The principle of food sovereignty draws its origins from the document "Food Sovereignty: A future without hunger" drawn up in 1996 by the international movement of farmers, called "Via Campesina" and even if it has undergone various adjustments and modifications,[18] it still can be summarised as the right of persons, communities and Member States to define and determine their own (national) food and agricultural system, as well as that of implementing policies that favour their domestic agricultural produce in both local and national markets. The concept was then assimilated by social organisations and civil and non-governmental organisations that believe that it is instrumental in fighting the problems that lead to the abandonment of small-scale farming and to hunger and poverty in rural areas. Paradoxically it is the small food producers who suffer hunger: the majority of the population in food deficit lives in developing countries especially in rural areas and is directly or indirectly connected to agriculture.[19] According to the supporters of food sovereignty this paradox is explained by structural adjustments imposed by the World Bank and the International Monetary Fund, as well as the international agreement on agriculture specified in April 1994 within GATT, as the outcome of the so-called Uruguay Round, launched in Punta del Este in 1984 and concluded in 1994, with the creation of the World Trade

[17]See Wittman et al. (2010), p. 1 ss. Many of the reflections that follow in the essay, arise from the reading of the papers of the international conference entitled "Food Sovereignty: A Critical Dialogue", in Yale University, on September 14–15, 2013. Some of the papers discussed at the Yale food sovereignty conference have been published in the Journal of Peasant Studies (JPS) in May and June 2014. A selection of these have been formally published in advance and are now available online—for free. These include papers by Henry Bernstein, Jack Kloppenburg, Jennifer Clapp, and Madeleine Fairbairn. Access is free of charge from the *Journal of Peasant Studies* website.

[18]Patel (2009), p. 666.

[19]Patel (2009), p. 667.

Organization. Structural adjustment measures laid down there in have led to the privatisation and the reduction of state support for the agricultural sector in developing countries while leaving the small farmers in a vulnerable situation. The liberalisation of the agricultural sector in developing countries has, in fact, resulted in national markets being flooded with very cheap agricultural produce imported from industrialised countries. The conviction that global food security could be achieved through the control of the private sector only in agriculture[20] has led to the concentration of power in the agricultural process and in trade, the creation of international oligopolies and finally to market failure.[21] The structural adjustment policies combined with economic liberalisation have led to a worsening of the conditions of production for farmers in developing countries. Often it has become impossible for farmers to compete with the low prices of subsidised imports of agricultural products and has pushed the peasants to poverty or the abandonment of their farms.[22] The concept of food sovereignty aims instead at ensuring the survival and prosperity of small food producers and farmers who have been largely neglected or excluded from wider development processes.

The Via Campesina argues that small farmers should play a more dominant role in agricultural policies: this objective can be achieved only if local communities have a better access to, and control over, productive resources and a greater political and social influence.[23] The concept of food sovereignty raises the elementary question of which type of food production, agriculture and rural development should be pursued to ensure food security at national and global level. By way of exemplification, there are two contrasting models of agriculture: the first is that suggested by the concept of food sovereignty while the second is the neo-liberal free trade model. Food sovereignty maintains a model of small farmers who operate on a local basis, environmentally friendly and dedicated to sustainable agriculture. This is in contrast with the neo-liberal tendency to free exchange within an industrialised agriculture, based on intensive imports of chemical products and exports of agricultural produce, involving large farms and dominated by international corporations.

The basic principles of the concept of food sovereignty are generally understood in the same way by different groups who predominantly invoke and define it. This consistency is intentional between supporters of food sovereignty to obtain a stronger political say. Yet beyond the declaration of principles, there are many definitions of food sovereignty in circulation since associations and organisations integrate the concept in their work and provide their definition of food sovereignty.[24] The definition given in the declaration of Nye'le'ni (2007) can be considered as the most representative, since more than 500 representatives of organisations of farmers/ families of farmers and artisan fishermen, indigenous peoples, landless people, rural

[20] Murphy (2008), p. 3.

[21] IAASTD (2009), p. 466 ss.

[22] Altieri and Nicholls (2008), p. 472 ss.

[23] Desmarais (2008), p. 138 ss.

[24] Patel (2009), p. 663.

workers, migrants, shepherds and forest communities, women, young people, consumers and people from an urban environment and movements from more than 80 countries have agreed on this definition (NGO/CSO Forum for Food Sovereignty 2007). The Nye'le'ni definition enjoys strong support in civil society: the food sovereignty is the right of peoples to healthy and culturally appropriate food produced through ecologically sound and sustainable methods and their right to define their own food and agricultural systems. It makes those who produce, distribute and consume the food at the heart of food systems and policies rather than the needs of the markets and business. It defends the interests and the inclusion of the next generation. It offers a strategy to resist and dismantle the trends of commercial enterprise and diet and food regimes, and sees that the directions for food, agriculture, countryside and systems of fishing are determined by local producers. Food sovereignty gives priority to local and national economies and markets and allows farmers, farmer families' agriculture, artisan fishing, shepherds, pastures, food production and distribution and consumption based on environmental, social and economic sustainability. Food sovereignty promotes transparent exchanges that ensure income for all and the rights of consumers to control their food and nutrition. It ensures that the rights to use and manage our lands, territories, waters, seeds, livestock and biodiversity are in the hands of those of us who produce food. Food sovereignty implies new social relations free of oppression and inequality between men and women, peoples, racial groups, social classes and generations".[25]

2.3 Intangible (Incorporeal) Cultural Heritage (Only a Few References)[26]

Finally, as further proof of the cultural significance of food, there is the insertion of some culinary preparations in the incorporeal cultural heritage of humanity. It is necessary to briefly mention the latter concept that came into being with the Convention for Safeguarding Intangible Cultural Heritage approved unanimously in the 32° session of the General Conference in Paris on 17 October 2003. The International Convention for the Safeguarding of Intangible Cultural Assets considers the interdependence between intangible cultural heritage and tangible cultural heritage essential, as defined in the Declaration of Yamato. Safeguarding Intangible Cultural Heritage is defined as "protection". The term safeguarding contemplates those measures designed to promote the transmission of Intangible Cultural Heritage through generations, which are: identification, documentation, preservation, protection, promotion and valorisation. These are processes that involve research aimed at the identification of an intangible heritage, written, photographic, audio and visual documentation, which will guarantee the transmission of historical and cultural

[25]NGO/CSO Forum for Food Sovereignty (2007).

[26]For an exhaustive treatment of the subject see the essay of Di Benedetto, in this book.

heritage. The protection aims at preserving the places, the natural environment and landscape, i.e. the historical, cultural and social context that has produced and produces—a living entity—the cultural asset in question. Promoting and enhancing intangible cultural assets is attained through knowledge and preservation of heritage, using formal and informal types of education.[27]

The International Convention for the Safeguarding of Intangible Cultural Heritage provides broad categories of assets where individual cultural assets related to any cultural heritage can be identified. These can be either traditional or existing or both: oral traditions, languages, performing arts, social practices and rituals, knowledge and practices that relate to nature and the universe, the knowledge and skills of craftsmen and the spaces associated with them, everything that communities, groups and even individuals recognise as part of their cultural heritage. According to this document an intangible cultural heritage to be safeguarded is characterised by: its being transmitted from generation to generation; that it be constantly recreated by communities and groups in close correlation with the surrounding environment and with its history; that it endows communities, groups and individuals with a social and cultural sense of belonging; that it promotes respect for cultural diversity and human creativity; that it promotes the respect of human rights and the sustainability of development of each country.

As mentioned before, in the list of "assets" to be considered intangible cultural heritage there are also some culinary practices: in 2010 the Intergovernmental Committee established by Art. 5 of the same Convention of 2003 has decided unanimously to enroll in the Special Representative List of the Intangible Cultural Heritage of Humanity the Mediterranean diet, together with the traditional Cuisine of the Mexican state of Michoacan and French *Repas gastronomical*, thus recognising officially for the first time the cultural value of coking customs.[28] Moreover these gastronomic heritages also include some material instruments such as the utensils used for the preparation and consumption of food, in the awareness of what is necessary to protect the traditional crafts and the work of the craftsmen to sustain the economy of a territory.

The Mediterranean diet combines the eating habits of the peoples of the basin of the Mediterranean Sea (Italy, Spain, Greece, Morocco, Portugal, Croatia and Cyprus), consolidated during the centuries and remained unchanged until the nineteen-fifties, and goes well beyond a simple list of food but also relates to the culture of life and traditional, social and agricultural customs. The motivation for this insertion is extremely significant for its references to the cultural element inherent in food but also the link with the territory and the conservation of agrobiodiversity:

> The Mediterranean Diet is much more than simple food. It promotes social interaction, since the common meal is the basis of social customs and festivities shared by a given community, and has given rise to a considerable body of knowledge, songs, aphorisms, tales and legends. The Diet is founded on respect for the territory and biodiversity, and ensures the preservation

[27] And ratified by Italy on 27 September 2007.

[28] See Brulotte and Di Giovine (2014), p. 10.

and development of traditional activities and crafts linked to fishing and agriculture in the communities of the Mediterranean.[29]

It is important to emphasise that the insertion of gastronomic heritage in the UNESCO list is symptomatic of the symbolic value of food, and hence its close link with the history and culture of a territory and a people, but also of belonging to Humanity, as a "common good" of all, of which it is necessary to ensure conservation.

3 Constitutional Protection: The Interweaving Between the Protection of the Right to Food, the Protection of Biodiversity, of Agrobiodiversity and of Food Sovereignty

If the debate on the right to food at international level focuses on the concept of RtAF and on the demarcation of the principle of food sovereignty, in close relation with the exploitation of the cultural dimension of food as witnessed by the insertion of some gastronomic procedures in the list of the Intangible Heritage of Humanity, at national level the protection of the right to food is varied. At constitutional level, as can be seen from a survey carried out by the FAO in 2011[30] and reported in doctrine,[31] sometimes recognition is explicit, direct and guaranteed to all (for example in the constitutions of Bolivia and Ecuador); other times it is recognised for particular groups of the population (as in Mexico and South Africa); yet in others the right is included in other human rights (as in Brazil); in some cases the recognition is implicit in human rights in the broad sense (Armenia); in other cases it is a goal or a governing principle (Ethiopia); in others it is the result of the processing of case law (Ireland); finally sometimes the protection arises from the direct applicability of treaties on human rights that recognise the right to food (as for example in Argentina). There is no denying the fact that there is a tendency, established above all in more recent Constitutions and in those countries in which the problem of poverty and of the fight against hunger is particularly felt, to recognise an explicit protection of the right to food. But the next step is to understand if this recognition also extends to food diversity and if so, the extent and through which means.

Meanwhile you can observe that some constitutions that *explicitly defend* the right to food as a human right, also recognise the principle of food sovereignty: this is true of the constitutions of Bolivia (ratified in 2009), Ecuador (adopted in 2007) and Nepal (ratified in 2015). The Bolivian Constitution, in Art. 16, within the title devoted to fundamental rights, says "Every person has the right to water and food.

[29] Intergovernmental Committee for the Safeguarding of the Intangible Cultural Heritage (2010).

[30] Knuth and Vidar (2011).

[31] Bottiglieri Longhi (2014).

The State has the obligation to guarantee food security for all through healthy, adequate and sufficient food", while Art. 405 stipulates that "comprehensive, sustainable rural development is a fundamental part of the economic policies of the State, which shall prioritise its actions to encourage to communitarian economic undertakings and those of the group of rural actors, placing emphasis on food security and sovereignty...".

Art. 13 of the constitution of Ecuador reads "Persons and community groups have the right to safe and permanent access to healthy sufficient and nutritional food, preferably produced locally and in keeping with their various identities and cultural traditions. The Ecuadorian State shall promote food sovereignty".

The same article recognises the right to food both quantitatively and qualitatively and connects it to the "favor" for local production and the principle of food sovereignty.

The link between the fundamental human right (food), territorial component (protection of biodiversity-food sovereignty) and cultural component (protection of agro-biodiversity and cultural heritage culinary) is consolidated in these Constitutions by a series of previsions that concern both the protection of the environment, of sustainable development and of the minority population allocated on the national territory who preserve indigenous knowledge related to crops and culinary preparations steeped in cultural and religious elements.[32] In fact indigenous peoples' right to food is inseparable from their rights to land, territories, resources, culture and self-determination.[33]

A series of articles are contained in the Bolivian Constitution that promote the conservation of biodiversity and of agrobiodiversity (Art. 354), the sustainable use of natural resources (Art. 380) and support the principle of sovereignty also on agrobiodiversity, as emerges from a combined reading of Articles 381, 382 and 383. On the other hand, the same Constitution enshrines the recognition of indigenous peoples (Art. 171). They recognise, respect and protect on the mark of the Law, the social, economic and cultural rights of indigenous peoples that inhabit the national territory, especially those relating to their original communal lands, guaranteeing the use and improvement of sustainable natural resources, their identity, values, languages, customs and institutions their link with the territory and their use and improvement of natural resources is encouraged and finally their heritage of values, habits and customs, especially those related to the rapport with nature, are acknowledged.[34]

The Ecuadorian constitution declares that the conservation of biodiversity and of agrobiodiversity are a priority for the State and constitute a limit to productive

[32]For the concept of "indigenous peoples" see Martinez Cobo (1986/7). Indigenous peoples' right to food is recognised in United Nations Declaration on the Rights of Indigenous Peoples, Annex to Human Rights Council Resolution (2007).

[33]Food and Agriculture Organization of the United Nations. Centre for Indigenous Peoples' Nutrition and Environment (2013), p. X.

[34]Aguilar et al. (2010).

activities (see Articles 57 and 72) and explicitly recognises the concept of sovereignty of biodiversity (Art. 400) too. Also, in this case the theme of food and agro/biodiversity proceeds in parallel with the recognition of the rights of indigenous peoples: Art. 56 reads "The indigenous peoples, who are automatically defined as nationals of ancestral races, and the black peoples or Afroecuadorians, form part of the Ecuadorian state, unique and indivisible".

Finally, Art. 36 of the Constitution of Nepal, promulgated in 2015 after a troubled constituent phase, in part III, entitled "The Fundamental Rights and Duties" stipulates that "right relating to food: (1) Every citizen shall have the right relating to food. (2) Every citizen shall have the right to be safe from the state of being in danger of life from the scarcity of food. (3) Every citizen shall have the right to food sovereignty in accordance with law".[35]

Subsequently the constitutional text in Art. 51, which relates to state interventions, provides policies (let.g) "relating to protection, promotion and use of natural resources: (1) to protect, promote, and make environmental friendly and sustainable use of natural resources available in the country, in consonance with national interest and adopting the concept of intergenerational equity, and make equitable distribution of fruits, according priority and preferential right to the local communities, . . . (5) to preserves, promote, and make sustainable use of forests, wildlife, birds, vegetation and bio-diversity, by mitigating possible risks to environment from industrial and physical development, while raising awareness of the general public about environment cleanliness . . . (7) to adopt appropriate measures to abolish or mitigated existing or possible adverse environmental impacts on the nature, environment or biological diversity. Finally, the constitution recognises the minorities and their peculiarities, in terms of the right to retain their own culture (Art. 32). Right to language and culture:. . .(2) Every person and community shall have the right to participate in the cultural life of their communities. (3) Every Nepalese community residing in Nepal shall have the right to preserve and promote its language, scripts, cultures, cultural civilisation and heritage, or protect and develop languages, scripts, cultures, literature, arts, motion pictures and heritages of various castes, tribes, and communities based on equality and co-existence, while maintaining the cultural diversity of the country".

Despite this network of rules relating to the right to food, the conservation of biodiversity and the protection of cultural minorities situated in the territory, there is no noticeable connection between these different instances in the constitution of Nepal. Conversely, it does not appear that these safeguards are the result of an overall vision aware of the importance of the link between minority populations, territory, culture and food, as instead happens in the constitutions of Bolivia and Ecuador. Therefore, the text promulgated in 2015, although very recent and apparently attentive to the theme of the preservation of the environment, biodiversity and

[35]Nepal's Constitution of 2015 is available at https://www.constituteproject.org/constitution/Nepal_2015.pdf?lang=en.

sustainability in intergenerational terms, can fall into the mainstream of "ecologist"[36] constitutionalism that particularly sees the Andean countries protagonists of a real reversal of the cultural perspective regarding the relationship of man with nature. Probably the different approach is to be attributed to the different matrix of the minorities that have influenced the drafting of the constitutional texts: in the case of the Andean countries it is indigenous minorities, permeated with an autochthonous culture linked to the land and to its "care" by an ancestral ratio. In the second case the religious minorities, especially attentive to the recognition of their dignity within the state.[37]

4 Conclusions

On the right to food differentiated and multilevel safeguards converge because they are linked to either national or international warranty. The dialectic between RtAF and Food Sovereignty is emblematic, in an inclusive or alternative interpretation, to assess their potential with respect to the need to develop international policies for solving the problem of hunger at global level and to reduce poverty. According to Via Campesina (Forum of the World Food Summit of 1996), the right to food can only take place within a system that ensures food sovereignty and indeed, the proponents of the concept of food sovereignty believe that the latter, together with access to agro-cultural resources, constitutes the fundamental precondition for achieving the right to adequate food. In reality, the two concepts have common aspects but also differences: while the RtAF is a legal concept, that of Food Sovereignty is a political concept,[38] useful in the orientation of the initiatives of policy makers with respect to agro-cultural policy, with a particular focus on the protection of small food producers and farmers who operate at rural and local level. Moreover, one operates at international level while the other can intervene at national level.

For our purposes, the two concepts seem to be able to combine in a virtuous way ensuring both the right to food in terms of quantity and quality, so that the latter attribute includes also the cultural profile of foods, becoming both fundamental and in-transferable. In fact, both direct attention toward the qualitative element of food, which connects to the biodiversity and to agrobiodiversity. The first, as we have seen, has become enriched with meanings, definitively including the cultural element in the forming of human right; the second aims to protect local production at

[36]Dating back to Stone (1972) essay and its development: see Bosselmann (1992), Boyd (2012), May and Daly (2014), Jasanoff (2013), Kim and Bosselmann (2013), Kotzé (2012), Grear and Kotzé (2015) and Hudson (2015).

[37]See Parajulee (2000), p. 106.

[38]Haugen (2009), pp. 291–282.

national level, so-called "niche", often the prerogative of minorities, which is the last stronghold of resistance of traditional agricultural technique applied to biodiversity.

The considerations arising from the analysis of the concepts described above are useful in defining in more detail the right to food and to rationalise its complexity.

What has been said so far shows that: (1) the right to food as a "human" right attracts in its orbit a surplus of protection that does not affect quantity but quality. Food diversity (and even before that agrobiodiversity) becomes functional to the enjoyment of the right to food; (2) Consequently the cultural aspect of the right to food tends to be absorbed within human rights to the detriment of the component of an economic nature, in the first place the freedom of free trade. If we think back to one of the expressions used at the beginning: "the right to food is an economic law, indeed the most fundamental of economic rights" and the pairs of opposites identified by Spitz, it is clear that the affirmation of principles such as that of the RtAF and even more that of Food Sovereignty affect the balance between the opposing factors causing the nature of traded commodities to become recessive with respect to that of an in-transferable fundamental human right. These considerations call for a reflection on not only the immaterial but even the unavailable nature of cultural heritage food, both with reference to the production and to the consumption, by reason of the intimate relationship the right to food has when intended as a fundamental human right. This is an in-transferability to build, not only in theoretical terms, perhaps through linking up to the reflections on the "commons" but in legal terms, as has been seen to develop in the countries that have adopted the principle of food sovereignty.[39]

References

Aguilar G, Lafosse S, Rojas H, Steward Tje R (2010) Constitutional recognition of indigenous peoples in Latin America. Int Law Rev 2(2):44–96

Almerico GM (2014) Food and identity: food studies, cultural, and personal identity. J Int Bus Cult Stud 8:1–7

Altieri MA, Nicholls CI (2008) Scaling up agroecological approaches for food sovereignty in Latin America. Development 51(4):472–480

Anderson E (2005) Everyone eats. Understanding food and culture. New York University Press, New York

Beuchelt TD, Virchow D (2012) Food sovereignty or the human right to adequate food: which concept serves better as international development policy for global hunger and poverty reduction? Agric Hum Values 29:259–273

Bobbio N (1990) L'età dei diritti. Einaudi editore, Torino

Bosselmann K (1992) Im Namen der Natur: Der Weg zum ökologischen Rechtsstaat. Scherz, Bern

Bottiglieri Longhi M (2014) Il diritto ad un cibo adeguato. Profili comparati di tutela costituzionale e questioni di giustiziabilità In: Macchia P (ed) La persona e l'alimentazione. Profili clinici, culturali ed etico-religiosi. Atti del Convegno (Asti, 30 novembre 2012). Aracne, Roma, pp 217–260

[39]Ostrom (1991) and Mattei (2011).

Boyd D (2012) The environmental rights revolution: a global study of constitutionalism, human rights and the environment. UBC Press, Vancouver-Toronto
Brulotte RL, Di Giovine MA (2014) Edible identities: food as cultural heritage. Ashgate, Farnham
Desmarais AA (2008) The power of peasants: reflections on the meanings of la via Campesina. J Rural Stud 24(2):138–149
Di Giacinto M (2012) Per una storia del rapporti tra alimentazione e diversità culturali. Studi sulla formazione 2:177–181
FAO (2004) Biodiversità: la nostra alimentazione ne dipende. Roma. Available at ftp://ftp.fao.org/docrep/fao/006/y5418i/y5418i00.pdf
FAO (2015) Coping with climate change. The roles of genetic resources for food and agriculture, Rome. Available at http://www.fao.org/3/a-i3866e.pdf
Food and Agriculture Organization of the United Nations Centre for Indigenous Peoples' Nutrition and Environment (2013) Indigenous Peoples' food systems& well-being interventions & policies for healthy communities. In: Kuhnlein HV, Erasmus B, Spigelski D, Burlingame B (eds) Rome. Available at http://www.fao.org/docrep/018/i3144e/i3144e.pdf
Grear A, Kotzé LJ (eds) (2015) Research handbook on human rights and the environment. Edward Elgar
Haugen HM (2009) Food sovereignty—an appropriate approach to ensure the right to food? Nordic J Int Law 78(3):263–292
HRC (2007) Declaration on the Rights of Indigenous Peoples, Annex to Human Rights Council Resolution, 61/295. Available at http://www.un.org/esa/socdev/unpfii/documents/DRIPS_en.pdf
Hudson B (2015) Structural enviromental consitutionalism. LSU Law Digit Commons 21:201–216
IAASTD (2009) Agriculture at a crossroads. Synthesis report of the international assessment of agricultural knowledge, science and technology for development. Island Press, Washington DC
Intergovernmental Committee for the Safeguarding of the Intangible Culural Heritage (2010) Nomination file for inscription in 2010 on the representative list of Intangible Cultural Heritage of Humanity. Available at https://ich.unesco.org/en/Decisions/5.COM/6.41
Jasanoff S (2013) A world of experts: science and global environmental constitutionalism. Boston Coll Environ Aff Law Rev 40(4):439–452
Kim RE, Bosselmann K (2013) International environmental law in the Anthropocene: towards a purposive system of multilateral environmental agreements. Transnl Environ Law 2(2):285–309
Kittler PG, Sucher KP, Nelms MN (2012) Food and culture. Wadsworth, Belmont
Knuth L, Vidar M (eds) (2011) Constitutional and legal protection of the right to food around the world. FAO, Roma. Available at http://www.fao.org/docrep/016/ap554e/ap554e.pdf
Kotzé LJ (2012) Arguing global environmental constitutionalism. Transnl Environ Law 1 (1):199–233
Martinez Cobo J (1986/7) Study of the Problems of Discrimination against Indigenous Populations. UN Doc. E/CN.4/Sub.2/1986/7 and Add. 1–4. paras 379–38. Available at https://www.un.org/development/desa/indigenouspeoples/publications/2014/09/martinez-cobo-study/
Mattei U (2011) Beni comuni. Un manifesto. Editori Laterza, Roma-Bari
May J, Daly E (2014) Global environmental constitutionalism. Cambridge University Press, Cambridge
Mechlem K (2004) Food security and the right to food in the Discorse of the United Nations. Eur Law J 10(5):631–648
Michael A (2014) Edible identities: food as cultural heritage. Ashgate, Farnham England
Montanari M (2004) Il cibo come cultura. Editori Laterza, Roma-Bari
Monteduro M (2015) Diritto dell'ambiente e diversità alimentare. Rivista quadrimestrale di diritto dell'ambiente 1:88–131
Murphy S (2008) Globalization and corporate concentration in the food and agriculture sector. Development 51(4):527–533
NGO/CSO Forum for Food Sovereignty (2007) Declaration of Nye ́le ́ni. Nye'le'ni Village, Se'lingue ́(Mali). Available at http://www.nyeleni.org/IMG/pdf/DeclNyeleni-en.pdf

Ostrom E (1991) Governing the commons: the evolution of institutions for collective action. Cambridge University Press, Cambridge

Parajulee RP (2000) The democratic transition in Nepal. Rowman & Littlefield, Lanham

Patel R (2009) What does food sovereignty look like? J Peasant Stud 36(3):663–773

Pierri M (2015) Agrobiodiversity, intellectual property rights and right to food: the case of Andean Countries. In: Monteduro M, Buongiorno P, Di Benedetto S (eds) Law and agroecology. A transdisciplinary dialogue. Springer, Berlin, pp 451–470

Rodotà S (2012) Il diritto di avere diritti. Editori Laterza, Roma-Bari

Rozin P (1976) The selection of foods by rats, humans, and other animals chapter. Adv Study Behav 6:21–76

Spitz P (1985) The right to food in historical perspective. Food Policy (November): 306–316

Stone CD (1972) Should trees have standing? Toward legal rights for natural objects. South Calif Law Rev 45:2–54

Wittman H, Desmarais AA, Wiebe N (2010) The origins and potential of food sovereignty. In: Wittman H, Desmarais AA, Wiebe N (eds) Food sovereignty. Reconnecting food, nature and community. Food First, Oakland, pp 1–14

Food, Culture and Law: The Terms of an Evolving Relationship

Marco Brocca

1 Preliminary Considerations

Expressions such as "Tell me what you eat and I'll tell you who you are" and "man is what he eats"[1] are increasingly common in everyday language, reflecting an awareness and shared belief: that of the close relationship between food and culture, that is, the idea that wine and food products, together with their raw materials and techniques, as well as their sites of cultivation, processing and production, are representative of local areas. Better yet, they are elements that contribute to the cultural identity of a population. They are the result, and at the same time the representation, of various factors, such as climate, morphology, social and economic structure, scientific and technical knowledge of territories and their populations, and therefore, are important indicators of the evolution of human and natural history.

Specularly to the spreading of increasingly globalized and homologated eating habits, we can witness the emergence of the relevance of food diversity and the search for and preservation of territorially specific food products, because they embody the typical features and identity of the places themselves. Not only does this trend reflect the increasingly desired nutritional needs, linked to the safeguarding of one's health and quality of life (as it is undisputed that the most traditional and natural foods are the healthier ones) and the need to preserve the biological diversity linked to the majority of traditional products, but it also denotes the accrued

[1]The first is attributed to Jean–Anthelme Brillat–Savarin, *The Physiology of Taste* (1825), and will be taken up again, with greater interest by the philosopher Ludwig Feuerbach with the statement "man is what he eats", in *Natural science and the revolution* (1850), and then *The mystery of sacrifice, or man is what he eats* (1862).

M. Brocca (✉)
Department of Legal Studies, University of Salento, Lecce, Italy
e-mail: marco.brocca@unisalento.it

A. Isoni et al. (eds.), *Food Diversity Between Rights, Duties and Autonomies*,
LITES – Legal Issues in Transdisciplinary Environmental Studies 2,
https://doi.org/10.1007/978-3-319-75196-2_26

awareness that we are speaking of goods that are an integral part of the culture of a people and are resources that can add to the attractiveness of a territory, since they are an expression of their cultural identity. Think of the increasing phenomenon, the so-called food-wine tourism, where wine and food products are goods that can steer tourist dynamics in the same way that traditional tourist resources can (cultural heritage, scenery, areas of religious interest, etc.), indeed having a transversal capacity of assertion, as they can add to the other tourist resources that denote a certain territory, resulting in the branding of the full range of tourist products.

The law reflects these new demands in different ways. Among the most obvious are food brands, which are found on a supranational, national and regional level and "certify" the link between a product and its territory, classifying it in the market in terms of its origin. Consequently, brands evoke of many elements related to the territory that reverberate on the product.[2]

The perspective proposed in this chapter regards the legal recognition of the properly understood cultural value that distinguishes different food and wine products (and their production techniques). A significant contribution in this direction comes from international cultural property law, particularly the UNESCO Convention of 17 October 2003 "for the safeguarding of the intangible cultural heritage", which, as will be seen, will place a strain on national laws, which are traditionally anchored to 'state-centric' settings and to a 'tangible' concept of cultural heritage, as is in the Italian legal system.

2 Solicitation from the International Law

The Convention on Intangible Cultural Heritage in 2003 marking the completion of the path followed by UNESCO for the protection of cultural heritage was based on the awareness of the interdependence and cultural affinity between tangible and intangible assets, besides that between cultural and natural assets. UNESCO's attention towards tangible cultural assets, dating back to the 1970s, translated into the Paris Convention of 16 November 1972 concerning the protection of the world cultural and natural heritage, which led to the recognition—because of their "outstanding universal value"—of more than a thousand sites through registration on a global list and this model was followed 30 years later, for intangible cultural heritage by the aforementioned 2003 Convention. This category, poignantly depicted as part of cultural heritage (Art. 2, Para. 1), is defined in a very broad sense, which, to give some examples, includes, but is not limited to: (a) oral traditions and expressions, including language as a vehicle of intangible cultural heritage; (b) performing arts; (c) social practices, rituals and festive events; (d) knowledge and practices concerning nature and the universe; (e) traditional craftsmanship.

---

[2]It deals with what doctrine defines "territorial signs", in particular, see Lucifero (2011), pp. 321–421.

The system of protection is carried out on two levels, one of international standing, placed under the auspices of UNESCO, and the other a national source, entrusted to the individual member states. UNESCO deals with the registration of deserving property on the Representative List of the Intangible Cultural Heritage of Humanity (Art. 16) and on the separate list of the Intangible Cultural Heritage in Need of Urgent Safeguarding (Art. 17), promotes programs, projects and activities, both national and sub-national, for the safeguarding of the intangible cultural heritage (Art. 18), as well as guaranteeing interventions of cooperation and international assistance (Arts. 19–24), using a fund specially set up for the intangible cultural heritage (Arts. 25–28). Member states are required to take specific safeguarding measures, including the compiling of specific inventories (Art. 12), the promotion of scientific, technical and artistic studies (Art. 13, Para. 1, Lett. c) and the adoption of "appropriate legal, technical, administrative and financial measures" (Art. 13, Para. 1, Lett. d) for the protection and enhancement of intangible cultural heritage.[3]

It is immediately obvious that intangible cultural heritage is extremely heterogeneous, drawing from various knowledge and manifestations of common life, as evidenced by the above definition of the Convention structured in very broad terms and, furthermore, of a merely illustrative, but not exhaustive nature.

On the world list, which includes nearly 400 goods, appear different foodstuffs, or goods and practices related to agriculture and food.[4] Emblematic is the recognition of the Mediterranean Diet, which represents and summarizes the many elements that characterize the category examined here: as claimed by UNESCO "The Mediterranean Diet – derived from the Greek word *díaita*, way of life – is the set of skills, knowledge, rituals, symbols and traditions, ranging from the landscape to the table, which, in the Mediterranean basin, concerns the crops, harvesting, picking, fishing, animal husbandry, conservation, processing, cooking, and particularly sharing and consuming the cuisine. It is at the table that the spoken word plays a major role in describing, transmitting, enjoying and celebrating the element".

The interest, therefore, goes beyond the individual elements that characterize the diet, which are also all traditional products and are widely recognized as healthy foods, and concerns a certain "way of life" that unites and distinguishes a specific geographical area, that of the Mediterranean Sea.[5] This same interest inspires the recognition of the Japanese, Mexican and French cuisines and foods, or individual goods such as paprenjak (Croatian gingerbread) or Turkish keskek (dish consisting of meat and wheat), in which the significance of the intangible asset does not arise so much from a link to a specific food, ingredient or technique, but from the combination of these elements and the ability to conjure up ways and habits of preparing and

[3]For more on the Convention see, among others, Francioni (2006), pp. 701–703; Zagato (2008), pp. 27–70; Ubertazzi (2011), pp. 777–798; Oggianu (2012), pp. 5–11; Crosetti and Vaiano (2014), pp. 37–38; Accettura (2015), pp. 50–54.

[4]For more see: Mirri (2015), pp. 59–67.

[5]The Mediterranean diet is, in fact, attributable to a plurality of countries: Cyprus, Croatia, Greece, Italy, Morocco, Portugal, and Spain.

eating meals, as well as celebrations and holidays.[6] As far as the Italian state is concerned, we find the listing of the traditional cultivation technique of "vite ad alberello" (head-trained bush vines) of the community of Pantelleria[7]: the reference is not to the zibibbo, known sweet wine of Pantelleria, but to the particular agricultural practice of growing grapes, carried out by burying the plant in deep holes covered by the branches from pruning, to protect the plant from the wind and heat, which denotes the ability of local farmers to adapt their activities to the morphological and climatic characteristics of the place in which they live. In the same context can be found the proceedings underway for the insertion of Neapolitan pizza.

Confirming the affinity that characterizes both the intangible and tangible cultural heritage, we can also note the applicability of the 1972 UNESCO Convention (regarding monuments and sites forming the world cultural and natural heritage) to goods and realities concerning agribusiness, as was the case in Italy and the "Vineyard Landscape of Piedmont: Langhe-Roero and Monferrato" registered on the World Heritage List through the UNESCO decision of 22 June 2014, on the grounds that it is an "outstanding living testimony to winegrowing and winemaking traditions that stem from a long history, and that have been continuously improved and adapted up to the present day. They bear witness to an extremely comprehensive social, rural and urban realm, and to sustainable economic structures. They include a multitude of harmoniously built elements that bear witness to its history and its professional practices".[8] Here farming is noted for its ability to adapt to and, at the same time, to shape the territory making it "landscape", in a profound union between man and environment, culture and nature.

3 State Choices: The Code of Cultural Heritage

The UNESCO 2003 Convention is ignored by the Italian legislature, which, in 2004, fully reformed the sector of cultural heritage, with the approval of the Legislative Decree 42/2004 (Code of cultural heritage and landscape). Furthermore, at that time, the international act lacked the ratification law (which later intervened with Law n. 167/2007), but it is also true that the national text, in its greatest reform effort,

[6]For example, French cuisine has been recognized for the following motivation: It "is a customary social practice for celebrating important moments in the lives of individuals and groups, such as births, weddings, birthdays, anniversaries, achievements and reunions. It is a festive meal bringing people together for an occasion to enjoy the art of good eating and drinking. The gastronomic meal emphasizes togetherness, the pleasure of taste, and the balance between human beings and the products of nature. Important elements include the careful selection of dishes from a constantly growing repertoire of recipes; the purchase of good, preferably local products whose flavors go well together; the pairing of food with wine; the setting of a beautiful table; and specific actions during consumption, such as smelling and tasting items at the table."

[7]Other registered goods include: Opera dei Pupi (Sicilian puppet theatre), Canto a tenore (Sardinian pastoral songs), Traditional violin craftsmanship in Cremona, Celebrations of big shoulder-borne processional structures, Mediterranean diet.

[8]About this see: Lombardi (2009), pp. 389–397.

showed consideration and acceptance of the principles and models of other international conventions that had not yet been ratified. This is the case of the European Landscape Convention (Florence, 20 October 2000) ratified by Law 14/2006, whose guiding principles and criteria have inspired much of the legislation on the countryside and the UNESCO Convention on the Protection of Underwater Cultural Heritage (Paris, 2 November 2001), ratified by Law 157/2009, which is referred to expressly by the code and qualified as the reference standard for some profiles of the protection of archaeological objects found at sea.

The legislature did not make a similar choice for the area of intangible cultural heritage, almost as if it intended to avoid, or at least delay, taking sides on a fundamental issue of the matter, that regarding the objective part, which is, instead, in other legal systems, the focus of considerable debate and causes deep tension between traditional settings and reformist forces.

The international success of categories such as "cultural heritage of universal value" and "World Heritage" (UNESCO Convention of 16 November 1972), "intangible cultural heritage" (UNESCO Convention of 17 October 2003) and "cultural diversity" (UNESCO Convention of 20 October 2005), denotes increased awareness of universal value and, at the same time, identity of culture and, in this way, raises the "ontological" question of concept, before the system of cultural goods. This setting shows just how essential international regulation, that is, the process that has been significantly defined the "globalization of the regulation" of the sector[9] or the emergence of a "global cultural property law"[10] is. It is an unprecedented regulatory front, which brings with it new settings and options, such as the intangible (and not only tangible) nature of cultural assets, the impact of which is disruptive in a traditionally state-formed sector anchored to a "historicist and tangible"[11] concept of cultural heritage.

Italian cultural property law is permeated by the materialistic concept where the value of the good is inextricably linked to its physical substrate and materialness is a precondition for the legal classification of cultural heritage, and this option ideally ties the first sector laws to current law: so, Law 1089/1939 was the first organic law on cultural heritage, in which the element of materialness can already be seen from the title ("Protection of things of artistic and historical interest") and it distinguishes the typed categories of cultural goods. Such is the afore-mentioned current code of cultural heritage, in which materialness inspires and influences the definition of cultural goods enshrined in Art. 2, as evidenced by the reference to "immovable and movable property" and by the term "things" typed by arts. 10–11, the latter being provisions that confirm in the lists of categories of cultural goods the exclusive presence of substrate tangible goods. From this premise coherently follows a system of measures and instruments of protection involving all the materialness of the asset on which to intervene (maintenance, conservation, restoration, coercive custody, export limits, expropriation, etc.).

---

[9]Cassese (2012), p. 781.

[10]Casini (2010).

[11]Cassese (2012), p. 782.

The regulatory situation was not even scratched when the legislative reform in the 1960s by the Franceschini Commission[12] was proposed, although it had the merit of stimulating the conceptual evolution of materials. The proposed definition of cultural heritage as a "tangible witness holding the value of civilization" marks the intent to cut the notion loose from the aesthetic criterion of the law of 1939 in favor of a historicist criterion, but it also consolidates the element of materiality.

Yet in this path we can highlight at least two moments, one of scientific reflection and the other of legislative initiative, oriented in the direction of freeing cultural heritage from the bond of materiality. At the doctrinal level the contribution of M.S. Giannini on the status of cultural heritage[13] remains crucial, in which the author identifies the unifying and distinctive traits of cultural heritage, identified in culturalism, immateriality and advertising. The qualification of cultural goods as "intangible"[14] marks the awareness that the value, cultural therefore immaterial, is the element that characterizes and qualifies the good and prevails on the *res* itself, which becomes a secondary, almost indifferent aspect. Of course, in the reconstruction of Giannini the *res* is always there, because it supports the good, but the former does not, per se, constitute *res* if the legal good,[15] as is plastically evident for assets whose interest is 'relational', that is, if it stems from its connection to political, military or literary history. The reference to intangibility, dominating and making sterile the aspect of tangibility, obviously opens new regulatory scenarios and offers new perspectives for the study on the subject, which, however, the doctrine, and even more so lawmakers, have only partially been able to grasp.

On the legislative front, Legislative Decree 112/1998 is relevant (the so-called Bassanini decree on administrative decentralization) for its definition of cultural goods: they are "the ones that make up the historical, artistic, monumental, demoethnoanthropological, archaeological, archival, book and other heritage that are testimony of the value of civilization as identified by law", (Art. 148, Para. 1, Lett. a). The definition clearly echoes the one proposed by the Franceschini Committee, with a variation, significant for present purposes, related to the omission of the adjective "tangible" testimony of the value of civilization, which breaks the bond of the materiality of cultural goods and offers legitimacy for any future legislative intervention in the new direction.

The issue of the reconciliation of intangible assets, also called "volatile goods",[16] with the legal category of cultural goods became urgent after the approval of the

---

[12]Commission for the investigation of the protection and promotion of things of historical, archeological, artistic and landscape interest, established by Law 310/1964.

[13]Giannini (1976), pp. 3–38.

[14]Giannini (1976), p. 24.

[15]As found in literature by Morbidelli (2014), p. 2.

[16]This expression was introduced by historian A.M. Cirese to indicate those goods such as "chants or fables, festivals or entertainment, ceremonies and rituals that are neither mobile nor immobile, in that, to 'use' them more times, they must be *reproduced* or remade, which is much different than houses or chests or hoes whose further usage [. . .] does not require a *remaking* [. . .]. Volatile goods are both identical and changing and can be lost forever if they are not 'fixed' in lasting memories": Cirese (1996), p. 251; Cirese (2007) (author's translation).

UNESCO 2003 Convention ratification law and the outcome of this review would be the inclusion in the code of a specific provision (Art. 7-*bis*) by the corrective decree of 2008 (Legislative Decree 62/2008). Indeed, the wording of the new law[17] reveals the intent of the legislature to confirm, rather than revolutionize, the conceptual and regulatory framework of the code and the reference to international acts appears to be simply formal and essentially elusive[18]: "collective expressions of cultural identity" (because of the international notion of intangible cultural heritage) are relevant for the purposes of the code on the condition that they constitute "tangible testimony" and fall under one of the categories (all of tangible nature) of cultural goods under Arts. 10–11. The introduction of these two conditions obviously produces a damping effect on, if not a sterilization[19] of, the potential of the International Convention and places the code in contrast to the decisions of other legal systems, even some similar to our own such as the Spanish and Portuguese ones, in which the notion of cultural heritage, including both the tangible and intangible aspects, is put into law and consequently, categories are formed including both tangible and intangible goods.[20]

4 Other Internal Guidelines

The choice the code makes, which emphasizes the tangible nature of cultural heritage, seems to overlook—even if by doing so it actually emphasizes the contrast with—other legislative choices that give legal importance to goods of cultural interest, even when they are devoid of tangibility. Conversely, the choice made by the code is not exclusive and, as has been observed in legal literature, regulatory policies, both state and regional, are now of such intensity that today "the notion of a cultural good is shattered into a plurality of legal categories, so that, we must go from the single 'good' to cultural heritage".[21] We are in the presence of an "open concept"[22] of cultural heritage, which includes and is not limited to the one in the code, and this position is also supported by the case law of the Constitutional Court,

[17]According to which "The expressions of collective cultural identity contemplated by the UNESCO Conventions for the safeguarding of the intangible cultural heritage and for the promotion of cultural diversity, adopted in Paris, respectively on 3 November 2003 and 20 October 2005, are subject to the regulations of the present code where they are represented by tangible testimony and the prerequisites and conditions for the application of article 10 have been met".

[18]Tarasco (2008), p. 2261. This highlights the oxymoronical nature of the norm, because it gives importance only to intangible cultural goods that have tangible support: Bartolini (2014), p. 2; Mirri (2015), p. 62. Casini (2008), p. 1057, speaks of a lack of coordination between the two sources and of a "non satisfactory" integration.

[19]Bartolini (2013), p. 112.

[20]Tarasco (2008), p. 2261; Bartolini (2014), p. 4 both authors highlight this point.

[21]Bartolini (2013), p. 113. In theory, the data has been presented using the model of concentric circles, where the largest one includes cultural heritage, as a "product" of culture in the anthropological sense, which, by now, has acquired full legal importance, and the smallest one coincides with the notion of cultural heritage as in the code: also, Vitale (2010), p. 176.

[22]Morbidelli (2014), p. 5; Poggi (2003), p. 3.

called upon several times to resolve conflicts between the state and regions regarding the legislative competence of "the protection of cultural heritage" (Art. 117, paragraph 2, letter s of the Constitution) and "the promotion of cultural heritage" (Art. 117, Para. 3 of the Constitution). The orientation of the Court is as rigorous in reserving the State with the competence of legislation in the protection of cultural heritage through a strict regime currently drawn up in the code of cultural heritage (a regime characterized by the application of the historical-artistic constraints and the application of various conservation and protection measures), as it is in admitting that the State authority does not prevent the Regions from considering other cultural goods where the community involved recognizes special cultural value, without this entailing formal qualification as cultural heritage and the consequent application of regulatory constraints.[23]

A further glimpse at National sources allows us to observe laws dedicated to intangible cultural goods, often in conjunction with tangible goods. These mostly include organizational norms tied to the Ministry of cultural goods and activities and of Tourism: Art. 52, Para. 1, Law 300/1999 (organizational reform of the Government) gives the Ministry all the power over cultural assets, and so, even those after the regulatory functions defined by the code.[24] The profile emerges more clearly in the recent ministerial organization regulation (Prime Ministerial Decree 171/2014), as intangible cultural heritage is cited multiple times and always in conjunction with tangible heritage. Examples include: Art. 6, regarding the press office, which is responsible for, among other things, handling specific communication projects on intangible and tangible cultural heritage; Art. 11, Para. 2, Lett. m), which entrusts the Secretary General with the function of promoting the entry of new sites and new elements in the lists of tangible and intangible world heritage based on the preliminary investigation carried out by the competent directorates-general (index that intangible heritage is an element transverse to the areas of competence of the individual directorates-general); Art. 19, Para. 2, Lett. f), which assigns the directorate general of tourism with the development of programs and initiatives aimed at increasing the tourist offer focusing, particularly on the use of sites and elements from the UNESCO

[23]It follows that regions can, for example: introduce special legislation for the protection of commercial and craft activities that have historical, artistic, environmental value, by calling them "historic places", without this qualification overlapping that of the code, but giving shop owners the chance to access funding and other promotional tools (Constitutional Court, 28 March 2003, no. 94). Grant municipal urban planning the authority to identify the culturally relevant properties for the community with the purpose of establishing the allowable interventions and uses, since this competence does not involve the introduction of criteria for the identification of cultural property beyond that of the State, but instead allows, in the area of regional law, the local government to consider goods other than those bound under the state legislation, which are typical of that territory (Constitutional Court 16 June 2005, no. 232, in Giur. cost., 2005, 3, 1991. See, among others, Salvia (2012), pp. 107–108; De Giorgi Cezzi (2007), pp. 219–234) for more on the protection procedures of cultural goods in urban law; regulate those goods, non having not been recognized or declared of "cultural interest" as provided in the code, which present, even minimally, some "cultural" interest for a given community or territory (Constitutional court, 17 July 2013, no. 194).

[24]Bartolini (2013), p. 113; Bartolini (2014), p. 2 underlines this point, recalling the principle of uniqueness.

world tangible and intangible cultural heritage list; and Art. 35, Para. 1, which defines museums as "permanent non-profit institutions serving society and its development. They are open to the public and do research concerning the tangible and intangible evidence of humanity and its environment; they acquire it, preserve it, communicate it and exhibit it for purposes of study, education and enjoyment".

When speaking of intangibles, the territorial dimension is stronger than with other goods. It is in the territory that these goods come to life through the execution and representation of events that recur periodically (for example, poles, festivities and holy rites). The territory is, therefore, a sort of "nursery" for these goods,[25] the place in which they born, stratified and ultimately become representative of an identity, usually at a local level, but that can also become of national or international interest. In any case, these assets are part of the identity and heritage of the territory of origin. This is particularly evident for food products, in which the bond with the land is deep, because of the raw materials available, the cultivation techniques adapted to the morphology of the soil, the places of production and use/sales, all of which are now perceived as typical and identifying of the history and traditions of a place or a social group.

The strong tie to the territory is the reason for the focus of the regional legislation, even concerning the UNESCO Convention on intangible cultural heritage. Particularly significant is the implementation by the Region of Sicily with Decree Councilor for cultural heritage, environment and education 77/2005, in which the warning of the Convention is carried out through the direct application of Art. 117, Para. 5, Constitution, where the authority-duty to participate in the processes of decision-making and implementation of international acts is assigned to the regions.

With the decree, the region incorporated the policies and guidelines of the Convention and established a register of intangible heritage (REI), recently replaced by another register (the intangible heritage of the Sicilian Region - REIS),[26] which is structured in four "books", according to the nature of the assets, which are an expression of: (1) crafts, knowledge and techniques; (2) celebrations, festivals and rituals; (3) dialects, parlances and slang; (4) oral traditions and expressions; (5) living human treasures; (6) symbolic areas.

The formalization of the specific aspects describing the intangible goods that act as requirements for enrolment in the register is interesting since it shows the adoption of the claims of the demo-ethno-anthropologic disciplines. They must be goods that are important for the "shaping of the history of the community to which they belong", ontologically "identical and changing" and destined to be "socially shared and participated", "preserved" so as to be 'enjoyed' over and over again, "fixed in lasting memories", so as to avoid the loss of their value (Art. 3, Decree 5 March 2014).

The Lombardy region took a step in the same direction with Law 27/2008 ("Promotion of the intangible cultural heritage") which is admittedly inspired by the UNESCO Convention of 2003 and proposes to promote intangible cultural heritage through numerous initiatives deriving from the activity of identification of the elements; hence, the establishment of the register of intangible heritage of the

[25]Tucci (2002), p. 54.

[26]Decree of the Councillor for cultural heritage and Sicilian identity of 5 March 2014.

Lombardy region (REIL), which is organized along the lines of the UNESCO indications, in five sections: (1) oral tradition; (2) arts and entertainment; (3) rituals; (4) knowledge concerning nature; (5) technical know-how".

The idea behind the registers is that of the recognition of the elements of intangible cultural heritage, a kind of regional classification of assets, as a condition for their preservation and enhancement through a series of public interventions, ranging from the promotion of study activities and field research to the dissemination of knowledge by the organization or support of cultural events, to the publication of documentary sources and realization of communicative products, to the dissemination of best practices to promote conservation and transmission from generation to generation.

These regional registers include various goods of agricultural and gastronomic importance: the Sicilian includes, to give a few examples: the chocolate of Modica, the salting of fatty fish from Sciacca, the cycle of wheat and bread of Enna, the Favignana Bluefin tuna.

The substantial equalization of tangible and intangible cultural heritage is guaranteed by other regions through the laws on cultural heritage and activities, approved by virtue of the legislative competence given to them in Art. 117, Para. 3 of the Constitution for matters of "promotion of cultural heritage" and "promotion and organization of cultural activities", or in accordance with the statutory provisions for regions with special status. The Tuscany region made a decisive choice with Law 21/2010 ("Consolidated provisions on assets, institutions and cultural activities"), whose Art. 1, in defining the purpose and objectives of regional intervention, specifically mentions "the tangible cultural heritage of Tuscany, as well as the immaterial heritage".

Similarly, Sardinia Regional Law 14/2006 ("Rules on cultural heritage, institutions and places of culture"), aimed at the objective of "the protection, enhancement and enjoyment of the tangible and intangible cultural heritage of Sardinia" and Liguria Regional Law 33/2006 ("Consolidated Law on culture") indicates the function of "protection, conservation, enhancement and social use of the cultural heritage represented by the Liguria folk tradition both in its material or written manifestations, and in the verbal or auditory ones" (Art. 2, Para. 3, Lett. d). More recent is Lazio Regional Law 15/2014, ("Lazio cultural system: Provisions relating to the performing arts and cultural promotion") which defines the promotion of culture in terms of "activities aimed at promoting awareness and appreciation of both tangible and intangible cultural heritage of Lazio" (Art. 10, Para. 1).

This choice is also present in other regional laws, in which the expressions of intangible cultural heritage can be found in the category of "cultural activities". This is the case of the Marche Regional Law 4/2010 ("Norms on cultural goods and activities"), whose Art. 3, in defining the cultural activities, encompasses the "revival of the tradition and customs of the Marche and the promotion of the history and identity of the region, with particular attention to celebrations and famous people" (Lett. e). In other cases, tangible assets constitute an area of intervention in cultural activities; one example is the Molise Regional Law 5/2000 ("New norms on cultural promotion"), according to which the activities and cultural services promoted by the law concern, among other subject areas, those of historical-humanistic and popular traditions, visual arts, communication, music and dance (Art. 3) and another, Friuli-Venezia Giulia

Regional Law 16/2014 ("Regional standards in the field of cultural activities"), which includes among the policy areas on cultural activities the "enhancement of the historical memory" (Art. 4, Para. 1, lett. e).[27]

In the regional overview, we can also find regulatory policies, in the form of ad hoc laws or individual provisions included in laws of the broader content, referring to certain manifestations of intangible cultural heritage, such as the local linguistic heritage,[28] popular music and dances,[29] historical events,[30] or

[27]This regulatory trend gets legitimacy from the material relative to cultural activities in Art. 117, paragraph 3 of the Constitution. In this context the link between intangible cultural heritage and cultural activities should be made clear, in the sense that the former does not identify with the latter, but is, instead, one of the areas of intervention (a profile, however, which is not always evident in regional legislation). As clarified by Sciullo (2011), p. 24; Bartolini (2013), p. 111), in the wake of the Constitutional Court (according to which cultural activities "include all activities related to the development and dissemination of culture": Constitutional Court, 21 July 2004, no. 255; Constitutional Court, 19 July 2005, no. 285), intangible cultural goods are to be kept separate from cultural activities, because the latter are activities that are fundamental for the formation and dissemination of culture, while the former are themselves expressions and products of culture.

[28]This is found in the most traditional legislation, also based on the link to the constitutional data related to the protection of minor languages (Art. 6 Constitution). For an *excursus* of regional legislation see, among others: Gualdani (2014), p. 5; Chiarelli (2010), pp. 433–436.

[29]Traditional is the legal branch related to the support of band and choir musical activities: among others we can remember Abruzzo Regional Law 89/1984, ("Intervention for the enhancement of Abruzzo band and choral music heritage"), Valle d'Aosta Regional Law 5/1986 ("Regional intervention for band activity and for the implementation of music courses for choir, instruments and band"), Piedmont Regional Law 49/1991 ("Norms for the support of formative activities in the sectors of band, choir and instruments in associations, schools and musical institutions in the Piedmont region"), Emilia-Romagna Regional Law 23/1992 ("Promotion of musical culture of bands and choir", repealed by Art. 12 of regional law n.13, 5 July 1999),Veneto Regional Law 13/1995 ("Norms for the promotion of musical culture of choir and bands"), Umbria Regional Law 9/2004 ("Promotion of musical culture of bands and choir"), Basilicata Regional Law 15/2006 ("Safeguarding and promotion of Musical, Band and Choral culture in Basilicata"), Friuli-Venezia Giulia Regional Law 16/2014, Art. 28, Lazio Regional Law 15/2014, Art. 8. The most recent legislation extends the focus to popular and folkloristic music, of which it recognizes the "historical and cultural value", in that they are "expressions of cultural identity of peoples, as well as instruments for learning the culture and popular traditions and for social and economic development, in particular, territorial tourism" (Lazio Regional Law 15/2014, Art. 9, Para. 1–2; similarly, Art. 28 Friuli-Venezia Giulia Regional Law 16/2014). Particularly advanced in this direction is Apulia Regional Law 30/2012 ("Regional interventions for the protection and promotion of the music and dances of the oral tradition"). It is specifically dedicated to music and folk dances and presents itself as a law implementing the UNESCO Convention for the Safeguarding of the Intangible Cultural Heritage: the law provides for the establishment of an appropriate regional register in which non-profit subjects, in any legal form, who perform music activities and folk dance can enroll. It provides financial assistance to such persons, as well as to those local authorities that promote the realization of specialized multimedia archives and libraries or the organization of festivals, gatherings and similar entertainment initiatives and to specialized publishers and record labels.

[30]Consider, for example, Umbria regional law 16/2009 ("Rules governing historical events"), which qualifies historical events as "expressions of historical and cultural heritage of the regional community" (Art. 1, Para. 1), and Molise Regional Law 12/2005 ("Provisions concerning the historical and cultural events authorized by the Region pursuant to Art. 3 paragraph 1 of Law n. 189, 20 July 2004"), aimed at promoting the historical and cultural events that are recognized by the

rather, local contexts that include both tangible and intangible cultural elements, such as ecomuseums.[31]

From this regulatory scenario we can get a glimpse of a generally common model regarding the functions and the specific institutions, which involves the identification/cataloging of items of cultural heritage, the recognition/accreditation of the workers in the sector, the planning of measures, the granting of subsidies, loans and other benefits in favor of the parties who guarantee the preservation, development, monitoring and control of the use of funds, as well as of the implementation of the interventions. In responsibilities, the "strong" role of the region emerges, also in administration, to which the above functions are allocated, except in the case of a different organizational design that highlights, with different shapes and intensities, local authorities.

When speaking specifically about goods in the agrifood sector, we can see some laws that have been dedicated to particular products or production methods: this is the case of Calabria Regional Law 59/2012 ("Recognition of the historical technique for 'Moscato al Governo di Saracena' as a cultural asset of Calabria"), which qualifies as "intangible cultural heritage" "the artisan and food and wine production methods documented through accredited and/or passed on historical sources, even if only oral, characterized by uniqueness and exclusivity and are that are likely to be definitively lost" (Art. 1), and of which Art. 2 provides examples, including the method of production of Moscato Passito di Saracena; or Sicily Regional Law 25/1998 ("Measures for traditional tuna fishing") aimed at "safeguarding and recovering the cultural value of tuna fishing with the traditional system of fixed traps, as per local traditions"; as well as Sardinia Regional Law 4/2016 ("Provisions for the protection of the bread making and typical baked goods of Sardinia"), which is

region as of "historical and unique importance to the local communities concerned", and Art. 11 Para. 1 of Lazio Regional Law n. 15/2014 which promotes "the creation of historical re-enactment events in all their forms of artistic expression whose aim is the enhancement of the territory's history."

[31]The Ecomuseums constitute and new widespread form of museum "aimed at preserving, communicating and renovating the cultural identity of a community. They consist in an integrated project to protect and promote a territory that is geographically, socially and economically homogeneous and that produces and contains resources, natural landscapes and tangible and intangible heritage elements" (Art. 1, Para. 2, Friuli-Venezia Giulia Regional Law 10/2006, "Institution of Ecomuseums in Friuli-Venezia Giulia"). From the regional laws it emerges that tangible and intangible regional goods are equivalent, making up the building blocks of the eco museum in equal parts (Piedmont Regional Law 31/1995 "Institution of Ecomuseums in Piedmont", Art. 1; Friuli-Venezia Giulia Regional Law 10/2006, Arts. 1–2; Lombardy Regional Law 13/2007, "Recognition of ecomuseums for cultural promotion and promotion of local traditions for environmental, scenic, cultural, touristic and economic purposes", Art. 1; Umbria Regional Law 34/2007, "Promotion e regulation of ecomuseums", Arts. 1–2; Molise Regional Law 11/2008, "Institution of Ecomuseums in Molise", Art. 1; Tuscany Regional Law 21/2010, Arts. 16 e 19; Apulia Regional Law 15/2001, "Institution of Ecomuseums in Apulia", Art. 1; Veneto Regional Law 30/2012, "Institution, regulation e promotion of Ecomuseums", Art. 2; Calabria Regional Law 62/2012, "Institution of Ecomuseums in Calabria", Arts. 1 and 4; Sicilia Regional Law 16/2014, "Institution of Ecomuseums in Sicily", Art. 2).

aimed at protecting and enhancing the bread and other baked specialties of the Sardinian tradition; or Apulia Regional Law 20/2008 ("Constitution of the regional wine/olive oil cellar") aimed at promoting Apulia wines and olive oil, also to ensure "the restoration and enhancement of traditions and the re-building of the economic and cultural identities of the territories of wine and oil" (Art. 1, Lett. c).

For other areas, the regions relied on matters under their legislative competences, either concurrent or residuary (Art. 117, Para. 3–4 Constitution) such as: commerce, agriculture or tourism, except where there are transversal clauses or matters strictly under State competence (including the protection of economic competition, environmental protection, civil order) or limits coming directly from Communitarian or International law (Art. 117, Para. 1–2).[32]

Consider, for example, the tourism sector in which the wine and food products are increasingly regarded as tourism resources. Important examples in this regard can be seen, among others, in the Venetian regional model, deriving from Regional Law 11/2013 ("Development and sustainability in Venetian tourism"). Here, wine and food products are expressly listed among "tourism resources" along with other goods traditionally considered to be of importance for tourism, that is, cultural assets in the strictest definition (cultural, historical, religious goods and places; cities of art, etc.) and environmental-landscape heritage (sea, mountains, lakes, rivers, protected areas, etc.). This combination shows the awareness that the food and wine component is a relevant factor in the orientation of tourist dynamics, as they are expressive and attractive elements of the territory. Among the projects to be implemented under the law we find one called "The Academy of the Venetian haute cuisine",[33] object of an institutional agreement between the region and regional union of chambers of commerce, whose aim is the promotion and enhancement of Venetian enogastronomy and typical products in regional tourism and is divided into a plurality of initiatives ranging from the organization of courses and production of information material to the collaboration with associations of Italian and foreign restaurateurs and chefs, to the assistance provided to agri-food businesses and hotel restaurants in scouting and accessing European and international financing.

At the intersection of Agriculture and Tourism material lies the agri-tourism institute whose "dual nature"[34] stems from state regulations (Art. 3 Legislative Decree 228/2001; Law 96/2006) that incorporate the new organization into the traditional scheme of the agricultural enterprise, sanctioning a close relationship between agri-tourism and agricultural activities and a criterion of prevalence of the latter over the former and, at the same time configures the new company as a means

[32]This fact is particularly evident in the matter of "agriculture", in which there is a great amount of transversal and diverse interfering material, besides an intense penetrating supranational source of law: on this question see: Germanò (2003), pp. 117–194.

[33]Resolution of the Veneto Regional Executive, 27 May 2014, no. 806.

[34]Ferrucci (2006), p. 63; Ferrucci (2007), pp. 680–681.

of promoting the territory, also through the promotion of local food products and related food traditions.[35] The latter profile is emphasized by regional legislation, both preceding and succeeding national law; among the most recent are: Sardinia Regional Law 11/2015 ("Norms regarding agritourism, fish/fishing tourism, teaching farms and abrogation of Regional Law n. 18, 1998") which declares, among its aims, that of "supporting and encouraging typical products, products of local quality with short production chains and traditional local food and wine products" (Art. 1, Lett. e). About the latter, the Sardinian law confirms that "meals, food and drinks offered to the public, preferably having a short chain of production, are the expression and promotion of traditional typical local food and wine and unique territorial food culture" (Art. 4, Para. 1), and so, they must be produced on the farm itself, that is, primary regional products, regional products with a biological brand, DOP, IGP, IGT, DOC or DOCG, products from a regional origin and source certified by the collective brand of quality control guaranteed by the Region, traditional regional wine and food products found on the national list (Art. 4, Para. 2), with the further condition that the sum of the products must be at least 85% of the total annual product used (Art. 4, Para. 3).

The definition of the product type and high percentage of the quota for home-produced/regional products are constant subjects for intervention of regional legislation (the 85% quota is found, for example, in Umbria Regional Law 12/2015, "Consolidated text on the subject of agriculture" and in Piedmont Regional Law 2/2015 "New dispositions on the subject of agritourism") and they make up an important *in melius* application of the bind introduced by national law which limits itself to mentioning "a prevalence" of home-produced and local farm products and a "preference", for the rest of the products used, for products coming from artisan food producers in the area or from similar contiguous areas of nearby regions.

At the crossroads between tourism and agriculture we can also find the legislative trend aimed at recognizing and enhancing "roads", "paths" and "itineraries" focused on the promotion of typical food and wine products of a specific area: wine, oil, milk, truffle, etc. In this way, it combines the safeguarding of the landscape, guaranteed through the preservation of the shape given to a territory by traditional agricultural practices, with the other typical aim of policies regarding landscape that of valorization, focused on increasing the awareness and enjoyment of goods. Tourism becomes a privileged means for the pursuit of these objectives and, at the same

[35]The aim of "supporting and promoting local products, quality production and related food traditions" is specifically enshrined in (Art. 1, Lett. f) of Law 96/2006 and is consistent in listing the various activities that can be offered by the agritourism business among which are included those of "distributing meals and drinks mostly consisting of their own products and products from farms in the area, including alcoholic drinks and spirits, with a preference for local products characterized by DOP IGP IGT DOC and DOCG or included in the national list of traditional food products" and organizing tastings of agrifood products, including wine bars "(Art. 2, paragraph 3, letters b-c) elements that also laid out in the model outlined by Art. 3 Decree 228/2001.

time, for triggering local economic development. The choice of territorial marketing, particularly interesting since it is both cultural and economic,[36] translates into norms that propose benefits and financial contributions for various projects: from the placing of signage to making the various itinerary roads safe, from the creation or adaptation of information centers to marketing through various advertising channels, from the promotion of cultural activities to the training of specialized workers.

On a normative level, we find a first series of local and regional initiatives, which are then transposed and formalized by Law 268/1999, "Regulations of the wine routes" (and ministerial decree n. 781500/2000; concerning the definition of the requirements and minimum standards), whose model has been taken, fortified and applied to other areas by the regions.[37]

5 Reflections on Cultural Heritage Law

The protection of the Intangible Cultural Heritage parallel to that of the material heritage enshrined at an international level, the "open" nature of the notion of cultural heritage highlighted by the doctrine, copious regional legislation for the protection and enhancement of intangible cultural heritage (including the explicit declaration of being on equal footing with tangible goods), and the support of the constitutional Court in favor of a number of legal descriptions of cultural heritage are all strains of such an intensity that they can no longer be ignored by the State. As seen, the code would prefer not to take sides on the point, a sign of a context still holding on to the "material" nature of cultural heritage, as evidenced by the new,

[36] Ferrucci (2011), p. 241.

[37] Among the regional initiatives see: Umbria Regional Law 38/1999 ("Regulation of wine routes in Umbria"), Emilia-Romagna Regional Law 23/2000 ("Regulation of wine and food itineraries in Emilia-Romagna"), Abruzzo Regional Law 101/2000 ("Regulation of wine routes in Abruzzi"), Veneto Regional Law 17/2000 ("Institution of wine routes and other typical products of Veneto"), Friuli-Venezia Giulia Regional Law 21/2000 ("Regulation of the labeling of agricultural products of Friuli-Venezia Giulia which are not GMO, for the promotion of traditional agrifood products and the realization of wine routes"), Lazio Regional Law 21/2001 ("Regulation of routes for wine, olive oil and typical traditional agrifood products"), Trento Provincial Law 19/2001 ("Regulation of agritourism, wine routes and 'tasting' routes"), Tuscany Regional Law 45/2003 ("Regulation of the routes of wine, extra virgin olive oil and agricultural products and quality foods", modified, most recently by regional law n. 4, 6 February 2013), Molise Regional Law 50/2005 ("Institution of routes of wine, oil, truffle, milk and flavors of Molise"), Liguria Regional Law 13/2007 ("Regulation of itineraries of the flavors and smells of Liguria, of regional wine cellars, and intervention in favor of widespread hospitality"), Piedmont Regional Law 29/2008 ("Identification, institution and regulation of rural districts and quality agrifood districts and modifications of regional law n. 37, 12 May 1980 ("Regional wine cellars, wine shops and city wineries, ethnographic-enological museums, wine routes"), Friuli-Venezia Giulia Regional Law 22/2015 ("Regulation for the realization, identification and valorization of 'Wine and flavor routes' of Friuli Venezia Giulia"). For further information on national and regional law, see, in particular, Strambi (2006), pp. 204–238; La Torre (2010), pp. 573–578; Cristiani (2011), pp. 258–268.

essentially elusive, provision Art. 7-*bis*. Yet, the system offers several models, which can even be taken from other sectors, to address the issue. Consider the environmental code (Law decree 152/2006) which divides environmental protection between the State and the Regions on the base of principle of greater environmental protection by the Regions (Art. 3-*quinquies*), which, applied to the case in point, would lead to placing the area of intangible cultural heritage under the protection of cultural assets of State competence, with the possibility for the Regions to act as reinforcement. Or, instead, consider the criteria of territoriality in cultural interests, cited several times by the Constitutional court, by which goods of national interest should be the responsibility of the State and those of regional/local interest under regional government.

In any case, intervention of the state administration is essential (the Ministry of Cultural Heritage), which, in this way, would recover a pre-eminent role in relation to the intangible cultural heritage, currently outlined in the new rules of organization and traditionally abdicated in favor of other subjects (Italian national Commission for UNESCO or other ministries, such as agricultural policy or health). The Ministry would be called on to identify the intangible assets of significant public interest (using models like the ones in the code of cultural heritage for tangible goods),[38] verifying the insertion of the good in the culture of a territory, and thus its unique character, but also its relevance at national level, leaving the regions with the eventual task of strengthening the protection or considering goods of regional/local importance. There would not even be interference with other administrations, which could be imagined in the agri-food sector, because the ministry of cultural heritage would not be called upon to assess the genuineness of the good, the taste or organoleptic characteristics which infringe on the responsibilities of other ministries, but it should only verify the cultural value with respect to the history of a people and its territory.[39]

Remaining on an organizational level, the topic of intangible cultural heritage leads to further reflection. We have already considered the strong bond these goods have with their territories, even closer than tangible goods, because volatile goods require constant repetition and re-execution, especially by the people of said territories. This fact highlights the willingness of the sector for the virtuous application of the principle of subsidiarity, in the dual meaning, vertical and horizontal. Local entities can actively participate both in the procedures of identification of cultural goods and in the other functions of protection and promotion and the same role can be attributed to private parties, singly or in the form of associations.[40] The case of the Mediterranean diet is emblematic. There are many associations involved in the

[38]The function of the Ministry in terms of identifying public interest, and so, selection of legally relevant goods (a mechanism that would be autonomous and additional compared to others selected by territorial bodies, as in the case of regional and local registries of intangible heritage) would be a valid representation of the role that law can have in overcoming the risks of panculturalism underlying the proposals for the widening of the notion of cultural heritage.

[39]Gualdani (2014), p. 4.

[40]Gualdani (2014), pp. 5–6.

promotion and dissemination of the diet, through the promotion its nutritional and/or cultural value.[41]

The profile is welcomed by the 2003 UNESCO Convention that sanctifies the principle of social participation: "the participation of communities, groups and relevant non-governmental organizations" (Art. 11, Lett. b) and "the widest possible participation of communities, groups and, where appropriate, individuals that create, maintain and transmit such heritage" (Art. 15) are cited in the State's definition of the measures of safeguarding intangible cultural heritage. Moreover, the level of involvement of the collectivity in identifying goods, starting from the initial phase of the presentation of the candidacy, is an important criterion for UNESCO for the enrollment of the good in the representative List (ch. I-2, R.4 of the operative directive for the fulfillment of the Convention of 19 June 2008 and successive modifications). If you look at the profile of administrative functions, the same UNESCO Convention offers further food for thought. It gives a quite ample transversal notion of "safeguarding', because in it there is not only a list of the different activities inherent to the conservation, but also includes typical profiles of other functions such as valorization. In fact, for "safeguarding" it intends those "means and measures aimed at ensuring the viability of the intangible cultural heritage, including the identification, documentation, research, preservation, protection, promotion, enhancement, transmission, particularly through formal and non-formal education, as well as the revitalization of the various aspects of such heritage" (Art. 2, Para. 3). A definition which, evidently, shows, on one hand, the vast gamma of activities that make up protection; and, on the other, the continuity, if not the overlapping, in the areas of protectionism and enhancement. For this reason, it stands out compared to the code of cultural heritage, in which the institutes of protection are all based on the assumption that they concern tangible goods and there is a clear distinction between protection and valorization, a result, in fact, of the Constitution (Art. 117) that distinguishes between protectionism and enhancement of cultural heritage for the purposes of legislative competence.[42]

As can be observed, the UNESCO Convention deals with the essential profiles of cultural heritage law from a new perspective. The attention inevitably returns to the preliminary question of the legal definition. The one provided by the code of cultural heritage, wrapped up in the element of materialism, seems reductive and its widening to include intangible cultural heritage is desirable, if not deserved, to fully implement international law. After all, it would run in the wake of other legal systems that have long since opted for more definitions of cultural heritage that include both tangible and intangible goods. Materialness cannot be the discriminating factor, the reason for isolating intangible goods, but rather should be an element that is useful for modeling the techniques and instruments for protection, which, as has been

[41] Mazzuoli (2015), p. 97.

[42] Regarding the distinction between the functions of protection and valorization of cultural heritage found in the code of cultural heritage which reflects the Constitutional note and its reciprocal interference see: Sciullo (2011), p. 60; Barbati (2007), p. 78.

observed, cannot be reduced to the ones indicated in the code. There should, at least, be awareness on the part of the code that its approach is not exclusive[43] and that, along with the code's definition, there exist other, equally dignified definitions open to pluralistic statutes and unified by the "sustainable lightness of the intangible value of cultural heritage goods (tangible and intangible)"[44]; that intangibility, which Giannini, back in the 1970s, defined as a distinguishing and unifying feature of cultural heritage.

References

Accettura B (2015) Valorizzazione del patrimonio culturale e nuovi modelli per lo sviluppo dei territori. Jovene, Napoli

Barbati C (2007) Art. 6. In: Cammelli M (ed) Il codice dei beni culturali e del paesaggio. Mulino, Bologna, pp 74–82

Bartolini A (2013) Beni culturali (diritto amministrativo). In: Enciclopedia del diritto. Annali VI. Giuffrè, Milano, pp 93–132

Bartolini A (2014) L'immaterialità dei beni culturali. Aedon. Riv di art e dir on line 17(1):1–7. http://www.aedon.mulino.it/archivio/2014/1/bartolini.htm. Accessed 30 May 2016

Casini L (2008) Le nuove modifiche al codice dei beni culturali e del paesaggio. Gior di dir amm 14(10):1057–1068

Casini L (2010) La globalizzazione dei beni culturali. Mulino, Bologna

Cassese S (2012) Il futuro della disciplina dei beni culturali. Gior di dir amm 18(7):781–782

Chiarelli R (2010) Profili costituzionali del patrimonio culturale. Giappichelli, Torino

Cirese AM (1996) I beni demologici in Italia e la loro museografia. In: Clemente P (ed) Graffiti di museografia antropologica italiana. Protagon, Siena, pp 249–262

Cirese AM (2007) Beni volatili, stili, musei. Gli ori, Prato

Cristiani E (2011) Codice del turismo e valorizzazione delle produzioni agro-alimentari: un'occasione mancata? In: Goldoni M, Sirsi E (eds) Il ruolo del diritto nella valorizzazione e nella promozione dei prodotti. Giuffrè, Milano, pp 253–269

Crosetti A, Vaiano D (2014) Beni culturali e paesaggistici. Giappichelli, Torino

De Giorgi Cezzi G (2007) Il diritto all'identità minore. Beni culturali e tutela degli status. In: Studi in onore di Leopoldo Mazzarolli, vol III. Cedam, Padova, pp 219–234

Ferrucci N (2006) L'agriturismo fra tutela dell'impresa e valorizzazione del territorio. Quad dei Georg 18(5):63–68

Ferrucci N (2007) L'agriturismo: icona della trasversalità dell'agricoltura tra impresa, alimentazione, ambiente e paesaggio. Riv di dir agr 86(4):679–683

Ferrucci N (2011) Il paesaggio agrario tra Convenzione europea del paesaggio, Codice Urbani e normativa agraristica. Dir e giur agr, alim e dell'amb 19(4):240–244

Francioni A (2006) Beni culturali (dir. int.). In: Cassese S (eds) Dizionario di diritto pubblico, vol I. Giuffrè, Milano, pp 690–704

Germanò A (2003) La "materia" agricoltura nel sistema definito dall'art. 117 Cost. Le Reg 31(1): 117–194

[43]Hence the admonition of authoritative doctrine, according to which the future discipline of cultural heritage "will not be able ultimately to provide a single definition of cultural property, but will have to accept different ideas, each functional to various needs of the community", Cassese (2012), p. 782 (author's translation).

[44]Bartolini (2014), p. 6 (author's translation).

Giannini MS (1976) I beni culturali. Riv trim di dir pubb 26(1):3–38
Gualdani A (2014) I beni culturali immateriali: ancora senza ali? Aedon. Riv di art e dir on line 17 (1):1–8. http://www.aedon.mulino.it/archivio/2014/1/gualdani.htm. Accessed 30 May 2016
La Torre ME (2010) Attività turistica e categorie contrattuali. In: Degrassi L, Franceschelli V (eds) Turismo. Diritto e diritti. Giuffrè, Milano, pp 539–578
Lombardi R (2009) La candidatura dei paesaggi vitivinicoli del basso Piemonte a "territorio patrimonio dell'Umanità": brevi note a margine della tutela dei paesaggi rurali. Nu aut 18(2–3): 389–397
Lucifero N (2011) La comunicazione simbolica nel mercato alimentare: marchi e segni del territorio. In: Costato L, Germanò A, Rook Basile E (eds) Trattato di diritto agrario. Il diritto agroalimentare, vol III. Utet giuridica, Torino, pp 321–421
Mazzuoli C (2015) Il ruolo del settore privato nella valorizzazione del patrimonio culturale immateriale: il caso della Dieta mediterranea. In: Germanò A, Strambi G (eds) La valorizzazione del patrimonio culturale immateriale di interesse agricolo. Atti del Seminario (Firenze, 21 aprile 2015). Giuffrè, Milano, pp 97–108
Mirri B (2015) La tutela UNESCO del patrimonio culturale immateriale e il riconoscimento delle 'diete' e 'cucine' come patrimonio dell'umanità. In: Germanò A, Strambi G (eds) La valorizzazione del patrimonio culturale immateriale di interesse agricolo. Atti del Seminario (Firenze, 21 aprile 2015). Giuffrè, Milano, pp 59–67
Morbidelli G (2014) Il valore immateriale dei beni culturali. Aedon. Riv di art e dir on line 17(1): 1–9. http://www.aedon.mulino.it/archivio/2014/1/morbidelli.htm. Accessed 30 May 2016
Oggianu S (2012) La disciplina pubblica delle attività artistiche e culturali nella prospettiva del federalismo. Giappichelli, Torino
Poggi AM (2003) Verso una definizione aperta di "bene culturale"? (a proposito della sentenza n. 94/2003 della Corte costituzionale). Aedon. Riv di art e dir on line 6(1):1–3. http://www.aedon.mulino.it/archivio/2003/1/poggi.htm. Accessed 30 May 2016
Salvia F (2012) Manuale di diritto urbanistico. Cedam, Padova
Sciullo G (2011) I beni. In Barbati C, Cammelli M, Sciullo G (eds) Diritto e gestione dei beni culturali. Mulino, Bologna, pp 21–51
Strambi G (2006) Le strade del vino, dell'olio e dei sapori: il quadro giuridico di riferimento. Riv di dir agr 85(2):204–238
Tarasco AL (2008) Diversità e immaterialità del patrimonio culturale nel diritto internazionale e comparato: analisi di una lacuna (sempre più solo) italiana. For amm CdS 7(7–8):2261–2287
Tucci R (2002) Beni demoetnoantropologici immateriali. Antrop mus 1(1):54–59
Ubertazzi B (2011) Su alcuni aspetti problematici della Convenzione per la salvaguardia del patrimonio culturale intangibile. Riv di dir intern 94(3):777–798
Vitale C (2010) La fruizione dei beni culturali tra ordinamento internazionale ed europeo. In: Casini C (ed) La globalizzazione dei beni culturali. Il Mulino, Bologna, pp 171–197
Zagato L (2008) La Convenzione sulla protezione del patrimonio culturale intangibile. In Zagato L (ed) Le identità culturali nei recenti strumenti Unesco. Un approccio nuovo alla costruzione della pace? Cedam, Padova, pp 27–70

Food Diversity and Legal Protection of "Made in Italy" Label

Rosa Calderazzi

1 Foreword

An in-depth analysis into the relationship between Made in Italy and food diversity arises from the observation that they are closely related, both directed to protect the quality and the characteristics of a specific product and its intrinsic link with the origin, and with natural and human factors, or even more and in general, with the food (and cultural) heritage of a country.

In the era of economy and market globalization, which is associated with a delocalization of the production, the safeguard of the product peculiarity arises as a very important problem: typical products are characterized by an exact localization, since they are closely linked to the area, and thus needing a strong protection to be extended also to international markets.[1]

The label Made in Italy aims to identify geographically, not only the origin of the product, but also its processing and Know-how. This concept does not exhaust itself with the subjects of "IGP" or "DOP", which do not represent general tools of protection of the food diversity; rather, it is connected to several legal aspects involving the mark profile, the geographical area, the disloyal competition, the labeling obligations, the counterfeiting, and more in general, the protection of the intangible cultural heritage, intended as the expression of a whole country.

In the traditional view, Made in Italy is to be considered as a tool in the campaign against counterfeiting, which provokes extensive damages to Italian economy

[1] Galli (2004) highlights the need of a "global" protection acting not only in the country of origin of the product, but also in other markets; De Filippis (2013), Jannarelli (2006), and Borghi (2015).

R. Calderazzi (✉)
Department of Economics, Management and Trade Law, University "Aldo Moro" of Bari, Bari, Italy
e-mail: rosa.calderazzi@uniba.it

A. Isoni et al. (eds.), *Food Diversity Between Rights, Duties and Autonomies*, LITES – Legal Issues in Transdisciplinary Environmental Studies 2,
https://doi.org/10.1007/978-3-319-75196-2_27

(in addition to image damages), but also closely related to the protection of the Community mark and to the upgrading of the geographical origin of the product.[2]

At this stage, a different perspective is provided by linking Made in Italy to the food diversity profile and by evoking the concept of the product peculiarity with that of food diversity.[3]

The analysis of the relationship between food diversity and protection of Made in Italy displays a special interest based on dual aspect: on one hand, it allows to demonstrate the legal relevance of the concept of food diversity, since—as it could be argued—Made in Italy rules imply and, at the same time, strengthen its existence and, on the other hand, it may open the way to an alternative interpretation of the concept of Made in Italy compared to the traditional one.

2 Relationship Between Food Diversity and Made in Italy

The protection of the label Made in Italy originates from the awareness of the existence of a product specificity, as well as from the variety and diversity compared to other products coming from different areas, and from the need to protect its intrinsic characteristics, peculiarity, and therefore the origin.

It implies a strong connection between food and area that leads to the identification of a food diversity needing legal protection. Therefore, recognizing and increasing the protection of Made in Italy implies the extolment of food diversity and the subscription of a legal acknowledgement.

To verify the accuracy of such assertion it is necessary to examine the concept of Made in Italy, starting from a couple of considerations: firstly, it is characterized by a very strong suggestive power, so that it is considered a true and proper mark, but also the most imitated and forged; secondly, in Europe, a traditional opposition exists to the force on the indication of the origin of the product and on marks of origin.[4]

The rules on Made in Italy is the resultant of a difficult equilibrium between the European rules denying the importance of the foodstuff origin and the Italian appeal to promote the legal valorization of the origin of the product.[5]

In Italy, rules are operative and the penal code (second book) addresses a heading to the crime against industry and trade with a consistent risk of retention or penalty (Art. 517 p.c.) for whoever put up for sale or into movement intellectual achievements or manufactured items with "Italian or foreign names, brands or trademarks acting at misleading the purchaser on the origin, source or quality of the work or of

[2]Vanzetti and Di Cataldo (2012), Vanzetti (2002), Masi (1998), and Spolidoro (2014).

[3]Concerning the concept of food diversity, Monteduro (2015) reconstructs food diversity as a synthetic concept, in the sense that "he summarizes, collects and incorporates in the same concept, legal values that, at least until now, were separated and distant within the rules and the doctrinal processing". About food law, Germanò (2011).

[4]D'Angelo (2014) and Sangiorgio (2015).

[5]Albisinni (2011a), Costato (2006), Libertini (2010), Ranaldi (2014), Canfora (2010), and Albisinni (2012).

the product", and (Art. 517 *quater* p.c.), more in detail, for the counterfeiting or alteration of the geographical indication or the name of origin of agricultural and food products. As unraveled by the law, penal rules have been set for the general interest both for the consumers freedom and good faith and for the protection of the manufacturers from the illegal competition.[6]

Where labeling and food quality is concerned, a law (February 03, 2011, n. 4) has been issued to guarantee the consumer a complete and correct information on the characteristics of commercialized foodstuff, both as processed, partially transformed or unprocessed, and to strengthen prevention and repression of alimentary frauds.

The current law (Art. 4, subparagraphs 1, 2) obligates to write on the product label, the indication of place of origin or source and, accordingly to the European Union rules, the eventual use of ingredients containing genetically modified organisms in any phase of the food chain from the initial place of production to the final consumption. Moreover, it must be reported that for "unprocessed foodstuff, the indication of the place of origin or source regards the country of production. Conversely, for processed foodstuff, the indication regards the place where the last substantial processing has occurred together with the place of growing and farming of the prevalent feedstock used for food preparation or production".

Even the Consumer Code (Art. 2) acknowledges to the consumer the protection of health, safety and quality of products and services, a satisfactory information and a fair advertising, as fundamental rights.

The law has fixed that "concerning agricultural and food products, the subject reported in Art. 4, subparagraph 49, law n. 350, December 24, 2003 (protection of "Made in Italy") must be interpreted in the sense that the origin of the product is defined by the geographical origin independently from the localization of the processing and exclusively for the products reporting the mark DOP (protected denomination of origin) or IGP (protected geographical indication), attributive of peculiarity and quality guarantee, whereas for all the other agricultural and food "generic" products that are devoid of such marks, to define the origin, it is necessary to refer to the criteria reported in Articles 23 and 24 of the European Customs Code".[7]

On the Community front, the approach to the problem seems to be different, at least at first attempt: the EU does not deny, in itself, differences among products, above all if they are based on objective and measurable characteristics; by contrast, it seems to consider that the recognition—to a specific food—of a quality connotation disregarding from the real characteristics of the product as linked to intangible factors such as the origin, may limit the free movement of goods in contrast to what reported in TFEU (Art. 28).[8]

[6]As in Penal Supreme Court of Cassation November 13, 2007, n. 2003.

[7]Recently, Court of La Spezia, January 23, 2015.

[8]Albisinni (2011b), Giacomini (2011), and Borghi (2013).

In this context, the Regulation n. 1169/2011[9] was issued (it concerns the supply of food information to consumers and modifies the previous Regulations n. 1924/2006 and n. 1925/2006, and repeals the directives 87/250/CEE, 90/496/CEE, 1999/10/CE, 2000/13/CE, 2002/67/CE and 2008/5/CE, and the Regulation n. 608/2004). Together with the Implementing Regulation n. 1337/2013 (fixing the procedures to apply the Regulation n. 1169/2011 regarding the indication of the country or the place of origin of fresh, chilled or frozen pork, sheep, goat and bird meats), with the delegated Regulation n. 1155/2013 (which modifies the Regulation n. 1169/2011 regarding the information as to the absence or reduced presence of gluten in food) and with the Regulation n. 952/2013 (introducing the Customs Code of the Union that will be applied starting from May 01, 2016), it represents the last stage of the European legislator concerning food information.

The Regulation n. 1169/2011 (Art. 26) sets out that "Indication of the country of origin or place of provenance shall be mandatory where failure to indicate this might mislead the consumer as to the true country of origin or place of provenance of the food, particularly if the information accompanying the food or the label as a whole would otherwise imply that the food has a different country of origin or place of provenance".

The Regulation n. 952/2013 (Art. 60) sets out that "1. Goods wholly obtained in a single country or territory shall be regarded as having their origin in that country or territory. 2. Goods the production of which involves more than one country or territory shall be deemed to originate in the country or territory where they underwent their last, substantial, economically-justified processing or working, in an undertaking equipped for that purpose, resulting in the manufacture of a new product or representing an important stage of manufacture".

From a more careful reading of the Community legislation, it emerges profiles of analysis not necessarily presuming a net contraposition between Italian needs and Community decisions: in fact, beyond the explicit prevision of the indication of origin or the place of origin, whenever its omission creates confusion, in the "whereas" of the Regulation n. 1169/2011 it is reported that, because of the perceived need by the European Union to assure high level of consumer protection, the free movement, not of the product itself but of safety and healthy food, represents an essential feature of the internal market, to contribute in a significant manner to the health and welfare of citizens, but also to the realization of their social and economic interests.

Therefore, the free movement of goods is a value to be protected, provided that they are healthy and safe products.[10]

Moreover, to obtain a high level of protection of the consumer health and to assure the right to information, it should be ensured that the consumers are properly informed about the food they consume: information becomes a junction tool between the consumer protection and the protection of competition, and the

[9]Capelli (2014).

[10]Canfora (2012, 2014).

consumer choice can be influenced by considerations of health, economic, social and ethical (whereas 2, 3).[11]

If, therefore, it becomes necessary to allow the consumer in making informed choices about the food they consume, it could be more interesting to confer a different sense to the intentions of the Community legislator, in a direction not necessarily contrasting the purposes of the Italian legislator: since the need of the consumer protection is expressed by the warranty information that make him more informed and responsible, it appears confusing to distinguish between objective characteristics of the product and its origin or source, because they are all used to reach the same purpose: "to prevent any practices that may mislead the consumer" (Regulation n. 1169/2011, 4th whereas).

Also the European Parliament resolution (January 14, 2014) deems that the indication of the country of origin on the label, although does not represent per se an instrument to fight food fraud, it could contribute to guarantee a better traceability of the whole food chain, as well as a more stable relationship between suppliers and meat processing firms, a greater care to choose suppliers and products by food business operators, and a greater reliability of information for consumers, to restore the trust of the latter.

On the other hand, this way of interpreting is highly consistent with the regulatory framework of the Treaty on the Functioning of the European Unit (Art. 169), which sets out that, "in order to promote the consumer interest and to assure a high level of consumer protection, the Union contributes to protect health, safety and economic interests of the consumer as well as to promote his right to information, education and organization to safeguard his interests".

3 The Promotion of Food Diversity Through the Instrument of Made in Italy

The foregoing considerations allow to interpret in a different perspective the relationship between the Italian law on Made in Italy and the Community guidelines: origin and quality of products are not opposites terms, rather they represent different variations of a single aspect, that of protecting the consumer about foodstuff, namely the right to the protection of human health.

Affixing on a food product the label Made in Italy helps to satisfy the right to information, constantly recalled by the latest European Regulation, and contributes also to protect health, safety and economic interests of consumer, and to promote information, education and organization to safeguard his interests, as reported in TFEU (Art. 169).

Also, Regulation n. 1151/2012 dealing with quality systems for agricultural products and foodstuff exalts the quality and variety of agricultural production as

[11] Vettorel (2014), Di Cataldo (2014), and Costato (2008, 2014).

important strength and competitive advantage for Union producers, as well as an integral part of its living cultural and gastronomic heritage.

Health, safety, economic interests of the consumer, the right to information, education, organization to safeguard own interests, are all aspects aiming to be a full right of the individual feeding and that reach their fulfillment through a recognition of the product origin. If you share this approach, that is, ultimately, the need to highlight the peculiarity of the product, by reasons of: (a) agro-biodiversity (different products are different because of the diversity of the used raw materials); (b) diversification of production techniques (due to the practices, knowledge, abilities, instruments belonging to the intangible cultural heritage); (c) variety of areas (where the food becomes a territorial factor, expressing its identity), the protection of Made in Italy poses as the preferred route for the emergence food diversity.

The close correlation between the protection of Made in Italy and food diversity allows, also, to contribute to the wider debate about the function of Made in Italy.

The right to feeding is considered by our Constitutional Court as a fundamental right, linked to the protection of the inalienable core of the dignity of each human person and to irrepressible constitutional values: the Constitutional Court (judgment n. 10/2010) has recognized that the constitutional rules allow to bring that of food in the "social rights" of which the legislator must bear.

Traditionally, the protection of Made in Italy was counted among institutions to protect competition, linked to notion of product, as a tool enhancing the territorial origin of the products, to realize a competition on the market.

The same right to feed was mostly focused on the notion of food: the Regulation n. 178/2002 (Art. 2) that sets out the principles and the general requirements of food law, establishes the European Food Safety Authority and fixes procedures in matters of food safety, defines the food product as 'food' (or 'foodstuff') means any substance or product, whether processed, partially processed or unprocessed, intended to be, or reasonably expected to be ingested by humans".

Focused the right to feed on the food, the entire law revolves around goods susceptible to economic exploitation, rather than on the rights of the food consumer.[12]

Even when a subject of the consumer protection was strengthened, this came as a protection of food production and marketing but not of the right to feed, attention has been focused on the stuff, the product, and not on the person and his needs. The previous quoted Regulation n. 1151/2012 highlights the diversity of quality products according to the positive effects for the rural economy: among the purposes pursued by the Regulation, the following are set out: (a) fair competition for farmers and producers of agricultural products and foodstuffs having value-adding characteristics and attributes; (b) the availability to consumers of reliable information pertaining to such products; (c) respect for intellectual property rights; and (d) the integrity of the internal market. Thus, it continues to protect the intellectual property and the values of competition, to facilitate the international trade.

[12]As, critically, Perfetti (2014, 2015).

However, the protection of the food market, as well as the free movement of goods, can receive legal recognition only if they comply with the provisions aimed at protecting human health: as the Constitution orders (Art. 41) the economic initiative "cannot be conducted in conflict with social usefulness or in such manner that could damage safety, freedom, and human dignity". Thus, it imposes the primacy of human health protection compared to that of free enterprise and market.[13] If a similar setting is shared, a different perspective even of the protection of Made in Italy can be grasped: the legal value of the territorial origin of the product becomes much more and not only a means of market competition, protection of competition and of goods, rather a tool allowing to reconstruct with a further piece the protection of consumer, his right to health, safety and quality of products and services, by providing a proper information.

In this sense, the link between Made in Italy and food diversity helps to strengthen such a thought: the strong link between food and land, as underlined by both concepts, allows to exploit natural resources and to enhance the variety and diversity of food, becoming a factor of human development, since diversity helps to ensure food security, a free (but protected) movement of goods on the market, a variety and adequate quantity of food allowing to focus the human beings and their needs at the center of the market.

References

Albisinni F (2011a) Il made in Italy dei prodotti alimentari e gli incerti tentativi del legislatore italiano. Agriregionieuropa 25:43–62

Albisinni F (2011b) Etichettatura dei prodotti alimentari. Dir Ali Merc Sic

Albisinni F (2012) La comunicazione al consumatore di alimenti, le disposizioni nazionali e l'origine dei prodotti. Riv dir agr 1:68–92

Borghi P (2013) Il diritto agrario e alimentare come laboratorio del diritto dell'economia: il ruolo del giurista. In: Rook Basile E, Carmignani S (a cura di) Sicurezza energetica e sicurezza alimentare nel sistema UE. Profili giuridici e profili economici. Giuffrè editore, Milano, pp 291–299

Borghi P (2015) Gli sviluppi dell'agricoltura in Italia. In: Sviluppo agricolo, protezione del territorio e produzione agroalimentare in un ambiente tutelato. Santarcangelo di romagna, Maggioli editore, pp 17–25

Canfora I (2010) La disciplina della concorrenza nel settore agricolo dopo il Trattato di Lisbona. Riv dir agr 1:689–717

Canfora I (2012) La responsabilità degli operatori alimentari per le informazioni sugli alimenti. Riv dir agr 1:114–138

Canfora I (2014) Quando la Corte costituzionale è più realista del re. Sugli ostacoli alla libera circolazione delle merci derivanti da leggi regionali che valorizzano i prodotti locali. Riv dir agr 2:32–42

Capelli F (2014) Il Regolamento (UE) n. 1169/2011 e le sue guide spirituali. Riv dir ali 2:13–25

Costato L (2006) Il Primo convegno europeo di diritto alimentare e l'emergere progressivo di una vasta ed articolata legislazione comunitaria e nazionale in materia di cibi. Dir giur agr ali amb 1:5–6

[13]Perfetti (2015).

Costato L (2008) L'informazione dei consumatori: postrema frontiera della CE. Riv dir ali 4:1–7
Costato L (2014) Prodotti agricoli ed alimentari e studio del diritto alimentare. Riv dir ali 3:1–3
D'Angelo D (2014) Prodotti agroalimentari e Made in Italy in etichetta. Agriregionieuropa 39
De Filippis F (ed) (2013) L'agroalimentare italiano nel commercio mondiale. Specializzazione, competitività e dinamiche. Tellus, Rome
Di Cataldo V (2014) Liti fra birrai, marchi e denominazioni di origine: evoluzioni del linguaggio e mutamenti dei prodotti. Giur comm 4:610–621
Galli C (2004) Globalizzazione dell'economia e tutela delle denominazioni di origine dei prodotti agro-alimentari. Riv dir ind 2:60
Germanò A (2011) Il diritto dell'alimentazione tra imprenditori e consumatori che chiedono e legislatori e giudici che rispondono. Dir giur agr ali amb 11:653–660
Giacomini C (2011) L'etichettatura Made in Italy dei prodotti alimentari: uno strumento a sostegno di strategie competitive per il mercato. Agriregionieuropa 25:39
Jannarelli A (2006) "Agricoltura" e "tutela della concorrenza" nel nuovo art. 117 Cost. Riv dir agr 1:31–54
Libertini M (2010) L'informazione sull'origine dei prodotti nella disciplina comunitaria. Riv dir ind 1:289–311
Masi P (1998) Il marchio collettivo. In: Marasa G, Masi P, Olivieri G, Spada P, Spolidoro MS, Stella Richter M (eds) Commento tematico della legge marchi. Giappichelli, Torino, pp 71–75
Monteduro M (2015) Diritto dell'ambiente e diversità alimentare. Riv dir amb 1:88–130
Perfetti LR (2014) Principi della disciplina pubblicistica dell'alimentazione. Premesse ad un diritto amministrativo dell'alimentazione. Riv dir agr 1:3–20
Perfetti LR (2015) Cibo: quando i diritti sono in conflitto. Agg soc 1:550–559
Ranaldi V (2014) Il confronto tra Stati membri ed Unione Europea in materia di Ogm nella giurisprudenza nazionale e comunitaria. Dir comm intern 4:1042–1043
Sangiorgio M (2015) La tutela del "Made in Italy" in due recenti sentenze della Cassazione: tra illecito penale e sanzione amministrativa. Riv dir ind 3:173–186
Spolidoro M S (2014) Reti di imprese e diritti di proprietà industriale. Riv dir ind 4/5:245–265
Vanzetti A (2002) I marchi nel mercato globale. Riv dir ind 1:91–108
Vanzetti A, Di Cataldo V (2012) Manuale di diritto industriale. Giuffré, Milano
Vettorel A (2014) L'indicazione obbligatoria relativa al Paese d'origine o al luogo di provenienza degli alimenti: quale informazione? Riv dir alim 2:26–29

The Right to Food and Food Diversity in the Italian Constitution

Michele Troisi

1 Introduction

"It is necessary to contrast the economic globalisation with the globalisation of solidarity. It is necessary to take advantage from the expo experience in order to think on the right to food, striving to see behind the *faces*. A hidden presence, but that actually shall be the real protagonist of the event: the *faces* of the men and women who are hungry, and that get sick, and even die, from a too poor or harmful alimentation.(. . .) My hope is that this experience will allow entrepreneurs, merchants, scholars, to feel involved in a great project of solidarity: to feed the planet in the respect of every man and woman who live there and in compliance with that natural environment, of that territory, which has been given to us".[1]

"The right to food is the right of every person to have regular access, permanent, free, either directly or by means of financial purchases, to quantitatively and qualitatively adequate and sufficient food, corresponding to the cultural traditions of the population to which the consumer belongs and be able to ensure a psychic and physical life, individual and collective, free of fear, fulfilling and dignified".[2]

I wanted to start this work with two quotes, expressed at different times, from two very different personalities: Pope Francis when greeting for the opening day of Expo 2015, and Jean Ziegler, as the UN special rapporteur on the right to food, in 2011.

Both contributions point out why, albeit from different cultural assumptions, they summarize in a similar manner the rights, the values involved, the interests that underlie the right to food: nourish, respecting; guarantee a right, protecting the area where materially is expressed.

[1]Bergoglio (2015).

[2]Ziegler (2004), p. 49.

M. Troisi (✉)
Department of Legal Studies, University of Salento, Lecce, Italy

A. Isoni et al. (eds.), *Food Diversity Between Rights, Duties and Autonomies*,
LITES – Legal Issues in Transdisciplinary Environmental Studies 2,
https://doi.org/10.1007/978-3-319-75196-2_28

In these statements, however, an essential profile for this research is subtended: the necessary distinction between the right *of* food that has the task of studying the rules of production and marketing of food and the right *to* an adequate food. If the right to food governs *how* the right should be produced and distributed, the right to adequate food rule *on what* the human being has the right to access.

The right to adequate alimentation, *rectius* the right to adequate food, comes always before the right of food, which is, therefore, serving in respect to the first.

But these "extra order solicitation" are especially useful because of them the researcher can make good use to wonder about the nature and degree of protection that the right to food has in their own Country, starting from its constitutional foundation.

In fact, the correlation between the right to food and the Constitution is *in re ipsa*: the first looks like a premise of the second, just like the right to life. The right to food is, thus, an aspect of the "right to a dignified existence," which seems to be not only essential, but also prejudicial, because, in hindsight, "the right to be able to use the rights".[3]

Not by chance a recent study shows how, according to a survey conducted by FAO in 2011, the right to food is enshrined at constitutional level in more than one hundred Constitutions.[4] Sometimes it is recognized explicitly (and is emblematic that this happens mostly in poor countries), and sometimes implicitly by referring to concepts such as the right "to an adequate standard of living", "well-being", "the development"; other times, the right to food is protected in a *mediated* manner by virtue of the recognized value in international treaties that foresee it.[5]

The Italian Constitution falls within the latter ones, for the postponement, operated, from the first paragraph of Art. 117, to the "constraints deriving from EU and international obligations".

But such location of our Charter, as will be seen, does not satisfy completely.

2 The Right to Food in a *Laborist* Constitution

In the Italian Constitution there is no explicit reference to the right to food.

Following the constitutional amendment of 2001, there was only the introduction, in Art. 117, III° c. Const., of the *alimentation* matter, whose ownership is shared between State and Region; but the effects of such entry have been disappointing. Indeed, the transversal character of the matter has made and makes it difficult to

---

[3]Ruotolo (2011), p. 403.

[4]Pizzolato (2015), p. 135.

[5]Ibidem.

identify its boundaries[6]: it connects to a number of expertise areas, from which often, in these 14 years of implementation of the Constitutional novella, it was compressed, struggling to establish itself as an autonomous title of competency.[7]

Because of its own nature, in fact, the nutrition comes in contact with other materials contexts: some of the exclusive competence of the Regions (agriculture, industry, trade); other of exclusive competence of the State (competition, determination of the basic level of benefits relating to civil and social rights), other of concurrent competence (health protection).

Well, with its rulings, the Constitutional Court has often attracted the discipline concerning the nutrition in the context of health protection. An example is the decision no. 104 of 2014, in which the Court is invested by the question of constitutional legitimacy of the law of the Valle d'Aosta region, which subordinates the exercise of trade activity of the food product sector to the possession of specific professional requirements.

The constitutional judge rejects the question, on the assumption that "these requirements appear functional to ensure that those who carry out activities in the food industry sector are provided with specific training and experience to the evident purpose of safeguarding the health of consumers in a sensitive and fundamental sector such as that of food".

It seems, then, that to the food matter, incident—by its nature—on other matters, has been given a *recessive* trait, by which it has been absorbed by other title of competencies.[8]

This phenomenon is due just to the circumstance that there is no explicit recognition of the right to food among the fundamental principles of our Charter.

Nevertheless, placing our Constitution among those that protect the right in question by virtue of the only appeal to supranational sources *former* Art. 117 Cost., appears a reductive choice.

The right to food, in fact, exist in the *soul* of our fundamental Charter.

It seems difficult to dispute, in fact, the profound relationship between the right to food and human dignity, the latter considered as "cornerstone" (using the famous expression of Giorgio La Pira) of the entire constitutional edifice.

From this perspective it would seem absurd, therefore, to consider the right to food as a right of "new generation", given its inherent coessentiality compared to the Constitution.

But why, then, unlike other Charts, there is no clarification of the right to food in the Italian Constitution?

Indeed, the absence is because of the choice, on the part of Constituents, to build our *welfare* system on the right to work, as a preventive remedy and the general issue

[6]Napoli (2013), p. 401.

[7]Vivaldi (2015), p. 239. About an overlapping of legislative competences, created by an unhappy reform, see at least: Caretti and Tarli Barbieri (2012), D'Atena (2013), Martines et al. (2012) and Mangiameli (2013).

[8]Vivaldi (2015), p. 238.

of poverty and social exclusion. The right/duty to work sanctioned by Art. 4 Const., is in fact the foundation of our republican order, to which the Art. 1 assigns an "force-idea" (using an expression note of Mortati).[9]

Democracy based on work constitutes the conception that is the basis of the whole constitutional architecture, in which the work is considered not only as a tool for the personality affirmation, but also as a source of material and spiritual realization of the person.[10]

In the constitutional perspective, therefore, the right to food exists and must be considered and guaranteed by the right to work, through the right to remuneration (Art. 36 of the Constitution.), to providential protection (Art. 38, paragraphs 2 and 3),[11] the right to social assistance (Art. 39 of the Constitution. 1 paragraph).

3 The "Discovery" of the Right to Food

Well, until a few years ago, the legislature, the doctrine and constitutional jurisprudence had never felt the need to make explicit this right.

Then, the serious economic crisis that began in 2008 (in which we are still immersed) pushed the operators of the right to explicitly intervene on this profile.

Emblematic has been the case of the institution of the so-called social card, which was followed by the Constitutional Court's judgment no. 10/2010.[12]

Particularly, Article 81 of Decree-Law n. 112 of 2008, had set up "a special fund aimed at meeting the priority needs of food nature [and subsequently also energy and sanitary needs of the poor citizens]" (paragraph 29), by regulating the funding (paragraph 30) and stating that, "in view of the extraordinary tensions faced by the food prices, in order to help the weakest segments of the population in a state of special need and demand for these", "is granted to the residents of Italian nationality who are in the greatest economic hardship" "a Purchase Card for the purchase of such goods and services, with burden borne by the State" (paragraph 32).

The regions who complained that the law has disciplined a matter of social policy, of the exclusive regional jurisdiction, the Court replied considering permissible the State intervention as "necessary in order to *effectively* ensure the protection of persons who, pouring in conditions of extreme need, boast a fundamental right" that, as "strictly relating to the protection of the irreducible core of the human person", especially in the presence of an "exceptionally negative economic situation," "must be able to be guaranteed throughout the national territory in a uniform,

---

[9] Mortati (1972), p. 26.

[10] Vivaldi (2015), p. 243. See also Cantaro (2008).

[11] Violini (2006).

[12] On such decision see the contributions of: Anzon Demmig (2010), Longo (2010) and Ruggeri (2010).

appropriate and timely manner, through a consistent and appropriate regulation purposive".

But which is the law qualified as 'fundamental' that, in this case, the Court considers should be protected at the cost of a derogation to the ordinary division of legislative competences?

It is the "right to achieve the essential services to relieve situations of extreme need - especially food". A right to which it corresponds—as it can be always read in the sent. n. 10/2010—"the duty of the State to determine the qualitative and quantitative characteristics, if the lack of a such provision might prejudge it". A right that the Court gets from the "fundamental principles of Articles 2 and 3, co. 2 Const., Art. 38 Cost. and Art. 117, co. 2, letter. m, Const.".

Well then, to the purpose of this research I find this judgment useful because it has three key points: (1) Firstly, in terms of the nature of the right in question, the Court has explicitly defined the right to food as a "fundamental right linked to the irreducible core of the human person"; (2) furthermore, in terms of the protection of this, as of other fundamental rights, the Court explained that the endorsement of a social law can not be divorced from its effectiveness; (3) Finally, for the exceptional nature of the discharge of duty in question solely from the state: in the final paragraph, in fact, the Court clarifies that, upon termination of "the economic situation that has required a social policy intervention extended to direct delivery of providence, it can not be separated from the participation instruments of the Regions, taking care also to ensure the full implementation of the principle of loyal cooperation, in observance of the distribution of competences defined by the Constitution".

4 The Right to Food and Principle of Solidarity

If, therefore, the right to food is *a very essential right* that, under normal conditions, all entities that make up the Republic shall guarantee, it seems appropriate to reflect on the scope of this *duty*.

This is certainly not the place to investigate the complex issues relating to constitutional duties[13]; we just dwell on some profiles that have connections with this research.

First of all it must be recalled that, in the more or less extensive catalog of constitutional duties, there are some that undoubtedly have direct effects on the right to food: think of Art. 4 Const., according to which it is a duty to contribute, through own work, to the material and moral welfare of society, or Art. 30 Cost. according to which the parents have the duty to "maintain" their children.[14]

---

[13]On the issue see at least: Balduzzi et al. (2007) and Lombardi (2002).

[14]For a classification proposal, see Pizzolato and Buzzacchi (2008), p. 326 ss.

Moreover, from a dogmatic point of view, it appears appropriate to emphasize that the issue of the duties find its natural declination in the principle of solidarity: it is a cardinal-principle of the legal order, which, reflected on our theme, recalls the duty to ensure over time, between generations, access to food and sustainable development.[15]

In fact, on closer inspection, the principle of solidarity seems ontologically created to communicate with the right to food: not only because the *right* to feed can not be fully configured without being coordinated with a *duty* to feed, but also for the vocation of this principle to hold together the two aspects of the right to food, or what implies public solidarity/responsibility and what it implies private solidarity/responsibility, according to the logic of interdependence between duties.[16]

Consider, in this regard, Art. 30 Cost., where different duties dialogue in the two paragraphs: it is the duty of the parents to "maintain" (also from a nutritional point of view) the children (horizontal solidarity); but in the alternative, if the parents do not have the ability, it becomes the duty of the Republic to ensure that maintenance ("public" solidarity).

Dwelling on the vertical solidarity, or public, it would seem to configure, in this field, a kind of *food solidarity*,[17] understood as a duty of the Republic (in all its articulations) to (to paraphrase the Art. 3 of the Constitution) remove all obstacles to equality between citizens in accessing adequate food, preventing the full development of the human person.[18]

This principle of food solidarity seems express itself in two aspects: (1) the duty *to render effective* the right to food, (2) the duty *to protect* it in all its declinations.

5 The Duty to *Render Effective* the Right to Food. The "Vital Minimum"

The sore point for the right to food is its concrete effectiveness; is, to paraphrase Bobbio, the "need to move from the abstract recognition of a right to the guarantee of its effectiveness" in respect to the Republic.[19]

Well, the duty to render effective the right to food requires that the Republic (not just the State) facilitate, with all kinds of positive action (legislative and administrative), access to adequate food especially the most vulnerable groups.

It was previously stated that our Constituent has designed an *welfare* state in which all rights (including food) rotate around the work; and now that work is going through a time of deep crisis, measures to guarantee the right to feed themselves

[15]Giuffrè (2002), Pezzini and Sacchetto (2005) and Tondi della Mura (2010).

[16]Shue (1984), p. 83 ss.

[17]Bottiglieri (2015), p. 246.

[18]Ibidem.

[19]Bobbio (1992).

seem to merge into those against poverty, to ensure the "vital minimum", as an essential core of the right to social assistance (Art. 39 Const. 1 paragraph).

And without doubt, in fact, that today the right to food recalls the theme of *social inclusion*: as has been authoritatively stated by the provisions on the rights and duties, emerges a constitutional design that requires continuous action aimed at the transformation of society, in which the foundation of the duty of the public authorities in the struggle against poverty and social exclusion "can be said to be strengthened by specific constitutional commitment to the removal of inequalities, clearly expressive of the need to place man in a position to exercise the rights and then to carry out his personality".[20]

In this sense, the introduction of a structural measure of income support could be understood as a concrete means of implementation, through legislation, of a true and proper right to food.

As it is known, on the issue of "the vital minimum", the doctrine has recently taken to question: for people who find themselves in fragile condition, that is the essential content of the right to social assistance, of which the legislator must take charge by making choices that are characterized by the principle of reasonableness.[21]

In fact, the reading of Art. 38, paragraph I, operated by relying on Articles 2 and 3 Const., leads, simultaneously, to the need to implement this right in ways that safeguard the principle of fulfillment of the mandatory duties of solidarity, to which is intimately linked the recognition of fundamental rights. Conversely, "a supplement to the income that places conditions of reciprocity seems to respond better to the idea of constitutional citizenship, both in terms of rights and in terms of duties".[22]

In argument, it is interesting to note that the German Constitutional Court has given its decision unequivocally that it is the duty of public authorities to guarantee citizens the right to an existential minimum. Invested by the issue of the constitutional legitimacy of the Hartz IV law (related to services on the labor market), the German Court stated that the principles of human dignity and the State of welfare is obtainable that the guarantee of a minimum subsistence is a fundamental right, addressed to ensure the necessary material conditions to existence and a minimum of participation in social, culture life of the country.

A right that is not subject to the availability of the legislature and that must be honored by the legislator who is responsible for the realization and the constant updating of the measures, adapting each time the performance "to the level of development of the community and the existing conditions of life".[23]

However, the doctrine has emphasized that the principle expressed by the German court is worth *a fortiori* in the Italian case, where the foundation of the duty of the public authorities in the struggle against poverty is reinforced by the specific constitutional commitment to the removal of inequalities clearly expressive of the

[20]Ruotolo (2011), p. 403.

[21]Vivaldi and Gualdani (2014).

[22]Tripodina (2013).

[23]Thus BVerfG, 1 BvL 1/09 vom 9.2.2010, cleverly commented by Ruotolo (2011), p. 406.

need to place man in the condition to exercise rights: this translates into a specific policy that the public authorities must necessarily implement to support the income and thereby indirectly to render effective the right to food.[24]

6 The Duty *to Protect* the Right to Food. Food Diversity

It has been said that the concept of "food solidarity" brings with it, for the Republic, also the *duty to protect the right to food.*

Well, the first point clarify is that the protection of this right should not be implemented in a typical assistenzialistic model of a certain idea (clearly ideological) of the welfare State, which confuses the intervention with the *mortification* of individual initiative, the pursuing of substantive equality with the *homologation*.

On the contrary, the protection of the right to food demands above all the respect and protection of the material and cultural conditions that allow, in different territorial realities, the operation and development of local agri-food systems, enhancing not only farm structures established therein, but I would say especially the people of flesh and blood (with their culture and tradition background), "the faces" who insist on those agricultural structures.

The protection of the right to food as the protection of human dignity, also passes through this aspect, then.

In essence, far from detecting it as a right of health care content to be able to assert against the "public" with economic welfare policies, the protection of the right to food is configured as a duty aimed at conforming the action of local institutions in view of the defense of "diversity", intended as peculiarities of the territory; territory intended not only from a material point of view, but also, and I would say especially, immaterial.

For these reasons, the right to food is underlined as a wholly singular fundamental right: if on one hand, as seen above, it includes within itself an universalistic demand that concerns the single individual who has "*the right to a dignified life*", on the other hand, in its practical implementation with respect to different socio-economic realities, is configured as "*the right of food to be protected*"; a right that demands to be conjugated to the territoriality and therefore with multiple social structures present in the area that generate it.

And it is in this sense, then, that acquires significant profile in the protection of "food diversity". This specific scope of protection, as it will be seen shortly, is the synthesis of multiple diversities: it is a value-system, in which converge and jointly compose several legal values of primary constitutional status (independence and territorial differentiation, landscape, cultural heritage, crafts, health and environment).

[24]Ruotolo (2011), p. 408.

The strength of these core values is combined and concentrated in food diversity, thus giving to this last, at least in the eyes of the interpreter, a role of absolute importance in the legal ordering.[25]

6.1 Food Diversity and Culture

A first legal matrix of food diversity is to be found among the fundamental principles of our Charter.

The reference is Art. 9 of the Constitution. That in the first paragraph reads: "The Republic promotes the development of culture." However, any pluralist democracy places at the base of itself "cultural pluralism as a model in which human cultures are not each other's enemy, but in which the predominant one poses the extinction of weaker cultures as a problem to fight and to solve".[26]

This concept is well suited to the cultural dimension of food. The task of the Republic, that is, to protect and promote the development of cultural pluralism in matters of food; this aspect implies, first of all, the duty to protect the diverse food cultures of their individual, as social formations.

This means, however, also a duty of the Republic to promote forms of integration of the different food cultures that coexist on the Italian territory.

Likewise, a foundation of food diversity is to be found in the second paragraph of Art. 9 Const., according to which the Republic "protects the landscape and the artistic heritage of the Nation".

This constitutional protection of the *landscape*, in fact, has been declined by the legislator in the sense that the landscape assets are those "constituents expression of historical, cultural, natural, morphological and aesthetic values of the territory" (Art. 2.3 Decree. N. 42/2004); furthermore, the landscape "is the expressive territory of *identity* whose character derives from natural and human factors and their interrelations" (Art. 131). However, the combination between the constitutional provision and the punctual normative production, provides protection to the so-called *food landscapes.*

It is about "places" that were characterized by the dynamic interrelation between human, natural and cultural forces: some of these landscapes are better known and valued because of the recognition, formal or informal, of the quality of specific food products that characterize them (just think of the added value of the brand PDO or PGI); in other cases are the social formations, specifically dedicated to food, that allowed to give to places of production of regional and traditional foods an appropriate recognition (think of *Slow food* principals). These realities the Republic (taken as a whole) has the task to protect and enhance.

[25]Monteduro (2015), p. 37.

[26]Betzu (2008), p. 72.

The other subject protected by the Republic, *ex* Art. 9, second paragraph of the Constitution, *the historical and artistic heritage of the nation*, has obvious connections with food diversity. It recalls the complex issue of cultural heritage, which over time has become a large value and inclusive of different types of goods.

For the purposes of this research, will be enough to emphasize that, if before with that term were considered the traditional categories of "artistic things", "antiquity", "things of historical interest", now this term is interpreted in the light of Article 810 of the Civil Code, which defines the goods as things that can be subject to rights. Such norm, in fact, points out that not all entities (tangible or intangible) can be considered, juridically, "goods": such are just the things to which the law attributes importance. The concept of good, therefore, constitutes a suitable qualification to assume the special ability of a thing to satisfy a man's interest (the so-called *utilitas*). Nor should it necessarily be an economic interest. The good in fact, can express extremely different values and interests, just as in the case of cultural heritage.

During the normative stratification, the legislator has shown, in fact, to appreciate and to attribute legal significance to an extremely diverse range of cultural goods in relation to the different interests they can express by aiming to meet the need of "culture" of a given society or mankind or in so far "testimony having civilization value" (this the felicitous expression of the "Franceschini commission").[27]

The relevance of these interests has greatly enriched and diversified over time and, starting, originally from goods of an historical, artistic and archaeological expression, has reached to *"ethno-anthropological goods*" (explicitly mentioned and protected first by the "Unique text on cultural heritage", the d., lgs., October 29, 1999, by the then-called "Urban code" of 2004–2008).

With this expression, the legislator intended to protect that category of goods belonging to the national cultural heritage which are distinguished by being "linked to local cultures and the lives of ordinary people and which constitute an expression of the traditions subject of anthropologists study". This heritage represents in itself the peculiarity that differentiates it from the historical, artistic and archaeological heritage, involving not only movable and immovable goods, but also and most of all the so-called intangible goods. Thus, in addition to sites, buildings, objects, have gradually rise to the object of attention and appreciation also civilization expressions such as language, dialects, music, rituals, symbolic practices, proverbs, food.[28]

It is precisely for this evolution of doctrine and legislative that this concept of cultural heritage is relevant for the purposes of our research: the protection of the right to food, in its declination of food diversity protection, it would seem to have its own cover both constitutional and legislative.

It is true, however, that the concept of the protection of the intangible good has always created more than some concern, because of the "volatility" of this cultural expression: the Constitutional Court has, therefore, repeatedly stressed the need of

[27]Crosetti (2014).

[28]Fiorillo and Silverio (2017), p. 27 ss.

the connection of the intangible good with its material support[29]; the same "Urbani Code" seems to have moved this more "materialistic line", given the continuous reference to a *res,* to a good.[30]

In fact, the Italian legal tradition is based on the concept of cultural emergency protection inasmuch "thing", material and physical object: to apply the same procedures for the protection of a painting, like at a party or to a language, appears very difficult.

In fact such problematic reliefs would seem only partially apply to *food diversity*; as it is easy to understand, in fact, we talk of a good that is not only intangible and, therefore, characterized by a certain abstraction and volatility.

The food, in fact, is a middle way between the tangible and the intangible good: of the latter has the intangible heritage, of the first has the *res*, the thing.

Food diversity, therefore, is an intangible cultural good that, since it is directly linked to the food-matter, has its own special protection under both ordinary and constitutional level.

6.2 Food Diversity and Territory

A second and fundamental matrix, legal and social, of food diversity is represented by the diversity of the territories of origin and their heirs communities and food custodians.

The point is crucial, because it shows how the "food heritage" are related not only to a specific terrain as a material space portion, with its particular physical-chemical characteristics, climatic and agronomic, but also and especially in a territory intended as a place for settlement and life of a particular community in a long time span.[31]

The relationship between the territorial diversity of the regions and differentiation of food products, which is an expression, manifested in the fact that the territories are demanding a right to self-represent their "food identity"[32] through their different typical products, to defend roots of their food culture models and to value their food diversity as instrument of territorial development (think of topics such as enogastronomic tourism).[33]

This territorial/identity dimension based on collective auto-determination of self-feeding models hinges, in legal terms, in constitutional principles of primary importance, such as the principle of autonomy (Art. 5 and 114 of the Constitution) and the principle of differentiation (Art. 118 Constitution).

---

[29]Ibidem.

[30]Ibidem.

[31]Monteduro (2015), p. 33.

[32]Ibidem.

[33]Ibidem.

It is interesting to note how these principles have been the subject of a constitutional amendment in 2001; well, at the base of this reform, there had been a socio-cultural debate precisely on the phenomenon of *re-emergence of the territory*, as an area of social and political relations with greater significance and characteristics and own needs.

It happened, in fact, that a number of contributing factors, such as the relief gained from the territorial dimension in economic relations, the fall of the social guarantees conquered in the experience of State performance and the subsequent mounting of a widespread perception of insecurity in the community, as well as the appearance of an "*identity question*" as a response to the failure of some integration policies, favored slowly "reactive phenomena compared with a cosmopolitan vision of social ties", imposing the territory as a place of emergence of more limited interest, but of greater density and consistency.

The very concept of "territory", which in its national declination has always tended to the *homogenization of local conditions*, resulted gradually in the aspiration to manifest their own rules and traditions. Faced with such phenomena the State appeared (and continues to appear) too uniform in its centralism to give recognition to the many differences that move on its territory and which are not reflected only in a political representative type.

Therefore, is the emergence of a related phenomenon, that is, the "rurality question", that did take to the rural areas a new "strategic importance", in which the development goes through the territories as knots in a network system made of interrelations, of additions, of complementarity. Each of these new communities, of these unique characteristics demanded (and demands) a special differentiated and specific protection by the legislator or otherwise by the closest administrator.

In this light seems to be placed the concept of "rural district", coined by Art. 13 paragraph 1 of the legislative decree 228/2001 (which was followed by a number of regional laws), "are defined rural districts, local productive systems (...) characterized by homogenous historical and territorial identity deriving from the integration between farming and other local activities, as well as the production of particularly specific goods and services, coherent with the traditions and natural and territorial potential".

From the socio-cultural background was then fed the boost that led to the reform of the 2001 Constitution, with a reversal of the pyramid that saw the autonomies in first place and the State at the last (although these hopes did not materialize).[34]

A reform that enhances the meaning and the autonomy of the territory that, in a subsidiary optical, restarted from the place, the nearest community to the citizen.

And so the food diversity gets a systematic protection: the "new" Article 114 (which establishes an *equal constitutional dignity* of all organizations-starting from the territory, which constitute the Republic) and 118 (which establishes the principle of subsidiarity, but also that of differentiation), interpreted systematically with the aforementioned Article 9 (relating to the cultural heritage, including the

[34]Among many Mangiameli (2013).

intangible one) and Art. 2, can expand the scope of the right (inviolable) to food; it is also included the value of identity diversity of the territory and the respect of different culinary cultures that generate it.

The Italian Constitution, although it was written nearly 70 years ago, placing the human person (not in abstract, but concretely as he is, with his needs) at the center of all, once again proves to be extraordinarily timely.

It places, in fact, a barrier to an "economic globalization" that tends to standardize and to "depersonalize" the food, reducing it to an anonymous commodity.

Moreover the barrier is represented precisely by that Art. 2 of the Constitution, by that dignity of the human person, inside of which is "covered" the knowledge-based specialties and flavors that make the history and the identity of each community and allow it to open to others. To be protected, then, it is the "territorial capital", within which fundamental it is precisely the "human capital" than their work.

Ultimately, the Constitution protects the right to food, preserving food diversity as cultural heritage of the people working in a given territory.

Under similar profiles seems to be demanding the European Union that, in the European Parliament resolution of 14 January 2014, 2013/2098(INI), called "on the regional brand: towards best practice in rural economies", advocates a coordination of goods (food and non-food) and quality services anchored in the specificities of each territory and particularly of its heritage (historical, cultural, geographical and so on).

This mark is intended to trigger a territorial valorization process that holds together the products and services within a perspective of identity, even compared to meta-economic parameters such as the environment, quality of life, economic and social cohesion and solidarity.

7 The Role of Regions and Local Authorities

It was previously stated that the right to food is a "fundamental right linked to the irreducible core of the human person" that, under the double profile analyzed, the Republic has the duty to implement and protect.

It is also stated, however, that these are tasks that weigh on all the articulations of the Republic.

It is then necessary to wonder about space and mode of action of Regions and local institutions.

Some "broken down" protection attempts are not convincing and seem to be related more to a "protectionist" logic than that of valorization.

For example, the Lazio Region had introduced a "regional" brand of quality designed to mark certain agricultural and food products for promotional purposes of the agriculture and gastronomic culture of the territory. The Court in its sentence no. 66 of 2013 declared the illegitimacy of that law, as capable to address the consumer preference towards products assisted by the brand than other homogeneous, of different origin. Such a user misuse would have product on the free movement of goods—at least "indirectly" or "in power"—those restrictive effects that the regional legislature is inhibited to pursue based on community ties, ex Art. 117 of the Constitution.

Different initiative has been that of some municipalities protagonists of the so-called "Anti-Kebab ordinances".[35] For example, the resolution of the Board of the City of Bussolengo in the province of Verona that, motivated under Article 34 l.r. Veneto 29/2007 (governing exercise of the activity of administration of beverages and foods), forbids the opening of kebab in the old town and within 150 mt. by religious and places of worship, hospitals and nursing homes, school buildings of any degree, cemeteries.

Another example is provided by the resolution of the City Council of Prato that, in the "implementation" of the Art. 110 of l.r. Toscana 28/2005 (Consolidated Law concerning trade, food and drinks administration), has identified in the "incompatible activities" within the area identified as PUA (Pedestrian Urban Area), craft activities of food cooking, such as kebabs and similar, also exercised also as a secondary activity compared to that prevailing, different from those "traditional".

Even Lucca City Council has moved in this direction: with resolution no. 12/2009, "in order to safeguard the culinary traditions and architectural, structural, cultural, historical and furnishing typicality", prohibited in the old town, "the activation of the administration exercises, whose activity is attributable to different ethnic groups", requiring also that the menu of all commercial establishments must have at least one local dish, prepared exclusively with "commonly recognized products" typical of the province of Lucca.[36]

The *ratio* given of such measures is the protection of "typical" enogastronomic products or of historical centers such places designated of "typical" and "traditional" enogastronomy, invoked as a justification to restrict, explicitly or implicitly, the sale of kebab as a product itself, as it is considered not "safe" and not "typical" or "traditional".

But the concept of "enogastronomic typicality" to be protected, lends itself to more than a few censorship: first, it does not seem certainly protectable by measures that, placed at the lowest level of the hierarchy of sources, intend to sensibly affect on legal interests of constitutional level such as freedom of private economic initiative (even with the limits of Art. 41 paragraph 2 of the Constitution), private property (even with the limits of Art. 42, paragraph 2 of the Constitution), the protection of competition, a matter attributed to the legislative competence of the State by Art. 117, paragraph 2, letter. *e*) Constitution (the notion of which, according to the Constitutional Court, reflects that which operates within the EU, with normative acts provided with primacy over national law, "regulatory intervention, the antitrust regulations and measures to promote an open market and in free competition" sent. n. 11/2004).

[35]Magrassi (2010), p. 325.

[36]Ibidem.

However, theoretically taken, these measures have raised suspicions that the true end pursued is not much the protection of "traditional products", but the corporate protection of commercial "local" businesses, if not a real marginalization in negative of certain determinable operators categories.

This profile, however, has nothing to do with the protection of food diversity, with the protection of "human and territorial capital" that creates the food of which was discussed in the previous paragraphs.

And then it is necessary to move from the abstract protectionist phase of the good, to that of "incentive" and valorizing of *what-the-good-represents;* that is, what the good "means" what "lies behind" the good, in terms of traditions, of knowledge, of culture.

Only in this way could be given proper implementation to a constitutional value of *food diversity* that "lives" within a *subsidiary State.*

In this respect are to be welcomed the initiatives designed to protect the agricultural heritage that are increasingly characterizing the regional regulations. They, starting with the fundamental principle of the integrity of the person, place in interaction environment, food, environment and culture: think of the spread of the so-called open-air museums or eco-museums dedicated to food.

On the same level, at the local level, think of the resolution no. 2/2009 of Prato City Council that, instead of producing a measure "anti," proposed a "pro" act. It adopted a regulation concerning trade based on a points system for the authorization of the activities of administration of foods and drinks in the old town, which recognizes an important score to the exercises that insert in the menu certain units of local products.

These measures are certainly not enough, but they can serve to dictate the way for local administrators or regional legislators: the protection of food diversity is another face of the right to food and should be protected with valorization systems (and thus publicizing) of the area that produces the food.

8 Conclusions

In front of this evolutionary scenario, it is possible to draft only partial final remarks.

It is no longer under discussion, in fact, the "constitutional cover" the right to food: although not explicitly foreseen, the Constitutional Court, as seen, has clarified it to be a "fundamental right linked to the irreducible core of the human person" and to the *laborist* principle that underlies our Charter.

It is not even in discussion the subject on which burden the corresponding constitutional duty to protect and guarantee this right: there are all entities of the Republic to have this important task of implementation and protection.

On *how* this right to food must be guaranteed, however, there are significant developments on the part of doctrine, jurisprudence and legislators.

An evolution that concerns the "two faces" of "food solidarity", both from a point of view of effective protection towards the weak subjects (the minimum wage is now on the political agenda of the central government and many regional governments), and from a point of view "of the right of food to be protected".

As it was possible to analyze the concept of food diversity, it is slowly emerging and is inextricably tying—in the constitutional fabric related to the territorial autonomies—even the "right to regional cultural identity"[37] and "the right to development of local food heritage".[38]

We are talking of synergistic concepts that feed each other.

In this perspective, as mentioned, also is placed the "landscape", as "the territory expressive of identity, whose character derives from the action of natural, human factors and their interrelations" (Art. 131, L. decree 42/2004): in it, it becomes apparent the close connection between local enogastronomic traditions and agricultural landscape, being the second, at the same time, the visible result of the molding action of the first over the centuries, and the prerequisite for them to be perpetuated in the future. Hence the rooting in Art. 9 of the Constitution of such dimension—identitary and collective of the territorial food diversity.[39]

From this point of view is to be welcomed that the Ministry of Agriculture, Food and Forestry Policies has adopted a special decree (the Decree of April 9, 2008, entitled "Identification of Italian agri-food products as an expression of the Italian cultural heritage") with which it has established that the traditional Italian food products contained in the lists referred to in the Ministerial decree of 18 July 2000 containing "the National List of Traditional Foodstuffs"(*PAT-Prodotti Agroalimentari Tradizionali*) "are an expression, as well as of inventiveness, ingenuity, and socio-economic evolution process of the Italian local collectivities, even of the traditions and culture of the regions, provinces and generally to the widespread community on the Italian territory, and as such must eligible for protection and preservation by the Italian institutions", as "an expression of the Italian cultural heritage".

Equally worthy of being mentioned is the initiative of the Abruzzo region that has recently proceeded, first among all the regions, by including in the statute a reference to the right to food. The Article 7-bis, in fact, reads: "The Region promotes the right to food and to proper nutrition, intended as the right to have regular, permanent and free access to quality food, sufficient, healthy and culturally appropriate, that guarantees the physical and mental fulfillment, individual and collective, required to lead a dignified life". "To achieve the objectives referred to in paragraph 1 - continues the second paragraph -, the Region contributes to favor and determine measures to combat malnutrition, both in the form of denutrition and that of overweight and obesity, the fight against waste, especially of food, and climate change, as key aspects of the right to health, even in its specification of the right to a healthy environment and it supports information and awareness activities".

---

[37] Ruggiu (2014), p. 486.

[38] Conte (2014), p. 461.

[39] Monteduro (2015).

Even with all its limitations on the effectiveness of the statutory provisions,[40] I believe it is to be greeted favorably the fact that the right to food, in the two analyzed profiles, has found an explicit recognition in a statutory charter: the "right to have access to food" and that such food has be "of quality, healthy and culturally appropriate".

I think this is the way that the regional legislature (which has seen greatly increased its range of action after the reform), but also local authorities and the State itself (in a subsidiary optic) should go for a full protection of a right that, as is easy to deduce from the various profile analyzed, has to do with the accomplishment of democracy itself.

References

Anzon Demmig A (2004) La Corte condanna all'inefficacia giuridica le norme programmatiche degli Statuti regionali ordinari. Giur Cost 6:4057–4068

Anzon Demmig A (2010) Potestà legislativa regionale residuale e livelli essenziali delle prestazioni. Giur Cost 1:155–164

Balduzzi R, Cavino M, Grosso E, Luther J (eds) (2007) I doveri costituzionali. La prospettiva del giudice delle Leggi. Giappichelli, Torino

Bartole S (2005) Possibili usi normativi delle norme a valore meramente culturale o politico. Le Reg 1-2:11–13

Bergoglio J (2015) Discorso di saluto alla cerimonia di inaugurazione di Expo 2016. https://w2.vatican.va/content/francesco/it/messages/pont-messages/2015/documents/papa-francesco_20150501_video messaggioinaugurazione-expo-milano.html

Betzu M (2008) Art. 9. In: Bartole S, Bin R (eds) Commentario breve alla Costituzione. Cedam, Padova

Bin R (2005) Perché le Regioni dovrebbero essere contente di questa decisione. Le Reg 1–2:15–19

Bobbio N (1992) L'età dei diritti. Einaudi, Torino

Bottiglieri M (2015) Il diritto al cibo adeguato. Tutela internazionale, costituzionale e locale di un diritto fondamentale "nuovo". POLIS Working Papers 222:5–548

Cammelli M (2005) Norme programmatiche e statuti regionali: questione chiusa e problema aperto. Le Reg 1–2:21–25

Cantaro A (2008) Lavoro (diritto al). Dizionario sistematico di diritto costituzionale. Il sole24ore, Milano

Caretti P (2005) La disciplina dei diritti fondamentali. Le Reg 1–2:27–29

Caretti P, Tarli Barbieri G (2012) Diritto regionale. Giappichelli, Torino

Conte L (2014) Autonomie territoriali e cultura. In: Morelli A, Trucco L (eds) Diritti e autonomie territoriali. Giappichelli, Torino

[40]On the problematic legal effect of certain norms of the regional statutes, see the judgements of the Constitutional Court nn. 372, 378-379/2004 and, among many of the doctrine reactions, at least: Anzon Demmig (2004), p. 4057; Bartole (2005), p. 11; Bin (2005), p. 15 ss.; Cammelli (2005), p. 21 ss.; Caretti (2005), p. 27 ss.; Cuocolo (2004), p. 4077; D'Atena (2005), p. 2 ss.; Dickman (2005), p. 2 ss.; Falcon (2005), p. 31 ss.; Ferrara (2005), p. 2 ss.; Lippolis (2005), p. 2 ss.; Mangia (2004), p. 4085 ss.; Pastori (2005), pp. 35–36; Pizzetti (2005), p. 37 ss.; Rinaldi (2004), p. 4090 ss.; Rossi (2005), p. 50 ss; Ruggeri (2005), p. 41 ss.

Crosetti A (2014) I beni demo-etno-antropologici: origini e parabola di una categoria di beni culturali. Dir e soc 2:355–388
Cuocolo V (2004) I nuovi statuti regionali fra Governo e Corte costituzionale. Giur Cost 6:4047–4057
D'Atena A (2005) I nuovi statuti regionali ed i loro contenuti programmatici. Available in www.issirfa.cnr.it
D'Atena A (2013) Diritto regionale. Giappichelli, Torino
Dickman R (2005) Le sentenze della Corte sull'inefficacia giuridica delle disposizioni programmatiche degli statuti ordinari (nota a Corte cost. 2 dicembre 2004, n. 372 e 6 dicembre 2004, n. 378 e 379. Available in www.federalismi.it, n. 1
Falcon G (2005) Alcune questioni a valle delle decisioni della Corte. Le Reg 1–2:31–34
Ferrara A (2005) Chi ha paura degli statuti regionali? Available in www.federalismi.it, n. 1
Fiorillo M, Silverio S (2017) Cibo, cultura, diritto. Mucchi, Modena
Giuffrè F (2002) La solidarietà nell'ordinamento costituzionale. Giuffrè, Milano
Lippolis V (2005) Le dichiarazione di principio negli statuti regionali. Available in www.federalismi.it, n. 1
Lombardi G (2002) Doveri pubblici (diritto costituzionale). Enciclopedia giuridica. Giuffrè, Milano
Longo E (2010) I diritti sociali al tempo della crisi. La Consulta salva la social card e ne ricava un nuovo titolo di competenza statale. Giur cost 1:164–182
Magrassi M (2010) Le c.d. "ordinanze anti-kebab". Le Reg 1–2:325–332
Mangia A (2004) Il ritorno delle norme programmatiche. Giur Cost 6:4068–4073
Mangiameli S (2013) Le Regioni italiane tra crisi globale e neocentralismo. Giuffrè, Milano
Martines T, Ruggeri A, Salazar C (eds) (2012) Lineamenti di diritto regionale. Giuffrè, Milano
Monteduro M (2015) Diritto dell'ambiente e diversità alimentare. Riv quad dir amb 1:88–131
Mortati C (1972) Scritti sulle fonti del diritto e sull'interpretazione. Giuffrè, Milano
Napoli C (2013) Le competenze degli enti locali in tema di cibo. Riv dir agr 3:395–404
Pastori G (2005) Luci ed ombre della giurisprudenza costituzionale in tema di norme programmatiche degli statuti regionali. Le Reg 1–2:35–36
Pezzini B, Sacchetto C (eds) (2005) Il dovere di solidarietà. Giuffrè, Milano
Pizzetti F (2005) Il gioco non valeva la candela: il prezzo pagato è troppo alto. Le Reg 1–2:37–40
Pizzolato F (2015) Il diritto all'alimentazione. Agg soc 2:131–141
Pizzolato F, Buzzacchi C (2008) Doveri costituzionali. Digesto delle Discipline pubblicistiche – Aggiornamento, Utet, Torino
Rinaldi E (2004) Corte costituzionale, riforme e statuti regionali: dall'inefficacia giuridica delle norme programmatiche al superamento dell'ambigua distinzione tra contenuto 'necessario' e contenuto 'eventuale'. Giur Cost 6:4073–4085
Rossi E (2005) Principi e diritti nei nuovi statuti regionali. Riv dir cost:51–96
Ruggeri A (2005) La Corte, la denormativizzazione degli statuti regionali e il primati del diritto politico sul diritto costituzionale. Le Reg 1–2:41–48
Ruggeri A (2010) Livelli essenziali delle prestazioni relative ai diritti sociali e ridefinizione delle sfere di competenza di Stato e Regioni in situazioni di emergenza economica (a prima lettura di Corte cost. n. 10 del 2010). Available in www.forumcostituzionale.it
Ruggiu I (2014) Identità culturale. In Morelli A, Trucco L (eds) Diritti e autonomie territoriali. Giappichelli, Torino
Ruotolo M (2011) La lotta alla povertà come dovere dei pubblici poteri. Alla ricerca dei fondamenti costituzionali del diritto a un'esistenza dignitosa. Dir pub 2:391–424
Shue H (1984) The interdependence of duties. In Alston P, Tomasewski (eds) The right to food. Martinus Nijoff, Leiden
Tondi della Mura V (2010) La solidarietà tra etica ed estetica. Riv AIC, 00:1–14
Tripodina C (2013) Il diritto a un'esistenza libera e dignitosa. Sui fondamenti costituzionali del reddito di cittadinanza. Giappichelli, Torino
Violini L (2006) Art. 38. In Bifulco R, Celotto A, Olivetti M (eds) Commentario alla Costituzione. Giappichelli, Torino

Vivaldi E (2015) Dall'alimentazione alla sicurezza alimentare. Il cammino della materia nell'ordinamento interno ed i possibili sviluppi connessi all'approvazione del d.d.l. di riforma costituzionale A.C. 2643. Riv dir agr 2:235–249
Vivaldi E, Gualdani A (2014) Il minimo vitale tra tentativi di attuazione e prospettive future. Dir soc 1:115–151
Ziegler J (2004) Dalla parte dei deboli. Il diritto all'alimentazione. Tropea, Milano

The manufacturer's authorised representative in the EU is Springer Nature Customer Service Centre GmbH, Europaplatz 3, 69115 Heidelberg, Germany. If you have any concerns regarding our products, please contact ProductSafety@springernature.com

Printed and bound by CPI Group (UK) Ltd, Croydon, CR0 4YY
01/10/2026
02225834-0018